Pesticide Chemistry and Bioscience

The Food–Environment Challenge

Pesticide Chemistry and Bioscience

The Food–Environment Challenge

Edited by

Gerald T. Brooks
University of Portsmouth, UK

Terry R. Roberts
JSC International Ltd, UK

RS•C
ROYAL SOCIETY OF CHEMISTRY

The proceedings of the 9th International Congress on Pesticide Chemistry: The Food–Environment Challenge held at the Queen Elizabeth II Conference Centre, Westminster, London on 2–7 August 1998.

Special Publication No. 233

ISBN 0-85404-709-3

A catalogue record for this book is available from the British Library

Published by The Royal Society of Chemistry,
Thomas Graham House, Science Park, Milton Road,
Cambridge CB4 0WF, UK

For further information see our web site at www.rsc.org

Printed and bound by Bookcraft (Bath) Ltd

Preface

During the last 30 years, the IUPAC sponsored Congresses of Pesticide Chemistry have provided an enthusiastically supported four-yearly forum for the presentation of the latest developments in this increasingly complex subject. The 1998 Congress is the last of the series in the present century. The scope of the Congress has increased greatly over the years, particularly to satisfy increasing environmental concerns. This development is summarised in the theme of the 1998 Congress: The Food–Environment Challenge, which reflects the problems that await solution in the next millennium.

The papers in this volume are derived from the plenary and symposia lectures presented at the Congress and are sectionalised into the eight main topics of the Congress, arranged in the traditional order, ranging from Chemical Synthesis, through Mode of Action and Resistance related topics, to Regulation and Risk Assessment. They demonstrate the remarkable changes that are taking place in approaches to crop protection, in response to the above challenge.

In addition to these papers, the Congress included more than 1100 poster presentations, with associated poster discussion sessions. Most significant is the now routine application of the techniques of molecular biology/biotechnology in all areas relevant to 'Pesticide Chemistry'. Together with concurrent advances such as combinatorial chemistry, these techniques are bringing a revolution in our understanding of the life processes involved and in our ability to develop complex, environmentally acceptable strategies for weed, pest and disease control.

Acknowledgements

IUPAC and The Royal Society of Chemistry express their gratitude to the speakers, participants and others who worked hard to ensure a successful Congress and especially to the following industrial companies and other organisations who have generously given financial and other support to the Congress:

AgrEvo UK Ltd, UK

BASF AG, Germany

Bayer AG, Germany

Dow AgroSciences, UK

Du Pont Agricultural Products, USA

JSC International Limited, UK

Monsanto Europe, Belgium

Novartis Crop Protection, Switzerland

Novartis Foundation, UK

Rhône-Poulenc Agro, France

Society of Chemical Industry, UK

Sumitomo Chemical Co, Japan

Witco OrganoSilicones Group, Switzerland

Zeneca Agrochemicals, UK

BBSRC, *etc.*

Committees for the Ninth International Congress of Pesticide Chemistry

Executive Committee

Mr John R Finney (Chairman), Battelle
Professor John A Pickett (Vice-Chairman), IACR-Rothamsted
Dr Gerald T Brooks, *Pesticide Science,* SCI London
Dr Patrick J Crowley, Zeneca Agrochemicals, UK
Professor Alan L Devonshire, IACR-Rothamsted
Dr Andrew Farnham, IACR- Rothamsted
Dr John F Gibson, The Royal Society of Chemistry
Dr Derek W Hollomon, IACR-Long Ashton
Mrs Penny A Mohamed, The Royal Society of Chemistry
Dr Terry R Roberts, JSC, UK
Mrs Paula J Whelan, The Royal Society of Chemistry

Scientific Programme Committee

Mr John R Finney (Chairman)
Members of the Executive Committee, and
Dr Stella Axiotis (Rhône-Poulenc Agro, Lyon, France)
Ms Elizabeth Behl (US-EPA)
Dr Hanspeter Fischer (Novartis, Switzerland)
Professor Franklin Hall (Ohio State University, USA)
Mrs R Hignett (UK Ministry of Agriculture, Food and Fisheries)
Professor Norio Kurihara (Kyoto University, Japan)
Professor Steven Ley, University of Cambridge, UK
Mr Patrick Mulqueen (Dow AgroSciences, UK)
Dr Klaus Naumann (Bayer AG, Germany)
Dr Kenneth Racke (Dow AgroSciences, USA)
Professor Atta-ur-Rahman (University of Karachi, Pakistan)
Mr Michael Skidmore (Zeneca Agrochemicals UK)
Dr Yoshiyuki Takimoto (Sumitomo Chemical Co., Japan)
Professor James Thomas, University of Manchester, UK
Mr Terry Tooby (JSC, UK)
Professor Peter Usherwood, University of Nottingham, UK
Dr John Unsworth (Rhône-Poulenc Agriculture, UK)

Advisory Committee

Dr H. Frehse (Germany),
Professor R Greenhalgh (Canada)
Dr P. Kearney (USA)
Professor J. Miyamoto (Japan)

Contents

Plenary Lectures

Synthesis and Structure–Activity Relationships

Delivery

Natural Products

Mode of Action and Resistance

Metabolism

Environmental Fate

Residues in Food and the Environment

Regulation and Risk Assessment

Plenary Lectures

How Can Technology Feed the World Safely and Sustainably?

David A. Evans

ZENECA AGROCHEMICALS, FERNHURST, HASLEMERE, SURREY GU27 3JE

1 INTRODUCTION AND BACKGROUND

I was delighted to be asked by the Chairman of the Executive Committee, John Finney, to provide one of the opening addresses for this meeting to be entitled, "How can technology feed the world safely and sustainably?" One of the great advantages of working in an industrial context is that one is influenced by other members of multidisciplinary teams who have complementary skills. I have been taught by our patent attorneys that the key approach to answering any question is to ensure that you fully understand it. Accordingly, I have dissected the question which I have been posed into individual words and will discuss the underlying meaning of each to provide some background to the full meaning of the question as a whole.

1.1 How

This begs the question of the impact of various technologies on the world's food supply. With regard to sufficiency, the World Bank estimates that ca. 90% of the required increase in food production will come from yield increases on existing acreage.

Technology will need to make vital contributions to:

- protecting yield - by control of weeds and pests
- increasing yield - through agronomic effects and by provision of cultivars which optimise production of the useful parts of plants and
- improving yield - by enhancing the composition of plant products eg. oils, proteins, nutrients relating to specific food needs

At a more detailed level, advances in technology will assist the whole process of provision of crop management products from invention through to market.

1.2 Can

The two key questions here are whether it will be technically feasible to feed the burgeoning world population in coming decades, and whether political choices will encourage or hinder the adoption of such technologies.

A combination of technologies has provided a 2% annual global increase in yields over past decades. It is tempting to simply assume by extrapolation that this rate of increase can be sustained over coming years, or even accelerated, but this can only occur through sustained innovation. Figure 1 illustrates that comparatively little of the total land area of the world is currently under the plough and the area made available for crop production increased very slowly in the period 1985-1995. During the same period however population increased by 16%. The food production index showed an increase of 22.1%, reflecting an increasing agricultural intensity to meet the consumer requirement for better food variety and quality[1].

The World 1985-1995

Total land area = 13.0 billion ha

	1985	1990	1995	% Change 1985-1995
Arable and permanent crops (billion ha)	1.44	1.46	1.48	2.8
Population (billion)	4.89	5.28	5.68	16.1
Food production index	90.7	100.8	110.8	22.1

Source : FAO Yearbook, Vol 50, 1996

Figure 1 *Land in agricultural production*

Figure 2 illustrates the gain in the yields of various crops in different geographic locations from the 1950's to the 1990's. It can be seen that gains have been substantive across a broad range of crops and countries[2]. The success of modern farming methods has confounded the gloomy predictions of world starvation which were so common in the 1950's and 60's.

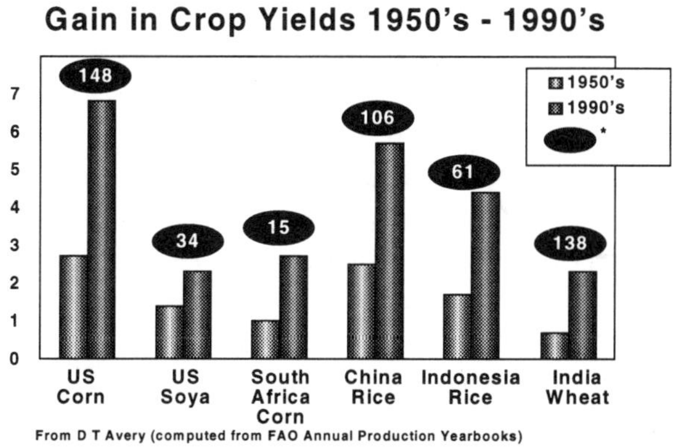

*Acres saved (million) versus 1950's yield

Figure 2 *Gain in Crop Yields 1950's - 1990's*

The continuance of this positive trend will have several dependencies. The successful progress of the last half century has been largely based upon science and technology applied to agriculture. It is therefore of concern that, in real terms, the funding of agricultural research is declining. Taking the US as an example, federal and state government spending on agricultural research has declined by approximately 30% in the last 25 years, and that in the private sector by 14%. In the early 1990's, the combined public and private spend in the US was approximately $5 billion, dwarfed by farming subsidies of $100 billion per annum. Furthermore, the spend on agricultural research appears minuscule alongside the three trillion dollar food industry which is supported by agriculture. It is likely that the current upsurge in the funding of research in biotechnology will reverse this trend, at least temporarily, but there must be concern that the more traditional, yet fundamental areas of agriculture and associated environmental issues will be neglected. Indeed, governments of developed countries with very adequate food supplies are presently questioning the current levels to which agricultural research (and the industry) are supported.

Even if we assume that advances in technology will enable us to meet food demands, there must be a question as to whether political forces will permit this to happen. In Figure 3, I quote the concerns of Norman E. Borlaug, recognised as one of the fathers of the Green Revolution for which he was awarded the Nobel Peace Prize. I believe that this quotation succinctly summarises the current situation[3].

"Twenty seven years ago, in my acceptance speech for the Nobel Peace Prize, I said that the Green Revolution had won a temporary success in man's war against hunger. I now say that the world has the technology - either available or well-advanced in the research pipeline - to feed a population of 10 billion people. The more pertinent question today is whether farmers and ranchers will be permitted to use this new technology. Extremists in the environmental movement from the rich nations seem to be doing everything they can to stop scientific progress in its tracks."

Norman E. Borlaug, Plant Tissue Culture and Biotechnology, 1997, **3**, 126

Figure 3

The perception of science and scientists by the public in very many developed countries is negative, especially in the context of pesticides in the environment and dabbling with Nature by the genetic modification of plants and hence food. Clearly, as scientists we have not sold our message. The benefits provided by technology are taken for granted but the risks are considered unacceptable, however minute they may be. Some pressure groups have now raised their suspicions around chemical pesticides and genetically modified crops to the level of an ideology and outright opposition is voiced. Moratoria are proposed, irrespective of the outcome of science-based risk assessment, and constructive dialogue seems impossible. In many countries, the political response is already leading to damaging legislation with no heed being paid to providing solutions to root causes of problems. It is now imperative for the crop management industries to address the issue of public perception more proactively and with greater skill than in the past.

1.3 Technology

The last five years have seen greater changes in technology than in the previous fifty. For half a century, organic chemistry has provided a reliable mainstay for crop protection, but its prominence is being challenged by exciting developments based upon biotechnology. For the first time, growers have a real choice of technology. Furthermore, the creativity of the scientists and technologists who support the industry continues unabated, and at present there is a positive prognosis for delivery of sufficient appropriate technology to feed the world for decades to come.

Recent advances in technology will provide the major focus for this paper, and will be addressed in Section 2 below.

1.4 Feed

Conventionally, crop protection chemistry has been considered as part of the chemical industry. More recently, we have all come to understand that we form a key part of the food provision industry. This has demanded that we understand the complex interrelationships with the other parts of the food provision chain and that we come to terms with new influences beyond the farm gate. Accordingly, dialogue with processors, food companies and retailers is intense. The strategies of several of the agrochemical majors is to participate by vertical integration into the food provision chain, whereas others such as Zeneca choose to rely upon partnerships. Whichever, we are climbing a steep learning curve with regard to the key customer values which motivate the aspirations of the downstream food companies. This has required an immense amount of adaptation and has required assimilation of a battery of new skills and concepts. Furthermore, consumer attitudes to food are highly variable by geography, and often run counter to trends in globalisation.

One notable feature of the food companies is that they are huge - Figure 4 illustrates the 1995 sales figures for the top five companies, taken from annual reports to shareholders. As such, they dwarf the sales of the individual crop protection/management companies. Given their position opposite the consumer, and their proximity to the regulatory affairs debate, it is not difficult to imagine why partnerships with the food companies are highly valued. It is noteworthy that several of these companies have an enormous global reach - Nestle's food sales for example are greater than food sale purchases in countries such as Mexico, Spain, Holland, Belgium and Australia.

The technological revolution which is presently impinging upon the crop protection industry has significant relevance to the food companies downstream in that cutting-edge long term science has the potential to impact the basis of competitiveness in the industry.

Top 5

Company	$ Sales Food (Billion)
Nestle	42.3
Phillip Morris (Food only)	35.9
Unilever (Food only)	24.2
PepsiCo (USA)	17.9
Danone	14.4

Figure 4 *Sales revenues of the top five food companies (1995)*

1.5 World

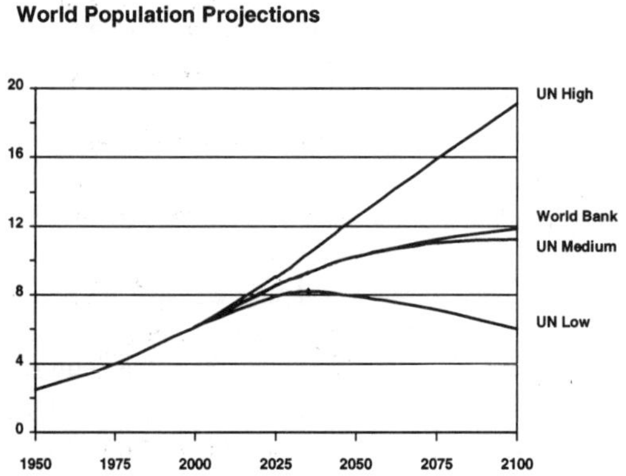

Figure 5 *World population*

Figure 5 illustrates estimates of world population growth provided by several agencies[4]. Up until 2040, there is general agreement on a rise in population to approximately 10 billion. We must regard population growth as inevitable, given that the females have already been born who will bear the population increase shown in the first part of the graph. As mentioned above, quality and variety requirements will ensure that demand for food will outstrip population growth. This suggests that we need to provide for roughly a tripling of food requirements by the middle part of the next century. This scenario provides the boundary conditions for two basic choices - either the same amount of land has to be three times more productive, or we put three times as much of the world's surface into agricultural production. Inevitably, the outcome will be a hybrid of these factors, but the balance between them is open to question. A detailed discourse on this topic is provided by Dennis Avery's thought-provoking book (1995).[5] In Figure 6, I have extracted some data from this volume in relation to world population growth. A commonly held myth is that providing more food to under-developed countries will lead to a population explosion. The data provided in Avery's treatise indicate that producing more food appears to lead to lower birth rates. In a range of very different countries, increased food provision via yield enhancement bears a correlation to a reduction in births per female. We must of course recognise that related factors, both social and educational, also play their part.

World Population Growth 1950 - Present

Producing more food leads to lower birth rates:

Country	Crop	Yield increase %	Population increase %	Births / female (1950)	(1995)*
USA	corn	152	89		2.1
France	wheat	195	38		1.8
Indonesia	rice	160	142	5.5	2.4
Chile	corn	x4	130	4.0	2.1
India	rice	>x2	149	5.8	3.1
China	rice	150	114		1.9

D. T. Avery, 1995

Figure 6 *World population growth in relation to food yield increase and births per female. *(Timescales are approximate and not strictly comparable, although data for individual countries are internally consistent)*

1.6 Safely

The crop protection industries are amongst the most highly regulated in the world. There is a highly effective and conservative regulatory framework for pesticides albeit with departures from science-based risk assessment being evident in several systems. The regulation of genetically modified foods and organisms is regionally heterogeneous, leading to differential rates of adoption of key technologies. The major outlier is the EU where the regulatory process has lacked transparency and decisiveness. The situation here is highly politicised, partly as a result of several recent food scares and the consequent emergence of highly motivated pressure groups.

The R&D companies recognise the imperative of providing technology which is benign both to the environment and to users, growers and the public at large. Indeed, the companies have responded to regulatory pressures by introducing tests at the early screening stages of product development which navigate the inventive effort towards products which have the key characteristics which satisfy regulators and users alike.

The following talk by Professor Sir Colin Berry entitled "Caution, precaution or indemnity? The cost of different approaches to regulation" will, I am sure, provide illumination on the current status of standards in regulation.

1.7 Sustainably

To be sustainable, food provision technology must satisfy the following criteria:

- agronomy - reliance on effective systems
- the environment - must conserve the natural balance
- society - must provide net benefits to communities
- economics - must provide rewards for all necessary components of the food product chain

As such, the concept of sustainability is extremely far reaching, and can be analysed under numerous headings.

One aspect which causes public concern is the effect of agricultural inputs on soil fecundity. An important contribution has been made by the Rothamsted group who demonstrated that repeated applications of five crop protection chemicals over a 20 year period produced no measurable effect on the yield of spring barley, taken as an indicator. No deleterious effects on crop productivity were observed, nor could significant differences be found in microbial processes in soil. Furthermore, no pesticide residues could be detected in soil 17 months after the last experimental treatment[6].

A beneficial contribution to sustainability can be found in the use of non-selective herbicides in no till or minimum till regimes. As such the use of 'chemical hoes' can obviate the need for mechanical ploughing, thus retaining moisture and soil structure. This is particularly important in dry land agriculture in order to protect topsoil from erosion, either by wind or storm water.

Another important, but often neglected, aspect of sustainability is provided by energy use. Comparison of technologies and products based upon relative energy consumption can be quite illuminating. As an interesting example, a study at Exeter University estimated the energy required to produce, process and distribute various foods to the household[7]. Figure 7 compares the energy (MJ) required from non-photosynthetic sources to provide one MJ of energy as food intake.

Ratio* of embodied energy (MJ)
required to provide 1 MJ food intake (nearest integer)

Rice	3
Peas	6
Chicken	7
Beef	8
Apples/pears	11
Wine	20
Salad vegetables	45
Coffee	177

Figure 7 *Comparison of embodied energy (defined as the energy required during the entire life cycle to bring a product to a household, not including the photosynthetic component) to the calorific value of various foods*

On this basis, some meats are amongst the least energy demanding foods with regard to production since they are not highly processed, grown locally, and are highly calorific. Conversely, coffee is highly processed, grown remotely, and provides little food energy. On this analysis, judicious choice of diet can contribute to alleviating the greenhouse effect. It is recognised by the authors that their data are necessarily imprecise and that several potentially confounding factors are aggregated, but the example does illustrate that comparisons of technologies must be made across a wide range of parameters in order to provide a total picture. Single factors taken in isolation can mislead.

I should like to conclude this background section by referring to the 1997 Bawden Lecture of Dennis Avery[8]. There is reference to the balance which must be struck between land productivity and total area under the plough. Avery states that between 1960 and 1992 agricultural yields have essentially tripled on the approximately 6 million square miles in agricultural production. This rate of productivity has been the key to meeting the food demands of a burgeoning world population. However, if there had been no yield improvements since 1960, a further 10-12 million square miles (roughly the land area of the US, EU and Brazil combined) would have been needed for agricultural production to provide the same amount of food. Avery concludes that high yield agriculture is the solution not the problem for wildlife and the environment by preserving wildlife habitat from the plough.

Doubtless the debate will continue to rage on whether effective alternatives can be found to current intensive technologies. In the next section, I intend to provide a realistic assessment of currently available technologies, and to highlight recent advances which have changed the face of the crop production industry.

2 SCIENCE AND TECHNOLOGY - FUTURE IMPACT ON CROP PROTECTION AND FARMING

2.1 Crop Protection Beyond 2000

Chemical pesticides will remain a keystone component of crop protection for several decades to come. This statement is frequently disputed by those who believe that more recently introduced technologies such as genetic manipulation will totally dominate the future. However, there are at present no real alternatives on the horizon to herbicides for weed control. Given that weed control accounts for approximately half of the crop protection market, chemical treatments, albeit based upon different molecules, will be essential for the foreseeable future. Furthermore, development of resistance by pests will mean that new toxophores will be highly prized. This is especially the case for both insect and fungal control. The R&D-based companies have invested heavily in technology for invention of superlative new compounds and these, taken together with careful application regimes, will undoubtedly provide sustainable methods.

I now wish to request that you imagine a chemical entity which:

- worked at 500 mg/ha (= 0.018oz/ac)
- offered exceptional control (rainfast, broad-spectrum, flexible application window pre and post activity) with superior crop safety
- favourable ecotoxicology (non-oncogen, reduced risk product)
- cheap to make on a small dedicated pilot plant (or even in the laboratory)
- range of formulations to suit all growing conditions

Clearly, it is going to be immensely difficult to invent a molecule with these demanding characteristics. However, all of these properties have been discovered to date, albeit in different molecules. The trick therefore is to invent a single molecule featuring all of these properties. Whereas this is recognised as a massive task, there is some new help available such as:

- functional genomics to identify targets
- high throughput screening, featuring these targets, to detect leads
- combinatorial chemistry to feed the screens and to optimise leads

Each of these will be taken in turn, together with related aspects of the research and development chain.

2.2 Functional Genomics

In its simplest form, ascribing the function to a gene promotes the opportunity either to change plant constitution (eg. high oil, low phytate) or to identify targets for chemical intervention in the biochemistry of an organism (eg. enzyme inhibition). In its most powerful form, functional genomics allows identification of the genetic sequences which are being expressed during specific biochemical processes and relates them to function. For example, this will allow identification of the genes which are responsible for the process of grain filling. Alternatively this technology can be used to determine which enzyme is being inhibited by a pesticidal molecule. Similarly, information can be provided on biochemical events such as xenobiotic metabolism, mode of action, mechanisms of resistance and the quantitative importance of a particular target site. The technology becomes particularly potent when comparisons are made between wild type organisms (whether plants, insects or fungi) and altered varieties eg. mutants. Thus, identification of a cultivar with a favourable property (eg. high oil), can lead to elucidation of the genetic events responsible for this property. The impact of the genomics revolution on agricultural research is heavily dependent on information technology which allows manipulation of very large data sets (bioinformatics).

Whereas this technology is presently in its infancy, there has been massive investment in the area, both public and private, in recent years and immense progress is being made. This has been made possible by intense activity in genome sequencing, again both in public and private organisations. There is no doubt that advances in this field, and parallel progress with proteomics, will change the way that research is done within the industry to a remarkable extent. Already today a paradigm shift has taken place in the pattern of R&D spend within the major crop protection companies. In the context of chemical invention, the identification of novel target sites and their validation is a key objective. Such high priority targets will then be included in high throughput screens to allow exposure to a diverse range of chemical challenges. Such methodology is now becoming standard in the industry as a preferred method of lead generation.

2.3 High Throughput Screening

For many years, it was usual in the conventional agrochemical industry that companies synthesised ca. 5,000-20,000 new chemical entities in their laboratories and submitted these together with purchased compounds to effective in vivo screens to test for herbicidal, insecticidal, fungicidal and plant growth regulatory activity. Typically, compounds were synthesised in 200mg-1gm amounts in order to provide sufficient material for biological testing on whole organisms. More recently, attention has been paid to developing customised high throughput screens requiring extremely small amounts of sample eg. 2mg for testing against a wide range of organisms. From our own experience at Zeneca, we have developed high throughput micro-screens based upon standard 96 well microtitre plates, featuring in vivo tests as well as those based upon enzymes/protein assays. In our laboratories the rate of screening has been raised from ca. 10,000 units in 1996 to well over 100,000 in the current year. Again, this has occasioned a paradigm shift in lead discovery philosophy. The maxim today is to focus upon "what you find" rather than agonise over "what you miss".

The competitive advantage afforded to a company by such screening methodology depends on numerous factors. Firstly, the effort expended in method development will be rewarded by highly relevant and customised screens. Secondly, the data reduction techniques applied to the massive amount of data generated will be crucial, as will the analysis of the data points emerging from these screens. Finally, the diversity of chemical inputs which can be achieved for these screens is perhaps the most important feature and forms the topic of the next section.

2.4 Chemical Diversity - Feedstocks for the Screens

The following sources are available to ensure that the screens are fed with a wide range of chemistries.

- in-house synthesis
- compound exchange with third parties
- compound libraries, combinatorial chemistry
- robotic synthesis
- natural product extracts (broths, plants)

In-house synthesis will increasingly become a relatively minor component of feedstocks for the screens. Miniaturised screens requiring only small sample sizes will facilitate compound exchange with third party companies whose collections contain small sample sizes (eg. pharmaceuticals). It is now possible to purchase a significant amount of chemistry requirements externally eg. through boutiques providing compound libraries based upon combinatorial methods. Advances in automation mean that synthesis using robots can be applied widely, and especially to the repetitive synthetic sequences which typify close analogue synthesis. Finally, Nature provides an unending supply of diverse chemical entities, and small companies are appearing which can provide tailored extracts of fermentation broths and plants in quantities appropriate for screening.

2.5 Lead Optimisation

Several strategies for providing leads for further optimisation have been described above. To complete the picture, mention must be made of lead generation by following and improving upon known biological activity (eg. from competitor patents) and by rational design of potent inhibitors of biological processes in pests and weeds. The comparative failure of the rational design process to meet expectations is due to the inherent complexity of achieving efficacy in vivo. Whereas it is frequently possible to design inhibiting molecules, success also requires:

- survival of the inhibitory molecule in the application environment and as a deposit on the pest surface (sunlight, temperature, rainfall, micro-organisms)
- efficient entry into the plant/pest
- efficient translocation to the site of action (including survival of metabolic destruction by the host organism)
- no adverse effect on the crop
- no deleterious effects in the environment (eg. soil micro-organisms, insects, mammals etc)
- cost effective manufacture
- success in the market place

It is clear that the molecular properties which are required for potent inhibition will not be optimal for the other processes mentioned above. Thus we are faced with a trade off of molecular properties, explaining why the introduction of a successful molecule usually requires so many different starting points, and a lengthy process of structural optimisation.

Techniques in combinatorial chemistry and robotic synthesis can be equally relevant to lead optimisation as well as lead generation strategies. Accordingly, there is presently a major trend to externalise the chemistry spend within the major agrochemical companies by forging alliances or agreements with the numerous contract houses which have established themselves in the recent past.

2.6 Environmental and Toxicological Parameters

It must be recognised that there are many customers for the agrochemical R&D process and its end products. These range from the distribution chain leading to the end user farmer and grower customer, through to the environmental agencies that represent the public in terms of the safe and ethical use of agrochemicals. It is essential that the agrochemical majors form a scientific dialogue with the regulatory agencies. By and large, this is now firmly in place in many countries of the world with technical advocacy becoming an increasingly important aspect of R&D activities. The reality is that the agrochemical industries are now amongst the most highly regulated in the world. As mentioned above, the regulations covering exposure to crop protection chemicals and their residues are stringent and large safety margins are employed. The development spend for a typical crop protection agent will be in the region of $50-100m, the majority of which is dedicated to ensuring safety to operators, the public and the environment. It is vital that research programmes are positioned to avoid areas of potential difficulty. In this context, early stage screening will incorporate indicative tests for favourable environmental properties (eg. soil persistence, leachability). The application of genomics to toxicology offers promise in the future for indicative tests for specific toxicological alerts.

Several regulatory authorities are now providing a tiered system of regulatory requirements in which "safer pesticides" received accelerated registrations. Given that time to market is a vital commercial parameter for agrochemicals, it is clear that research work focused upon identifying safer molecules will be significantly rewarded.

2.7 Fast-tracking

As mentioned, time to market is one of the key features which dictates the ultimate profitability of an agrochemical product. In addition to early market entry, the availability of more lengthy patent protection is a key competitive element. Accordingly, many companies have re-engineered their development chain in recent years, with consequent and significant improvements. However, it is perhaps more important to examine the research process and particularly the phase in which product candidates are fully characterised. Fast-tracking in this area will require parallel pursuance of activities which were hitherto taken stepwise. Departure from robust sequential decision making causes major discomfort in many organisations and deep seated change is required with processes, organisation and culture in order to gain sustained success. Amongst these changes are:-

- a move to a more risk-taking culture - fewer data than previously will be available to underpin decisions. Parallel pursuance of activities means that failure in the earlier stages is more expensive than hitherto.
- significant resource needs to be invested in method development to provide early tests as indicators of key produce attributes eg. soil persistence to underpin early risky decisions.
- an innovative managerial focus is required on the processes involved in product R&D. This inevitably de-emphasises function (eg. Chemistry, Biology) based management in favour of process-focused multi disciplinary project teams. Removal of traditionally-powerful functional baronies is non-trivial in any organisation - no less so in the agrochemical industry.

2.8 The Benefits of Potency

Apart from the obvious benefit provided by potency which can be reflected in cost per treatment (eg. low variable product cost contributing to available gross margin) there are several consequential benefits which can include:

- small manufacturing plant (waste, effluent burden, energy input etc)
- lower bulk transportation costs
- greater flexibility in formulation type (eg. tablets) and lower adjuvant costs (sometimes)
- smaller packs; more convenient application methods

Thus, it is clear that leads which hold the promise of high potency will be highly prized. Again, the advent of new bioscience will assist in identifying target sites which are capable of exploitation to provide powerful but selective crop protection agents.

2.9 Formulation Technology

Significant advances have been made in recent years with regard to formulation technology in parallel to the excitement raised in new compound invention. The industry has tended to move away from liquid formulations to solids, spurred on by considerations of safety to users. Elsewhere, the advantages provided by micro-encapsulated formulations have been recognised viz. controlled release, hazard reduction and mixture compatibility. The use of water soluble bags has also become important, based upon advantages such as unit dosing, convenience in use and reduction of exposure hazard.

In addition to considerations of safety, it has also been well recognised that formulation technology provides a potential route to enhancement of bioperformance. Detailed studies of the effects of adjuvancy on observed activity have provided major gains in efficacy.

2.10 Precision Agriculture

Mention must be made of the startling progress made in recent years by the application of global positioning systems and geographic information systems to the application of chemical treatments to crops. Using such technology, irregularities across terrain (eg. pH, organic matter, nutrients) have been accurately mapped to within a few metres. In turn, boom sprayers with adjustable outputs are being developed to deliver an optimal but varied dose across the crop. Adoption of this technology has been variable, with the US being the most advanced to date.

Thus, the technology and the information management systems associated with precision farming have the potential to transform practical agriculture over the next two decades and play an essential role in maintaining the industry's licence to operate through more accurate timing and application of product.

3 BIOTECHNOLOGY IN AGRICULTURE - A PARADIGM SHIFT

The impressive progress in crop protection chemistry described above has been matched by advances in the harnessing of bioscience to meet agricultural needs. The first transgenic crops were introduced into a few major markets in the mid-1990's with notable success. Whereas biotechnology-based solutions will not be relevant to all pest, disease and weed problems, there will be many applications which find favour with farmers and growers. In addition, transgenes will be employed to improve both the quality (eg. flavour) and agronomics (eg. stress tolerance) of crop plants. The farming of transgenic crops will dramatically change the basis of competition within the crop protection industry. This will bring changes in traditional channels of trade, terms of business and marketing strategies.

Public perception of genetically modified crops presents a most interesting picture. Whereas several groupings hail biotechnology-based agriculture as a new green revolution, other oppose its use as an interference with Nature. Beyond the farm gate, the food processors and retailers, particularly in Europe, reflect a nervousness over the outcome of this issue and presently favour food labelling to provide the consumer with choice.

3.1 Application of Biotechnology

It must be emphasised that the impact of biotechnology on crop production will benefit both chemically and genetically-based technologies. Biotechnology is key for supporting

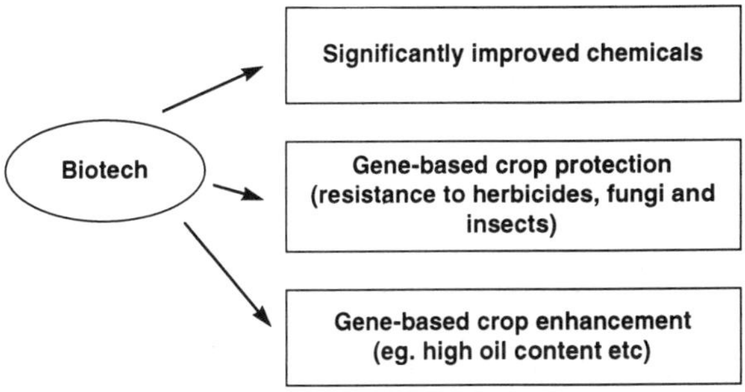

Figure 8 *The Impact of Biotechnology on Agriculture*

discovery and validation of novel target sites, as described above. Additionally, very significant potential lies with the generation of crop varieties with increased pest resistance, and of varieties resistant to chemical herbicides. Perhaps the most exciting role for biotechnology is in the manipulation of crop quality (eg. high oil, low phytate, high nutrient). In Figure 9 an estimate of the pace and quantum of uptake of gene-based technologies is provided.

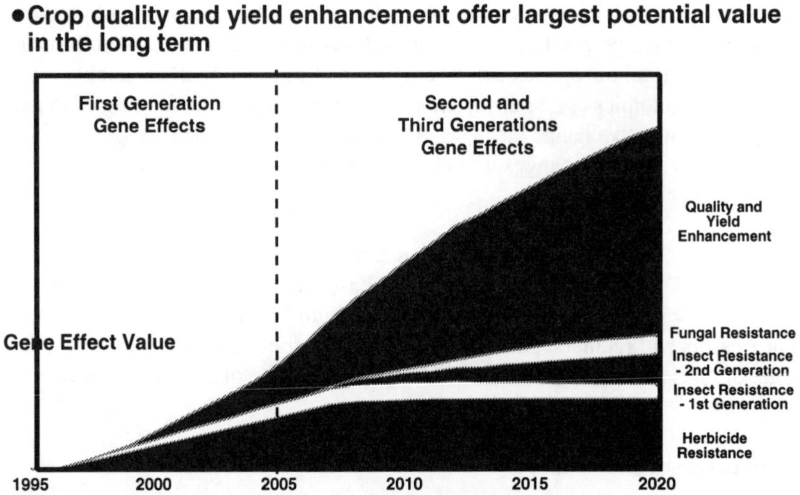

Figure 9 *Biotechnology: Quality and Yield Enhancement*

Crop quality and yield enhancement are likely to offer the largest potential value in the longer term. Research in this area is already well advanced and examples of this include:

- high oil, modified starch, high protein crops etc
- modification of flavour and sweetness
- agronomic effects eg. cold tolerance, drought tolerance
- post-harvest benefits eg. anti-sprouting, anti-bruising
- enhancement of beneficial components eg. nutraceuticals

There are many excellent examples available which describe the success of the use of transgenic plants in the context of crop protection. These include disease resistance, resistance to insect pests and the engineering of selectivity to herbicides which would otherwise damage the crops. Descriptions of recent successes will be provided elsewhere at this meeting. Looking to the future, the incorporation of multi-gene constructs into plants holds the promise of delivery of powerful broad spectrum effects, tailored to specific crop plants.

3.2 Crop Business Systems

In Figure 10, five essential components are illustrated which require to be in place to provide a commercial product in a crop of interest:

- effect genes - to provide required traits eg. herbicide tolerance, fungal control, high oil etc.
- enabling technology - eg. promoters, selectable markers and particularly methods for transforming crops with foreign DNA.
- freedom to operate - intellectual property position, availability of licenses and agreements.

- access to leading germplasm - whereas proof of concept of technical effects can be obtained in model plants, commercial success will require expression of traits in elite crop varieties.
- value protection and extraction - establishing the commercial systems for getting reward eg. by licence fees or seed premiums.

As can be seen, the route from laboratory bench to market from a gene effect is very different, and in many ways more complex and less certain, than for chemicals.

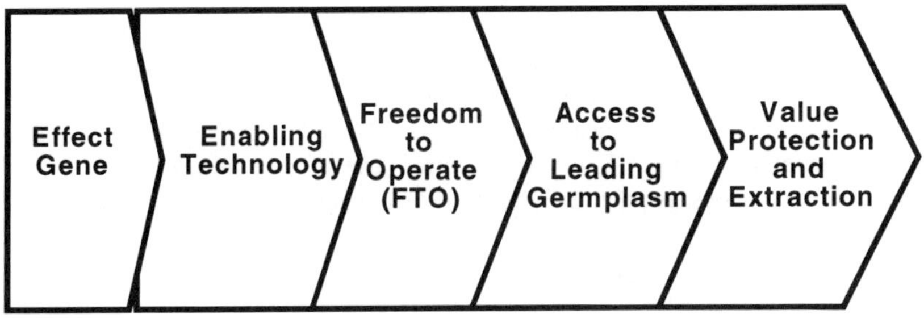

Figure 10

3.3 Genetically-modified Produce

One example of the commercialisation of a genetically-enhanced crop is provided by the launch of Zeneca tomato puree derived from genetically-modified tomato in the UK. By down-regulation of the pectin-degrading enzyme polygalacturonase a higher yield of thicker puree is produced. The puree was launched in 1996 by two retail supermarket chains in the UK following regulatory approval. The success of the product (it outsells the conventional variety) has been based upon three key factors:

- provision of information - a major thrust to provide the consumer with information both in store and via scientific literature
- a statement of benefit - one important feature here is that higher yield of paste allows cost savings to be passed on to the consumer
- consumer choice - most importantly, the cans of puree are clearly labelled as "produced from genetically modified tomatoes", and the unmodified product is made available on the same shelf

4 BIOLOGICAL CONTROL AGENTS

Amongst the many interpretations of the above title, I shall concentrate upon the use of living organisms to control pests and weeds. Whereas the use of biological control agents in control measures is a potentially elegant and specific solution to agricultural problems, the technology has not yet lived up to its earlier promise, except in special situations. Undoubtedly, this is partly due to the relatively low levels of resource which have been expended on R&D, but also there are some inherent difficulties which must be overcome. Amongst these are:

- the characteristically narrow spectrum of biological control agents lends itself well to inclusion in integrated pest management programmes, but this is accompanied by the difficulty of providing complete control in practical and dynamic situations.
- formulation stability - preparations of biological control agents need sufficient shelf life to be accommodated in commercial operations in a wide range of climatic conditions.
- sensitivity to environmental conditions - this leads to variability in response. To compete with characteristically robust chemical treatments, and emerging treatments based upon biotechnology, reliability of control is paramount.
- manufacture - although technically tractable, the costs of manufacture can nevertheless be detrimental to margins. However, the current use of cell cultures (eg. for production of insecticidal viruses) is potentially able to solve this problem.

In the light of the foregoing, it can be appreciated that biological control agents have found it difficult to compete commercially with chemical treatments in broad acre crops. However there has been significant success in some specific applications (eg. pests of glasshouse crops) and in well orchestrated government-sponsored programmes. In particular, biological control agents can play a major part in integrated pest management systems. Excellent reviews on this topic have recently appeared. A review by van Emden et al gives significant coverage to this topic[9].

5 THE FOOD PROVISION CHAIN

Mention has been made earlier of the position of the recent crop management industries in the food provision chain, and the sheer scale of the latter has been emphasised. With regard to the UK alone, the agri-food chain:

- provides 9% of the GDP
- employs well over 1 million people
- enjoys weekly consumer expenditure (1996) on food in the UK totalling well over £1 billion
- is a major exporter
- is amenable to technological input

Source: Agri-Food Chain Research Steering Group (1995/6)

In Figure 11, a simplified schematic showing how plant products reach the user is displayed. It is recognised that there are many additional steps in the full chain but in essence the story starts with a gene and ends with the purchase of food by consumers in retail outlets. This situation has been colloquialised as 'gene to gut', 'dirt to dinner' or 'pipette to plate'. Whereas the traditional focus of agrochemical usage has been at the stage of plant growth (with additional involvement in seed treatments and post harvest treatments) the influence of plant biosciences pervades the whole chain. Thus, the effect of genetic manipulation at the start of the chain will be felt throughout the whole process.

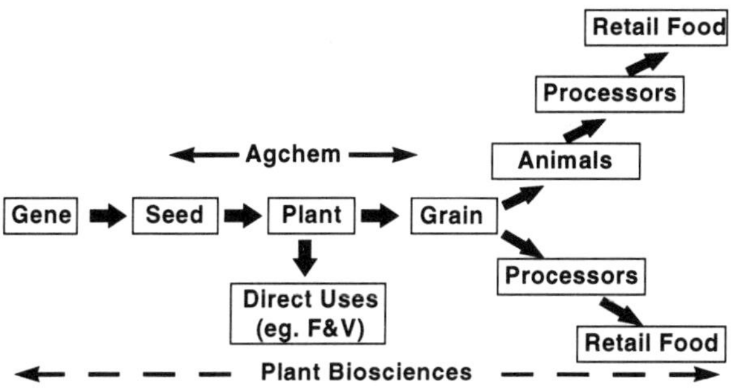

Figure 11 *Getting Plant Products to the User*

In Figure 12, this chain is reanalysed in terms of the influences which are presently occurring, and suggests that this will lead to a significant redistribution of value. The crop protection industry is presently experiencing an increase in specification of allowable inputs to crop regimes, leading to increasing integration along the chain. For the foregoing reasons, the technology suppliers to the chain are engaging in this integration by one of two mechanisms. Firstly, several of the major companies are seeking to establish collaborative partnerships with downstream companies, and secondly others have chosen to purchase key representative companies within the chain.

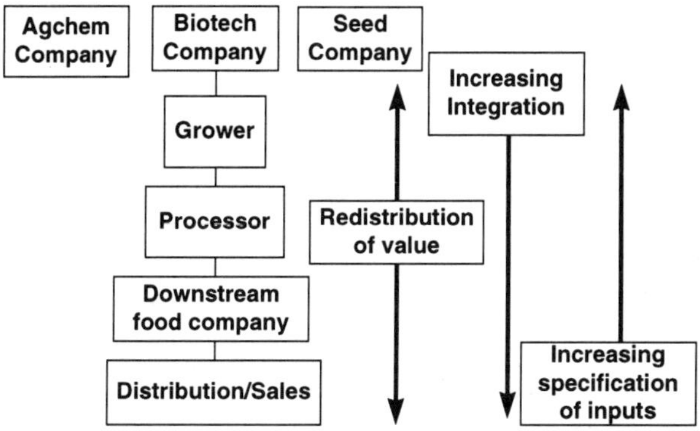

Figure 12 *Vertical Integration*

5.1 Enhancement of Value

It is my personal view that current changes in the structure of the industry will lead to enhancement of value in the future. Many of today's broad acre crops are traded commodities. The introduction of new traits through genetic manipulation will convert these commodities to speciality crops. This will necessitate that such specialities have preserved identity and thus can be tracked throughout the value chain. Furthermore, speciality crops by their very nature demand segmentation strategies, and segmentation in turn will be the engine of value capture. This enhanced value will derive from two sources. Firstly, reward will be provided for the value of the introduced traits. Secondly, a more valuable crop will justify and bear a greater amount of crop protection input. Thus, I believe that the conversion of commodities to specialities will lead to enhanced value, whether for food, feed or fibre (Fig. 13).

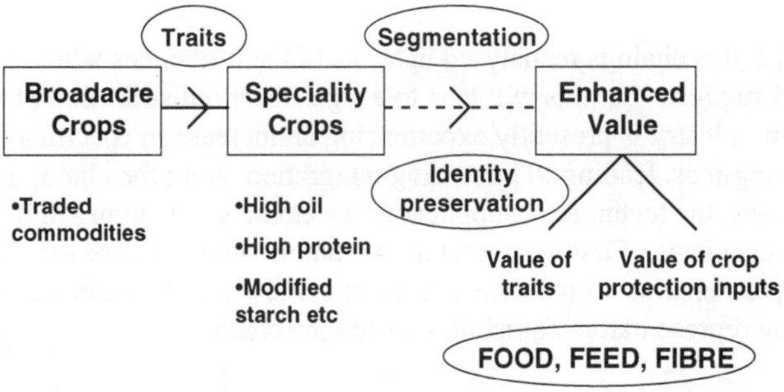

Figure 13 *Enhancement of Value*

5.2 Identity Preservation

As mentioned above, the production of speciality crops will demand that the identity of a particular crop product will need to be preserved throughout the production chain. In addition to the context of value extraction, identity preservation will be required to support the regulatory system and to allow for consumer choice. For many crops, present systems of distribution will require significant modification to promote preservation of identity. Nevertheless, these logistical problems can, and will, be solved.

In Figure 14, comparison is made between the present day situation of limited segregation, and future scenarios. Taking corn as an example, genetic manipulation will enable the growing of corn varieties directed towards specific animal feed lots, thus providing differentiated feed for eg. pigs, ruminants and poultry. Similarly, designer starches will be provided in planta by genetic modification.

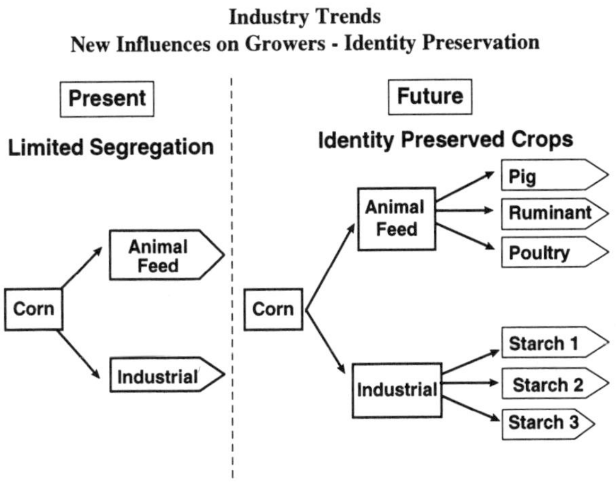

Figure 14 *Industry Trends : New Influences on Growers - Identity Preservation*

6 THE FUTURE

Given the unprecedented amount of change that we have experienced in our industry in the past few years, forecasting the future requires immense courage. To date, the industry has been revolutionised by transgenic expression of a small number of genes.

An enticing prospect for the future rests with our increasing ability to manipulate secondary metabolism. Indeed, it is secondary metabolism which provides the essential differentiation between the vast variety of plants in the world. Once we are able to shift whole secondary metabolism pathways between plant species, then we shall be close to gaining exquisite control over the constituents which plants produce. Once this has been achieved routinely, it is not too fanciful to regard fields of crops as flexible chemical factories (Fig. 15).

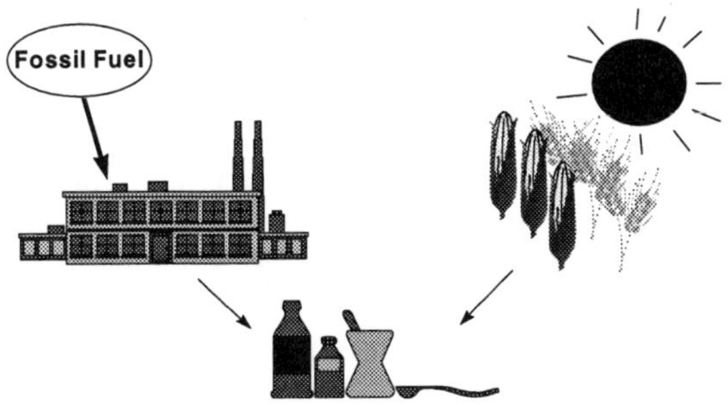

Figure 15 *Fields as Chemical Factories*

This leads to very different scenarios for the chemical production economy. The prime energy source will be sunlight, rather than the present reliance on fossil fuels. Manufacturing processes will experience the greatest change - with combine harvesters working alongside synthesis plants, and chemical tank farms alongside grain silos.

The key that will be most instrumental in bringing about this change stems from the genomics revolution. It is vital that we keep our scientific minds open to the immense prospects for change that now greet us, and to recognise that business as usual is not an option. Early successes are already to hand. Pharmaceutically-valued proteins can now be produced in plants as can biopolymers, useful as plastics. Nutrient-dense foods are being researched, as touched upon earlier in this paper. The production of customised chemicals in plants from genetically-modified enzymes is a little way off, but is undoubtedly technically tractable.

7 CONCLUSION

The coming years promise to be nothing if not fascinating. I believe that the imperative of providing a safe supply of food will dictate that pragmatic solutions will be reached on a world-wide basis, albeit with some countries adopting outlying positions. There will be a migration towards integrated crop management systems which optimise the best available technologies, including effects delivered by chemicals, genes and biological control agents. These changes will be catalysed by progress in the current revolution with information management. In addition to changes brought about by technologies such as precision agriculture and electronic commerce, we must recall that our licence-to-sell documentation, incorporating both efficacy and environmental data, is a rich source of information. The present state of technology relevant to crop production, taken together with advances which are evident in our research laboratories, predicts that we shall meet the challenges of the future. I therefore conclude that integrated crop management systems which optimise technology solutions will ensure that the global population is fed safely and sustainably for decades to come.

References

1. FAO Yearbook, Vol 50, 1996 - data extracted from several sections
2. Extracted from FAO Yearbooks 1950-1996
3. N. E. Borlaug, *Plant Tissue Culture and Biotechnology*, 1997, **3**, 126
4. Source: World Bank and United Nations projection data, quoted in *Saving the Planet with Pesticides and Plastic*, D. T. Avery, Hudson Institute, Indianapolis, 1995, p5
5. ibid, chapter 3, pp46-60
6. R. H. Bromilow, A. A. Evans, P. H. Nicholls, A. D. Todd and G. G. Briggs, *Pesticide Science*, 1996, **48**, 63-72
7. D. A. Coley, E. Goodliffe and J. Macdiarmid in *Energy Policy*, Elsevier Science Ltd, 1998, **26**, 455-459
8. D. T. Avery, *Proc. Brighton Crop Protection Conference - Weeds*, 1997, **1**, 3-16
9. H. F. van Emden and D. B. Peakall, *Beyond Silent Spring - Integrated Pest Management and Chemical Safety,* Chapman & Hall, 1996

Caution, Precaution or Indemnity? The Cost of Different Approaches to Regulation

Sir Colin Berry

DEPARTMENT OF MORBID ANATOMY, THE ROYAL LONDON HOSPITAL, LONDON E1 1BB

The dictionary definitions of the terms of the title are given as **Caution** - the taking of heed, **Precaution** - the exhibition of prudent foresight, and **Indemnity** - the provision of protection against contingent hurt. It is evident that we should be cautious with compounds we have grounds to believe are dangerous, and that we should take precautions where our information is incomplete. However, it is far from clear that a system could be devised which will guard against all contingencies, or that such a system would be desirable. It is illustrative to consider these various approaches to regulation against the reality of risks to which the population are exposed. Regulation is intended to confer benefit and must impose costs, but to regulate without regard to cost may impose secondary risks which confound the purpose of the exercise.

The range, quantity and quality of food available to us has improved beyond recognition in the last 40 years and this has imposed its own problems. Lutz and Schlatter[1] and Hart et al[2] have argued convincingly that natural or additive carcinogens do not explain food related carcinogenic risks - the problem is one of overnutrition. Nevertheless there are clearly problems with food production, but almost without exception these problems are microbiological and occur in food processing, preparation and handling (examples include E.coli, Salmonella, Campylobacter and enteric viruses.[3] Despite this evidence, it is the chemicals in food including food additives, pesticides and preservatives - all of which are used to prevent some of the problems described above - that appear to cause anxiety.

So my first emphasis might be that while most of us would accept that it is mandatory to use caution in the evaluation of pesticides, and precaution in their use, these constraints on behaviour should be directed against real targets. It is difficult to understand why natural toxicity is preferred to synthetic safety, sometimes with very damaging consequences - consequences which are generally ignored by the media. There are many recent examples, some involving deaths. A large number of cases of renal failure and a very significant number of transitional cell carcinomas have been induced in Belgium from a Chinese herbal dieting remedy. The dietary supplement L-tryptophan was withdrawn following extensive use in the USA because of the sometimes fatal eosinophilia-myalgia syndrome. This essential amino-acid occurs at low levels in some foods and may need to be supplemented in special feeding regimes. As a "natural" product L-tryptophan was taken by faddists for insomnia, the premenstrual syndrome and stress reduction; some EMS victims were taking 17 gms/day; this is an example of the absurd but widely held idea that if a little of something is good for you a lot must be better. Other examples include the renal toxicity of dietary germanium supplements.[4] No pesticide has had an effect of this kind.

Many compounds normally present in the diet are both essential in small amounts and dangerous in large quantities - selenium is an example;[5] it is surprising that no-one has started a scare about Soy proteins.[6] The recommendation of a diet distorted in any way by a conviction that the change is beneficial should be made only after it has been carefully evaluated. The concept of dose related effects is almost entirely lacking in response to food scares - hence the reciprocal of the idea noted above - if a vast amount of a compound does harm in testing, a little must cause some problems. Curiously, this belief is combined with the idea that if a little is necessary (for example, with vitamins) a lot must be better, despite the fact that ample evidence exists of the poisonous characteristics of a number of vitamins.

I believe that it is irrational to expect the non-trained population to undertake rational analyses of risk; this is not meant to be a patronising comment but is intended to emphasise the fact that most of us have had to be trained in rational thought (that is what PhD's are for). We do not use this mode of thought in regulating our day-to-day activities, or no-one would buy lottery tickets. This can be illustrated with some simple probability theory on the distribution of birthdays in this audience which enables us to see how misuse of probability theory let us down in "beef on the bone".

So if it is not reasonable to expect members of the public to carry out risk assessments, even isolated and directly relevant assessments such as those used in thinking about your diet, what do we do? There is little point in producing numbers which explain very little to the uninformed. Most consumers will think in terms of risks from single foods rather than diets and not consider their drinking water at all; paradoxically most will be worried about occasional or intermittent low-level intakes where the problems that concern them have usually been identified by experiments involving high life-long **daily** exposures in animals - perhaps the most important point in producing effects. Public concern appears most often to be related to the long term effects of compounds ingested at very low doses whereas operator and acute exposures will bother most toxicologists. Even this is not certain; the data are often supplied by pressure groups and have been collected to support causes and the issues may relate to non-pesticidal uses of compounds such as the organophosphates (in sheep dips, or in the Gulf war, for example).

It is dangerous and misleading in trying to indemnify against all possible harm; this may lead to irrational acts even in those who believe they are rational. Lest I appear to be criticising other regulators only, let me comment on some actions of the ACP. A number of products containing a moss killer were not permitted for garden use on the basis of the precautionary principle, because of the risk of damage to eyes. The compound was used in a formulation of very high pH - clearly dangerous. It is almost as dangerous as oven cleaner (pH a unit higher - but this is a log scale). No incidents of injury had been reported or could be found which had involved the use of non-agricultural pesticide products; in the month we considered more data at the ACP there was a newspaper report of an oven cleaner being sprayed on a number of people in a bus queue and of damage to the skin of a baby's face in this incident (JAN. 1998). On this basis, total body protection should be worn when cleaning the oven !

Your industry is being affected by the fact that irrational concerns about health are of increasing concern to a healthy population. Why this is so interesting; many feel it is a form of self-indulgence and the population at risk clearly differs from the rest of us.[7] What Barsky and Borus[8] have called **somatisation** - the reporting of somatic symptoms that have no pathophysiological explanation - has been recently characterised. These authors have described how changes in social attitudes have reduced the tolerance of the public to mild symptoms and benign infirmities. The threshold for seeking medical attention has been lowered and discomforts and isolated symptoms are identified as diseases by patients. These authors write in a general context as psychiatrists, but part of the problem in our area of concern is that, as Wessely[9] has pointed out, attributions are not neutral. Those affected by pollutants are generally referred to as "innocent" victims" but if the same symptoms are attributed to a psychological problem or psychiatric disorder there is substantial blame attached by society and guilt is experienced by the individual. In a situation of uncertainty there are many compelling psychological reasons why individuals may prefer environmental explanations for their ills. Once this is done there are ample sources of reinforcement as Sutherland[10] has pointed out; a desire for attention, the comfort of being able to abdicate responsibility for one's failings by ascribing them to an illness, the luxury of receiving therapy, the excitement added to a humdrum life by weird beliefs about one's past, the propagation of stories produced by hysterics by the mass media, the attempts by therapists to push their clients into revealing multiple personalities, non-existent sexual abuse and even alien abduction in the mistaken belief that such revelation will affect a cure.

In considering these difficult issues we must rely on science to be rational. And what is it about science that helps? Science delivers data that are reproducible in temporal and geographical terms, in different laboratories at different times - or in different populations if an epidemiological linkage is what is sought. If a hazard cannot be characterised in some reproducible way its establishment as a risk is problematic. Bad epidemiology has been damaging when masquerading as science; if six factors are analysed and the confidence intervals are set at 95% (they usually are) then the probability of finding a false positive is one in four - it should be the case that studies concentrate on very few factors and require very low P values to be significant.

It will only be possible to ensure that there is a well considered and appropriate public response to environmental problems if there is an informed public debate. This, in turn, requires that there is an examination of the relative value of the various kinds of evidence presented to them. In a number of instances, notably with 2,4,5-T, careful examination of evidence by the courts has rejected wild assertions from pressure groups, the same is true for daminozide. These events get recycled[11] and the continuing presentation by pressure groups of discredited data confuses the issues.

My final example is the presentation of issues to the public about the use of radiation in food processing. The activity of pressure groups has caused real harm and one might argue, many deaths.[11] There is undoubted public anxiety about radiation in all sorts of forms, but it is ironic to note that recent food poisoning incidents with, let me emphasise, a number of deaths - might have been prevented had irradiation been used a an effective method of disease prevention.

Wildavsky[12] has posed the important question of "why we are not all dead?" He makes the point that the benefits of risk aversion and risk taking should be considered together; those changes in diet, assumed by many to be benefits, are of course risks - and we will not know their effects for a long time. Very small risks are compounded with time and then become very large, in fact impossibly so. Mumpower[13] has pointed out that as the actuarial data suggest a decrease in the overall level of risk in society, the rate of risk reduction from unconsidered factors must exceed the additional risks being introduced by new chemicals or practices. Why is cancer of the stomach disappearing? It might be reasonable to associate this change with pesticide use and the disappearance of mycotoxins from food; indeed, some have suggested a link with changes in methods of preservation of food, but these are not useful suggestions; there are no good data. The important point is that we do not attempt to assess, in any significant way, the factors which are making our daily life safer. The erroneous conclusions imposed by the mishandling of small risks, the methodology of risk evaluation used by some regulatory systems combined with the use of estimated data about food intakes and various other worst case assumptions produce nonsensical estimates of risk. This *de minimus* thinking allows figures to assume a life of their own and may persist in the face of new information. The problem lies in our failure to include benefits in our assessments.

In the presence of confrontation rather than rational debate, money is spent on irrational responses in ways that fail to resolve our problems. Recent legislation in the United States of America has addressed the problem of the quality of information used in the courts in medical malpractice cases and it would be valuable if similar constraints were placed on some aspects of the environmental debate. In commenting on the debate about phenoxyherbcides and soft tissue sarcoma Gough[14] has said "The Swedish studies have been widely cited and will probably will be cited for years to come. People who cite these studies in arguing for a link between dioxin and soft tissue sarcomas will have to ignore the evidence that doesn't support the link, but that can be done."

Real catastrophes happen; one related to a source of current public concern, the processing of uranium. Uranium is a rather common element, commoner than arsenic, antimony, iodine, mercury, silver, cadmium and about one sixth as abundant as lead - all well known toxins for man. We each have about 60 micrograms of uranium in our skeleton. The radioactivity from this source is not of great significance and the element is regulated mainly by virtue of its toxicity as a heavy metal. The plant at Fernald, Ohio dealt with uranium as part of US national defence operations and was dealing with large amounts of the element. In 1984 a filter failure occurred and an estimated 124 kg of uranium as dust was released into the atmosphere. Following this release an investigation into the previous history of the plant was made and a truly remarkable story emerged[15] with estimated releases of a total of 179,000 kg of uranium in the years between 1951-1987 and 6500 kg thorium between 1962 and 1978 together with extensive contamination of the Miami river. This appalling history quite properly lead to a study of health effects in the exposed population both within the plant and around it. A number of conservative estimates were made (it was assumed that people spent all of their time outdoors) and particle size was considered in a way which probably inflated estimations of dose, but the estimated dose to a person living closest to the plant and downwind of it was such that they received approximately 40% of what the average US citizen receives from natural sources in a year from the plant. In the same period radon exposures from the plant were around 1000 less than those from natural sources. Using the US Environmental Protection Agency (EPA) linear model for risk assessment one discovers that the loss of life expectancy for the exposed individual is 6 days - compared with 2000 days for smoking a pack of cigarettes per day.

It cannot be supposed that this degree of environmental pollution is in any sense tolerable, but what followed the identification of the problem was irrational. Bear in mind the magnitude of the risk estimate; an independent study suggested that the loss of life in the exposed population was an excess of 7 deaths (0.2 per year) from the 34 years of exposure. Cohen[15] points out that for comparison the air pollution from a typical large coal burning electric power generating plant results in an estimated 25-100 excess deaths per year.[16] Although these figures should also be taken with a pinch of uranium they indicate that the subsequent expenditure of around $1.1 billion in clean-up costs and almost $80 million in health care activities was foolish; it would in no way affect outcomes. However, that part of the money spent on improving the plant ($356 million) was clearly a necessary expense.

There is an unconsidered view that regulation increases safety and in an obvious sense this is true - building regulations are a clear example. Regulation based on caution (an adequate overbuild in supporting girders, say) or precaution - ensuring that fire escapes will allow evacuation in unlikely combinations of circumstances, are clearly prudent. However, to ensure that a building is constructed so that no-one ever comes to harm in it is impossible (in a fire they might fall down the fire stairs and certainly should not use the lift; all buildings cannot be single story). This is a trite example but the principle is an important one, regulation does not generate safety. No regulation should be formulated without risk and cost benefit analyses; these should provide the basis for rational expectations about potential harm and not seek to adopt solutions which will ensure that no individual human being, vertebrate or invertebrate, protozoan, plant or soil structure is altered.

References

1. W. K. Lutz, and J. Schlatter, J. *Carcinogenesis*, 1992, **13**, 1191.

2. R. W. Hart, D. A. Neumamm and R. T. Robertson (eds) "Dietary Restriction: Implications for the Design and Interpretation of Toxity and Carcinogenicity Studies", ILSI Press, Washington DC, 1995, p. 396.

3. Anonymous, 'Communicable diseases control and epidemiology. Introduction'. *World Health Statistics Quart*, 1992, **45**, 166.

4. T. Matsusaka, M. Fujii, T. Nakano, K. Okada, K. Okagawa, Y. Furod, K. Morizumi, K. Sato, H. Morita, S. Shimomura and S. Saito, *Clin. Nephrol*, 1988, **312**, 219.

5. T. H. Jukes, *Nature*, 1995, **376**, 545.

6. K. D. R. Setchell, L. Zimmer-Nechemias, J. Cai and J. E. Heubi, *Lancet*, 1997, **350**, 23.

7. G. H. Hall, W. T. Hamilton and A. P. Round, *J R Coll Physicians Lond*, 1998, **32**, 44.
8. A. J. Barsky and J. F. Borus, *JAMA*, 1995, **74**, 1931.
9. S. Wessely, *Trans Cult Psych Res Rev.*, 1994, **31**,173.
10. S. Sutherland, *Nature*, 1997, **388**, 239.
11. Health Which? "Pesticides in fruit & veg", June (1998) pp 8.
12. A. Wildavsky, "Searching for safety", Transaction Books, New Brunswick (USA) and London (UK) p. 40.
13. J. Mumpower, *Risk Analysis*, 1986, **6**, 437.
14. M. Gough, "Phantom Risk:Scientific inference and the Law", K. R. Foster, D. E. Bernstein, and P. W. Huber (eds), MIT Press, Cambridge,MA, 1993, p. 249.
15. B. L. Cohen, "Phantom Risk: Scientific Inference and the Law", K. R. Foster, D. E. Bernstein and P. W. Huber (eds), MIT Press, Cambridge, MA, 1993, p.319.
16. R. Wilson, S. D. Colome, J. D. Spengler and D. G. Wilson, "Health Effects of Fossil Fuel Burning", Ballinger, Cambridge, MA, 1980

Synthesis and Structure–Activity Relationships

Chirality in Agrochemistry: An Established Technology and its Impact on the Development of Industrial Stereoselective Catalysis

Gerardo M. Ramos Tombo*[1] and Hans U. Blaser[2]

[1] FUNGICIDES RESEARCH, NOVARTIS CROP PROTECTION AG, CH-4002 BASEL, SWITZERLAND
[2] CATALYSIS & SYNTHESIS SERVICES, NOVARTIS SERVICES AG, CH-4002 BASEL, SWITZERLAND

1 INTRODUCTION

Chirality is a property related to molecular symmetry. Symmetry considerations have become fundamental for a deeper understanding of molecular structure, reactivity and biological activity.[1]

Although biological processes are conducted in a chiral environment, chirality is not a requirement for the biological activity of small molecules. However, in those cases in which the bioactive molecule contains one or more stereogenic centers, the desired biological activity often strongly depends on a given absolute configuration.[2,3]

The advantages of using stereochemically pure (or enriched) agrochemicals have been recognized for many years. Owing to practical limitations, only a few substances (mainly natural products or their derivatives) were developed as single stereoisomers until the mid 80s. The dramatic improvements over the past 20 years in the synthesis and analysis of enantiomerically enriched compounds enabled the investigation of stereoisomerism as one of the many determinant factors of the biological properties of crop protection chemicals.[3]

Hence, it has become a major task for contemporary agrochemists, to provide routes of access to single stereoisomers of chiral target molecules, thus making the evaluation of their corresponding biological profiles and other related properties possible. In many cases, the development of economically feasible synthetic processes for their large scale preparation is decisive for the commercial future of a chiral agrochemical.[4,5]

1.1 Available Synthetic Methods for Introducing Stereogenic Centers[6]

The *resolution of racemates* is probably still used most often despite the fact that the yield of the desired enantiomer is at best 50% per cycle. If the undesired isomer cannot be recycled, it must be disposed of. Similar problems occur when applying stoichiometric chiral reagents or chiral auxiliaries. This does not occur if *chiral building blocks*, isolated from natural products or produced by fermentation can be used, since they are incorporated in the final product and nature has already created the desired absolute stereochemistry. However, for larger scale applications it is not always possible to find the suitable starting material, for chemical and economical reasons. Therefore, *enantioselective catalysis* with either biocatalysts or chiral chemical catalysts is now being applied more frequently and has the best future perspectives because the chiral auxiliary is required only in sub-stoichiometric quantities. The present paper reports on the opportunities and problems associated with the use of chiral metal complexes as enantioselective catalysts for industrial purposes. Also some specific examples of biocatalytic processes will be discussed. However, it must be stressed that there are so many factors that influence the economical and ecological quality of a chemical process that no single approach is able to meet all requirements and every method mentioned above can be the most suitable one for solving a particular problem (Table 1).

Table 1 *Applicability of the different stereoselective synthetic methodologies in the developmental phases of an agrochemical principle*

Phase	*Scale*	*Methods*			
		Separation	*Chiral Pool*	*Biocatalysts*	*Chemical Catalysts*
Lab	g	√	√	-	-
Field	kg	√	√	-	?
Production	< 100 t/y	-	√	√	√
	> 100 t/y	-	√	?	√

1.2 Progress in Enantioselective Chemical Catalysis for Industrial Applications[7]

Whether a synthetic route containing an enantioselective chemical catalytic step can be considered for the preparation of laboratory quantities or for the manufacture of a particular product is usually determined by the answer to two questions:

- Can the costs for the over-all manufacturing process compete with alternative routes?
- Can the catalytic step be developed in the given time frame?

This means that several factors determine the technical feasibility of an enantioselective catalytic process:

- The *enantioselectivity* for which >80% values can be acceptable for agrochemicals.
- The *maturity* of a catalytic method (good procedures, reliability, known scope and limitations) is an important point for the synthetic chemist, both in the discovery phase and in development.
- The *catalyst productivity*, given as turnover number (ton) or as substrate/catalyst ratio (s/c), determines catalyst costs. Tons ought to be >50,000 for large scale or less expensive products. Catalyst re-use increases the productivity.
- The *catalyst activity*, given as turnover frequency (tof, h^{-1}) affects the production capacity. Tofs ought to be >500 h^{-1} for small and >10,000 h^{-1} for large scale products.
- *Availability and cost of the catalyst*: chiral ligands and many metal precursors are expensive and/or not easily available. Typical costs for chiral diphosphines are 100–500 \$/g for laboratory quantities and 5000–20,000 \$/kg on a larger scale. Only few ligands are available commercially.

• The *development time* can be a hurdle, especially if an optimal catalyst has to be developed for a particular substrate (substrate specificity) and/or when not much is known on the catalytic process (technological maturity).

For most other aspects such as catalyst stability and sensitivity, handling problems, catalyst separation, space time yield, poisoning, chemoselectivity, process sensitivity, toxicity, safety, special equipment, *etc.*, enantioselective catalysts have similar problems and requirements to nonchiral catalysts.

As shown in Table 2, not many enantioselective catalysed transformations fulfill all of the above criteria. This is true despite a very considerable effort both in academia and in industry.

Table 2 *State of the art and evaluation of useful enantioselective catalytic transformations*[7]

Transformation	*ee range (best)*	*scope*	*maturity lab*	*maturity production*
Hydrogenation of enamides	80-95 (99)	broad	yes	yes
Hydrogenation of α,β-unsaturated acids	80-95 (99)	medium	yes	yes
Hydrogenation of unfunctionalized C=C	60-75 (96)	narrow	?	no
Hydrogenation of (functionalized) C=O	80-90 (96)	broad	yes	yes
Hydrogenation of N-aryl imines	60-80 (90)	narrow	?	yes
Hydrogenation of N-alkyl imines	60-80 (96)	narrow	?	no
Hydrogenation of N-acyl-hydrazones	70-90 (96)	narrow	yes	no
Hydrogenation of cyclic imines	90-95 (99)	medium	yes	no
Epoxidation of allylic alcohols	90-95 (98)	broad	yes	yes
Epoxidation of C=C	80-95 (97)	broad	yes	?
Dihydroxylation of C=C	85-95 (99)	broad	yes	?
Hydroformylation of C=C	60-80 (97)	narrow	?	no
Hydrosilylation and hydroboration of C=C	80-90 (97)	narrow	?	no
Michael addition reactions	70-90 (99)	narrow	?	no
(Hetero) Diels–Alder	80-95 (99)	narrow	?	no
Aldol reactions	80-90 (98)	narrow	yes	no
Au-aldol reaction	90-95 (97)	narrow	yes	?
Addition of ZnR_2 to RCHO	90-95 (99)	narrow	yes	no
Formation of cyanohydrin	90-95 (98)	narrow	?	no
Cross coupling	70-90 (95)	narrow	?	no
Nucleophilic allylic substitution	70-90 (>95)	narrow	yes	no

Of special importance for the development of stereochemically enriched chiral agrochemicals in Novartis [c)] were the following advances in enantioselective catalysis:

[c)] Novartis, a life sciences company formed by the merger of Ciba-Geigy and Sandoz in 1996

1.2.1 Ru-binap catalysts. This catalyst system was developed by Noyori and coworkers in the early 1980s and turned out to be one of the few generally applicable enantioselective catalysts that gave very high ees for unsaturated acids, enamides, unsaturated alcohols and, most important for us, substituted ketones (Figure 1).[1,8,9] Even though the catalyst activity of these complexes is often relatively low, several technical applications have been claimed.[10]

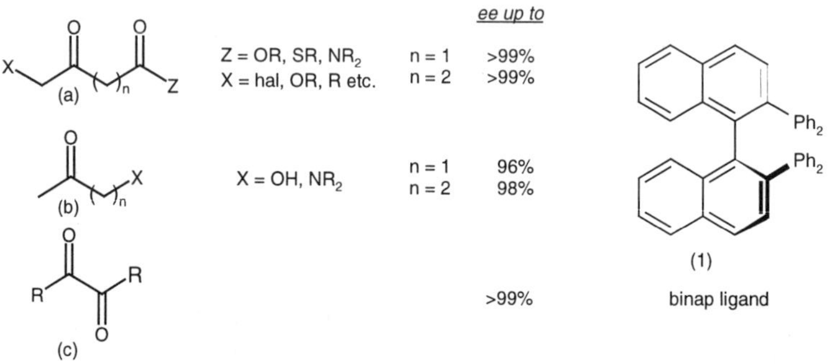

Figure 1 *Best enantioselectivities for the hydrogenation of substituted ketones using Ru-binap catalysts*[10]

1.2.2 Heterogeneous Pt-cinchona catalysts. These catalysts, originally developed by Orito, were investigated and developed to technical maturity within Novartis.[c)11] The catalysts are highly selective only for α-keto acid derivatives (Figure 2) but several research teams are working on possible extensions.

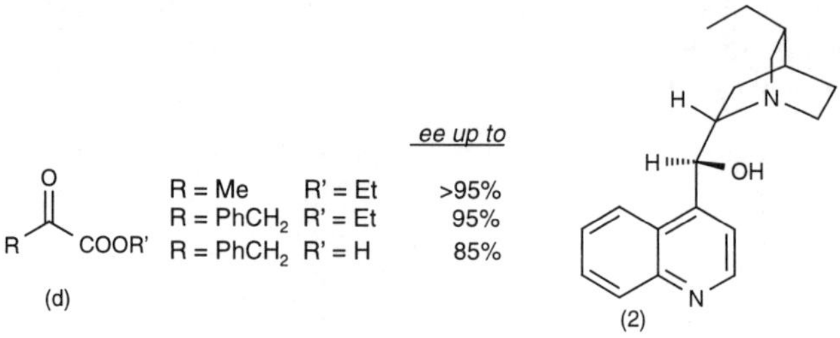

Figure 2 *Best enantioselectivities for the hydrogenation of α-keto acid derivatives using 10,11-dihydrocinchonidine (2) as modifier and 5% Pt/Al_2O_3 as catalyst*

1.2.3 Ir-diphosphine catalysts. The hydrogenation of C=N bonds is one of the more neglected areas of asymmetric catalysis. The need for an efficient synthesis of chiral amines, a important functional group in many chiral agrochemicals, motivated us to make a very strong and extensive effort to find effective catalysts for this relevant transformation.[12]

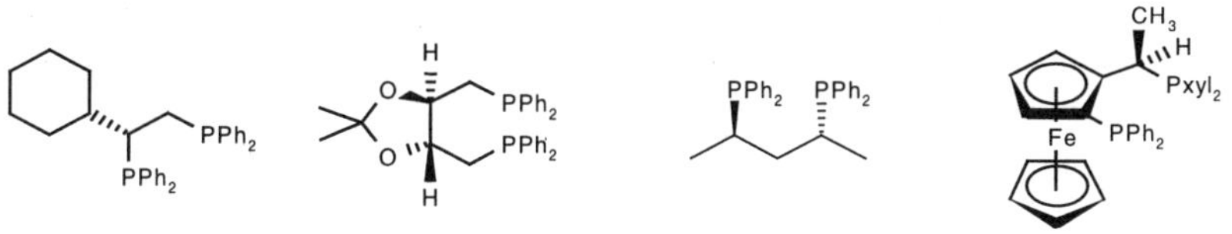

(3) cycphos (4) diop (5) bdpp (6) xyliphos

Figure 3 *Important ligands used for the enantioselective hydrogenation of N-aryl imines*[12]

Table 3 *Performance of different catalytic systems for the enantioselective hydrogenation of N-Aryl imines*[12]

Catalyst system	*best ee*	*best ton*	*best tof* (h^{-1})
Rh - cycphos	69	670	15
Ir - diop - iodide	70	10,000	800
Ir - bdpp - iodide	84	500	40
Ir - xyliphos - iodide - acid	81	1,000,000	150,000

2 CHIRAL CROP PROTECTION AGENTS

A rough statistical analysis of the entries of the Pesticide Manual 11th. Ed.[13] shows that products containing stereoisomerically enriched active ingredients are an established technology (Figure 4): approximately 19% of all entries are (carbon)-chiral compounds. 5% of the listed compounds are commercialized as single or enriched isomers. Perhaps more illustrative is an analysis of the 25 world market leading products (Table 4). These compounds represented a market volume of about 9.5 billion US$ in 1996, *i.e.* 30% of the total world market at that time (Wood Mackenzie, 1996). Six of these compounds are chiral, with a sales volume of 2 billion US$; 3 of them are being substituted by single isomers or mixtures of enhanced stereoisomeric composition.

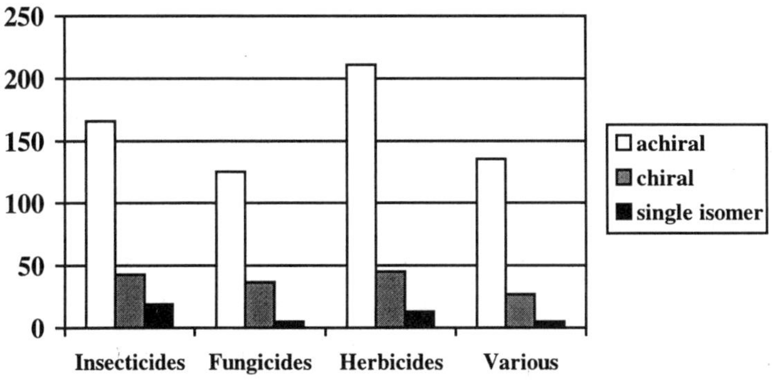

Figure 4 *Distribution of (carbon)-chiral compounds among the entries of the Pesticide Manual 11th Ed.*[13]

Table 4 *Top 25 world market leading crop protection products. In italics the chiral compounds. The star indicates compounds being substituted by single isomers or enriched stereoisomeric mixtures.*

Herbicides	*Insecticides*	*Fungicides*
Glyphosate	Chlorpyriphos	Mancozeb
Atrazine	Imidacloprid	Chlorothalonil
Metolachlor *	*Cypermethrin**	*Propiconazole*
Imazethapyr	Carbofuran	*Metalaxyl**
2,4-D	*Methamidophos*	
Paraquat	Monocrotophos	
Trifluralin	Endosulfan	
Pendimethalin		
Dicamba		
Nicosulfuron		
Isoproturon		
MCPA		
Bentazone		

A detailed analysis of the correlation between biological activity and stereochemistry of the active principles shows clearly that the situation is far from being black and white.[14] Typical situations frequently met are the following:

1. The stereoisomers have complementary biological activity.
2. All stereoisomers possess nearly identical quantitative and qualitative biological activity.
3. The stereoisomers have qualitatively similar activities but quantitatively different potencies.
4. The stereoisomers have qualitatively different biological activities (the desired biological activity resides in one stereoisomer).

It is not our purpose to go through an exhaustive review and analysis of cases. A literature search narrowly focused on agrochemical relevant queries gave more than 700 references of scientific articles and roughly 200 patent references dealing with chirality in crop protection for the time frame of 1993–1998. We rather want to discuss the impact of the technology development on the industrial enantioselective synthetic methodology, based on a few examples in which the authors were directly involved.

2.1 Chiral Dioxolanes

The following examples will illustrate two aspects: the preparation of chiral diols and hydroxy acids as well as stereoselective acetal formation.

One of the most important triazole fungicides in the market is Propiconazole (8) (Tilt®), (Scheme 1) a systemic foliar fungicide for broad spectrum disease control in cereals, bananas, turf and other food and non-food crops.

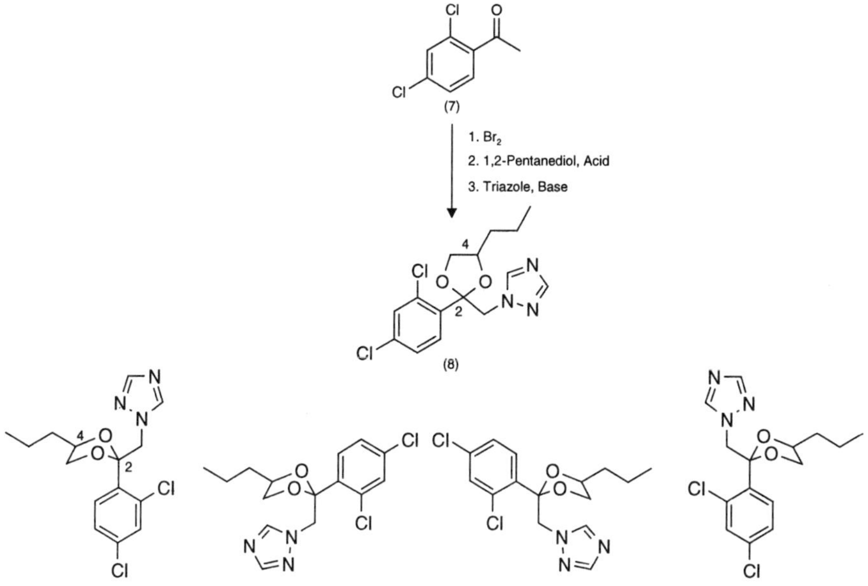

2S,4R -(8) 2R,4R -(8) 2S,4S -(8) 2R,4S -(8)

Scheme 1

The stereoisomers with the 2S absolute configuration were shown to be more efficient *in vitro* inhibitors of ergosterol biosynthesis in *Ustilago madis.*[15] After this finding, large amounts of all 4 stereoisomers were required for their *in vivo* biological profiling. The preparation of enantiomerically pure 1,2-pentanediol and a stereoselective formation of the dioxolane ring were key steps towards a suitable large scale approach to the stereoisomers.

The enantioselective hydrogenation of 1-hydroxypentan-2-one (9) using a Ru-binap complex demonstrates the broad applicability of this catalyst to carbonyl hydrogenations (Scheme 2). The high substrate/catalyst ratio (>1000:1) makes the process very attractive for large scale production. The (R)-catalyst produces (2R)-pentanediol (R)-(10); (2S)-pentanediol (S)-(10) can be prepared using the (S)-catalyst.

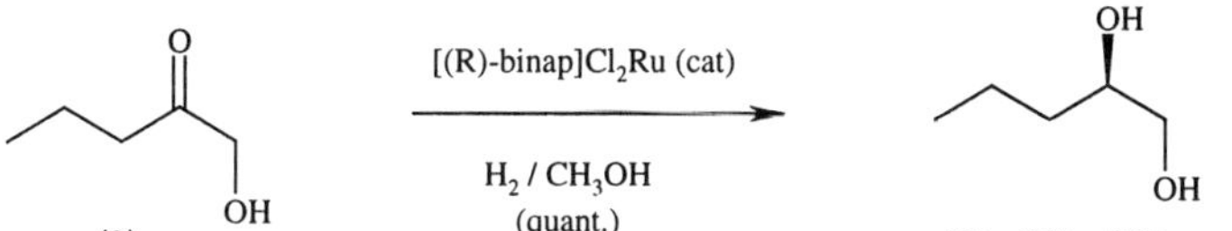

(9) (R) - (10) 95% e.e.

Scheme 2

An alternative enzyme catalysed process based on the stereoselective hydrogenation of 2-oxo-pentanoic acid (11) using lactate dehydrogenase (LDH) and recycling the required NADH cofactor with formate dehydrogenase (FDH) afforded enantiomerically pure hydroxy-valeric acid (12) (Scheme 3).[17] The process can be performed either batchwise or continuously, using an enzyme membrane reactor. L-LDH was used for the synthesis of (2S)-hydroxyvaleric (S)-(12) acid and D-LDH for the (2R) (R)-(12) enantiomer. Reduction of the hydroxy-acid with borane generated *in situ* gave the desired diol in high yield.

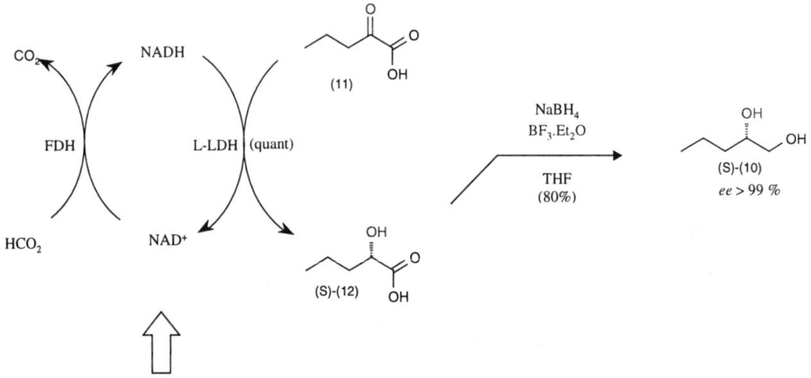

ENZYME MEMBRANE-REACTOR

volume: 270 ml
reaction time: 100 h
conversion: 95%
ee: >98%
yield: 417 g/l.d

L-LDH: L-Lactate dehydrogenase from bovine heart, spec. enzyme consumption 194 U/kg
FDH: Formate dehydrogenase from yeast, spec. enzyme consumption 2000 U/kg

Scheme 3

The formation of acetals is generally catalysed by protons, but several other Lewis acids, including a number of transition metal complexes have been reported as being active catalysts for this reaction.[18] The equilibrium of this reversible reaction is determined by the nucleophilic addition of the alcohol to the carbonyl group. The idea to control the stereochemistry of the dioxolane-ring formation gave rise to a series of experiments which afforded novel, mild and very efficient acetalization catalysts. The concept was based on the utilization of rhodium(III) complexes of sterically very demanding terdentate phosphine ligands (Scheme 4).[18,19]

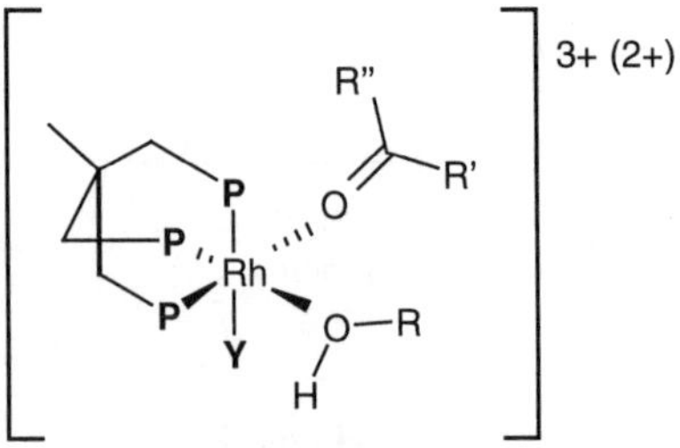

(e)

for *triphos*, **P** = PPh_2 , **Y** = solvent or (anionic ligand)

Scheme 4

Some examples of the use of $[Rh(CH_3CN)_3(triphos)](OTf)_3$ as catalyst (s/c: 1/2000) are given in Scheme 5.[19]

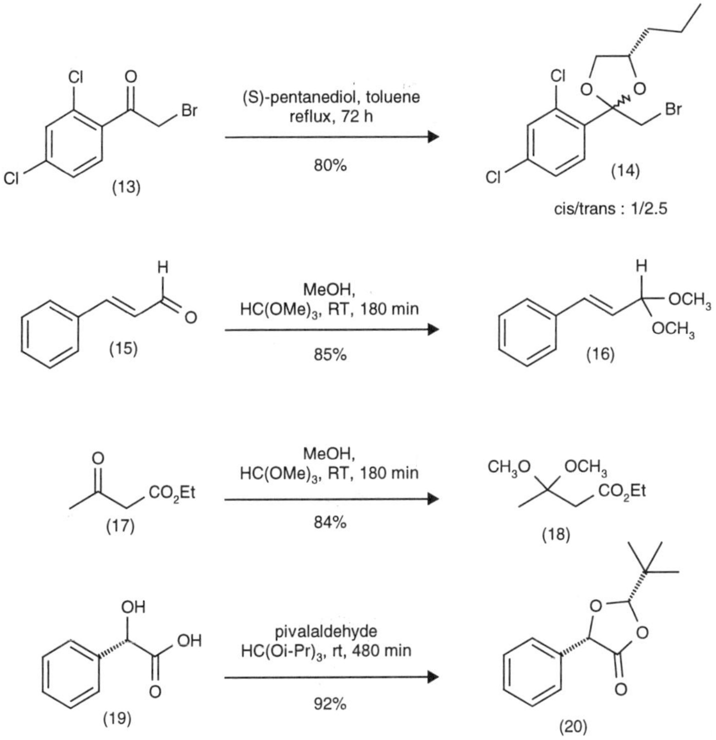

Scheme 5

Unfortunately, the stereoselectivity of the acetalization reaction using Rh-triphos complexes did not fulfill the requirements for the synthesis of the isomers of propiconazole. Thus, the single propiconazole stereoisomers were prepared in large scale by selective crystallization of the nitrates of the corresponding diastereomeric acetals.[20]

Field testing of the single isomers clearly showed that the different pathogen species discriminate between each isomer. However, the fungicidal activity was not constant across the spectrum of target organisms, each individual stereoisomer showing a somewhat different activity profile (Figure 5). As a consequence, all stereoisomers are needed for a strong activity across a broad spectrum of pathogens, thus contributing to the extraordinary commercial success of propiconazole.[20]

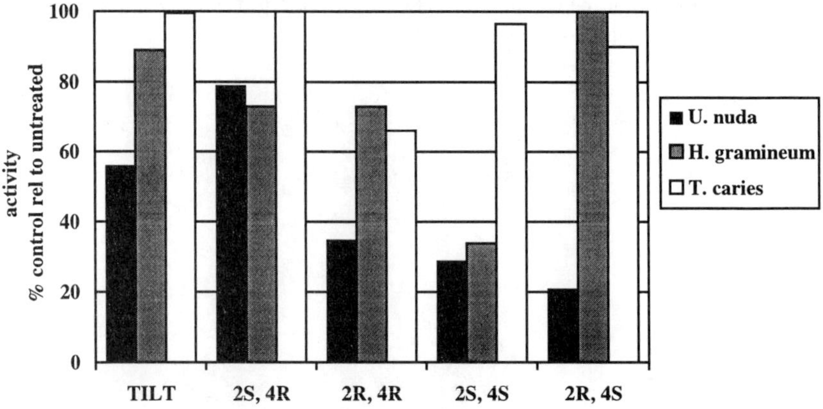

Figure 5 *Activity of propiconazole and its stereoisomers against U. nuda, H. gramineum and T. caries on seed treatment (rates: 2.5g ai per 100 kg seed) in wheat.*

Interestingly, the Rh-Triphos catalysed acetalization which failed in the example discussed above was very useful in the enantioselective synthesis of (26), a chiral experimental herbicide.[21] The chemo-enzymatic approach, combining the enantioselective enzymatic preparation of the mandelic acid analog (22) with the diastereoselective Rh-triphos catalysed dioxolanone (23) formation and subsequent stereoselective chloroalkylation and fragmentation to (26) is shown in Scheme 6.

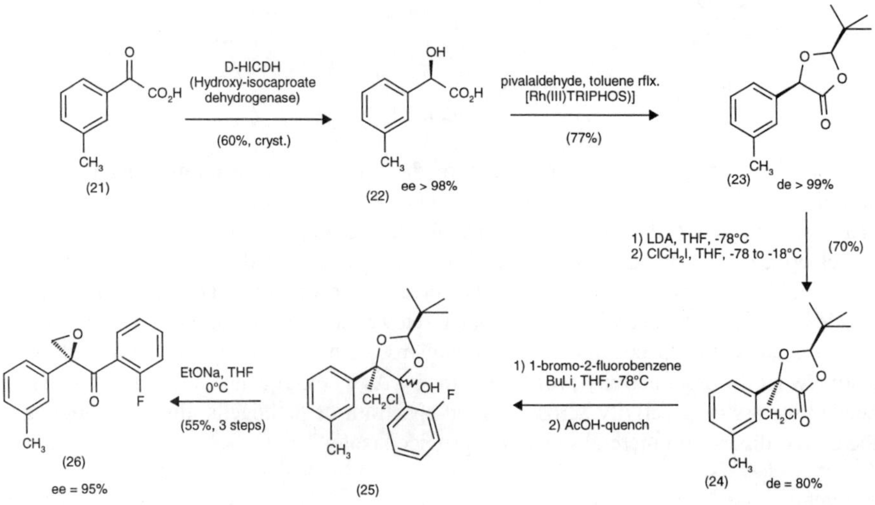

Scheme 6

Another relevant class of 1,3-dioxolane derivatives showing insect growth regulating activity are the juvenile hormone (JH) mimics (30) and (31) (Scheme 7), which contain 2 stereogenic centers in the dioxolane ring.

A major research breakthrough in the development of JH mimics was the replacement of the unstable epoxy sesquiterpene unit in natural JH (f) by a phenoxy–phenoxy moiety (27). Further modifications led to (28) and to the market product (29)[23] and the experimental insect control compounds(30)and (31).[24,25]

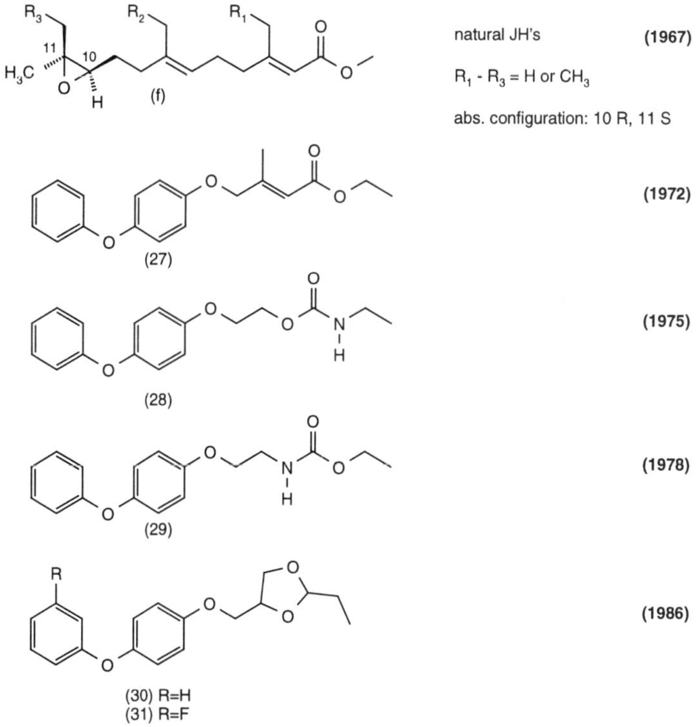

Scheme 7

The chiral glycerol derivative (34) was the pivotal compound in the chosen approach to the target structures. Besides the classical chiral building block approach (Scheme 8) and a catalytic process using a Ru-binap catalyst (Scheme 9), a non conventional stereoselective crystallization gave the most promising results for a large scale synthesis.

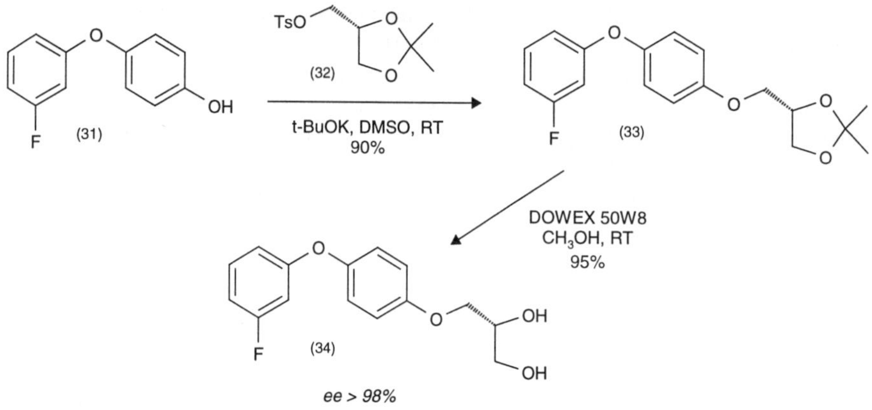

Scheme 8

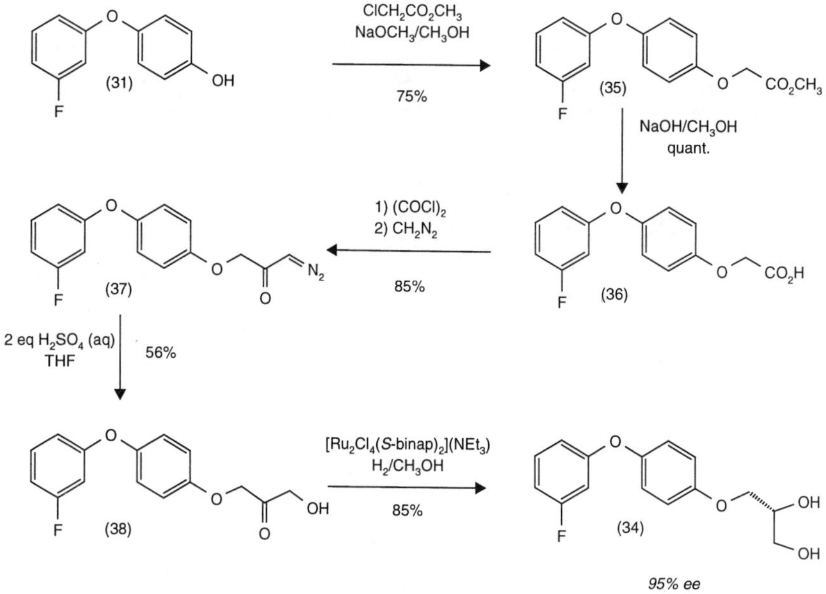

Scheme 9

In analogy to described procedures[26] using (1R)-camphor-10-sulfonamid (39) as chiral auxiliary, one diastereomeric acetal was formed preferentially upon acetalization with glycerol. The major product was isolated in diastereomerically pure form by crystallization. In spite of intensive optimization attempts a full conversion grade was never attained. However, the mixture of the minor diastereomeric acetals could be equilibrated under the reaction conditions to a mixture containing *ca.* 60% of the desired stereoisomer. The target diol (34) was then prepared using standard procedures. The chiral auxiliary (39) was quantitatively recovered, thus fulfilling the requirements of a large scale process (Scheme 10).

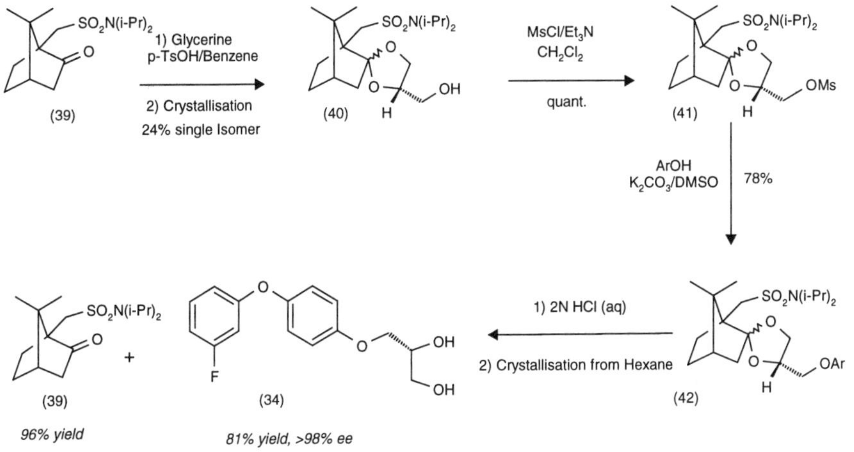

Scheme 10

Preliminary biological testing of isomeric enriched mixtures obtained by chromatographic separation indicated that the most active isomers have a *cis* substituted 1,3-dioxolane ring. Thus, our efforts were focused on a stereoselective, irreversible acetal formation. The use of a Rh-tripodal phosphine catalyst gave the *cis* isomer with good stereoselectivity (Scheme 11).

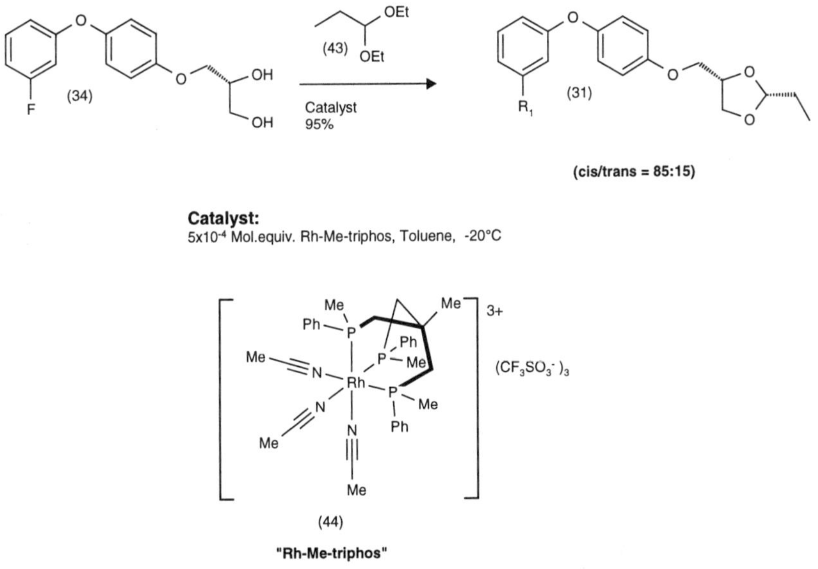

Scheme 11

In the end, the face-selective, heterogeneous hydrogenation of the appropriate ketene-acetal (47) was the best solution to this problem (Scheme 12).

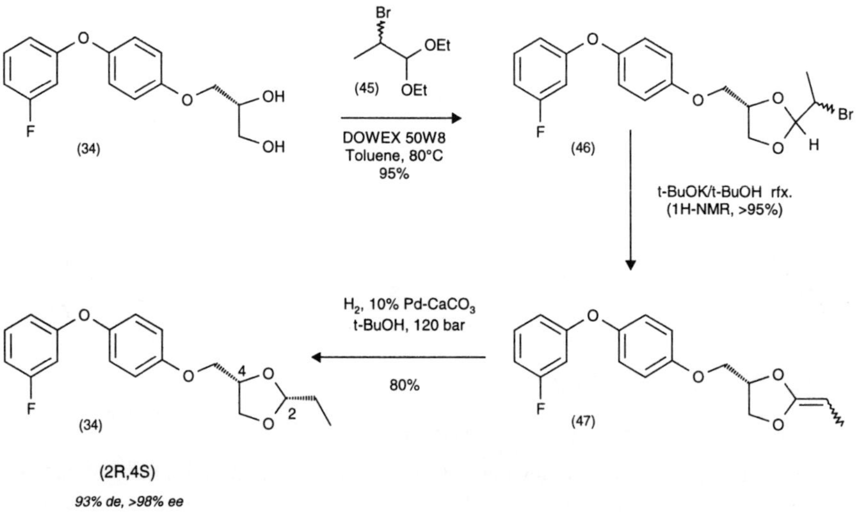

Scheme 12

The biological activity of the isomers of the dioxolane (34) was assessed *in vivo* on the cockroach *Nauphoeta cinerea*. The single isomers dissolved in olive oil were injected into the pre-adult stages. The evaluation of the juvenoid activity was based on five different morphological characteristics which are known to be under juvenile hormone control. The results clearly indicated that all stereoisomers have juvenoid activity but that the configuration at C(4) in the dioxolane ring seems to play a major role in the potency of the effects. In this context, it was also demonstrated, that there is no epimerization at C(2) at biological pH values (Figure 6).[25]

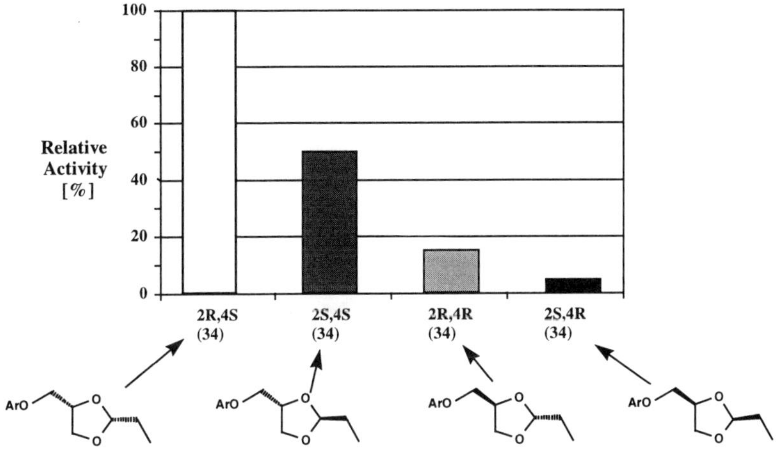

Figure 6 *In vivo activity of the dioxolane stereoisomers against Nauphoeta Cinerea*[25]

2.2 Chiral Acylanilides

Chiral acylanilides play a predominant role among the commercially successful chiral crop protection agents. Metolachlor (47), the active ingredient of Dual® one of the most important grass herbicides for use in maize, is structurally closely linked to metalaxyl (49), an important active ingredient in products used for the management of plant diseases involving pathogens of the *Oomycetes* class. From the research point of view these products are also related. Replacement of the alkoxyalkyl substituent in a metolachlor analogue by an ester group led to the compound (48), which exhibited fungicidal as well as herbicidal activities (Scheme 13). These activities were clearly associated with the absolute configuration of the compound (48) as illustrated in Figure 7.[27] The work on this compound was discontinued once the biologically more powerful analogue metalaxyl (49) was discovered and it was demonstrated that the racemate shows no phytotoxic effects.[28]

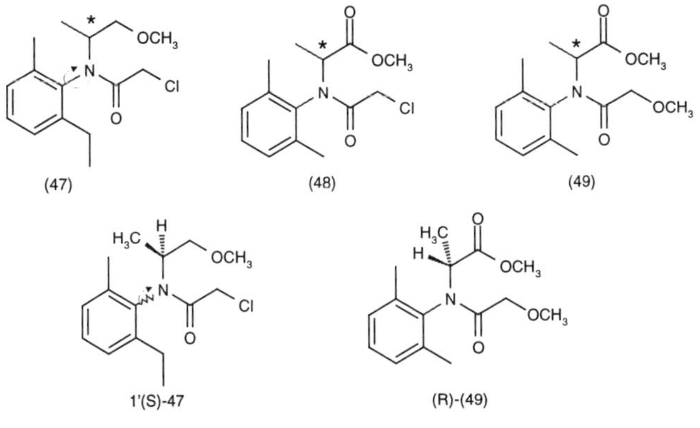

Scheme 13

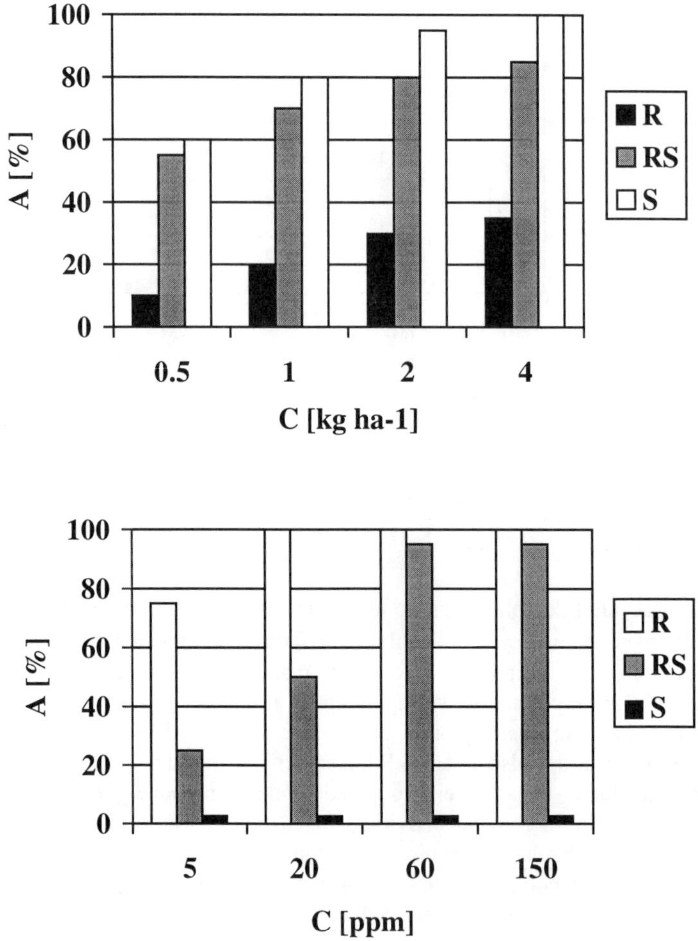

Figure 7 *Biological activity of 48 and its stereoisomers. Top: mean herbicidal activity on eight grassy weeds, pre-emergent application. Bottom: fungicidal activity against grape downy mildew*[27]

Because of the stereogenic center at C1' and the atropisomerism due to hindered rotation around the aryl carbon-nitrogen bond (marked with an arrow in Scheme 13), metolachlor consists of 4 stereoisomers. In 1982 it was found that the two 1'S isomers (epimeric at the C,N axis) provide most of the biological activity.[29] The challenge of finding a production process for the desired isomers was enormous, mainly due to the lack of a suitable technology, the volumes of production and the existing, extremely efficient production process for metolachlor. From the beginning, the efforts were concentrated on catalytic approaches, using the imine (50) formed *in situ,* which is also the intermediate in the non-stereoselective process (Scheme14).

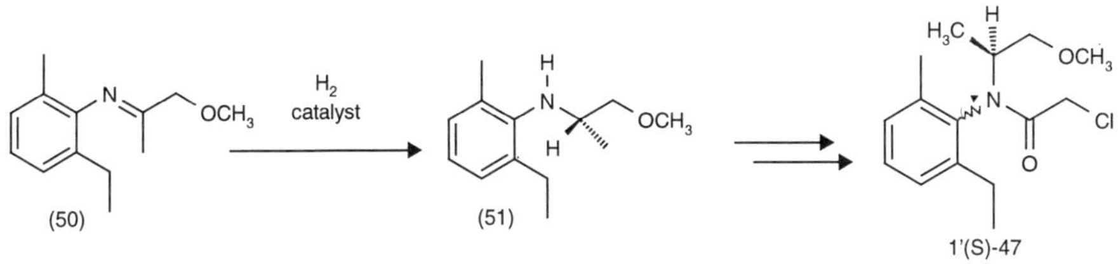

Scheme 14

During more than 12 years of uninterrupted research work, numerous approaches involving different substrates as well as catalytic systems were screened.[12] Two related projects were, among many other contributions, decisive for the development of the technology which in the end allowed the problem to be solved.

1. The hydrogenation of sterically hindered imines with iridium catalysts (instead of the classical Rh catalysts), although giving only moderate enantioselectivity, allowed us to understand and develop the system from the point of view of catalytic performance. (Scheme 15)[30]

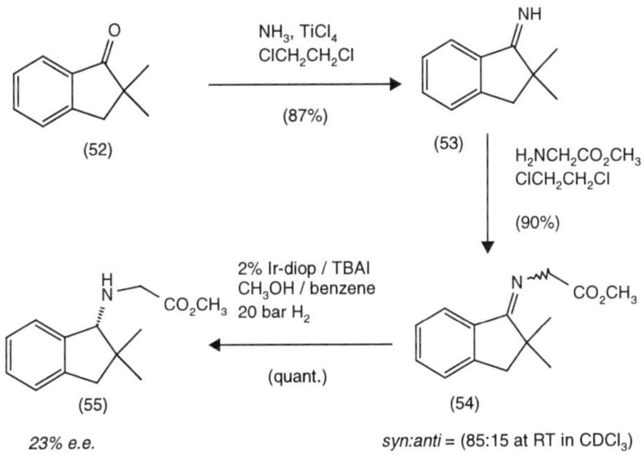

Scheme 15

2. The development of a technically feasible synthesis of ferrocenyldiphosphine ligands in the context of a multi kg-scale synthesis of the antidepressant levoprotiline (62), (Scheme 16).[31]

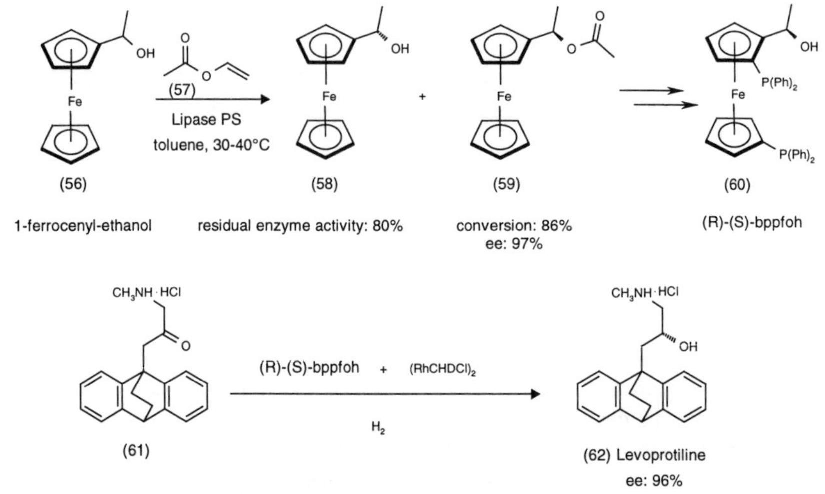

Scheme 16

An extensive screening of ligands, solvents, additives and an optimization of the reaction conditions led to a process for the hydrogenation of the imine (50) which sets a new standard for the application of enantioselective homogeneous catalysts (Scheme 17).[12] At 80 bar hydrogen at 50 °C, in presence of acid and iodide and generating the catalyst *in situ* from $[Ir(cod)Cl]_2$ and the ferrocenyldiphosphine ligand xyliphos (63) with a substrate to catalyst ratio of 1,000,000, complete conversion of the imine (50) was reached within 2–3 h with an initial tof exceeding 1,800,000 h^{-1} and ee values around 80%. The results were confirmed in a 1000 litre pilot loop reactor. On the basis of these results a production process was developed and is in operation since the end of 1996. Dual Magnum®, with (S)-metolachlor of 80% ee and presenting the same biological effect with about 35% less environmental load, was introduced to the market in 1997.

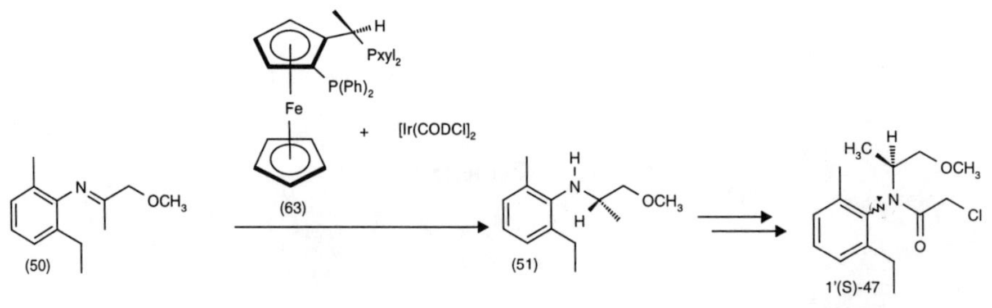

Scheme 17

Because of our interest in getting access to the pure (R)-enantiomer of metalaxyl[32] (Mefenoxam) the catalytic hydrogenation of the corresponding imine (64) was intensively studied.[33]

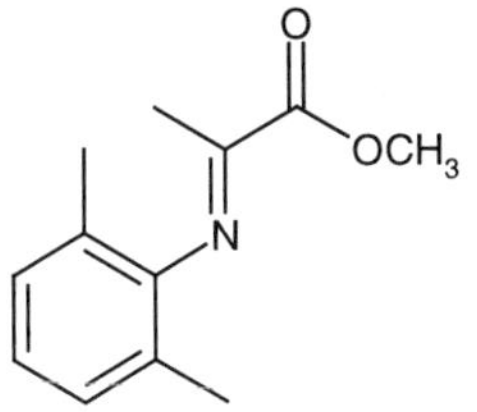

(64)

Scheme 18

To our disappointment, only poor ee values (< 30%) and very low catalytic activities were achieved with Ir catalysts generated *in situ* in analogy to the (S)-metolachlor process. Interestingly, the catalytic approach through hydrogenation of an enamide intermediate, which did not work in the metolachlor case, was more successful in this case. The enantioselective hydrogenation of N-aryl substituted enamides was actually developed during the enantioselective preparation of the stereoisomers of clozylacon (70) (Scheme 19), an experimental acylanilide fungicide for soil application against *Oomycetes.*[34]

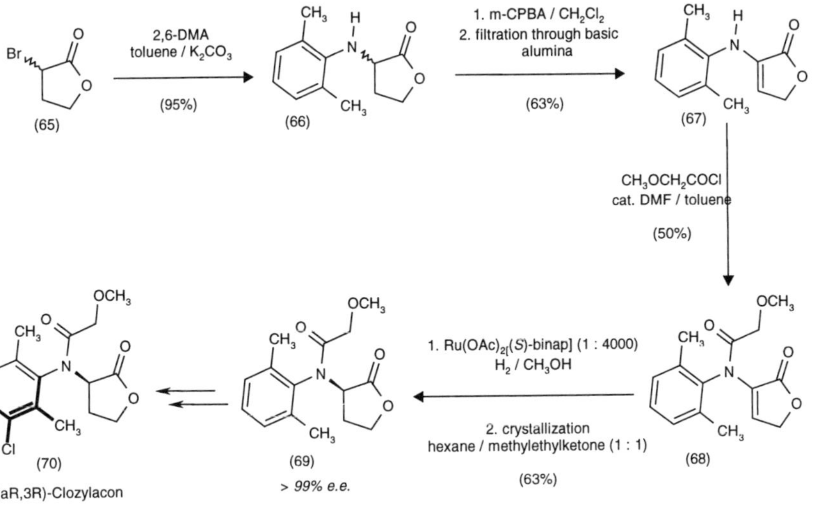

Scheme 19

In the case of the enamide (74) (Scheme 20) the effects of the temperature (20–60 °C) and hydrogen pressure (1-60 bar) on the enantioselectivity were strongly dependent on the nature of the ligand. They were almost negligible with (R,R)-Me-duphos but very pronounced with ferrocenyldiphosphine ligands. Optimization and scale-up experiments with the catalyst $[Rh(nbd)_2]BF_4$/(R,R)-Me-duphos (75) improved the turnover number to 50,000 at 10 bar H_2 and 60 °C. The ee of 95% and the tof of 5200 h^{-1} were clearly above the specified minimum limits. The enantioselective catalytic enamide hydrogenation is technically feasible for a large scale production process for Mefenoxam, (R)-49.

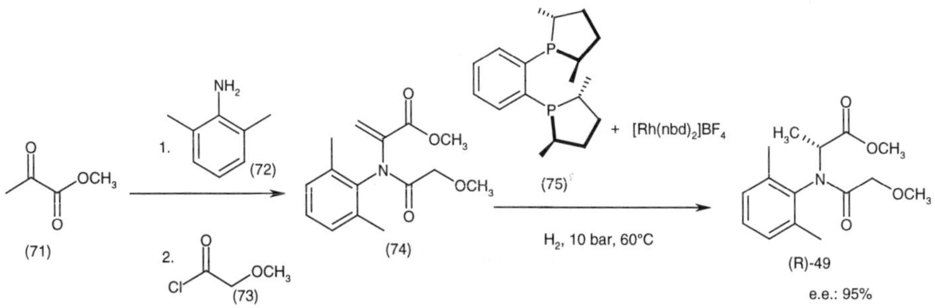

Scheme 20

3 CONCLUSIONS

1. The need for efficient synthetic methods has led to a rapid development of asymmetric catalytic methods, both with biocatalysts and with metal catalysts.
2. Most chiral catalytic systems are very substrate specific, *i.e.* the catalyst performance can vary strongly even for small changes in the structure of a starting material. As a consequence, a successful method can not necessarily be transferred from one target molecule to an other one even of the same class, as shown for the optimal synthesis of the 3 acyl anilides inTable 5.

Table 5 *Comparison of different enantioselective approaches to chiral acyl anilides*

Molecule	*Enamide hydrogenation*	*Imine hydrogenation*	*S_N2-Substitution*
Metolachlor	negative?	Ir-xyliphos	negative
Metalaxyl	Rh-duphos	negative	ok
Clozylacon	Ru-binap	-	ok

3. Since more and more catalysts (enzymes, ligands) will be available commercially for testing and the scope and limitations of an ever increasing number of reaction types are being investigated, chances for a successful application of catalytic methods for the synthesis and manufacture of agrochemicals will increase.

References

1. E. Heilbronner and J.D. Dunitz, 'Reflections on Symmetry', VHCA, Basel, 1993.
2. H. Frank, B. Holmsted and B. Testa (Eds.) 'Chirality and Biological Activity', H.R. Liss, Inc. New York, 1990.
3. E.J. Ariens, J.J.S. van Rensen and W. Welling (Eds.), 'Stereoselectivity of Pesticides', Elsevier, Amsterdam, 1998.
4. H.P. Fischer, H.-P. Buser, P. Chemla, P. Huxley, W. Lutz, S. Mirza, G.M. Ramos Tombo, G. Van Lommen and V. Sipido, *Bull. Soc. Chim. Belg.*, 1994, **103**, 565.
5. A.N. Collins, G.N. Sheldrake and J. Crosby (Eds.), 'Chirality in Industry I and II', John Wiley, Chichester, 1992 and 1997.
6. R.A. Sheldon, 'Chirotechnology', Marcel Decker, Inc., New York 1993.
7. H.U. Blaser, B. Pugin and F. Spindler in 'Applied Homogeneous Catalysis by Organometallic Complexes', B. Cornils and W.A. Herrmann (Eds.), Verlag Chemie, Weinheim, 1996, p. 992.
8. R. Noyori, *Chemtech*, 1992, **22**, 366
9. R. Noyori, 'Asymmetric Catalysis in Organic Synthesis', John Wiley & Sons Inc., Chichester 1994, p. 16 and references therein.
10. H. Kumobayashi, *Recl. Trav. Chim. Pays-Bas*, 1996, **115**, 201.
11. For a review see: H.U. Blaser, H.P. Jalett, M. Müller and M. Studer, *Catalysis Today*, 1997, **37**, 441.
12. F. Spindler, B. Pugin, H.P. Jalett, H.-P. Buser, U. Pittelkow and H.U. Blaser, *Chem. Ind.*, (Dekker) 1996, **68**, 153.
13. C.D.S. Tomkin (Ed.), 'The Pesticide Manual', 11th Edition, BCPC, Farnham, 1997.
14. G.M. Ramos Tombo and D. Belluš, *Angew. Chem. Int. Ed. Engl.*, 1991, **30**, 1193.
15. E. Ebert, W. Eckardt, K. Jäckel, P. Moser, D. Sozzi and C. Vogel, *Z. Naturforsch. C*, 1989, **44**, 85.
16. M. Kitamura, T. Okkuma, S. Inone, N. Sayo, H. Kumobayashi, T. Ohta, H. Takaya and R. Noyori, *J. Am. Chem. Soc.*, 1988, **110**, 629.
17. M.J. Krim and G.M. Whitesides, *J. Am. Chem. Soc.*, 1988, **110**, 2959.
18. J. Oh, B. Schmid, L.M. Venanzi, G. Wang, T.R. Ward and G.M. Ramos Tombo, *New. J. Chem.*, 1990, **14**, 495 and references therein.
19. J. Oh, B. Schmid, L.M. Venanzi, G. Wang, T.R. Ward and G.M. Ramos Tombo, *Tet. Lett.*, 1989, **30**, 6151.
20. G.M. Ramos Tombo; R. Nyfeler and J. Speich, '7$^{th.}$ IUPAC Int. Congr. of Pestic. Chem.', Book of Abstracts, Vol. I, H. Frehse, E. Kessler-Shcmitz, S. Conway (Eds.), Hamburg 1990, p. 129.

21. T. Früh, G.M. Ramos Tombo, *Synlett*, 1994, 727.
22. F. Karrer, S. Farooq in 'Regulation of Insect Development and Behavior', F. Sehual, A. Zabza, J.J. Menn and B. Cymborowsky, Eds., Wroclaw Tech. Univ. Press, Wroclaw, 1981, Part I, p. 289.
23. U. Fischer, F. Schneider and R. Zurflüh, U.S. Pat. 4.215.139 (17.3.1978), F. Hoffmann-La Roche AG.
24. F. Karrer, U.S. Pat. 4.971.981 (19.1.1987), Ciba-Geigy Ltd.
25. F. Karrer, H. Kayser, H.-P. Buser and G.M. Ramos Tombo, *Chimia*, 1993, **47**, 302.
26. C.-Y. Hsu, Y.-S. Lin and B. J. Chang, *Tetrahedron Asymmetry*, 1990, **1**, 219.
27. A. Hubele, W. Kunz, W. Eckhardt and E. Sturm, 'Proc. 5$^{th.}$ Int. Congr. Pestic. Chem.', Kyoto, 1982, J. Miyamoto and P.C. Kearney, Eds., Pergamon, Oxford, 1983, Vol. 1, p. 233.
28. H. Moser and C. Vogel, '4$^{th.}$ Int. Congr. Pestic. Chem.', Zürich, 1978, Abstract Volume, II, 310.
29. H. Moser, G. Rihs, H.-P. Sauter and B. Böhner, 'Proc. 5$^{th.}$ Int. Congr. Pestic. Chem.', Kyoto 1982, J. Miyamoto and P.C. Kearney, Eds., Pergamon, Oxford, 1983, p. 315.
30. F. Spindler, B. Pugin and H.U. Blaser, *Angew. Chemie,* 1990, **102**, 561.
31. H.U. Blaser, R. Gamboni, G. Sedelmeier, E. Schaub, E. Schmidt, B. Schmitz, F. Spindler and Hj. Wetter in 'Approaches to pharmaceutical process development', K.G. Gadamasetti, Ed., Marcel Dekker, NY, to appear in 1998.
32. C. Nunninger, G. Watson, N. Leadbitter, H. Ellgehausen, 'Proc Brighton Crop Prot. Conf. Pests and Dis. 1996', BCPC, Farnham 1996, Vol. 1, p. 41.
33. H. U. Blaser and F. Spindler, *Topics in Catalysis*, 1997, **4**, 275.
34. H.-P. Buser, B. Pugin, F. Spindler and M. Sutter, *Tetrahedron*, 1991, **47**, 5709.

Herbicide Discovery in the 21st Century – A Look into the Crystal Ball

James V. Hay

DUPONT AGRICULTURAL PRODUCTS, STINE-HASKELL RESEARCH CENTER, PO BOX 30, NEWARK, DE 19714, USA

When asked to speak at this Congress on the status of chemical weed control and herbicide research, my initial thoughts were that I would discuss all the new and exciting herbicides, their attributes and their impact on world agriculture. After additional consideration, I arrived, however, at the conclusion that presenting just another review of this type is not what we in herbicide research really need as we approach the new millennium. I, therefore, decided to take this opportunity to attempt to describe, at least from my perspective, some of the changes that herbicide research may take, and in some cases must take, in the coming years. But in order for us to have a clear view of where we are going, we need to know where we are. So, before gazing into the crystal ball to see what herbicide research may look like in the future, I would like to take a few minutes to describe what I see as the status of herbicide research in 1998.

If we only look at the surface of herbicide research today we would conclude that it is healthy, highly productive and innovative. We would see that across the industry, new herbicides are continually being discovered and moving into development to replace, or to complement, older less effective herbicides. We have developed new high throughput *in vitro* and *in vivo* screening techniques that have enabled us to test a larger and larger number of compounds while requiring a smaller and smaller quantity of the test samples. Transgenic, herbicide resistant crops have entered the marketplace and have had a major impact.

However, let us take a closer, more detailed inspection at herbicide research over the last several years. I believe that you will agree that the picture is somewhat different from the view on the surface. In fact, from this closer study we should be quite concerned about some aspects of our current herbicide research.

Let's look first at the new herbicides that have been introduced in the last 5–7 years. As Pillmoor[1] first pointed out during the last meeting of this Congress, and as reiterated by Pallett[2] at the 1997 Brighton Conference on Weeds, the vast majority of new herbicides introduced since 1991 have well known sites of action. This fact was further confirmed by the new herbicides introduced at the 1995 and 1997 Brighton Conferences (Figure 1). Of the 13 new herbicides introduced at these Conferences six were acetolactate

- 6 Acetolactate synthase inhibitors
- 3 Protophorphrynogen oxidase inhibitors
- 1 *p*-Hydroxyphenylpyruvate dioxygenase inhibitor
- 2 N-Acetylglutamate kinase inhibitors (?)

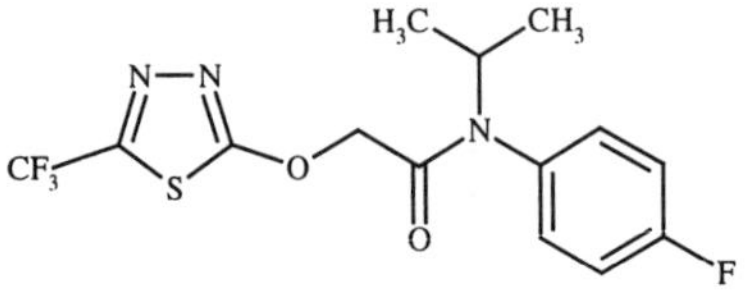

BAY FOE 5043

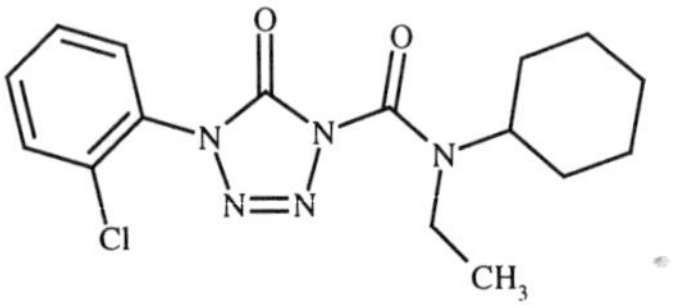

Fentrazamide

Bayer

- New, unreported site of action (?)

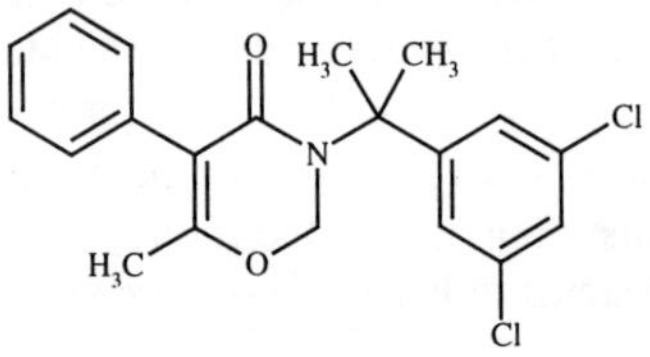

Oxaziclomefone

Rhone-Poulenc

Figure 1 *Classes of herbicides announced at 1995 and1997 Brighton Crop Protection Conferences*

synthase inhibitors, three were protoporphyrinogen oxidase inhibitors, one was a *p*-hydroxyphenylpyruvate dioxygenase inhibitor, while the oxyacetamide, FOE 5043[3] and the tetrazolinone, fentrazamide,[4] both from Bayer appear to have the same site of action as alachlor and mefenacet.[5] Only one of the new herbicides introduced, the oxazinone, oxaziclomefone from Rhône-Poulenc[6] for use in rice, appears to have new, unreported sites of action.

Moreover, if one reviewed the patent literature for the past several years, one would also find that most of the new patent applications continued to be directed to herbicides whose sites of action are those noted by Pillmoor and Pallett. Let us take a closer look at these herbicidal sites of action (Figure 2). The first examples of acetolactate synthase inhibitors, the sulfonylureas, were discovered nearly 25 years ago. Other classes of ALS

HERBICIDE CLASS	DISCOVERY DATE
• Acetolactate synthase inhibitors	mid 70s–mid 80s
• Protoporphyrinogen oxidase inhibitors	late 70s
• *p*-Hydroxyphenylpyruvate dioxygenase inhibitors	early 70s–early 80s
• Phytoene desaturase inhibitors	late 70s

Figure 2 *Discovery dates of classes of herbicides announced at 1995 and1997 Brighton Crop Protection Conferences*

inhibitors, imidazolinones, triazolopyrimidine sulfonamides and pyrimidinyl benzoic acids, were discovered in the 1980s. Similarly, the first examples of protoporphyrinogen oxidase inhibitors, such as azafeniden, oxadiargyl and isopropazol (proposed common name) recently announced at the Brighton Conference, were synthesized in the mid-seventies. Research on *p*-hydroxyphenylpyruvate dioxygenase inhibitors, such as sulcotrione and isoxaflutole, has been an extremely active research area by CPC companies in the 1990s. But in fact the first examples of *p*-hydroxyphenylpyruvate dioxygenase inhibiting herbicides, the benzoylpyrazoles, were prepared in the early 1970s. Similarly the benzoylcyclohexanediones class of *p*-hydroxyphenylpyruvate dioxygenase inhibitors were first reported by Stauffer in the early 1980s. Based on the number of published patent applications in the last several years covering novel classes of phytoene desaturase inhibiting herbicides, it is evident that there has been a resurgence in research on this class of herbicides that was first discovered 18–20 years ago.

Yes, our chemists have been very creative and innovative in their discovery of novel herbicide structures in these older classes. They have utilized the newest synthetic methodologies that have allowed them to synthesize compounds previously inaccessible. Even with the new discoveries noted above, the fact remains that much of the recent herbicide research has been focused on chemistry that was first discovered 15–25 years ago. Despite the fact that at least 37 different species of weeds have developed resistance to acetolactate synthase inhibitors,[18] this class of herbicides continues to be one of our most active areas of research. Have we become so comfortable in our research in these fruitful areas of chemistry that we are reluctant to explore new areas of chemistry because of the low probability of finding new active classes of herbicides? We cannot continue down this path. We must move away from the chemistry with older sites of action and be willing to accept the risks involved by venturing into unknown areas of chemistry. Are we waiting for others to discover the new areas of herbicides, and then after their patent application publishes, try to carve out a proprietary position for ourselves?

This is not to say that these are the only areas of herbicide research we are exploring. We have discovered novel structural classes of herbicides with new sites of action (Figure 3). Examples of are the triazole phosphonates that inhibit imidazoleglycerolphosphate dehydratase,[7,8] hydantocidin,[9] a natural product inhibitor of adenylosuccinate synthetase;[10–12] the imidazole carboxylates[13] and the heteroaryl carbinols[14] that block plant sterol biosynthesis by inhibiting obtusifoliol-14-α-methyl-demethylase, carbocyclic coformycin,[15] a inhibitor of adenosine 5'-phosphate deaminase,[16] and inhibitors of anthranilate synthase.[17] However, no examples of these

- Imidazoleglycerolphosphate dehydratase

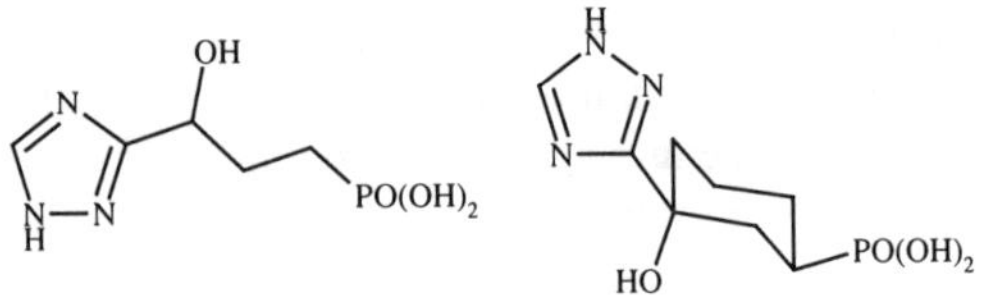

Triazole phosphonates

Novartis Zeneca

- Adenylosuccinate synthetase

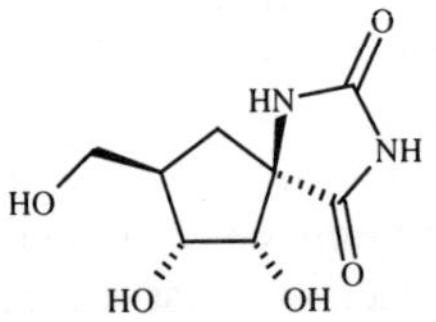

(+)-Hydantocidin

- Obtusifoliol-14-α-methyl demethylase

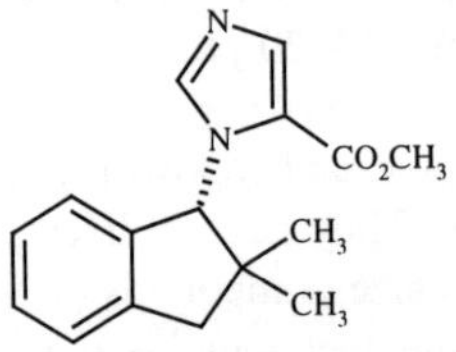

Imidazole carboxylates

Novartis

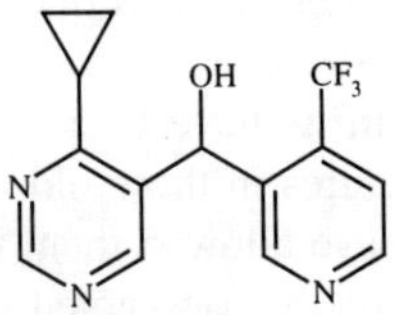

Heteroaryl carbinols

Zeneca

- Adenosine 5'-phosphate deaminase

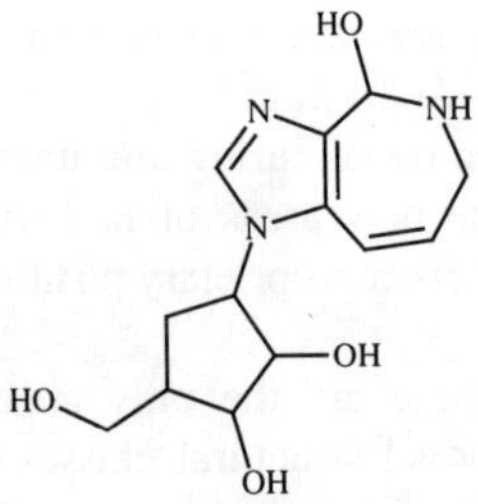

Carbocyclic coformycin

Figure 3 *New, recently discovered herbicide sites of action*

- Capacity to screen 100,000 compounds per year
- *In vitro* assays on specific enzyme targets
- Miniaturized whole plant screens

Figure 4 *New developments in screening for herbicide leads*

classes of herbicides with new sites of action have been reported to enter the commercialization process.

Let us next examine our approaches to screening for new herbicides (Figure 4). A common goal throughout the industry is to develop high throughput screens that have the capacity to test 100,000 compounds per year. This is a staggering number considering that most companies typically have only screened in the range of 10,000–20,000 compounds per year in the past. Most companies have had to revamp their biological screens to provide a greater capacity for testing this increased number of compounds. One approach many companies have taken is to develop and incorporate into the screening protocols new *in vitro* assays to detect inhibitors of specific enzymes. [A much more in-depth discussion of this topic is the subject of another session of this Congress.] Examples of these new enzyme assays include isoleucyl-tRNA synthetase,[19] adenylosuccinate synthetase,[20] adenylosuccinate lyase,[21] and the heavily researched *p*-hydroxyphenylpyruvate dioxygenase.[22] Another approach has been to develop new smaller *in vivo* screens utilizing a limited number of species that require much smaller quantities of test samples. These miniaturized screens are typically less costly to conduct, provide greater capacity and are used primarily to detect whole organism activity for discovering new leads.

With these new high capacity screens in place, the next issue we face is "What are the sources that will consistently provide 100,000 new compounds for testing each year?" Using traditional organic synthesis approaches, even the most skilled and productive chemist cannot be expected to synthesize more than 200–300 novel compounds per year of relatively noncomplex structures. So in-house traditional synthesis will not provide the 100,000 compounds per year needed to achieve our goal. Therefore we must look to alternate approaches to supply this large number of novel compounds.

One such approach is to acquire compounds externally. Most companies have established biological testing agreements with university professors, whereby samples from their research are purchased for biological testing. There are success stories from this approach (Figure 5). The nitromethylene class of insecticides can be traced back to a research agreement between Shell and Professor Henry Feuer of Purdue University.[23] The strobilurine class of fungicides, *e.g.* azoxystrobin and kresoxim methyl, are an example of commercial products that have evolved from university collaborations.[24] The origin of DuPont's fungicide famoxadone was a collaboration with Professor Detlef Geffkin while at the University of Hamburg.[25] The continued viability of universities as sources of a large number of new compounds is limited however. Most academic synthesis research programs are focused on a specific target molecule or the development of new synthetic methods. Typically in these research programs only very small quantities of samples are prepared. Thus, many novel compounds prepared in academic laboratories are not available in the amounts that companies need to obtain screening data at the whole organism level.

Product or Class	University Collaborator
• Nitromethylene insecticides	Purdue University
• Famoxadone fungicide	University of Hamburg
• Strobilurine fungicides	University of Kaiserslautern & Gesellschaft für Biotechnologische Forschung

Figure 5 *Commercial products/classes originating from collaborations with university professors*

Many companies have looked to the Former Soviet Union (FSU) as another source of collections of novel compounds for screening. Collaborations with Research Institutes have been established. Scientists in the FSU have formed brokerage companies offering to sell compounds to CPC companies for testing. Although successful discoveries from acquisitions of compound collections from the FSU are rare, one apparent success is the collaboration between Rhône-Poulenc and the Novosibirsk Institute of Organic Chemistry whose patent applications on herbicidal nitrone derivatives were recently published[26] (Figure 6).

As many of us have learned, however, the compounds obtained either through collaborations with Research Institutes or from brokers in the FSU are not without their problems. Sample purity and sample integrity have been a major problems. Structural diversity of the offered compounds is not as great as desired. Moreover, the same group of compounds may be offered time and time again through different brokers. Still another concern is sample exclusivity. How many of us are testing the same compounds in our screens? Even with these problems we most likely cannot rely on the FSU as source for a continued supply of novel compounds in the future due to the plight of the chemical sciences in these countries.[27]

The two sources with the most potential to provide a consistent and abundant supply of novel compounds are natural products and combinatorial chemistry. Not only can nature provide actual commercial products, such as glufosinate (phosphinothricin) and spinosad, but as pointed out by Duke,[28] compounds isolated from natural sources are most likely to give us a multitude of new sites of action on which we can focus research. A multitude of microbial metabolites that possess phytotoxic properties have been described by Stonard.[29] A more recent review by Pachlatko[30] summarized the status of research in natural products in crop protection and describes additional examples of naturally occurring herbicidal compounds of both microbial and plant origin. Although these natural products are often structurally complex and not readily synthesized, for example hydantocidin and carbocyclic coformycin as previously mentioned, they do give our chemists new lead structures on which they can establish research programs. Through their creativity the chemists can sometimes design structurally simpler compounds which contain all the key structural components that will enable the compounds to retain useful

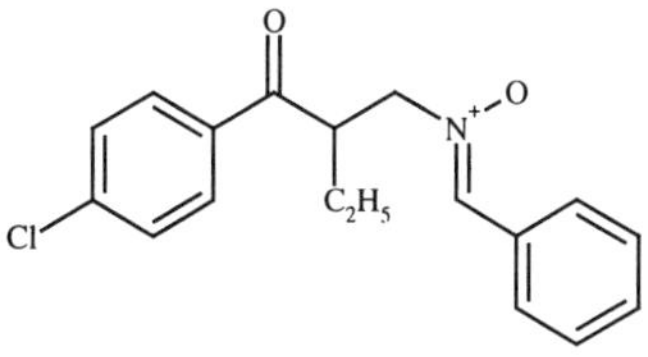

Figure 6 *Nitrone herbicides from Rhone-Poulenc/Novosibirsk research collaboration*

herbicidal activity. Using natural products as sources of leads, however, is problematic. From his research on carbocyclic coformycin, Pillmoor[31] has described some of the many difficulties often encountered in natural product research programs. Other issues that must be considered in natural product research programs for the CPC industry have been summarized by Rice and co-workers.[32]

Most CPC companies are turning to combinatorial chemistry and parallel synthesis to supplement their existing supply of compounds for screening. Some companies are developing in-house capabilities to synthesize libraries for screening. Others have established alliances with small combinatorial chemistry companies for both the synthesis and screening of libraries. Combinatorial chemistry does have the capability to provide large numbers of compounds for screening. To date there has been only one published report of the discovery of a new herbicide lead through a combinatorial approach. Parlow[33] has described the discovery of a herbicidally active pyrazole amide 1 from a library of 8000 amides and esters synthesized by a solid phase approach (Figure 7). Employing parallel synthesis techniques for the optimization program the research team from Monsanto ultimately discovered the pyrazole amide 2 that possessed a 4-fold increase in herbicidal activity.[34] Based on their bleaching symptomology, these herbicides were suggested to be phytoene desaturase inhibitors even though they do not contain the typical substituent patterns found in this class of herbicides.

Although the utility of parallel synthesis in lead optimization programs can be readily seen, the question remaining to be answered is "Will combinatorial chemistry play a major role in lead discovery for the crop protection chemical industry?" Whereas the pharmaceutical industry is unable to screen samples in the whole organism, they can continually rescreen the same compound library by developing new *in vitro* assays to discover leads for new therapeutic targets. In contrast, most CPC companies screen initially at the whole organism level, which by its nature embraces a multitude of targets, as well as in a number of *in vitro* assays against a specific target. Hence, once a compound library is screened and found to have no whole organism activity, the value of that library is diminished. Even if we find a library that demonstrates activity in a newly developed *in vitro* assay, since we have found previously that it was inactive in the whole organism we are faced with the formidable challenge of how to translate *in vitro* activity into *in vivo* activity. We must discover the modifications of the structure of the compound which imparts the correct physical properties that will enable the compound to (1) cross through the cuticle of the plant or insect, (2) be transported through the organism and (3) reach the

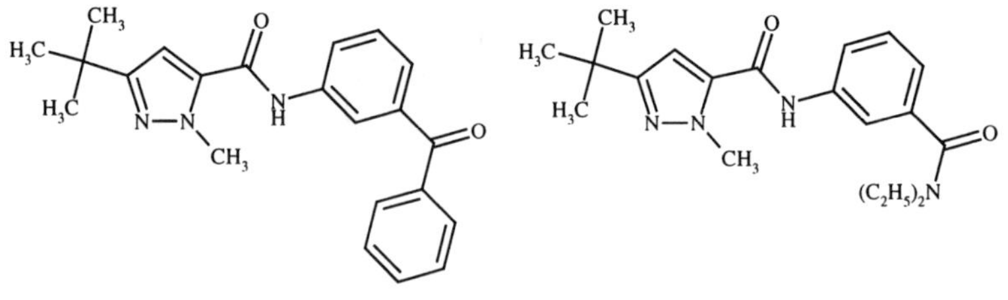

1 2

Figure 7 *Herbicidal lead discovered through combinatorial chemistry and parallel synthesis*

active site of the enzyme before being metabolized. Reprepping samples where *in vitro* activity is detected is another issue — is the likelihood of the sample having *in vivo* activity worth the cost and effort to scale up the synthesis.

In the past several years transgenic herbicide resistant crops have entered the marketplace. Maize varieties resistant to glyphosate, glufosinate and the imidazolinones have been developed. Similarly soybeans, cotton, canola and sugarbeets resistant to glyphosate, glufosinate or bromoxynil are now available to the grower. The number of hectares planted with these transgenic crops has increased each year since their introduction. According to the Wall Street Journal,[35] 50% of the cotton, 40% of the soybeans and 20% of corn planted in 1998 are transgenic, containing either herbicide or insect resistant traits. Research to develop transgenic wheat and rice resistant to glufosinate, bromoxynil, and sulfonylureas is in progress.[36,37]

Next, let us take a quick look at the current registration process for new active ingredients. Throughout the world registration requirements are becoming more stringent. More and more studies are being required by governmental regulatory agencies in response to public and political demands for food and environmental safety. In the U.S., the Food Quality Protection Act has had a major impact on the registration process. If the EPA considers a new active ingredient to be significantly safer and more effective, to have a new site of action, or even to be less expensive, it may accept the registration package under a "Reduced Risk" status and the review process may be accelerated. Otherwise, the length of time for the EPA to complete its review of the registration package and approve registration may approach 8 years. Also, the EPA has stated that it will accept only 5 registration packages per year from each CPC company for new active ingredients or label expansions. Similarly, the registration process has slowed in Europe and the rest of the world.

Now that we have taken a closer look at the current status of herbicide research today, what does the future hold for us? In the next few minutes allow me to gaze into the crystal ball and share with you what changes that I see the future holding for herbicide research. We can anticipate that major changes will occur. Some changes will be voluntary, others will be imposed upon us. Regardless of the stimulus, we must start now to plan for these changes (Figure 8).

- Use of transgenic herbicide resistant crops will increase
- Research on older, established sites of action must be refocused
- Chemical weed control will require continuing introduction of herbicide with new sites of action
- Natural products and combinatorial chemistry will be major focus for lead discovery
- Biological weed control will become an important tool
- Safer, more effective and environmentally friendly herbicides will be required
- Registration requirements will continue to become even more stringent
- Regulatory agencies will not register racemic mixtures

Figure 8 *Changes that will impact future herbicide research*

First, even though there is a reluctance in some countries to use transgenic crops as a food source, we can confidently assume that farmers will increase their use of transgenic herbicide resistant crops. Wheat and rice with traits providing resistance to the major herbicide classes will be commercialized, as will corn and soybeans that carry genes that impart resistance to multiple herbicides. What impact will this increased use of herbicide resistant crops have? It means that a farmer may use the same herbicide, or group of herbicides, season after season in the same fields regardless of the crop grown. Given this situation, we can expect that weed resistance to these herbicides may develop more rapidly and we will not be able to rely on these herbicides to provide effective weed control. Thus, a continuing supply of herbicides with new sites of action will need to be discovered and introduced to combat weed resistance.

To meet the need for herbicides with new sites of action, we must accept the fact that we must cease our research on herbicides with older established sites of action and refocus our efforts to the discovery of new sites of actions. The traditional chemical approaches we have used to discover new herbicide leads will in the future play an important, but reduced, role in our discovery efforts. In the coming years our approach to agricultural chemical research will evolve toward an approach similar to that used in the pharmaceutical industry. We will place an increased emphasis on combinatorial chemistry and natural products as sources of new herbicide leads because of their capability to provide a greater diversity of novel compounds. We will shift primarily to the use of *in vitro* assays to detect activity against target enzymes. When activity is observed, we will then have to rely on the creativity of our chemists to find the ways needed to translate *in vitro* activity into *in vivo* activity as I described earlier.

Not only will natural products serve as a source of compounds for the potential discovery of new herbicides, we can anticipate that biological weed control will grow in importance. For example, we have only just begun to realize how useful allelopathic agents may be as weed control tools either by plant–plant or plant–microorganism interactions.[38] Even with the negative experiences with biological weed control we have had over the past several decades and the challenges we will continue to encounter, biological weed control will most likely become an important tool in an integrated pest management approach. We can also expect that registration requirements will continue to become even more stringent than today as public and political pressure on regulatory agencies increases. The public will demand that only safer, more environmentally sound herbicides be registered. With this increased public pressure, biological weed control agents may be considered as "green" and may face a less stringent registration pathway. One specific registration requirement we can anticipate being implemented by agencies through the world is that racemic mixtures will not longer be registerable, only the active enantiomer, or diastereomer will be permitted to be registered.

In closing, I hope that I have been able to convey my perspective on both the current status of herbicide research, and probable changes we can expect to occur in the coming years that will impact our research. The changes we face will present many challenges for us, but some of the changes will provide great opportunities for us to advance herbicide research so that we will be able to aid the world farmers to provide an abundant and safe food supply for the world's growing population. If we have not planned for these changes and readied our research organizations to meet the challenges of the future, we will miss the opportunities these changes present, and thereby fail to meet our responsibilities to society, the growers of the world and ourselves.

References

1. J. B. Pillmoor, S. D. Lindell; G.G. Briggs, K. Wright, 'International Congress of Pesticide Chemistry, Options 2000', American Chemical Society, Washington, D.C., 1995, p 303.
2. K. E. Pallett, 'Proceedings of the British Crop Protection Conference—Weeds—1995', 1995, p 575.
3. R. Deege, H. Förster, R. R. Schmidt, W. Thielert, M. A. Tice, G. J. Aagesen, J. R. Bloomberg, H. J. Santel, 'Proceedings of the British Crop Protection Conference—Weeds—1995', 1995, p 43.
4. K. Yasui, T. Goto, H. Miyauchi., A. Yanagi, D.Feucht, H.Fürsch, 'Proceedings of the British Crop Protection Conference—Weeds—1995', 1997, p 67.
5. K. Tietjen, 9th International Congress of Pesticide Chemistry, 1998, Abstract #1C–0013
6. K. Jikihara, T. Maruyama; J. Morishige; H. Suzuki; K.Ikeda; A. Takagi, Y Usui, A. Go, 'Proceedings of the British Crop Protection Conference – Weeds – 1995', 1997, p 73.
7. J. M. Cox, T. R. Hawkes, P. Bellini, R. M. Ellis, R. Barrett, J. J. Swanborough, S. E. Russell, P. A. Walker, N. J. Barnes, A. J. Knee, T. Lewis, P. R. Davies, *Pestic. Sci.*, 1997, **50**, 297.
8. D. Ohta, I. Mori, E. Ward, *Weed Sci.*, 1997, **45**, 610.
9. M. Nakajima, K. Itoi, Y. Takamatsu, T. Okazaki, K. Kawabuko, M. Shindo, T. Honma, M. Tohjigamori, T. Haneishi, *J. Antibiot.*, 1991, **44**, 293.
10. D. R. Heim, C. Cseke, B. C. Gerwick, M. G. Murdoch, S. B. Green, *Pestic. Biochem. Physiol.*, 1995, **53**, 138.
11. R. Fonné-Pfister, P. Chemla, E. Ward, M. Girardet, K. E. Kreuz, R. B. Honzatko, H. J. Fromm, H.-P. Schär, M. G. Grütter, S. W. Cowan-Jacob, *Proc. Naatl. Acad. Sci. USA*, 1996, **93**, 9431.
12. D. L. Siehl, M. V. Subramanian, E. W. Walters, S.-F. Lee, R. J. Anderson, A. G. Toschi, *Plant Physiol.*, 1996, **110**, 753.

13. H.-P. Fischer, H.-P. Buser, C. Chemla, P. Huxley, W. Lutz, S. Mirza, G. M. Ramos Tombo, G. Van Lommen, V. Sipido, *Bull. Soc. Chem. Belg.*, 1994, **103**, 565.
14. S. Howard, D. L. Lee, H. L. Chin, D. B. Kanne, T. H. Tsang, D. P. Dagarin, T. H. Cromarte, E. D. Clarke, T. E. Fraser, T. R. Hawkes. M. P. Langford, 9th International Congress of Pesticide Chemistry, 1998, Abstract #1C–0013.
15. B. D. Bush, G. V. Fitchett, D. A. Gates, D. Langey, D. *Phytochem.*, 1993, **32**, 737.
16. J. E. Dancer, R. G. Hughes, S. D. Lindell, *Plant Physiol.*, 1997, **114**, 119
17 D. L. Siehl, M. V. Subramanian, E. W. Walters, J. H. Blanding, T. Niderman, C. Weinmann, *Weed. Sci.*, 1997, **45**, 628.
18. I. M. Heap, *Pestic. Sci.*, 1997, **51**, 235.
19. T. R. Hawkes, *WO 95/09927*, 1995.
20. S. L. Potter, US 5,519,125, 1996.
21. C. D. Guyer, E. R. Ward, US 5,712,382, 1998.
22. S. Sturner, L. M. Hirama, B. Singh, N. Bascomg, WO 98/04685, 1998.
23. S. B. Soloway, 214th American Chemical National Meeting, Division of Agrochemicals, Abstract # 003, September, 1997.
24. K. Beautement, J. M. Clough, P. J. de Fraine, C. R. A. Godfrey, *Pestic. Sci.*, 1991, **31**, 499.
25. M. M. Joshi, J. A. Sternberg, 'Proceedings of the British Crop Protection Conference—Pests and Diseases—1996', 1996, p 21.
26. C. W. Ellwood, A. Yakovlevich, WO 98/03478, WO 98/03479, 1998.
27. M. Freemantle, "Russian Science on the Rack", *C&EN*, 1997, **75**(51), 25.
28. S. O. Duke, F. E. Dayan, A. Hernandez, M. V. Duke, H. K. Abbas, 'Proceedings of the British Crop Protection Conference—Weeds—1997', 1997, p 579.
29. R. J. Stonard, M. A. Miller-Wildeman, 'Agrochemicals from Natural Products', Marcel Decker, New York, 1994, Chapter 6, p 285.
30. J. P. Pachlatko, *Chimia*, 1998, **52**, 29.
31. J. B. Pillmoor, *Pestic. Sci.*, 1998, **52**, 75.
32. M. J. Rice, M. Legg, K. A. Powell, *Pestic. Sci.*, 1998, **52**, 184.
33. J. J. Parlow, J. E. Normansell, *Mol. Diversity*, 1995, **1**, 266.
34. J. J. Parlow, D. A. Mischke, S. S. Woodard, *J. Org. Chem.*, 1997, **62**, 5908.
35. S. Kilman, S. Warren, "Old Rivals Fight for New Turf—Biotech Crops", *Wall Street Journal*, 1998, **231**(102), B1.
36. D. A. Chamberlain, R. I. S. Brettel, D. I. Last, B. Witrzens, D. McElroy, R. Dolferus, E. S. Dennis, *Aust. J. Plant Physiol.*, 1994, **21**, 95.
37. J. H. Oard, S. D. Linscombe, M. P. Braverman, F. Jodari, D. C. Blouiin, M. Leech, A. Kohli, P. Vain, J. C. Cooley, P. Christou, *Mol. Breed,*, 1996, **2**, 359.
38. F. A. Macias, D. Castellano, R. M. Oliva, P. Cross, A. Torres, 'Proceedings of the British Crop Protection Conference—Weeds—1997', 1997, p33.

A New Paradigm for Structure-guided Pesticide Design using Combinatorial Chemistry

Gregory A. Petsko[1], Dagmar Ringe[1] and Joseph Hogan[2]

[1] ROSENSTIEL BASIC MEDICAL SCIENCES RESEARCH CENTER, BRANDEIS UNIVERSITY, 415 SOUTH ST., WALTHAM, MA 02254-9110, USA
[2] ARQULE INC., 200 BOSTON AVENUE, MEDFORD, MA 02114, USA

1 INTRODUCTION

A revolution is sweeping through the pharmaceutical industry, driven by the simultaneous emergence of three new technologies.[1] Genomics is providing drug developers with thousands of new potential targets every month. High-throughput screening is enabling biologists to screen hundreds of thousands of chemical compounds in the time that would have been needed, a few years ago, to screen a few hundred compounds. And combinatorial chemistry is providing millions of new molecules every year, of unknown utility to be sure, but nevertheless a sheer quantity of novel compounds unprecedented in the history of chemistry.

The herbicide and pesticide industry has been slower to embrace these new technologies, but that appears to be changing. In this paper, we discuss briefly the reasons that adoption of these methods is desirable, summarise the types of combinatorial strategies that are available, and then describe how all this fits into structure-guided methods of lead compound design.

1.1 The Need

The pharmaceutical industry began to change the way it discovers drugs over a decade ago. In the U.S. today it takes, on average, the synthesis of 6,200 chemical compounds for every drug that makes it to the market. Even at the level of clinical trials, the failure rate is about 4 to 1. The time required to bring a drug to the public is about 13 years and the total cost is well in excess of $350 million, of which almost two-thirds is spent on lead compound discovery and optimization. So in the 1980s, pharmaceutical companies began to invest in technologies for "rational" drug design, in the hope that this would shorten the time-line required to get the compound into the clinic. Perhaps the best illustration of the success of this approach is the development of the protease inhibitor drugs that target the HIV protease; these compounds have literally transformed the treatment of AIDS in the developed world. But protein structure determination, which is at the heart of the "rational" - or, as we prefer to call it, "structure-guided" - approach,[2] is a time consuming and chancy business, and cannot always be employed anyway (50% of the targets for human drugs are G-protein-coupled receptors or ion channels, for which no structures are available). Moreover, this method allows one to *design* any molecule one likes, in a perfectly logical manner, but does not guarantee that said molecule, once designed, can be *synthesized* or that, if it can be, that it will bind to the desired target with the right affinity and have the desired toxicological and other properties required of a lead compound.

At the same time, the pharmaceutical industry was becoming painfully aware that it had not begun to exploit the available target space or, for that matter, the available classes of chemical compounds to interact with those targets. The approximately 6,000 known drugs do not even represent a few hundred well-separated chemical classes. And as for target space, the roughly 3,000 drugs that target human proteins hit fewer than 1,000 distinct targets. In other words, less than 1% of the human genome has been accessed thus far. When it comes to agents directed at non-human targets the diversity is even worse: of the thousands of herbicides, fungicides, insecticides, antimicrobial and antiviral compounds known, the number of distinct targets is only a few hundred.

1.2 One Solution

So a new paradigm has started to emerge in the pharmaceutical industry, driven by the need to overcome these limitations. This paradigm suggests that the problem of finding a new drug can be likened to the problem of finding a house in a city such as Tokyo that has few street signs and no house numbers. Random searching may get you there, but it is inefficient and costly and not guaranteed to succeed. *But if you knew the neighborhood and could knock on every door in that neighborhood, you could find the house.* In this new paradigm, structural techniques and/or high-throughput screening of large combinatorial libraries put you in the right neighborhood, while combinatorial chemistry allows you to knock on a huge number of doors. It is the thesis of this paper that this paradigm is transferable to the development of pesticides and herbicides.

2 FINDING THE RIGHT NEIGHBORHOOD

So how do we find the right neighborhood ? Structural information can define it quite precisely (for an example, see reference 2), but as we have just seen, such information will not always be available. And even if it is, there is a big gulf between having the structure of a drug target and knowing where on the surface of that target - which may resemble the Andes mountain range - a drug is likely to bind. Computational approaches for scanning protein surfaces have been developed, but these do not define the most likely binding sites with great precision. A few years ago, one of us (D.R.) recognized that this problem arose because in real life, the binding of any ligand to a protein is carried out in the presence of 55M of a competing ligand, namely water.[3,4] To find out how real ligands would bind under these circumstances, Ringe and coworkers developed an experimental method for mapping the binding surface of any crystalline macromolecule. One takes organic solvents, which represent functional groups on drug molecules, and soaks them into protein crystals. Solution of the protein crystal structure in the presence of these organic molecules reveals the binding sites, which can be used as the basis for rational drug design in a "connect the dots" kind of approach. The method is rapid (a solvent structure can be done in less than a week) and gives the desired information directly. An NMR analog of this procedure has been developed recently by Fesik and coworkers.[5]

Until recently, there was no method for defining the neighborhood in the absence of structural information. Now that has changed. Large libraries of small-molecule compounds, synthesized by combinatorial chemistry techniques, allow one to approximate the type of compound that will bind to the target structure even in the absence of structural data.

3 COMBINATORIAL CHEMISTRY

Combinatorial chemistry involves the production of every possible molecule from a basis set of modular units. It has had a number of incarnations. In its earliest form, it involved the synthesis of vast numbers of peptides and oligonucleotides. This was rapidly superseded by the synthesis on solid phase support of mixtures of small organic molecules. This incarnation is alive and well in many companies, but it has several limitations. Screening mixtures leads to a high percentage of ephemeral hits, and many screens are not amenable to solid supports. The range of chemistries that can be accessed is limited as well. So a new generation of combinatorial chemistry companies is emerging, wherein high-speed parallel synthesis of many thousands of pure compounds is carried out on robotics stations. These compounds can be delivered pure, in 96-well plate form, immediately ready for high-throughput screening. This approach has already yielded interesting new classes of inhibitors.[6]

With modern combinatorial techniques a single chemist can synthesize hundreds of thousands of novel compounds per year. This does not mean that the traditional role of the medicinal chemist is dead. Far from it: these compounds are not drugs, and the medicinal chemist will be more important than ever in the quest to turn them into drugs. What this technology means is that there are now tools available to access a much greater range of molecular space than ever before, to find new classes of compounds that target new classes of proteins in plant and pathogen genomes.

4 THE PARADIGM

So what should herbicide and pesticide discovery look like if it follows this model ? Figure 1 shows the process of finding a bioactive compound in the absence of combinatorial chemistry: powerful, yes, but limited in scope. Figure 2 shows the steps in this process where combinatorial techniques have already started to make a major impact. No arrowheads are drawn connecting things to combinatorial chemistry, because the flow of information is a two-way street. And the required diversity of any combinatorial strategy is inversely proportional to the amount of information available about the target. As the agrochemical industry enters the 21st century it cannot afford to remain tied to the techniques of the 20th. Genomics, high-throughput screening and combinatorial chemistry, combined if possible with structural information, are changing the shape of the pharmaceutical world. Truly modern herbicide and pesticide discovery must, in our view, adopt the same paradigm.

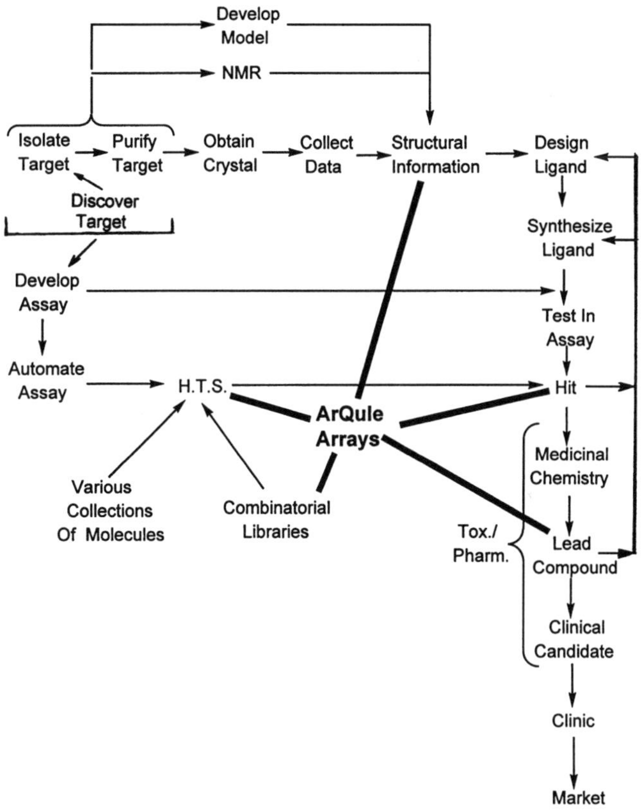

Figure 1 and Figure 2 *Figure 1 (see text) is the above scheme* <u>*without*</u> *the ArQule Arrays and connections (bold), which are applied to give Figure 2.*

References

1 G. A. Petsko, *Nature,* 1996, **384,** (6604 Suppl.): 7.
2. K. N. Allen, A. Lavie, G. A. Petsko and D. Ringe, *Biochemistry,* 1995, **34**, 3742.
3. C. Mattos and D. Ringe, *Nat Biotechnol.,* 1996, **14**, 595.
4. D. Ringe, *Curr Opin Struct Biol,* 1995, **5**, 825.
5. P. J. Hajduk, R. P. Meadows and S. W. Fesik, *Science*, 1997, **278**, 497.
6. E. Peisach, D. Casebier, S. L. Gallion, P. Furth, G. A. Petsko, J. C. Hogan Jr., and D Ringe, *Science,* 1995, **269**, 66.

Delivery

Impact of Biotechnology on Pesticide Delivery

B.D. Hammock[1], A.B. Inceoglu[1,6], W. Rajendra[1], J.R. Fuxa[2], N. Chejanovsky[3], D. Jarvis[4] and T.N. Hanzlik[5]

[1] DEPARTMENT OF ENTOMOLOGY, UNIVERSITY OF CALIFORNIA, DAVIS, CA 95616-8584, USA
[2] DEPARTMENT OF ENTOMOLOGY, LOUISIANA STATE UNIVERSITY, BATON ROUGE, LA 70803-1710, USA
[3] INSTITUTE OF PLANT PROTECTION, THE VOLCANI CENTER, BET-DAGAN 50250, ISRAEL
[4] DEPARTMENT OF MOLECULAR BIOLOGY, UNIVERSITY OF WYOMING, LARAMIE, WY 82071-3944, USA
[5] CSIRO DIVISION OF ENTOMOLOGY, INSTITUTE OF PLANT PRODUCTION AND PROCESSING, CANBERRA, ACT 2601, AUSTRALIA
[6] DEPARTMENT OF PLANT PROTECTION, ANKARA UNIVERSITY, 06110 ANKARA, TURKEY

1 INTRODUCTION

This chapter provides a brief overview of recent trends in the use of recombinant plants as delivery systems for pest control. With only two years of widespread use behind us, it already is clear that recombinant products offer tremendous benefit to the farmer, the consumer and the environment. This occurs as an expanding world population increases the demand for food. To complicate matters further there is a real increase in the purchasing power of a vast number of people who appear to want an improved and interesting diet. This process places a tremendous leverage on world food production, and it will make food much more valuable in the near future. If there is any hope of preserving wilderness and other natural areas, agriculture must become still more productive and intensive.[1] Using organisms as delivery systems will be a key technology in increasing the amount, quality and profitability of world food production. Regulation of biotechnology needs to consider both immediate risks to health and the environment by specific end products designed for use in agriculture, but also the risk to health and the environment which will result in not developing the technologies. Scientists need to be cognizant of the concerns of society over misuse of recombinant technologies in the long term and the immediate socio-economic disruption which will occur in the near future due to the tremendous success of recombinant delivery systems. The chapter ends with a more detailed evaluation of the status of recombinant baculoviruses for insect control.

In closing his chapter on biopesticide formulation at the last IUPAC meeting, Shieh stated, "I believe the technology used for creation of various transgenic plants that are resistant to viral disease and insect pests, along with herbicide resistant crops will be the noble delivery systems for crop protection in the 21st century."[2] The increase in sales of transgenic crops from \$75 million in 1995 to a projected value of \$1,835 million in 2001 represents a massive technological revolution which likely is in its infancy. Clearly the use of organisms to deliver pest control in agriculture is demonstrated.[3] In this article we will address briefly this revolution in the use of plants as delivery systems for pest control and touch on other approaches to deliver pest control with several organisms. If the impact of this technology on agriculture is as great as initial acceptance indicates, we must consider some of the socio-economic consequences of this revolution. The dramatic success of transgenics does not mean that these technologies can or should be used in isolation, rather it indicates that transgenic delivery systems will have a major role in integrated pest management. By no means are the possible consequences mentioned in the article exhaustive and several certainly are unlikely. However, it is hoped that by introducing them we will stimulate thoughtful discussion. Finally, the chapter will focus on a specific biological delivery system in the use of baculoviruses to deliver peptides and proteins for insect control.

David Evans gave our discipline its marching orders for the next decade in pointing out in the first chapter that to keep pace with the expanding population's need for food and of more significance the increasing demand for food of greater diversity and higher quality, we must employ a range of innovative technologies, including recombinant technologies. Failure to do this will result in the loss of our wilderness areas, wild lands and wetlands.[4] The food we produce must not only be abundant and of high quality but also profitable. Humanitarian efforts may solve short term hunger problems, but it is profitability of agriculture that delivers a high quality diet in a sustainable fashion. Colin Berry cautioned us that public and regulatory worries over both real and envisioned risk can stifle a technology. Against this dynamic background we find that many agricultural chemical companies are now seed companies. Thus, we are certain to see seeds, whole plants and other organisms as delivery agents for pest control systems. Practical development of this technology faces technical, regulatory, perceptive, political and economic problems. In spite of these barriers agricultural biotechnology is an unquestioned success. The recombinant products already on the market demonstrate that we are involved in a revolution in agricultural pest control, while the products announced for introduction over the next few years will alter dramatically how crops are produced worldwide.

The world pesticide market is mature with a growth of only 1–2% per year from an estimated base of $31 billion. Of this market biologicals contribute only about $250 million with 60% of this being microbial products largely based on *Bacillus thuringensis* (BT). Although biologicals represent a very small market, a variety of drivers are leading to a massive rate of increase of 25% per year in biologicals. Certainly one driver is the group of transgenic crops for insect control described below which underwent a 450% increase in acreage in a single year[5] and which are likely to undergo a still greater increase this year. In many cases these crops need to be supplemented by other pest control strategies. It would be attractive if softer treatments such as microbials could be used rather than broader spectrum synthetic compounds. In many cases the regulatory barriers to the development of microbial materials are lower providing an added incentive.

1.1 Plants as Delivery Agents for Pesticides

Although the potential for delivering pest control materials in recombinant plants has been recognized for a long period, the rapid implementation of this technology over the last three years has been staggering. In just a two-year period (1996–97) there have been over 10,000 field trials of new transgenic crops. In 1997, 2.8 million hectares of transgenic crops were grown commercially with 54% being planted for herbicide tolerance, 31% for insect resistance, 14% for viral resistance and approximately 1% for quality traits. This massive increase occurred in spite of demand exceeding the supply of transgenic seed in many cases.[5] The distribution of recombinant traits in planted crops changed dramatically from 1996 when herbicide tolerance accounted for 23%, 37% for insect resistance, and 40% for virus resistance. These data indicate how rapidly the field is changing in that there was no decrease in crops planted for disease resistance, but its relative importance declined due to the massive success of herbicide tolerance. Data such as these will soon be meaningless with many crops having multiple recombinant traits. The data are misleading about the future even now since they indicate that output or quality traits are a trivial component of the market while it is clear that they soon will dominate the field. One can anticipate that both stacked traits and output traits will have a major impact in the near future. It is likely that multiple crops will have several recombinant genes to preserve the yield from loss due to pests, genes to improve the crop qualitatively, and possibly even genes to improve yield. Recombinant plants are no panacea for agriculture but their economic benefit to the nations using them are clear. Owing to transgenics we likely are in the midst of a revolution in agricultural pest control at least as dramatic as the advent of DDT and other synthetic pesticides. It is having a profound impact on the pest control industry. However, the agricultural landscape has been dramatically altered since the 1950s. Compared with the 'green revolution' the changes wrought by biological delivery of pest control are likely to be very important but more incremental rather than revolutionary in agriculture as a whole.[6] The use of recombinant plants to deliver resistance to pests offers all of the advantages of classical plant breeding leading to innate resistance, and in addition recombinant technology facilitates more incisive use of specific traits, a wider selection of hosts producing useful traits, and much greater speed in generating useful varieties.[7] These statistics on transgenic crops underestimate the impact of biotechnology on agriculture since much classical plant breeding depends upon recombinant technology for high speed and precision. Although there are complaints the rapid international acceptance by the agricultural community is testament to the success of the technology. Below we will briefly discuss some aspects of insect and herbicide tolerance and then use these examples to bring up some questions regarding the technology.

1.1.1 Plants delivering insecticidal proteins. The clearest examples of plants as a delivery system are those producing proteins from the microbe *Bacillus thuringiensis* for insect control.[6,8] There are many conceptual advantages of the approach of using plants for delivery of pesticide chemicals. Of course the greatest benefit is delivery of the pesticidal agent directly to the pest without direct effects on nontarget organisms, risk to field workers or contamination of food or the environment. The use of the plant to deliver pesticides eliminates the problems of shelf life and field stability faced by the approaches to the formulation discussed above as well as providing on site biosynthesis of the selecting agent. Although the recombinant crops are being first designed for markets that can bear the cost of their development, one hopes that a spin off value will be improved agricultural productivity and profitability in developing countries. Recombinant products in agriculture have been criticized for helping first developing countries. As discussed below, this is not necessarily avoidable. With many beneficial technologies such as the development of antibiotics or immunization, we see the development and utilization first in developed countries, but the subsequent impact on developing counties has been enormous and largely beneficial. It also seems positive that approximately 25% of the commercial transgenic crops in 1997 were grown in developing countries.[5] It will be much easier to transfer pest control based on transgenic crops to a developing country than complex ecologically based management programs or industrial insecticides which depend upon a high technological level for their production.

An alternative or addition to the use of BT is the expression of proteins involved in chemical defense of plants. Recombinant technology allows one to move classical varietal resistance among different species of plants which allows exploitation of pest management strategies developed by the plant. Some varietal resistance is certain to be polyfactorial and polygenic, but there are resistance mechanisms that are due to single proteins. This is easily addressed using current recombinant technology. In many cases these proteins have a broader action than just BT; however, they also tend to be less active. Thus, they often slow the growth of a pest without actually killing it. This could be more useful in resistance management offering a more sustainable pest control, but it has not attracted industry in the way that single factorial genes that result in pest death have. The first generation products have been labelled as simplistic, short sighted, and entirely profit driven by critics of the technology. The products may be simplistic in concept, but by no means is their technological development simple. It also may not be socially bad for the first products to be simplistic in concept rather than having industry attempt grandiose projects designed to generate the ultimate sustainable engineered crop. The catastrophic rush to develop the 'Comet' as the first commercial jet aircraft or the rush to develop advanced nuclear power plants in response to political pressure provide testament to the fallacy of making massive technological leaps rather than incremental steps.[9] It probably is beneficial for society that our initial entry into recombinant technology has been pragmatic with projects likely to yield short term success and profits and products that fit readily into existing farming practice.[10] This concept is illustrated by the early claims in agriculture biotechnology that nitrogen fixing crop plants would shortly be universal. Although simple in overall concept, the technologies involved are enormously complex, and the nitrogen fixation projects gave way to goals that could be accomplished in the short term. For the technology is it important that a few of these recombinant products succeed, but by no means do they all need to survive commercially. If regulatory and technological barriers are not so high that vast units of society's capital for development go into single projects, it may have a beneficial effect on agricultural biotechnology for some of these approaches to fail. However, so much has been invested in the first few products society does need a success to sustain the technology.

Classical plant breeders have pyramided or combined genes to increase the level and generality of insect control. As with classical traits, it is important to remember that the recombinant traits could interfere with each other as well as with inherent resistant traits. However, their effects could also be additive or synergistic. This is obvious where protease inhibitors might stabilize protein toxins such as BT.[11] Stacked genes can be a resistance management tool. However, one can draw a parallel to this technology in the use of a tank mix of high levels of a variety of insecticides of differing modes of action as a resistance management tool. This approach has conceptual advantages but also many disadvantages.

As with many biological approaches the advantage of specificity offered by these technologies is also a limitation. The current resistant plants do not control all pests. The promise is that by controlling several key pests without destruction of natural biological control, that other more sustainable strategies can be used in IPM programs. At best the BT crops, for example, will form an ecological void waiting to be filled by another pest.

With both classical fungicides and insecticides the development of resistance in the target pest is well correlated with the commercial success of the control agent. If the material is successful, there is resistance. We can anticipate the same outcome from BT crops. It is unsettling to have a major component of plant biotechnology rely on a single set of related genes. The actual structural diversity among BT proteins in the open literature is alarmingly small. Our resistance management programs for BT crops are unproven, and Robert Metcalf once warned that, "the best resistance management program is a hairs' breath from the best system to induce resistance rapidly." Of course numerous laboratories are searching for new BT isolates and alternate proteins. There are some leads such as cholesterol oxidase for the control of boll weevil and a new group of weakly toxic microbial proteins from *Photorhabdus luminescens*.[12] Successes such as these are exciting but also caution that the search for supplements to BT genes will not be trivial since they must meet many criteria for success. Of greatest difficulty is that they must survive proteolysis in the insect gut as well as degradation by ingested plant chemicals and then penetrate or disrupt the insect's intestinal epithelium. In addition the materials must be potent, easily expressed and highly selective for pests. It has proven difficult to find new genes which meet these criteria.

Even with these negative factors which have and are being rigorously debated,[13-15] the impact of crops expressing the BT toxins has been enormous and largely positive. The crops developed to date delivering insecticidal toxins are simplistic in terms of using a single peptide designed to kill the insect pest. There are many more sophisticated approaches that could be used in the future, taking advantage of a large and complex fundamental literature on insect plant interactions as well as decades of classical breeding to yield host-plant resistance. The goal for both classical breeding and using recombinant delivery systems is protecting the plant and obtaining a high value crop rather than killing pests.[16] However, there is no doubt that we are in the middle of a worldwide experiment in which multiple crop plants engineered with a single gene will be used for delivery of a product highly lethal to specific pests. If it takes 50 to 100 years for technological advances to have a major impact on human life,[10] the apparently dramatic changes in agricultural biotechnology in the last two years are a mere preview of the revolution to come. However, if this hypothesis holds, then we can predict that the practical products resulting in our lifetime from recombinant DNA technologies can be predicted from contemporary data.

1.1.2 Plants with herbicide resistance. There are multiple successful ways to engineer plants to confer resistance to herbicides.[6,17] Herbicide resistance is not technically a delivery agent for pest control; however, this is an important and controversial concept with many parallels to crops showing direct resistance to pests. Weed control is the largest pest problem in agriculture, and herbicides represent the largest market for agricultural chemicals. By making plants resistant to a herbicide unprecedented weed control can be achieved.[3] Herbicide resistance is the dominant transgenic trait in agricultural now,[5] and as discussed below it best illustrates some of the socio-economic issues that will face the agricultural community immediately. It is likely that these plants will be regulated in the same way as plants with transgenes against pests and that they will be engineered in the same way. It is certain that herbicide resistance will, in many cases, be stacked with other traits in seed lines offered commercially. The availability of transgenic crops with herbicide resistance has already cut into the market for classical herbicides. It is likely that the diversity of materials available for pest management of weeds will drop as a result of transgenics, possibly to the point where only a handful of broad spectrum compounds remain.

1.1.3 Plants delivering antiviral, antifungal and antinematode compounds. The People's Republic of China first registered transgenic tobacco and later tomato resistant to viruses. Although transgenic approaches to control plant disease now lags behind BT crops and herbicide resistant crops, it is exciting in that it offers the potential to control diseases which previously were intractable.[6,18] Recombinant methods, in particular functional cloning,[19] also provide the technology to elucidate resistance traits in model or noncrop species and transfer the traits to agronomic species. They also will help elucidate fundamental mechanisms of resistance. Disease resistance in crops is likely to have many of the same problems as discussed for resistance to insect pests,[19] but in many cases molecular approaches are the first successful tools for preventing these problems. The technology is certain to have a positive impact on agriculture, and Monsanto for example is marketing potatoes with genes to control two viral diseases stacked with anti-insect proteins. As molecular techniques become more widespread and our understanding of mechanisms of pathogenesis increase,[20] one can hope for an increased diversity of tools. Several approaches are promising for control of nematodes with transgenic plants.[21,22] An exciting observation is that a gene originally isolated for control of the root knot nematode also confers resistance to an aphid.[23] Hopefully the serendipity that has provided numerous chemical tools will provide molecular tools as well. Unlike insect damage where control of the corn rootworm or cotton bollworm each represent massive markets, with pathogen control there are many smaller markets. Success here will require rapid molecular methods on the technical side, and low regulatory barriers on the policy side.

The pathogen area illustrates the value of basic studies on how plants distinguish self and nonself. An exciting and promising area which has been exploited in the first generation of resistant plants is pathogen-derived resistance.[24] There may be several mechanisms which confer this type of resistance, but in at least some cases it appears that translation of message is not needed for the resistance to develop. This RNA dependent resistance to viral infection may be due to homology dependent gene silencing.[25,26] As our appreciation of this area increases there may be the possibility of using master switch genes to develop broad spectrum yet durable resistance. Such a technical advance could address the problem of small markets raised in the previous paragraph. For instance a slight increase in the NPR1 gene in *Arabidopsis thaliana* results in the plant responding more strongly to several diverse pathogens. Since the NPR1 gene is only activated when the plant is challenged, one may confer resistance without draining the plant's resources.[27] As with more generalized resistance systems against insect attack based on allelochemicals, we need to know if the chemicals induced in the plant offer any risk to humans or domestic animals.

1.1.4 Plants as nonhost producers of insecticidal viruses. Hopefully the rapidly developing techniques in molecular biology will open the way to many innovative approaches to control insects and other pests. So far we appear to have few options, but as the general techniques become better established a diversity of innovative approaches hopefully will appear. A very exciting new concept has been advanced by Hanzlik and Gordon and is based on a small RNA virus in the Tetraviridae termed the *Heliothis* stunt virus.[28] In first instar larvae of *H. armigera* only 5000 virus particles are needed to stunt the insect completely so that it never matures. Since the virus only seems to infect gut cells and is mainly effective in early instars, production of sufficient quantities of the virus for practical insect control is very difficult. The virus consists of three genes on two virus particles enclosed in a capsid made of 240 copies of an identical 71 kDa viral protein. These workers have been able to produce biologically active virus outside of its normal host both in the baculovirus system and in plant cells. This raises the possibility that plants and possibly other organisms can be used not only to deliver peptide and protein toxins but also viruses for controlling insects or other pests. The viral capsid also can be used for the delivery of other genes which could be useful in insect control.

1.1.5 Recombinant technologies to enhance production of allelochemicals. To date there is no commercialization of plants that have been engineered to deliver allelochemicals leading to the plants' resistance to weeds, insects, or pathogens. However, most major companies in agricultural biotechnology have some effort in this area. There are certainly many reasons for this lack of commercial success. One factor is that the resistance of plants to herbivores is complex and not well understood even within the plant, much less in its ecological context.[29] The complexity of resistance mechanisms is a strength against adaptation of herbivores and a great advantage over the single gene approaches discussed above, but it presents both regulatory worries and scientific complexities. Most allelochemicals arise from complex synthetic pathways involving many genes and gene products. Not only does one need the multiple biosynthetic components in the plant, but they need to be coordinately expressed and the enzymes may need to be in specific locations within the cell. Even simple stacking of multiple recombinant genes in a crop plant has proven to be technically challenging. Toxin distribution within a plant is not constant with either plant tissue or with developmental stage, and this may have a major effect on insect control.[16] Many of the plant natural products active against insects and fungi are general toxins active on herbivores. The term herbivore may in fact include humans and domestic animals. Since many of the genes involved in allelochemical biosynthesis have broad substrate selectivity, it is difficult to predict all of the chemical products which could result from the manipulation of a series of pathways. There are multiple examples of allelochemicals interacting in plants in complex and unexpected ways to enhance or diminish control of pests.[29-31] Analysing the natural products in the resulting plants would be technically demanding even with advances in LC MS/MS technology. This uncertainty of products is also likely to result in additional regulatory concerns. On the positive side there already is technology developing to control biosynthesis of natural products from the pharmaceutical industry. There are also several companies focusing on the mechanisms to control allelochemical production in plants.

Another area of active research in private and public sectors involves the use of chemical signals from the plant and/or pest to induce resistance in plants. There has been some success in inducing resistance by the application of chemicals, cultural manipulation or inoculation.[29,32] Recently such induction has been shown to give an advantage to a plant in the field by reducing damage from both sucking and chewing insects.[33] As the mechanisms of induced resistance are better understood, it becomes attractive to exploit this using recombinant techniques to control production of allelochemicals at the level of gene regulation.[27] This approach offers many advantages but of course raises the possibility that the resistance agents the plants use may not be highly selective for the pest species. The lack of success with this approach using classical breeding also indicates that the approach will be complex. A simpler application of this concept is the expression of single genes under a wound inducible promoter. There have been several cases where this approach appears very promising, but it suffers from relying on a single trait.[29]

It is certainly attractive to think of a plant producing commercial amounts of highly potent but selective pesticides such as the natural pyrethrins for insect control. Simplistically this approach is attractive in a crop plant, and recombinant plants could represent an alternative source of these valuable natural insecticides as green factories. However, the number of enzymes involved in the biosynthesis of the natural pyrethrins is intimidating as is their regulation. Juvenile hormone, phytojuvenoids, and the phytoecdysones represent other materials of exceptionally high activity on insects and low mammalian toxicity. Once again their biosynthesis involves multiple coordinated steps. However, all of the above compounds are known to be produced naturally in plants, and most plants probably have the majority of enzymes available for the production of these materials. It is possible that in some crops only minor modifications of existing pathways could yield effective concentrations of these or other natural materials. Thus, we may see in the future multistep biosynthesis in plants of allelochemicals for pest control. However, biotechnology is already having a major impact on allelochemicals in plants. A variety of routine techniques in recombinant DNA technology have greatly accelerated the pace of classical plant breeding for resistance to plant pathogens and pests. Although the resulting crops are not engineered, they were developed quickly and incisively through the application of biotechnology.

1.1.6 Green chemistry for pest control agents. Biotechnology promises to make many industrial processes less expensive and less polluting. It may even open up previously inaccessible synthetic directions. Some industrial processes even for major classical pesticides such as Esfenvalerate employ a biological step. Such chemistry may make the use of materials such as pheromones much more economically attractive in pest management. There are no examples but one could envision with more sophisticated recombinant technology on site or near site production of a pesticide in a fermentation system where the fermenting organism acts as a formulating agent, or even a crop acting as a production system for a nematocide or insecticide. By no means is this trivial technology.

1.2 Concern over Transgenics

In spite of the numerous advantages and rapidly expanding use of these transgenic crops, it is not certain that public acceptance of foods containing recombinant pesticides will be universal. Many worries over transgenic crops including escape of foreign genes and human health effects have been presented in both rational and alarmist fashions, yet the products which now are being marketed certainly appear devoid of human health effects, and the positive environmental effects will far outweigh the negative. Along these lines Freeman Dyson in the 1976 Princeton hearings on recombinant technology cautioned that we should not stifle a technology because of unsettling scenarios far in the future.[34] Dyson cautioned that no public agency should have the authority to restrict research simply because the individuals in power are philosophically opposed to it. However, he also criticized scientists for having too short a view of the possible dangerous outcomes of their technologies. As with the Princeton experiments, the immediate risks of recombinant DNA research are adequately controlled, but the more thoughtful public is justifiably concerned of the long term potential of the technology and the abuse of this knowledge. Recombinant techniques will have a massive positive impact on agriculture and part of this impact will be the use of organisms as delivery agents for pest control. However, every successful technological advance has major consequences, and some of these will be unsettling and locally damaging.[35] Even now one can envision scenarios that could be disruptive in agriculture and several of these are outlined briefly below. We argue against trying to correct all the ills of capitalism by draconian regulation of recombinant technologies in agriculture. However, Nottingham cautions that the great potential for good in these recombinant technologies could be stifled by regulation if society perceives a rush for short term profits prompted by corporate greed.[36] We also could see over regulation if there is the perception that a national company is imposing its technology on another country or if a developed country imposes technology on a developing country. Neither the long term ethical concerns nor the socio-economic impact of a technology seem an adequate justification for restrictive regulation, but we will be remiss as a scientific community if we fail to address them.

Of the recent books on genetic engineering possibly the work of Ho[37] is among the most damning. She charges that current trends in biotechnology represent bad science walking with big business against the public good. She charges that many of the early engineering projects are unethical and exploitative, and she provides a check list of risks that may be associated with agricultural biotechnology. Much of her work mixes mysticism with science, and we personally feel that in the specific cases under consideration, experimental evidence challenges her general assertions. In other cases she attempts to correct the ills of capitalism and all modern technology by a proposed ban on biotechnology. However, some of these cautions in time will certainly prove true, and society must address them if the benefits justify the risks. In other cases the speculation clearly is unwarranted in the near term, but as Freeman Dyson cautioned, can her cautions become reality in the long term? Although the first round of biotechnology products appear very safe it does not follow that all future biotechnology products will be safe.

1.2.1 Industrial consolidation. One can conceive of many scenarios which will be initiated or accelerated by recombinant technology. One dramatic process clearly accelerated by the promise of biotechnology is the loss of economic entropy in the agricultural chemical industry. The acceleration of both horizontal and vertical integration of companies is already in progress as companies fuse horizontally to combine resources and fuse vertically to capture more of the potential profit in the seed to shelf movement of agricultural products. As we see chemical companies position themselves as biology companies and food providers, it is clear that biotechnology is having an impact on this consolidation. The size of the international research program on pest control is likely to be approximately proportional to the number of companies involved in pesticide production.[38] However, other technologies like precision farming certainly accelerate this process as well as social issues and regulation itself. Thus, loss of economic entropy is not a reason to single out agricultural biotechnology for special regulation. However, there is concern that loss entropy driven by economics of scale can result in economic and technical instability.

Large companies will initiate grand projects. In part this trend is driven by economic overhead in large companies and in part it is driven by human nature. Important people, and especially important committees, like large projects that bring notoriety. However, science often is most innovative with small projects, and certainly technology works best when as a world community we can experiment with many technologies, no one of which has so much of the planet's scientific capital invested in it that we cannot allow it to fail. Pioneering agricultural biotechnology has been risky and expensive. Large companies have developed numerous conservative products which look like they will be enormous successes in major markets. The very size of these companies dictates that they must only approach large markets. This situation provides opportunities for small innovative units to develop a diversity of agricultural biotechnologies under conditions where many of these technologies can be allowed to fail since investment in any one of them has not been large. If this process occurs we can expect even rapid progress in agricultural biotechnology with at least some of it addressing problems in developing countries. There is a dark trend however. This is that many of the small companies are purchased by large companies before their products have a chance to succeed or fail on their own. This process certainly is driven by a desire to limit competition, but probably even more by a desire of companies to control enabling technology in an area. If this trend continues we will see small companies and public sector scientists unable to develop the diversity of new biotechnologies that we need for a Darwinian process of economic selection. Even this trend may not be all bad. If freedom to operate is deigned to innovative individuals and companies in developed countries, there is an escape valve for innovation. The technical barriers to second and third generation products in agricultural biotechnology are not nearly so high as are barriers to the production of chemical pesticides. We could find that innovation simply moves to the developing countries who most desperately need improvements in agriculture. The dominant nations could find themselves in an intellectual backwater of their own making.

Output traits currently are only a small part of the agricultural biotechnology market. However, they are likely to be far more profitable than plants designed for delivery systems. If with consolidation industry begins to concentrate on these more profitable directions, we may be left with a private and public sector research community in agriculture too poorly staffed and funded to sustain this technological revolution in pest control. This could have a major impact if either of the two scenarios described below for BT or herbicide tolerant plants occur. In the early days following the discovery of DDT far thinking scientists such as Steinhouse, Newsome and Smith cautioned that we should not use synthetic pesticides as a crutch to allow us to forget hard learned lessons in the ecological interactions among crops and their pests. It is an exciting time as we are faced with integrating these new technologies into ecologically based pest management systems. Hopefully as a world community we will not again forget principles of pest management in this process.[39] We also can hope the optimism of Dyson that the theoretically possible dangers of genetic engineering will not be real is correct,[10] and go further to carry out research to insure that the fears do not become realities.

1.2.2 Resistance to BT crops. There certainly are many negative aspects to the use of recombinant plants in agriculture including the rapid development of resistance resulting from high selection pressure. As technological barriers drop one can anticipate the *Bacillus thuringiensis* genes to be used in a variety of crops. As when chemical pesticides were used on a spray schedule, we will have prophylactic treatment and continuous exposure of the pest to a selection agent with BT plants. If BT is effective, resistance of target species is a certainty, the only questions are how fast it will develop and if it will be economically prohibitive. Since the original BT gene is public domain, one can even envision that the first companies to have products could see BT resistance as a benefit if they have subsequent products to follow.

We could envision numerous scenarios associated with this resistance problem. For example in some areas of the world cotton production is effectively limited by devastating populations of *Helicoverpa armigera.* We can anticipate that BT will be moved into local cotton cultivars in these regions both by recombinant techniques and classical breeding. This BT cotton could lead to massive production of cotton in these regions, overwhelming the world market and driving many farmers and even nations out of the cotton business. Once out of the cotton business it is unlikely that an individual farmer or a nation will return to it. If BT crops reduce dramatically the market for synthetic insecticides on several major crops, companies will drop existing materials and shy away from developing new materials for the cotton market. If we accept as reasonable the plans to retard resistance in *Heliothis virescens* to BT cotton,[15] then we can predict from the same arguments that *H. armigera* is poised for development of rapid resistance. After several years of cotton production in new areas change the face of world cotton production, a failure of BT crops based on high levels of resistance in *H. armigera* with no classical chemicals to replace it could decimate cotton production. Once the path is closed on a technology it will be hard to open again. The world community could then find itself with no technological base to produce and register new cotton insecticides as well as no farmers with the resources to grow cotton. The above example may be far fetched; however, it is certain that changing technologies will alter dramatically the relative profitability of producing a crop by an individual farmer, region or even nation. In the wild scenario above it might not be bad in the long term for the world to move away from cotton production for numerous reasons, after all two companies have just announced technology to produce nylon precursors in plants.

1.2.3 Problems with herbicide resistance. Most plant input traits make the plant more competitive under some conditions. Since the definition of pest and in this case weed is an economic and not a biological definition[40] there is always the worry that such transgenic plants can become weeds. With the possible exception of rape seed this seems unlikely for the first generation of recombinant plants but a real possibility with many plants whose evaluation needs to be considered on a case by case basis.[35] The argument by some environmental groups that such engineering will encourage herbicide use clearly does not apply to the first generation products. These products all move agriculture from prophylactic treatment to treatment based on economic thresholds and from environmental questionable products to greener materials. However, one can argue that the products also move agriculture from a diversity of largely inexpensive products to a very limited number of expensive products. Most of the objections to herbicide resistance traits in first generation products can be effectively discounted.[35,37] However, this may not apply to future products and there is the caution that plants are known to have become weeds many decades following their introduction. There will be an inevitable reduction in the diversity of genotypes available as seed production is concentrated in the hands of a few companies capable of offering improved quality and pest resistance in the resulting crops. Ethical and legal questions have and are arising as companies restrict the use of other technologies over the top of their crops. The likelihood of these and other scenarios is debatable, but they should be debated and considered as the new recombinant technologies are realized. This process may help us to minimize deleterious socio-economic impacts of these exciting technologies.

There are other aspects of the technology that are certainties. The dramatic success of herbicide tolerant crops coupled with the loss of entropy in the agricultural chemical area and other factors associated with the ageing of pesticide products is certain to accelerate the removal of many classical herbicides from the market. This process will reduce the diversity of tools available to a pest management specialist. If practitioners of integrated pest management need a diversity of tools, this trend will be a setback for IPM. The appearance of resistance or expansion of weed varieties, such as nightshade, tolerant to herbicides paired with transgenic crops after most classical herbicides have been removed from the market could leave us vulnerable to massive yield loss. Since a variety of conservation tillage practices depend upon the availability of synthetic herbicides, loss of weed control would lead to a return to cultural practices causing loss of soil and loss of soil tilth. Once classical herbicides go off the market they are unlikely to be reintroduced to control the weed problems. The agricultural industry may not be able to respond with sufficient speed to replace these products either with new herbicides or new transgenic approaches.

1.2.4 Loss of biodiversity. The dramatic success of transgenic crops, consolidation of the agricultural industry, destruction of wilderness areas, reduction of public sector support of plant breeding and numerous other factors are reducing germplasm for agriculture. Is it ironic that at the very cusp of history when we can best exploit biodiversity for societal benefit, we are failing to train scientists in bionomics, losing global biodiversity, and losing agricultural germplasm.[41] Recombinant technology seems certain to accelerate loss of agricultural germplasm and the shift from training in bionomics to more reductionist careers. The loss of biodiversity is a commonly used argument against agricultural biotechnology.[6] This loss of biodiversity is both a very real technical issue as well as an emotional rallying call for environmentalism. However, it is a possible outcome of many economic drivers, not a direct and unequivocal result of agricultural biotechnology.

1.2.5 Agricultural biotechnology in developing countries. A major charge against agricultural biotechnology is that it is producing toys for wealthy nations and profits for large companies.[6,35–37] A friend with a large biotechnology company recently lamented that he joined to company to make the 'volkspotato' to feed the world but is engaged in making the 'yuppie potato' optimized for fast food franchises. As discussed above, it is common when technologies are paid for in developed nations that the technologies are first applied there. This trend did not negate the value of polio and small pox vaccines nor antibiotics in developing countries. The process also avoids the charge that new technologies are experimentally applied to societies that cannot defend themselves. The delay in itself is not bad, but if issues in agriculture specifically residing in developing countries are not addressed because of the pressures of capitalism, a wonderful humanitarian opportunity will be lost. Capitalism alone is likely to be slow to generate products that will aid developing counties at an affordable price, but the rate of development could be increased somewhat if there were good patent protection in these countries.[6] One hopes that activists can move beyond blocking questionable technologies toward a positive program where technologies which improve the environment and encourage social justice are advocated. Ignoring the technological needs of developing economies will serve us ill. There is an enormous gulf between rich and poor in some of these nations. In attempting to mitigate this problem in developing countries, we may gain insight to reducing the same economic dichotomy in developed nations.

Products specifically for developing countries are more likely to come from public than private research. As funding for public sector science is reduced and scientists are told to look for near market funding, applied science will increasingly be under the control of businessmen who are likely to give support to products that the affluent can buy.[9] This is of course counter productive for using agricultural biotechnology to solve problems in developing countries. Thus the very trends in the funding of Western applied science designed to make the scientist more responsive to human needs seems fated to insure that the scientists cannot be responsive to the needs of developing nations.

There are positive trends as well. Although multinational companies are unlikely to design products specifically for developing nations, some of the major products developed will be admirably suited for use in developing economies. Also as the major hurdles in the recombinant technology are over come by large science, the barriers decrease to applying these techniques to problems specific in developing countries. For example the production of many modern synthetic chemicals for agriculture requires a complex industrial infrastructure. Such an infrastructure is not needed for either discovery or production of many products in biotechnology. There is no shortage of drive or creativity in the science in many developing countries, and training of individuals is far easier to provide than an industrial infrastructure. The infrastructure of Borlag's green revolution is still in place and is well suited for implementation of advances in biotechnology. Finally as mentioned above, if developed nations cannot correct their problems of over centralization and a lack of freedom for innovators to operate, the developing nations can become the creative leaders in agricultural biotechnology.

1.2.6 Role of public sector in agricultural biotechnology. During the golden age of pesticides there was a strong public sector research component dedicated to agriculture.[42] This component provided fundamental research in support of industry, independent evaluation of technologies, assistance in implementation of new materials, and input on the changes in agriculture driven by technology. In most countries this public sector component has been reduced dramatically,[4,38,43] possibly caused by a reduction in the perceived value of agriculture with the voting public and the number of people actually involved in agriculture.[38] The minimal impact of public sector research in agricultural genomics and proteomics attests to this reduction. Thus, in a revolutionary period of rapidly evolving technologies in agriculture there are few scientists who will be viewed as unbiased by the public and thus will have the time to investigate and the credibility to evaluate the benefits and safety of new technologies. We can see an excellent example with herbicide resistant crops. There are examples of successful and profitable herbicide free agriculture; however, herbicides have been used in low till approaches to reduce input, erosion, water use and loss of tilth dramatically.[6] We very much need public sector research to integrate the use of herbicides, herbicide resistant crops and other technologies into sustainable programs for crop production and soil conservation.

Recombinant methods are certain to disrupt current agricultural practice as have all previous technological innovations.[6] There will be fewer public sector scientists to help agriculture adjust to these new tools. A heavy burden will thus fall on industrial scientists where they must go beyond innovating and developing new generations of technologies. More than ever before the private sector must carry out the underlying science. They are faced with much of the implementation research in biotechnology, and they must succeed in business. As cautioned by Dyson they also must be society's conscience regarding what technologies are best to develop while aiding in integration of these technologies to minimize disruption of world agriculture. Many factors contributed to the decision not to commercialize atrazine resistance by major companies, but an appreciation by industrial scientists of the public opposition to the technology and its likely impact on agriculture certainly were contributing factors.

1.2.7 Regulation of agricultural biotechnology. When there is public concern over a technology the legislative solution is to regulate it. In the short term such regulation has necessary societal benefit of reducing risk of damage from a technology, but the regulation is probably implemented to protect the reputation of the politicians against unforeseen problems. Thus, there is a political driver to regulate based on perception rather than science. We of course cannot and should not develop regulations for the unknown. Nor is it reasonable to regulate biotechnology and not other technologies just because we have an opportunity to do it early in its evolution.[6] However, possibly the greatest social contribution of the revolution in biotechnology was to empower the general public to voice their concerns early in the evolution of a technology. Regulation also can get out of hand where we strive for absolute safety as described elsewhere in this book for classical pesticides. Although large companies often complain about regulation, it also has the outcome of raising barriers for commercial success and thus eliminating competition from small and innovative units. Under conditions where we try to eliminate all risks of new technologies bureaucratic institutions fret over the remotely possible dangers of saying 'yes' to a technology without considering the certain negative impacts of saying 'no'. At best any bureaucracy tends to be inflexible and unable to accept that the future is an unpredictable moving target, yet a reasonable process for approval can serve to build public trust in a technology. For instance Henry Miller[44] charges that various bureaucracies, "have built huge, expensive, and gratuitous biotechnology regulatory empires preoccupied with negligible-risk activities, and have succeeded in protecting consumers only from enjoying benefits of the new technology." Every regulation has a killing effect on creativity. On the other hand in the last two years agricultural biotechnology has made unprecedented inroads into agriculture world wide. At least in developed countries, informed regulatory systems appear to be in place which will insure that nothing will be done with agricultural biotechnology which presents an overt risk to human or environmental health. The value of their reputation and concerns over liability provide another layer of insurance that major companies will be prudent with the products developed in the agricultural field.

Possibly we have a new group of benign and informed scientists involved in regulation who weigh the consequences of saying yes with saying no to agricultural biotechnology. We certainly need an enlightened process of evaluation of recombinant products in agriculture. Scientists have argued for product rather than process based regulation where recombinant products are accepted or rejected based on the characteristics of the product, not the technical process to arrive at that product. In the U.S. this concept seems generally accepted. Clearly with the first several rounds of genetically engineered crops, the high technical and regulatory barriers will restrict entry in developed countries to large companies. These companies have been judicious in selecting products both likely to be profitable and very unlikely to cause deleterious health or environmental effects. With future products society will need to consider them on a case by case basis. Reiss and Straughan[35] suggest a rational approach where we do not expect absolute guarantees of safety and strive to minimize risk and maximize societal benefit. With a positive selection trait such as herbicide resistance in a crop with close relatives that are weed species, one can anticipate that in a few cases the gene will escape. One possible way to regulate the process is to make the company that developed the transgene liable for control of such weed species. It certainly does not follow that if the first generation of recombinant products appears safe, that future products generated with the technology will be safe as well. A science based process of evaluation should provide a cost effective mechanism for evaluating a greater diversity of new products without stifling innovation and products for small markets. Hopefully we can maintain the trend where regulation provides a rational process for evaluation and implementation of agricultural biotechnology.

1.3 Baculoviruses as Delivery Agents for Pesticides

The success of recombinant plants makes it clear that organisms can be used for delivery agents. Plants offer many advantages, but there also are advantages to environmentally benign materials which can be applied as needed to a crop and which can be used on a variety of minor crops. Of these recombinant organisms, the furthest along in development are the insecticidal baculoviruses. Baculoviruses are double-stranded DNA viruses which are not known to infect any organism other than arthropods.[45] Their pathogenicity to insects, their high specificity for just several insects and their clear safety for vertebrates and plants have long made them targets for biocontrol efforts. Simple systems exist for their engineering and they are commonly used in producing recombinant proteins for research and biomedical applications.[46,47] These viruses are important components of natural insect control, especially among the lepidopterous insects which include many of our most destructive pests. Most baculoviruses are orally active and microencapsulated in a protein coat[48] which makes them more stable in the environment than many other microbes. This is an example of biological formulation which might be applied to other viral systems. Following ingestion, the protein coat or polyhedron is dissolved releasing viral particles. These particles invade gut cells where they must replicate and spread before they are shed at the next molt. Within the insect the virus spreads among cells using a different viral form termed the budded virus. Many of these viruses are capable of converting almost 50% of the biomass of the larva into virus particles which can then lead to an epizootic. Like BT, the viruses are attractive for biological control since they do not disrupt efficacy of parasites and predators.[49] However, they generally are more selective than BT which could offer advantages in not affecting butterflies for example in sensitive ecosystems. Natural baculoviruses are increasingly attractive for IPM both as a biological control agent and with inundative treatments. The use of baculoviruses in resistance management programs, to circumvent resistance to classical pesticides, and as supplements to transgenic crops is attractive. Conceptually one should consider separately the baculoviruses used as biological control agents which certainly are well adapted for this role and baculoviruses and other viruses which can be used as biological pesticides.[50] In this latter application there clearly are ways to use molecular techniques to improve the efficacy of the viruses. In this regard speed of kill of the viruses is critical for use as a biopesticide.

1.3.1 Development of recombinant viruses as biological pesticides. Although superb agents for natural biological control, natural baculoviruses have a major deficiency for use as biopesticides. The viruses have developed a sophisticated system to keep the host insect alive as a medium to produce additional virus. This attractive trait for natural control complicates the use of the viruses as biopesticides in ecologically based pest management programs. Thus, what is needed as a viral pesticide is a virus that leads to the rapid death of the host or a blockage of its feeding. There have been several recombinant approaches to achieve rapid kill with the virus. One elegant approach has been to remove a gene which adapts the virus to its host termed ecdysone glucose transferase.[51] The first approach evaluated was the insertion of insect hormones into viruses.[52] Although promising, this approach has not been widely used due to our limited knowledge of the insect endocrine system. It certainly has potential utility in demonstrating biological roles for neuropeptides but also in pest control.[53] Insect enzymes have been used successfully[54] along with other proteins to reduce feeding damage,[55] but the simplicity of using neurotoxins has dominated the field.[56,57] The wildtype and recombinant viruses appear to have a very similar spectrum of activity on lepidopterous larvae, but one can anticipate that the recombinant viruses may have slightly increased activity in some cases on semi permissive insects. Production, formulation, distribution and economic problems have also limited the use of wild type baculoviruses in insect control. The attraction of the quick kill recombinant viruses has led industry to address these problems, and the solutions hopefully can also be applied to wildtype baculoviruses and other microbials for use as biological control agents. There have been numerous recent reviews of this field.[41,55-61]

1.3.2 Future improvements in baculovirus efficacy. From quite early in the project it was possible to envision many improvements that could be made in recombinant baculoviruses.[62] Several of these improvements are in place in viruses now in field tests. Such improvements include expression of more potent toxins, use of earlier or better promoter systems, use of enhancers, and use of virus stocks which are inherently more active on the target species.[60] Improvements in separation technologies, mass spectral evaluation and sequencing will make it possible to mine even previously studied venoms for new toxins.[63,64] An improved understanding of structure activity relationships among peptide toxins may allow the design of synthetic genes for non natural peptides that are of increased potency and high specificity. Structural biochemistry has allowed, for example, the juvenile hormone esterase to be stabilized against degradation by the insect and increased its biological activity.[65] The same concept could be applied to toxins. The use of more potent toxins will facilitate the use of earlier and weaker promoter systems and help extend the technology to other viruses.[41] Synergistic combinations of toxins and dispersants are common in venomous animals and, exploitation of such natural systems for insect control is promising.[66] Co-infection of a host with two baculoviruses bearing pharmacologically complementary anti-insect toxins reinforce the potency of the virus. Another interesting extension of this concept is the engineering of such complementary toxins in a single virus.[67]

The baculovirus genome has an amazing plasticity that may allow farther sophisticated engineering. Thus, for example, insertional mutagenesis of the viral genome by placing a foreign gene (*i.e.* encoding for a potent toxin) is an approach that may lead to better viruses. The feasibility and advantages of such an approach have been provided recently,[68] by inserting a mite neurotoxin in the *egt* gene of a baculovirus disrupting it and conferring at the same time new toxic properties to the virus. Baculoviruses encode other gene products that modulate the viral infection in the host and potentially they are good targets for the rational design of more potent baculoviruses.

Finally, novel approaches are developing to extend the host range of specific baculoviruses. A few model systems have been a described.[69] Most of them aimed to extend and analyse the expansion of the host range of the *Autographa californica* nuclear polyhedrosis baculoviruses.[52,70–72] Interestingly, progress has been made and in two cases this goal was achieved.[52,71,72] However, it turns out that different genes were required to extend the host range of AcNPV to *Bombyx mori* and *Lymantria dispar,* respectively. Novel genes, required for AcNPV replication in different cell hosts, were isolated as well.[73-77] It is expected that this trend will lead to a better understanding of the baculoviral infection that will pave the way to expand the host range of other agriculturally important baculoviruses. It seems that this route is contributing already to basic science and manipulation of viral infections as well. It also illustrates a general principle in agriculture that the timeline between pure fundamental research and its application to practical problems is very short. A side but not less important benefit is that this approach may provide us with new tools to restrict the host range of baculoviruses in case that is needed (see also below). Fundamental information on the mechanisms of tolerance, resistance and host range may facilitate manipulation of host range, but also provide a fundamental understanding of viral specificity useful in registration and risk evaluation.[78–80]

If the first generation of viruses are successful it is likely to be commercially attractive to extend the host range to a wider variety of pest insects. This approach of course relies on the development of a great deal of fundamental information, and it comes with the disadvantage of generating viruses that are less specific. If regulatory barriers are not too high it may also be attractive to market combinations of recombinant and/or wildtype viruses tailored to pests on a specific crop or for use in an area.

It is likely that the recombinant viruses now under development will find practical use in agriculture. However, the value of this work could extend well beyond these first generation viruses. For example, these recombinant viruses illustrate a commercial value for proteins and peptides which can disrupt the development of insects even if they are not orally active. Many organisms utilize venoms which include synergistic mixtures of toxins, this approach could be used with recombinant organisms both as a resistance management tool and to increase efficacy. There is a plethora of venoms that can be mined for selective toxins, possibly of far greater potency than the current toxins. More potent toxins allow the use of earlier and weaker promoters and other viruses. The enhancement of baculovirus efficacy could be applied to a variety of other pest specific viruses and biological control agents, and some of the ancillary technology developed for the recombinant baculoviruses could be applied with advantage to a variety of wild type microbial agents for pest control. Although not a topic of this article, development of improved formulations for baculoviruses like other biologicals is critical. In some cases there could be formulation ingredients that have a direct effect on the interaction of the virus and insect.[81] Even in formulation and production, fundamental research with the virus and target are likely to have a major impact on the practical use of the viruses.

1.3.3 Efficacy in the field. In a series of studies from 1986–89 the behavior of genetically marked baculoviruses was evaluated[82] followed in 1993 by the first small scale field test of a virus with enhanced speed of kill.[83] In even the early tests the recombinant viruses demonstrated clear efficacy. Recent test data are from more realistic experimental designs, but data have not yet been published on the viruses likely to actually be marketed nor on large scale field tests. Using an Acal NPV expressing the AaIT toxin 10^{12} PIBs/Ha dramatically reduced square damage from corn ear worm and tobacco bud worm on Delta Pine Cotton. Although the control was lower than with Esfenvalerate at 17 g/Ha the lack of destruction of natural enemies led to the virus treated plots having control for a month after spraying. The virus also was found to act additively with BollGard cotton to control a mixed infestation of the above insects under conditions where the recombinant plant alone did not provide adequate control on its own.[84] Virus treatment gave slightly higher yield although less reduction of boll damage than treatments with a pyrethroid.

1.3.4 Potential for integration into ecologically based pest management systems. If success in pest management stems in part from the availability of a diversity of agents with different properties, the recombinant baculoviruses offer a great advantage. For smaller crops this will clearly require research and development in the public sector. The short persistence and high specificity of baculoviruses represents an alternative to more persistent and broader spectrum synthetic insecticides. The potential for integrating recombinant viruses into IPM systems in the near future is possibly more limited than the potential for the wild-type NPVs. The wild-type viruses have great versatility in IPM, due in large part to their long-term environmental persistence and efficient transmission (*e.g.* vertical transmission, which contributes to viral transport) in many cases. Thus, these wild-type NPVs can be used as short-term viral insecticides when rapid kill is not needed, but they also can provide season-long control (*e.g.* the *Anticarsia gemmatalis* NPV in

Brazil) or even permanent suppression of a pest in classical biological control (*e.g.* the *Gilpinia hercyniae* NPV in North America). These approaches to insect control integrate well with other natural enemies that attack the pest as well as with chemical insecticides. Pest suppression can even be enhanced by naturally occurring NPVs by altering cultural practices to favor viral populations already present, an approach which almost epitomizes concepts of IPM. These long-term approaches, which take advantage of certain strengths of the NPVs in pest management, are not acceptable with recombinant NPVs under the current regulatory climate. Persistence and efficient transmission are undesirable, because, until confidence is gained in the environmental safety of recombinants, persistence and transmission (transport) increase their environmental risks. Also with the first generation of quick kill viruses, the desired trait precludes effective transmission and persistence.

In those situations where an incisive reduction of a population of caterpillars without the disruption of a biological complex in a field is desired, the recombinant viruses offer a tremendous advantage. An example of this application is in early season cotton. Similarly in sensitive ecosystems such as forests both wildtype and recombinant viruses offer much higher specificity than either synthetic insecticides or BT. In contrast they are not attractive where a wide variety of phylogenetically distinct pests need to be controlled immediately. The quick kill viruses could be used to supplement BT plants to control pests lacking high susceptibility to the BT strains used or in a program of resistance management for transgenic crops. Thus, integration of recombinant NPVs into IPM initially will concentrate on situations in which environmental safety of the pest control agent is paramount (*e.g.* preserving populations of predatory arthropods) or in which the crop-pest system requires an alternative control agent. Examples of the latter might include crops in which pesticide registrations are not renewed and new chemistry is unavailable, or BT crops which require an alternative control method at times in order to hinder the development of pest resistance to the BT δ-endotoxins. Recombinant NPVs also might be useful in minor and "organically-grown" crops, although this would require research and development by the public sector due to small markets. If adequate formulation and production systems can be developed to make the recombinant viruses economically competitive with classical insecticides, pest management systems will need to be developed to take advantage of the maintenance of beneficial insects by the viruses if their true potential is to be recognized.

The recombinant viruses produce levels of toxins that reduce the coordination of pest insects before there are clear symptoms. This knocks the insects off of plants increasing the level of control beyond simple kill times.[85] The scorpion toxins used in the current generation of viruses bind to the same channel as pyrethroids and DDT. Thus, they represent a peptide version of this earlier generation of compounds. This raises the possibility of cross-resistance, but fortunately it appears that strains of *H. virescens* previously selected for resistance to pyrethroids become even more susceptible to the toxin AaIT. Thus, the recombinant viruses could have potential in resistance management programs for pyrethoids as well. It is likely that the quick kill viruses commonly will be used in combination with other insecticides to control a wider selection of insects. One can anticipate a variety of interactions, but fortunately with pyrethroids the interaction appears to be synergistic. The speed of kill of the recombinant virus is enhanced with a low dose of pyrethroids. The high doses of pyrethroids often used in cotton are driven by the need to control resistant caterpillars. By combining the pyrethroid with a baculovirus much lower doses of pyrethroid could be used reducing problems with disruption of biological control and environmental contamination.[86] It is likely that other interactions can be found with synthetic and biological control systems.

The chemistry of the crop plant is known to have a major influence on efficacy of agricultural chemicals so it is not surprising that the effects of plant chemistries on biological material such as baculoviruses can be even higher. The chemistry influences both the stability and efficacy of the viruses. For example in two plant species, the higher the peroxidase and polyphenyl oxidase levels the lower the mortality caused by baculoviruses. The situation was further complicated by the complexity of phenolics present in different plants. These data certainly indicate that one should integrate the use of some plant resistance traits with baculovirus use carefully.[87–89]

1.3.5 Environmental safety. Numerous wildtype viruses have been approved for use in agriculture in many countries.[60] The benefits of these selective and potent pest control agents clearly outweigh the risks. It remains to be seen if recombinant viruses will offer benefits which outweigh the additional regulatory burden of dealing with recombinant organisms. So far regulatory agencies in the U.S. appear pleased with the safety issues regarding the recombinant viruses, and several companies are moving recombinant materials toward the market. With the expression of toxins which are exceptionally selective for insects, under promoters selectively expressed in insects in viruses which largely are selective for species in a single family, and orally active in even fewer species, the recombinant viruses are far more specific in their action on insect pests than any synthetic insecticide or BT.

A major consideration with recombinant organisms is their environmental safety.[90] There have been sufficient disasters with classical biological control agents that the release of any organism in an environment novel to it should be approached with caution. One consideration is if the recombinant organism has a survival advantage over the endogenous organisms that it may compete with in its new environment. For instance when we place herbicide resistance or viral resistance genes in a crop plant, we are giving that crop plant a survival advantage over plants lacking this gene. Thus, we need to be very careful with releasing this germplasm because organisms containing a gene which provides a selective advantage (recombinant or not) could be difficult to recover once released into the environment. In contrast, with the recombinant viruses under consideration as biopesticides, recombinant technology has been used to make the viruses better biopesticides in inundative biological control. The viruses are not designed to compete with the wildtype viruses and in fact will rapidly vanish from the environment and be out competed with the wildtype viruses. Since there are advantages to mixing wildtype and recombinant viruses in terms of economy and insect control, one could even enhance this process of competitive replacement by spraying mixtures of viruses. The viruses which lead to quick kill of pests produce fewer viral progeny and thus are at a great survival disadvantage.[91,92] Although these viruses can be further crippled by a variety of techniques, the fact that the very gene which makes the recombinant viruses attractive to agriculture also gives them a massive competitive liability indicates that the viruses will be environmentally safe. Bergelson has cautioned that altered fecundity does not predict invasiveness, and that large scale field tests are needed for this evaluation.[93] This is a valuable caution; however, it is based on data from *Arabidopsis thaliana* with a resistance gene to chlorsulfuron. This is a situation where seed number indicates that the resistant genotype is at a very slight negative disadvantage, but the plant carries a gene which can give it a tremendous selective advantage under some conditions. In contrast the recombinant baculoviruses carry a gene which gives them a tremendous selective disadvantage under every condition of infection so far examined. Of course actual experiments are needed to settle this question. Ignoffo reported that the wildtype and recombinant baculoviruses were of similar sensitivity to ultraviolet light.[92] Fuxa compared two recombinant and a wildtype virus acting on *T. ni* feeding on collard plants in a greenhouse microcosm. The recombinant viruses failed to maintain a sustained high level of infection when used alone while the wild type virus did and the wild type virus out competed the recombinants for a niche in the greenhouse. These data suggest that it is very unlikely that quick kill viruses will persist in an agricultural ecosystem,[94] and similar results were obtained by American Cyanamid.[60] Wild type viruses often elicit a behavior in their host leading to spread of the virus including larvae moving near the top of the plant and liquefying to spread the virus. In contrast the recombinant viruses tend to knock larvae off plants at an early stage often as hard cadavers which do not disintegrate to spread virus.[83] This reduces feeding but could lead to a build up of recombinant virus in soil reservoirs. However, preliminary data indicate that the wildtype virus also is far more abundant in the soil than the recombinant viruses under microcosm conditions.[60,94] To estimate the contribution of a species to maintaining a virus in an ecosystem many factors need to be known including its likelihood of encountering the virus, susceptibility to the virus, negative effects of the recombinant on the species, production and propagation of the virus. For example a species which is highly susceptible to oral infection by the recombinant virus, allows the virus to replicate to high levels, but is not killed by the virus could maintain environmental populations at a higher level than anticipated. Although the data at present are on limited species and are qualitative, it seems safe to conclude that replication of the virus in the ecosystem will be minimal.

Baculoviruses are able to recombine among themselves to a low extent but there are many barriers that a species has to retain genetic integrity.[60] It is conceivable that a baculovirus engineered to express an anti-insect toxin could be released in an ecosystem containing larvae carrying another wild strain of baculovirus. Even in this case a putative emergent new recombinant virus, resulting from co-infection of one single larvae by both types of baculoviruses, will not have any selectional advantage. As outlined above, the mode of action of the toxin-expressing viruses has an initial advantage by killing the larvae fast, but also has a long term disadvantage because of the small number of viral particles resulting in the infected organism. In addition the larvae infected with the quick kill viruses usually are unable to disseminate the virus by melting as is in the case of the wild type counterpart.[85,95] To obtain experimental recombination even in the most favorable circumstances, molecular biologists normally employ genes with a strong positive selection pressure as a driver. In the case of recombinant viruses there is a strong negative driver that will force them quickly into extinction.[60]

Of greater significance than the recombinant gene is the consideration of the wildtype parent virus in an ecosystem. Since the recombinant could lose the lethal gene and effectively become a wildtype virus again, it would be wise to examine the potential ecological effects of the introduction of a wildtype virus lacking the recombinant gene into an ecosystem if the virus is not already there. An alternative is to develop recombinant baculoviruses based on endogenous viral genotypes for a region.

With the recombinant baculoviruses under development it is critical to view them as a biological pesticide.[50] There will be some environmental impact from their use because this is the purpose of applying any pesticide. Direct effects on the target population in a field is the purpose of the application and indirect effects on species relying on the target population for food are anticipated. These negative effects of recombinant baculoviruses should be weighed against alternative control strategies and the consequences of no pest control.[4] Nontarget effects depend upon both the sensitivity of the nontarget organism and its exposure to the recombinant virus.[90] The high specificity of the virus for target species combined with their crippled ability to reproduce in the environment greatly reduces the likelihood of nontarget effects compared with most other synthetic and biological pesticides. In general the recombinant viruses appear exceptionally safe for nontarget organisms.[56,60] A dramatic recent study came from cotton plots treated weekly with wildtype virus and an AaIT quick kill construct. No adverse effects were seen on nontarget arthropods, earthworms and arachnids in 18 non-lepidopteran families.[96,97]

1.3.6 Human safety and public acceptance. A large number of tests of numerous baculoviruses at high doses in many vertebrate species has failed to yield any negative effects.[60] Such tests are being repeated now by at least two companies with recombinant viruses. With high specificity of the virus coupled with high specificity of the toxin, the recombinant viruses can be assumed to be safe. The scorpion toxins being evaluated for use in the viruses also are undergoing extensive testing and multiple studies published to date indicate that the peptides are exceptionally selective to insects.[56,60,62] For example the scorpion toxins used in the recombinant baculoviruses are very selective for insect sodium channels. By comparison the α-cyanopyrethroids are derived from natural pyrethrins from flower heads. These compounds are thought to act largely through the voltage sensitive sodium channels in both insects and mammals.[98,99] The relative safety of viruses comes from high levels of safety at multiple steps in the causal chain leading to effects on target insects. In this regard we can compare them to these commonly used insecticides with a similar mechanism of action. These insecticides are generally regarded as safe for use in agriculture, but depending on the technique, species, and preparation used, the α-cyanopyrethroids such as Esfenvalerate are of similar activity on insect and mammal sodium channels and nerves while no activity has been found for the peptide toxins being used in field tests with recombinant baculoviruses. A combination of barriers of penetration and metabolism as well as careful use patterns allows Esfenvalerate to be used safely in agriculture, although it has a rat oral LD_{50} of 75 mg/kg.[100] The peptide toxin is produced in baculoviruses under a promoter which is inactive in mammalian cells in a virus that cannot effectively penetrate most mammalian cells and is not able to replicate in any mammalian cells. The virus cannot penetrate the mammalian gut, and it is enclosed in a protein matrix that does not dissolve readily in mammalian intestines. Testing by companies and some published data support the safety of the recombinants.[60,62,101]

1.3.7 Use in developing countries. The initial cost of developing and registering recombinant baculoviruses will be justified by targeting major markets in developed countries. However, one hopes that the technology can benefit agriculture in developing countries. It may be that the recombinant and wildtype viruses under investigation for major markets will fit into a technological niche which can benefit developing agriculture. It has been difficult to deliver effective chemical based pest control in some cases to developing countries. This situation, in some cases, can result from the dilemma that a country may lack the international currency to pay for compounds yet not have the industrial infrastructure to make the compounds themselves. The newer compounds with enhanced human and environmental safety generally are more expensive and result from increasingly sophisticated synthetic procedures, thus exacerbating the problem. Ecologically based pest management systems, even ones with a pesticide base, often rely on information and skills which farmers lack. The viruses could represent a valuable technology of intermediate complexity. If a central facility can send out inocula, the recombinant viruses could be produced *in vivo* by local industry. Both recombinant and wildtype viruses are likely to be initially expensive due largely to labor costs, although *in vivo* production does not necessarily rely on a complex industrial infrastructure. A lower labor cost in developing countries could reduce dramatically the cost of the virus preparations. The viruses offer great human and environmental safety while being simple to use since they are biological pesticides applied with existing technology. They of course have the disadvantage of not being broad spectrum chemicals, but there are likely cropping systems where a single insect or insect complex are the major problems leading to high levels of pesticide use.[56]

The wild-type viruses have been quite successful as viral insecticides in pest management in developing nations. This probably is due largely to inexpensive hand labor, to the cost of importing and using chemical pesticides, and perhaps to a relative lack of regulatory oversight in such countries. The best example is the *Anticarsia gemmatalis* (soybean caterpillar) NPV, which is applied to approximately one million Ha of soybean per year in Brazil. Other examples include the *Oryctes rhinoceros* (rhinoceros beetle) nonoccluded baculovirus on coconut in Pacific Islands; granulosis virus (GV) of *Erinnyis ello* (cassava hornworm) in Brazil; *Spodoptera frugiperda* (fall armyworm) NPV on maize in Brazil (*ca.* 20,000 Ha treated/year); various *Spodoptera* NPVs against *Spodoptera* spp. in China, India, Taiwan, Guatemala, and Thailand; *Heliothis armigera* NPV on various crops (primarily cotton) in China, Thailand, Vietnam, and India; *Heliothis assulta* NPV on tobacco, cayenne pepper, and tomato in China; NPVs of the semi looper complex (Noctuiidae: Plusiinae) on various crops in Zimbabwe and China; GV of potato tuber moth in Peru, Colombia, Ecuador, and Bolivia; *Cydia pomonella* (codling moth) GV in Argentina; and NPV of *Perigonia lusca* (hornworm) on tea in Paraguay. One hopes that the technologies developed in support of the recombinant viruses will improve the utility of the wildtype viruses, and that recombinant viruses can be developed to improve agricultural productivity and profitability specifically in developing countries.

2 CONCLUSION

With recombinant delivery systems we are embarking on a brave new venture in agriculture which will offer tremendous benefit to humans and the ecosystem. Any revolutionary technology will be disruptive at the socio-economic level,[35] and agricultural biotechnology will be no exception. Certainly we have not regulated other revolutionary changes in agriculture based on unpredictable social consequences. However, the argument against regulating the social and economic consequences of recombinant technology in agriculture does not indicate that companies and individual scientists are relieved of their obligation to minimize the social and economic impact of their technologies. As a scientific community in both the public and private sector we must strive to minimize such disruption and to emphasize the benefits over the problems. While advancing the immediate and safe benefits of this technology, the scientific community must have sympathy for public fears over the long term effects of misuse of recombinant technology. This sympathy does not translate as restrictive regulations that go beyond the immediate concerns of safety for humans and the ecosystem. In the seventeenth century Milton cautioned against banning freedom of the press and intellectual inquiry in England and condemned the imprisonment of Galileo that crippled innovation in Italy. In reference to this caution Freeman Dyson warns,[34] "Perhaps, after all, as we struggle to deal with the enduring problems of reconciling individual freedom with public safety, the wisdom of a great poet may be a surer guide than the calculations of risk-benefit analysis." A thoughtful and cautious approach and a sympathetic ear for short and long term concerns of the public should go a long way toward emphasizing the benefits of recombinant organisms as delivery systems.

3 REFERENCES

1. D. Evans, *Chemistry in Britain,* 1998, **34**, 20.
2. T. R. Shieh, 'Eighth International Congress of Pesticide Chemistry Options 2000', N. N. Ragsdale, P. C. Kearney and J. R. Plimmer, eds, American Chemical Society, Washington, DC, 1995, p. 104.
3. S. P. Briggs and M. Koziel, *Curr. Opin. Biotechnol.,* 1998, **9**, 233.
4. D. T. Avery, 'Saving The Planet With Pesticides and Plastic: The Environmental Triumph of High-Yield Farming', Hudson Institute, Indianapolis, IN, 1995, p. 432.
5. C. James, *ISAAA Briefs,* 1997, **5**, 31.
6. S. Krimsky and R. P. Wrubel, 'Agriculture Biotechnology and The Environment: Science, Policy, and Social issues', University of Illinois Press, Urbana and Chicago, IL, 1996, p. 294.
7. R. J. Cook and C. O. Qualset. 'Appropriate oversight for plants with inherited traits for resistance to pests', presented at: Institute of Food Technologists, 1996, p. 1
8. M. Peferoen, *Tibtech,* 1997, **15**, 173.
9. F. J. Dyson, 'Imagined Worlds', Harvard University Press, Cambridge, MA, 1997, p. 216.
10. F. J. Dyson, 'Infinite in All Directions. Gifford Lectures', Harper and Row, New York, 1985, p. 321.
11. A. M. R. Gatehouse and J. A. Gatehouse, *Pesticide Science,* 1998, **52**, 165.
12. E. Strauss, *Science,* 1998, **280**, 2050.
13. D. N. Alstad and D. A. Andow, *Science,* 1995, **345**, 1394.
14. R. T. Roush, 'Advances in Insect Control: The Role of Transgenic Plants.', N. Carozzi and M. Koziel, eds, Taylor and Francis, London, 1997, p. 271.
15. F. Gould, *Annu. Rev. Entomol.,* 1998, **43**, 701.
16. C. W. Hoy, G. P. Head, and F. R. Hall, *Annu. Rev. Entomol.,* 1998, **43**, 571.
17. A. Tsaftaris, *Field Crop Res.,* 1996, **45**, 115.
18. L. Herrera-Estrella, L. S. Rosales, and R. Rivera-Bustamante, 'Plant-Microbe Interactions', G. Stacey and N. T. Keen, eds, Chapman & Hall, New York, NY, 1996, p. 33.
19. B. J. Staskawicz, F. M. Ausubel, B. J. Baker, J. G. Ellis, and J. D. Jones, *Science,* 1995, **268**, 661.
20. K. E. Hammond-Kosack and J. D. Jones, *Plant Cell,* 1996, **8**, 1773.
21. V. M. Williamson and R. S. Hussey, *Plant Cell,* 1996, **8**, 1735.
22. G. Gheysen, W. Vandereycken, N. Barthels, M. Karimi, *et al.*, *Pestic. Sci.,* 1996, **47**, 95.
23. M. Rossi, F. L. Goggin, S. B. Milligan, I. Kaloshian, D. E. Ullman, and V. M. Williamson, *Proc. Natl. Acad. Sci. U S A,* 1998, **95**, 9750.
24. T. van den Boogaart, G. P. Lomonossoff, and J. W. Davies, *Amer. Phytopathological Soc.,* 1998, **11**, 717.
25. J. J. English and D. C. Baulcombe, *Plant J.,* 1997, **12**, 1311.

26. N. S. Al-Kaff, S. N. Covey, M. M. Kreike, A. M. Page, R. Pinder, and P. J. Dale, *Science,* 1998, **279**, 2113.
27. J. Cao, X. Li, and X. Dong, *Proc Natl Acad Sci U S A,* 1998, **95**, 6531.
28. T. N. Hanzlik and K. H. J. Gordon, *Adv. Virus Res.,* 1997, **48**, 101.
29. R. Karban and I. T. Baldwin, 'Induced Responses to Herbivory', University of Chicago Press, Chicago, 1997, p. 319.
30. S. S. Duffey and G. W. Felton, 'Naturally Occurring Pest Bioregulators', P. A. Hedin, eds, American Chemical Society Symposium Series, Washington, DC, 1991, p. 166.
31. M. R. Berenbaum and A. R. Zangerl, 'Plant Resistance to Herbivores and Pathogens: Ecology, Evolution and Genetics', R. S. Fritz and E. L. Simms, eds, University of Chicago Press, Chicago, IL, 1992, p. 97.
32. J. S. Thaler, M. J. Stout, R. Karban, and S. S. Duffey, *J. Chem. Ecology,* 1996, **22**, 1767.
33. A. A. Agrawal, *Science,* 1998, **279**, 1201.
34. F. J. Dyson, 'Disturbing the Universe', Harper and Row Publishers, New York, 1979, p. 283.
35. M. J. Reiss and R. Straughan, 'Improving Nature? The Science and Ethics of Genetic Engineering', Cambridge University Press, Cambridge, UK, 1996, p. 288.
36. S. Nottingham, 'Eat Your Genes, How Genetically Modified Food is Entering our Diet', Zed Books Ltd., London, UK, 1998, p. 208.
37. M.-W. Ho, 'Genetic Engineering Dream or Nightmare', Gateway Books, Bath, England, 1998, p. 304.
38. W. Klassen, 'Progress and Perspectives for the 21st Century', J. J. Menn and A. L. Steinhauer, eds, Entomological Society of America, Washington D.C., 1991, p. 43.
39. M. Kogan, *Annu. Rev. Entomol.,* 1998, **43**, 243.
40. L. D. Newsome, 'Annual Review of Entomology', R. F. Smith and T. E. Mittler, eds, Annual Reviews, Palo Alto, CA, 1967, p. 257.
41. B. D. Hammock and S. Maeda, 'Progress and Perspectives for the 21st Century', J. J. Menn and A. L. Steinhauer, eds, Smithsonian Institution, Washington, D.C., 1991, p. 77.
42. J. E. Casida and G. B. Quistad, *Annu. Rev. Entomol.,* 1998, **43**, 1.
43. N. N. Ragsdale and C. R. Curtis, 'Modern Fungicides and Antifungal Compounds: 11th International Symposium, May 14th-20th 1995: Castle of Reinhardsbrunn/Berg Hotel Friedrichroda, Thuringia, Germany.', H. Lyr, P. E. Russell and H. D. Sisler, eds, Intercept Ltd., Andover, Hampshire SP10 1YG, UK, 1996, p. 25.
44. H. Miller, *Bio/Technology,* 1993, **11**, 1075.
45. R. R. Granados and B. A. Federici, 'The Biology of Baculoviruses: Practical Application in Insect Control', CRC Press, Boca Raton, FL, 1986, p. 276.
46. G. E. Smith, M. J. Fraser, and M. D. Summers, *J. Virol.,* 1983, **46**, 584.
47. L. K. Miller, *Ann. Rev. Microbiol.,* 1988, **42**, 177.
48. G. F. Rorhmann, *J. Gen. Virol.,* 1986, **67**, 1499.
49. P. F. Entwistle and H. F. Evans, 'Comprehensive Insect Physiology Biochemistry and Physiology', L. I. Gilbert and G. A. Kerkut, eds, Pergamon Press, Oxford, 1985, p. 347.
50. B. D. Hammock, *Nature,* 1992, **355**, 119.
51. D. R. O'Reilly and L. K. Miller, *Science,* 1989, **245**, 1110.
52. S. Maeda, *Biochem. Biophys. Res. Commun.,* 1989, **165**, 1177.

53. P. W. Ma, T. R. Davis, H. A. Wood, D. C. Knipple, and W. L. Roelofs, *Insect. Biochem. Mol. Biol.,* 1998, **28**, 239.
54. B. D. Hammock, B. Bonning, R. D. Possee, T. N. Hanzlik, and S. Maeda, *Nature,* 1990, **344**, 458.
55. B. C. Bonning and B. D. Hammock, *Ann. Rev. Entomol.,* 1996, **41**, 191.
56. B. D. Hammock, 'Reviews in Pesticide Toxicology', R. M. Roe and R. J. Kuhr, eds, IOS Press, Amsterdam, The Netherlands, 1997, p. 1.
57. M. Gurevitz, N. Zilberberg, O. Froy, D. Urbach, E. Zlotkin, B. D. Hammock, R. Herrmann, H. Moskowitz, and N. Chejanovsky, 'Modern Agriculture and The Environment', D. Rosen and E. Tel-Or, eds, Kluwer Academic Publishers, Dordrecht, Boston, 1997, p. 81.
58. B. C. Bonning and B. D. Hammock, *Biotechnol. Genet. Eng. Rev.,* 1992, **10**, 453.
59. B. F. McCutchen and B. D. Hammock, 'Natural and Derived Pest Management Agents.', P. Hedin, J. J. Menn and R. Hollingworth, eds, American Chemical Society, Washington, DC., 1994, p. 348.
60. B. C. Black, L. A. Brennan, P. M. Dierks, and I. E. Gard, 'The Baculoviruses', L. K. Miller, eds, Plenum Press, New York, 1997, p. 341.
61. R. D. Possee, A. L. Barnett, R. E. Hawtin, and L. A. King, *Pestic. Sci.,* 1997, **51**, 562.
62. B. D. Hammock, B. F. McCutchen, J. Beetham, P. Choudary, E. Fowler, R. Ichinose, V. K. Ward, J. Vickers, B. C. Bonning, L. G. Harshman, D. Grant, T. Uematsu, and S. Maeda, *Arch. Insect. Biochem. Physiol.,* 1993, **22**, 315.
63. Y. Nakagawa, Y. M. Lee, E. Lehmberg, R. Herrmann, R. Herrmann, H. Moskowitz, A. D. Jones, and B. D. Hammock, *Eur. J. Biochem.,* 1997, 496.
64. H. Moskowitz, R. Herrmann, A. D. Jones, and B. D. Hammock, *Eur J Biochem,* 1998, **254**, 44.
65. B. C. Bonning, V. K. Ward, M. M. v. Meer, T. F. Booth, R. D. Possee, and B. D. Hammock, *Proc. Natl. Acad. Sci.,* 1997, **94**, 6007.
66. R. Herrmann, H. Moskowitz, E. Zlotkin, and B. D. Hammock, *Toxicon.,* 1995, **33**, 1099.
67. G. G. Prikhod'ko, H. J. R. Popham, T. J. Felcetto, D. A. Ostlind, V. A. Warren, M. M. Smith, V. M. Garsky, J. W. Warmke, C. J. Cohen, and L. K. Miller, *Biol. Control,* 1998, **12**, 66.
68. H. J. R. Popham, Y. H. Li, and L. K. Miller, *Biol. Control.,* 1997, **10**, 66.
69. S. M. Thiem, *Curr. Opin. Biotechnol.,* 1997, **8**, 317.
70. E. Gershburg, H. Rivkin, and N. Chejanovsky, *J. Virol.,* 1997, **71**, 7593.
71. C. J. Chen, M. E. Quentin, L. A. Brennan, C. Kukel, and S. M. Thiem, *J. Virol.,* 1998, **72**, 2526.
72. G. Croizier, L. Croizier, O. Argaud, and D. Poudevigne, *Proc. Natl. Acad. Sci.,* 1994, **91**, 48.
73. S. M. Thiem, X. Du, M. E. Quentin, and M. M. Berner, *J. Virol.,* 1996, **70**, 2221.
74. C. J. Chen and S. M. Thiem, *Virology,* 1997, **227**, 88.
75. X. Du and S. M. Thiem, *Virology,* 1997, **227**, 420.
76. R. J. Clem and L. K. Miller, *J. Virol.,* 1993, **67**, 3730.
77. A. Lu and L. K. Miller, *J. Virol.,* 1996, **70**, 5123.
78. B. A. Kirkpatrick, J. O. Washburn, E. K. Engelhard, and L. E. Volkman, *Virology,* 1994, **203**, 184.
79. J. O. Washburn, B. A. Kirkpatarick, and L. E. Volkman, *Nature,* 1996, **383**, 767.
80. L. Volkman, *Adv. Virus Res.,* 1997, **48**, 313.

81. J. O. Washburn, B. A. Kirpatrick, E. Haas-Stapleton, and L. E. volkman, *Biol. Control,* 1998, **11**, 58.
82. D. H. L. Bishop, P. F. Entwistle, I. R. Cameron, I. R. Allen, and R. D. Possee, 'The Release of Genetically-Engineered Microorganisms', M. Sussman, C. H. Collins, F. A. Skinner and E. Stewart-Tull, eds, Academic Press, New York, NY, 1988, p. 143.
83. J. S. Cory, M. L. Hirst, T. Williams, R. S. Hails, D. Goulson, B. M. Green, T. Carty, R. D. Possee, P. J. Cayley, and D. H. L. Bishop, *Nature,* 1994, **370**, 138.
84. J. N. All and M. F. Treacy, 'Proceedings Beltwide Cotton Prod. and Research Conference', P. Duggar and D. A. Richter, eds, Memphis, TN., 1997, p. 1294.
85. K. Hoover, C. M. Schultz, S. S. Lane, B. C. Bonning, S. S. Duffey, B. F. McCutchen, and B. D. Hammock, *Biol. Control.,* 1995, **5**, 419.
86. B. F. McCutchen, K. Hoover, H. K. Preisler, M. D. Betana, R. Herrmann, J. L. Robertson, and B. D. Hammock, *J. Econ. Entomol.,* 1997, **90**, 1170.
87. S. Duffey, K. Hoover, B. C. Bonning, and B. D. Hammock, 'Reviews in Pesticide Toxicology.', R. M. Roe and R. J. Kuhr, eds, Toxicology Communications, Inc., Raleigh, North Carolina, 1995, p. 137.
88. K. Hoover, J. L. Yee, C. M. Schultz, D. M. Rocke, B. D. Hammock, and S. S. Duffey, *J. Chem. Ecol.,* 1998, **24**, 221.
89. K. Hoover, M. J. Stout, S. A. Alaniz, B. D. Hammock, and S. S. Duffey, *J. Chem. Ecol.,* 1998, **24**, 253.
90. A. Richards, M. Matthews, and P. Christian, *Annu. Rev. Entomol.,* 1998, **43**, 493.
91. Y. Kunimi, J. R. Fuxa, and B. D. Hammock, *Entomologia Experimentalis et Applicata,* 1996, **81**, 251.
92. C. M. Ignoffo, C. Garcia, B. C. Bonning, R. Herrmann, and B. D. Hammock, *J. Kansas Entomol. Soc.,* 1997, **70**, 149.
93. J. Bergelson, *Ecology,* 1994, **75**, 249.
94. J. R. Fuxa, A. Inceoglu, S. A. Alaniz, A. R. Richter, L. M. Rilley, and B. D. Hammock, *Biol. Control.,* in press.
95. L. K. Miller, *J. Invertebr. Pathol.,* 1995, **65**, 211.
96. M. F. Treacy, J. N. All, and C. F. Kukel, 'New Developments in Entomology', K. Bondari, eds, Research Signpost, London, 1997, p. 57.
97. M. F. Treacy and J. N. All, 'Proceedings Beltwide Cotton Conferences', P. Duggan and D. A. Richter, eds, National Cotton Council of America, Memphis, TN, 1996, p. 911.
98. D. M. Soderlund and J. R. Bloomquist, *Ann. Rev. Entomol.,* 1989, **34**, 77.
99. J. R. Bloomquist, *Annu. Rev. Entomol.,* 1996, **41**, 163.
100. T. C. Sparks, J. J. Sheet, J. R. Skomp, T. V. Worden, L. L. Larson, D. Bellows, S. Thibault, and L. Wally, 'Proceedings of the 1997 Beltwide Cotton Production Conference', eds, National cotton Council, Memphis, TN, 1997, p. 1259.
101. D. Possee, M. Hirst, L. D. Jones, D. H. L. Bishop, and P. J. Coyley, *British Crop Protection Council Monograph 55. Opportunities for Molecular Biology in Crop Production.,* 1993, **55**, 23.

ACKNOWLEDGMENTS: The authors thank reviewers in both the public and private sector. Research was supported in part by USDA grant #97-35302-4406; USDA Risk Assessment Grant #98-33120-6435; BARD #IS-2530-95C and IS-2465-94; NIEHS Superfund # P42 PHS ES04699; 1 P30 ES05707; EPA #CR819658 and State of Texas Advanced Technology Program Grant #999902-039. We thank Martina McGloughlin and Kathleen Dooley for a critical review and Maeva Yao for editorial assistance.

Plant Protection – Current State of Technique and Innovations

Heinz Ganzelmeier

BIOLOGISCHE BUNDESANSTALT FÜR LAND- UND FORSTWIRTSCHAFT (BBA), MESSEWEG 11/12, D-38104 BRAUNSCHWEIG, GERMANY

1 INTRODUCTION

Plant protection products must be applied according to the rules of Good Agricultural Practice if plant protection measures are to be economical, with low losses, environmentally safe and efficient. To be able to come up to this demand when controlling pests in different crops, diverse plant protection equipment types are needed, such as field sprayers, fruit crop, vineyard and hop sprayers, pedestrian sprayers, seed dressing machines, granules applicators, fogging machines, etc.

About 60 % of the plant protection products used in Europe are used in field crops such as e.g. cereals, sugar beet and rape. They are mostly applied by field sprayers equipped with a horizontal spray boom. Some 28 % of the plant protection products are applied in fruit crops, vineyards and hops, normally by air assisted sprayers with vertical nozzle bars[1]. In the United States, some 80 % of plant protection products are applied in field crops and about 7 % in bush and tree fruit crops[2]. In this presentation, I shall survey the application techniques employed in field and tall growing crops.

2 CURRENT STATE OF SPRAYER TECHNIQUE

Manufacturers of sprayers have changed their product ranges in favour of bigger and better equipped implements in adaptation to the structural change in agriculture. Germany, which also regulates sprayers, has detailed statistics about this[3] (Fig. 1). The share of self-propelled and trailed sprayers has noticeably increased over the past few years from 1 to 6 % and from 19 to 29 %, respectively, while the proportion of tractor-mounted sprayers has decreased from 62 to 46 %. Spray tank volumes reach 6600 l, and the working width of some spray booms is up to 45 m. Among air assisted sprayers, types with axial flow fans predominate (56 %). The trend in fan types is towards such with a more horizontal air stream. Tank volumes of air assisted sprayers reach 4000 l, and fans produce an air flow rate of about 100 000 m^3/h.

Sprayers and air assisted sprayers for field crops	 3-point hitch	 mounted	 trailed	 self propelled	 line spraying	 among all: air assisted
Available sprayer types	95 (46%)	35 (17%)	61 (29%)	12 (6%)	5 (3%)	24 (12%)
Tank (l)	200 - 1500	600 - 4000	600 - 5000	1800 - 6600	100 - 300	600 - 4000
Spray boom (m)	7 - 24	8 - 36	12 - 45	12 - 36	4 to 12 rows	12 - 28
Pump (l/min)	58 - 240	94 - 2x225	94 - 2x225	138 - 1400	16 - 70	100 - 344
Tech. rest volume (% of tank capacity)	1.6 - 2.6	1.9 - 3.0	1.2 - 3.0	0.8 - 2.1	2.1 - 3.1	1.6 - 3.0

Figure 1 *Survey of the offer for sale of field sprayers in Germany*

3 WHAT ARE THE DEMANDS ON SPRAYERS TODAY?

Above all sprayers shall

- exactly dose plant protection products
- evenly distribute plant protection products on the treatment area
- achieve sufficient spray deposit on the target area
- cause only little ground deposit
- cause only little drift

3.1 Field Sprayers

The **dosage** is done with the help of relatively simple and sufficiently exact dosing facilities. Big sprayers are normally equipped with electronic dosing control systems which meet current requirements and considerably relieve the driver[4].

The **distribution** of the spray on the treatment area is determined by the lengthwise and crosswise distribution generated by the sprayer[5]. Dynamic measurements carried out on field sprayers have shown that the distribution varies a lot under practical conditions (coefficient of variation between 15 and 38 %), making technical improvements urgently necessary (Fig. 2).

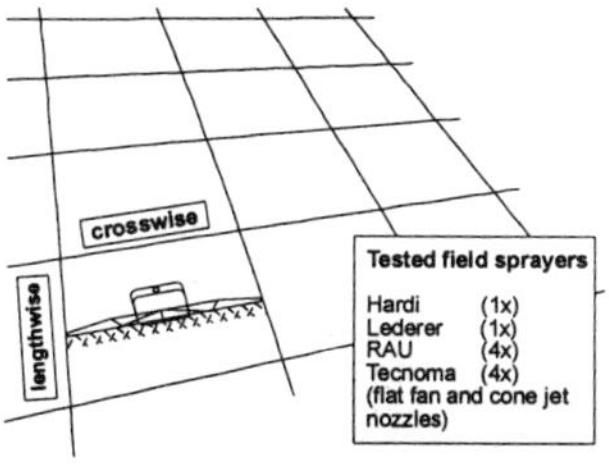

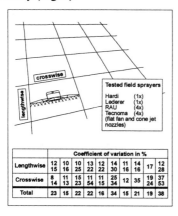

	Coefficient of variation in %									
Lengthwise	12 15	10 16	10 25	13 22	12 22	14 30	11 16	14 16	17	12 28
Crosswise	8 14	11 13	15 23	11 54	11 15	25 34	12	35	19 24	37 53
Total	23	15	22	22	16	34	15	21	19	38

Figure 2 *Quality of lengthwise and crosswise spray distribution by field sprayers under practical field conditions*

The **spray deposit** of target areas also varies greatly. The spray deposit is determined both by the distribution quality achieved by the sprayer and the density of the crop[6]. Still, the biological success of application by today's techniques is fairly good, even if depending on the measure, only part, or a small percentage, of the product applied reaches the actual target, such as ear, flag leaf and stalk base on a cereal crop. Current techniques allow only modest improvements in deposit, in particular in the depth of a crop (Fig. 3).

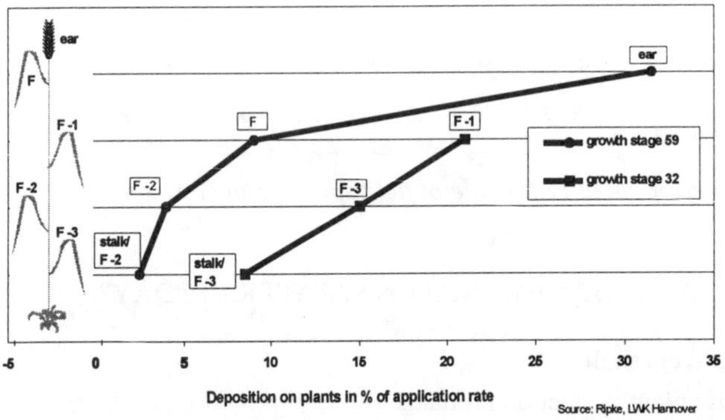

Figure 3 *Deposition of plant protection products on different parts of the plants in a winter wheat crop*

The **ground deposit** – The ground of the treatment area is usually not the target of treatment and should be covered with plant protection products as little as possible. Yet, at early crop growth stages the ground will be covered with more than 50 % of the spray applied[7]. This unwanted deposit declines to a few percent with progressing vegetation (Fig. 4).

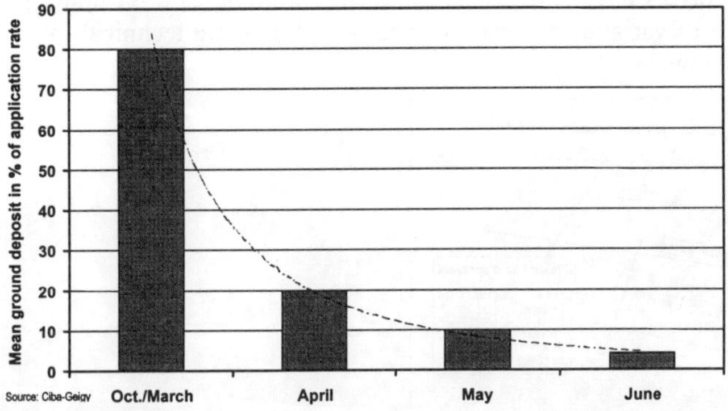

Figure 4 *Deposits of plant protection products on the ground as non-target area, in a winter wheat crop, as a function of vegetation*

Drift cannot be completely avoided. It declines asymptotically with growing distance and is essentially determined by the construction of the sprayer and by the crop[8]. The ratio of drift (related to field sprayers) occurring with the sprayers now in use may be stated as follows: field crops/vineyards/fruit crops/hops as 1/6/15/25 (Fig. 5). These drift values play a role in the risk assessment for the authorisation of plant protection products. As a result, authorisation of a product may be tied to instructions about observance of 'buffer zones', which means that the grower has to keep certain distances to surface waters when applying the product.

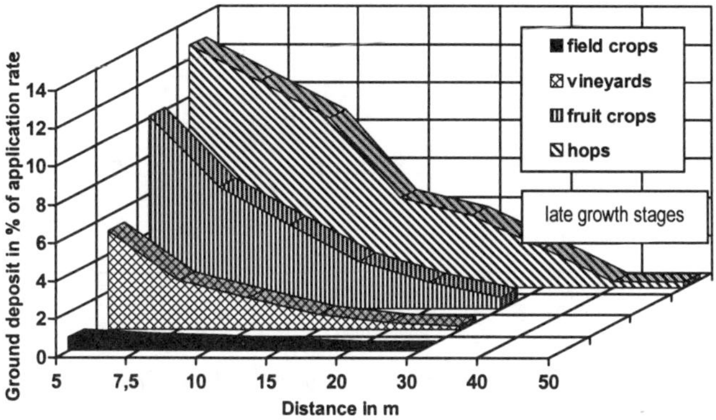

Figure 5 *Drift (ground deposit) of plant protection products in field crops, vineyards, fruit crops and hops to areas neighbouring the treatment area. The ground deposit is represented by the 95 % percentile. Sprayers and spraying conditions are in line with good agricultural practice.*

3.2 Air Assisted Sprayers

Air assisted sprayers are much less equipped with electronic controls, although with today's state of technique, electronic controls would also function smoothly under the difficult treatment conditions in tall growing crops.

The treatment of tall growing crops (vineyards, fruit crops, hops) is very difficult because of the specific manner of their cultivation, their height, and the large variety of crop plants and pests. This treatment is best made by sprayers with variable air assistance moving between the crop rows.

Today's air assisted sprayers mostly have axial fans generating a so-called 'soft airstream' (large volume, slow speed), but with strongly vertical direction[3]. Different measures are taken (additional fan cases, cross-flow fans) with the aim to generate a more horizontal airstream, which brings at least the same spray deposit with less drift. In the following I will show some examples describing the quality of today's air assisted sprayers with the help of the parameters of spray deposit, ground deposit, and drift.

It is always intended to achieve a dense and even deposit on the target area by spraying. Fig. 6 shows the situation with different air assisted sprayer types in commercial fruit cropping. It shows that an average deposit of about 15 % is reached, almost irrespective of the type of sprayer. The variability of deposit, characterised by the coefficient of variation, lies mainly between 40 and 60 %[9,10]. Contrary to our original expectations, both parameters did not vary much with different equipment. The results are based on fluorimetric measurements of leaf deposits, for which 40 leaves or more each were taken from the crowns of 130 trees as representative samples.

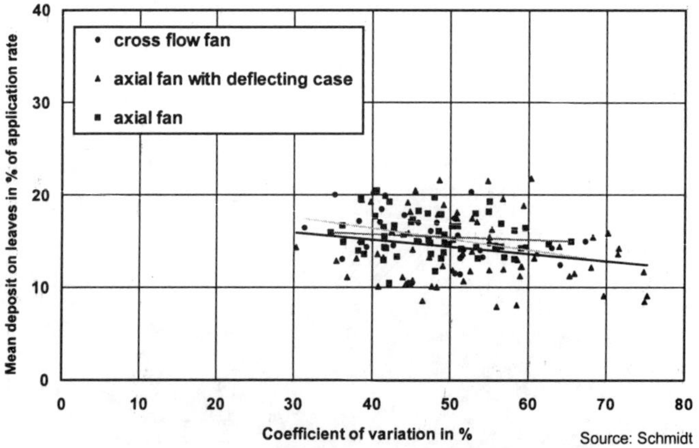

Figure 6 *Quality of deposit (mean deposit of spray and coefficient of variation) on leaves of apple trees with different sprayers*

Considering the total leaf surface of the sampled trees and ground deposit inside and outside the treated area, one can calculate the proportion of plant protection products which is deposited on the crop and that which is lost through sedimentation on the ground or through drift. Fig. 7 shows these proportions in fruit crops at initial bud break and at full leaf.

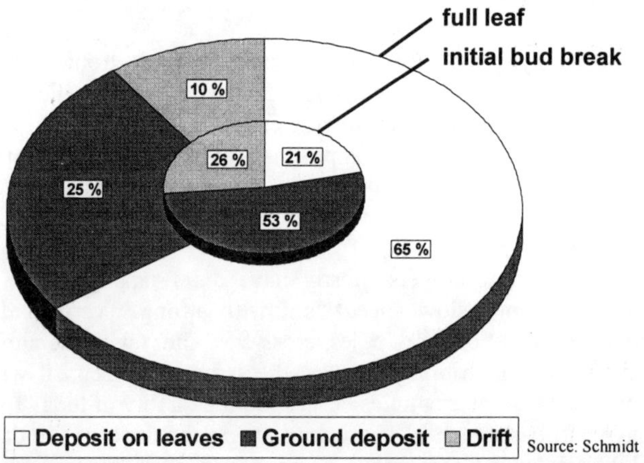

Figure 7 *Quantification of plant protection products applied by conventional air assisted sprayers with regard to target (leaves) and non-target (ground deposit, drift) areas in fruit crops*

It shows a high proportion of loss (to the ground and drift) in early treatments and a decline of such losses down to about one third as the actual target area (leaves, fruits) increases with progressing vegetation. In parallel, there is a decline in drift. For more findings concerning drift, I want to refer again to Fig. 5 mentioned previously. Technical measures to further reduce such losses will be described later.

4 WHAT PROGRESS CAN BE EXPECTED IN APPLICATION TECHNIQUES?

A great many of innovations are to be discerned in the field of application technique:

4.1 Agricultural BUS System (ABS)

Up till now, agricultural electronic controllers for tractors and implements have mostly been stand-alone systems. A tractor cab is normally encumbered with a multitude of buttons, dials and switches. Each system needs its own sensors, power supplies, boxes, cables and so on. The ergonomic and economical effects on, for example, visibility are great. To solve this problem activities have been started at the beginning of the 1980s to design and propose a universal BUS system, comprising a serial BUS, a man-machine interface and a connection to the farm computer[11, 12] (Fig. 8).

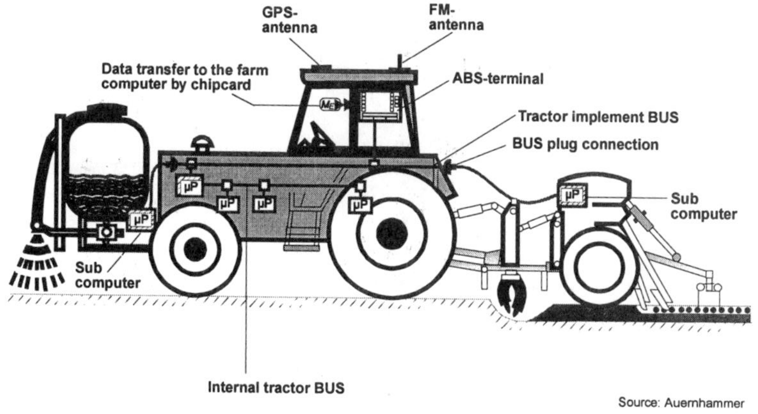

Figure 8 *The agricultural BUS system sets up electronic communication between tractor and sprayer and enables data transfer between mobile electronic equipment and farm computer*

This standardised communication system is comparable to the tractor three-point linkage, by creating clear interfaces and integrating them in the technique in the form of standardised hardware (plugs, cables, operation boards, chips) and software (way of communication, transmission rates, addresses, data, data formats). DIN standard no. DIN 9684/2 - 5 provides the necessary standard for the agricultural BUS system. It became valid at the beginning of 1998. A corresponding ISO standard (Standard ISO 11783 under development by ISO/TC 23/SC 19/WG 1) is expected in the years to come.

Several manufacturers of tractors and farm machinery already offer ABS. Introduction of ABS is still hindered by the high price of the necessary investments. BUS and global positioning systems (GPS) permit the design of efficient integrative systems for field site specific farming. GPS provide localisation and navigation by reporting place and time, while ABS sets up the necessary electronic communication between the tractor and the sprayer and enables the data transfer between the farm computer and mobile electronic equipment.

4.2 Site Specific Application

With the given heterogeneity of agricultural land, the progress of sensor technique for the identification of the state of crops, image analysis, and the navigation and control of vehicles opens up new possibilities to respond to local differences in nutrient supply or weed infestation[12,13]. Weed control is the purpose of about 60 % of all applications of plant protection products.

Ideas concerning the use of site specific application centre around using image analysis fitted to the spraying unit for identification of weed species and density (Fig. 9). But really practicable techniques for automatic identification of weeds in cereals with the help of image analysis systems will not be available in the next few years. The current safety of identification and processing speed are not sufficient to adjust spraying technique in the field in real time.

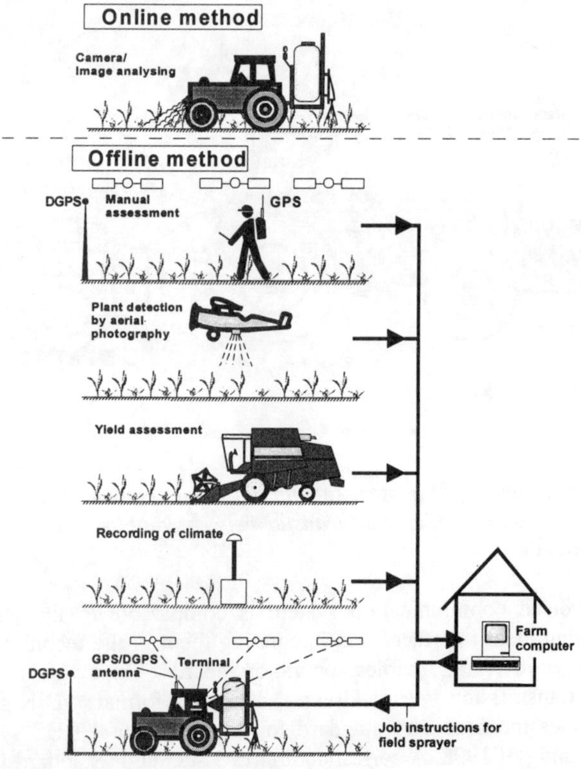

Figure 9 *Methods of site specific application of plant protection products and of assessing heterogeneous areas*

Site specific application means that the field must be divided into (as many as wanted) homogenous patches. Because of the small size of these homogenous patches, counting must be tightly screened to obtain useful weed assessments. Useful offline weed-assessment methods are manual counting and all partially automated photo-detection and analysis methods[14]. The use of aerial photography for the recognition of weeds is confined to the photometric detection of differences in the weed density in agricultural areas[15]. Application of this method is naturally limited because there are too few days for flying; with good visibility, the photos have too little resolution, and there is no interpretation with regard to farming problems. Local yield determination in combine harvesters has reached a technically advanced standard, and several systems to measure yields are on the market now. Still, the combine harvester also counts a high potential of weed seeds, which is not yet registered, collected or destroyed separately. Electronic facilities registering places of weed infestation and their specific degrees of infestation and providing this information for specific local weed assessments could facilitate local weed control measures.

Equally, information on the current and preceding weather and the growth stage of the crop are important and must be permanently updated.

Instructions for how to proceed in individual patches of the field are worked out at the farm computer on the basis of geographical information systems (GIS). For that, the field in question is overlaid with a grid of a width defined by the farmer, e.g. 50 x 50 m. Within that grid, individual decisions or proportionately differentiated application decisions can be automatically allocated. Job instructions for sprayer control are transmitted to the tractor terminal by chip cards. Tractor computers designed to carry out patch spraying must contain software extensions for satellite location and site specific dosing. Differentiated dosing instructions are realised during the spraying passages.

The dosage can be adjusted by various technical facilities on the sprayer. The simplest way of controlling the dose is to change the spraying pressure and/or the speed. The disadvantage of that method is an involuntary change of the spray drop size. This leads to bad deposit when the droplets are too large, or to high drift when the droplets are too small. The dose variation allowed by this method is ± 25 to ± 30 % around a suitable mean.

Dosage by direct injection is more complicated. It means that the liquid plant protection product is fed directly to the spray water, either immediately before the sections of the spray boom or before the nozzles. This method offers unlimited variation of dosage, but it is technically complicated and has not yet been practised on a large scale to a good effect. Cleaning of such systems may also pose a problem.

Model calculations about the advantages of site specific treatment have shown that from an economic point of view, scaling of the herbicide dose only makes sense if there are only few levels. It is also not necessary to subdivide fields into patches for treatment smaller than a sprayer's working width by 50 m passage length. Experience with differentiated herbicide application has shown that economic advantages can only be gained when the proportionate cost of the patch spraying does not exceed 20 DM/ha. Manual weed countings and assessment on the basis of partially automated photo detection are still intolerably expensive in terms of money and labour.

Practicable techniques of site specific application will come onto the market in the next few years. Acceptance of such techniques will largely depend on how fast simple weed assessment methods and sensors will be developed, and how practicable injection systems of multiple products will be.

The specific knowledge of many farmers about the condition of their fields, the use of positioning techniques, and relevant GIS processing software, however, have already allowed mapping of field patches to take place and the realisation of roughly differentiated herbicide use.

4.3 Sensor Controlled Sprayers

Adjustment of the liquid output of nozzles to the variance of target areas with the help of optical sensors has been practised in crop protection in many ways[16]. The sensors fitted to the sprayer are able to detect green plants or plant parts (target areas) against non-target areas (the ground or gaps in the crop), and nozzles linked with the sensors are switched on or off so that the spray is almost solely deposited on the target areas. The sensors -- so-called green detectors -- function by the fact that green plants strongly absorb ambient light in the range of 630 to 660 nm because of their chlorophyll and strongly reflect light in the range of 750 to 1200 nm. There is nearly no light reflection by non-target areas (ground, gaps in the crop). The sensors register reflection in the red (R) and near-infrared (NIR) range by photo diodes and control the nozzles on the basis of the ratio of the two degrees of reflection[17,18] (Fig. 10). Sensors working with reflection in the infrared (IR) range only are also very promising. After first results of the research about opto-electronic sensors for plant detection were obtained in the U.K. about 20 years ago, and later in Australia and Canada, such sensor techniques have been successfully applied for non-selective plant detection in a number of fields (Fig. 11).

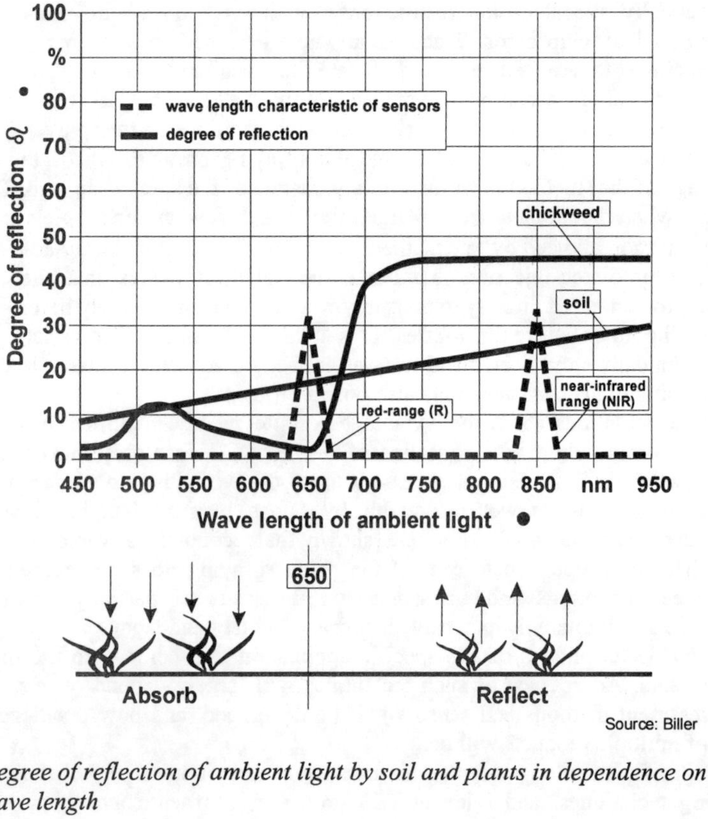

Figure 10 *Degree of reflection of ambient light by soil and plants in dependence on light wave length*

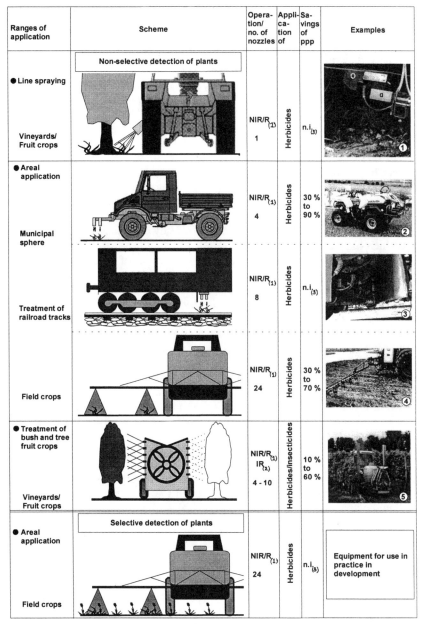

Ranges of application	Scheme	Operation/ no. of nozzles	Application of	Savings of ppp	Examples
● Line spraying Vineyards/ Fruit crops	Non-selective detection of plants 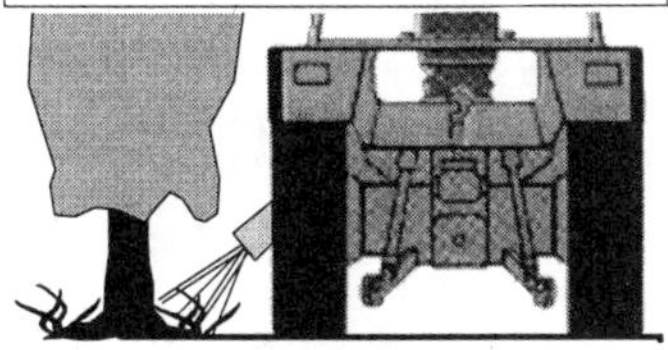	NIR/R(1) 1	Herbicides	n.i.(3)	
● Areal application Municipal sphere		NIR/R(1) 4	Herbicides	30 % to 90 %	
Treatment of railroad tracks		NIR/R(1) 8	Herbicides	n.i.(3)	
Field crops	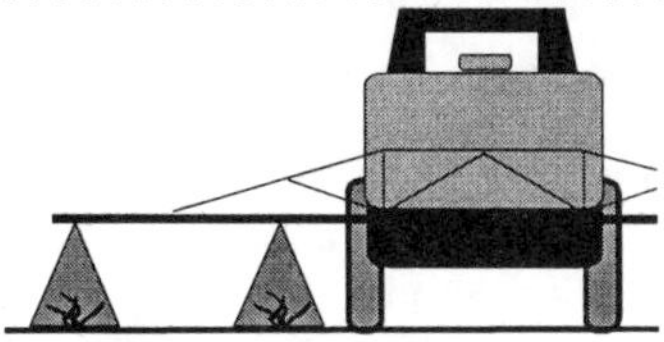	NIR/R(1) 24	Herbicides	30 % to 70 %	
● Treatment of bush and tree fruit crops Vineyards/ Fruit crops	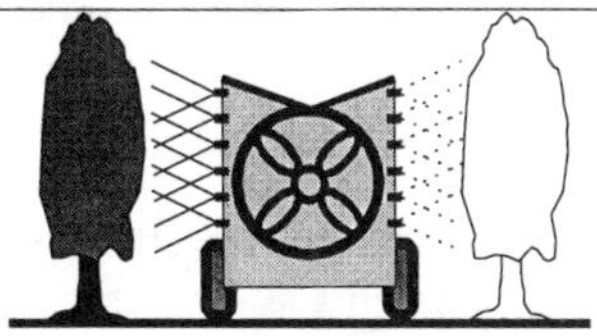	NIR/R(1) IR(2) 4 - 10	Herbicides/Insecticides	10 % to 60 %	
● Areal application Field crops	Selective detection of plants 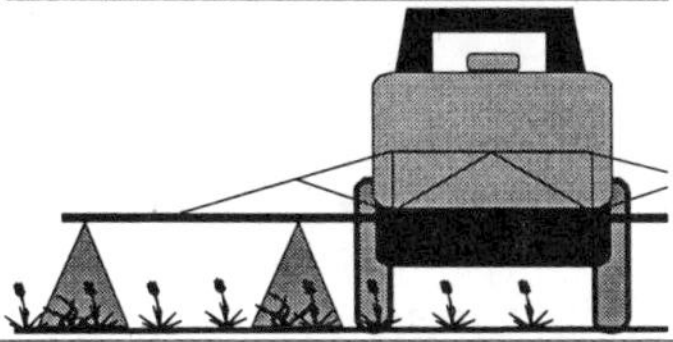	NIR/R(1) 24	Herbicides	n.i.(3)	Equipment for use in practice in development

Figure 11 *Examples of use of opto-electronic sensors for plant detection during the application of plant protection products.*
1) Control based on reflection in the red (R) and near-infrared (NIR) range
2) Control based on reflection in the infra-red (IR) range
3) No information

Pictures: ① *Uhl, W.;* ② *Werkfoto Douven;* ③ *Osteroth, H.-J.;* ④ *Biller, R. H.;* ⑤ Bäcker, G.

Sensor techniques are successfully applied in some regions in **line sprayings**, that is in spraying rows of fruit trees or the ground under vine rows to control sporadic weeds.

In **areal application** in farming, municipal sphere or on railroad tracks, sensor techniques also aim to apply plant protection products in a concentrated manner on locally occurring weeds. While Germany allows the use of plant protection products in public amenities only in exceptional cases, experts expect a great potential of savings in plant protection products from sensor-controlled spraying on railroad tracks. This savings potential will be environmentally relevant.

Conventionally managed farms have only little use for field sprayers with sensor-controlled nozzles. So far, this technique only makes sense with black fallow or sod seeding. In these cases, it is estimated that between 30 and 70 % of plant protection products can be saved compared to aerial application[17,18].

Vine and fruit growers have also applied sensor techniques. After several years of experience, it is estimated that savings in plant protection products amount to 10 to 60 %, drift is reduced by half, on average, and the ground contamination in the treated area is reduced by up to 75 %[19,20]. Several manufacturers already offer sensor-controlled vineyard sprayers which have successfully undergone BBA testing. The sensor technique of fruit crops sprayers still had to be adapted to the larger dimensions of trees. Improvements have been made and are being tested in practice.

In field cropping, it is now the aim to improve optical sensors so that they can be used for selective weed detection. Different reflection by weeds and weed grasses and by crops is the basis for operating nozzles depending on the occurrence and density of weeds. If sensors can be improved so that they do not only recognise green plant parts but also differentiate between the green of crops and that of weeds and weed grasses, this would be a real breakthrough for site specific application of plant protection products in conventional arable farming. The problems of site specific application of plant protection products, which come from time-consuming manual or computer-aided photo-registration of the local infestation situation and have allowed only off-line application to date, could be solved by online opto-electronic, selective plant detection. Results of relevant research and trials are very promising[21]. Other approaches proceed from marking the crop plant, for instance by inherent fluorescence, to allow sensors to differentiate between weeds and crop[22].

Improvement of sensor techniques for fruit crops beyond the current standard of switching on and off nozzles depending on gaps in the crop is aimed at regulating the nozzles' liquid output depending on the density of foliage[23].

4.4 Recycling Sprayers

The two contrary aims in spraying tall growing crops are to achieve an optimum distribution of the chemical on all parts of the plant while spraying as little product as possible. Additional precautions need to be built into the equipment. One possibility is to use a recycling unit. Several types are available, which collect or reflect the droplets to avoid their spreading into the environment. They have been developed initially for use in vineyards, but newly-developed types are now also available for fruit crops. They are generally successful in avoiding losses, but the practical handling of such units is difficult and their manoeuvrability is very limited, particularly while turning at the end of a row and in hillside areas [24,25].

Fig. 12 shows the known basic types of recycling units. The tunnel principle is mostly used in vineyards, with or without air assistance, but units with air assistance are more successful in ensuring a sufficient deposit. The collection principle can be combined with conventional axial blower types as well as with tangential blowers. Reflecting units are mostly combined with tangential blowers. The circulation principle uses a circulating air stream. Tunnel and circulating units treat the row completely in one run whereas collecting and reflecting units need often a second treatment of the same row, particularly when the canopy is fully developed. Tunnel units have already been tested by the BBA and accepted for 5 years. However, the introduction of such recycling units on the market is lagging behind.

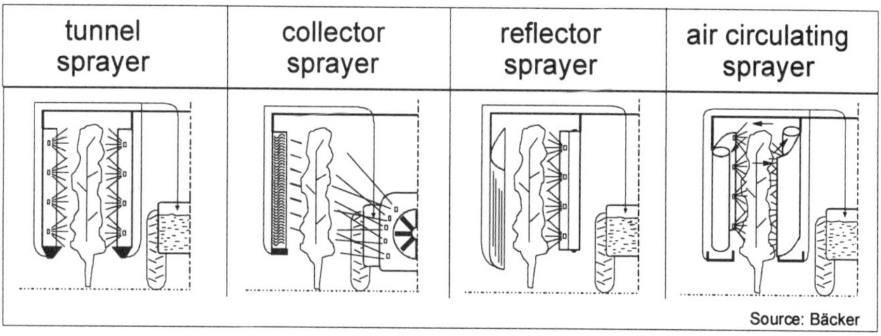

Figure 12 *Basic types of recycling units for air assisted sprayers*

These devices bring annual average savings in plant protection products of 30 to 40 %. Fig. 13 shows the benefits of recycling sprayers in vineyards and fruit crops with regard to ground deposit and drift to the environment.

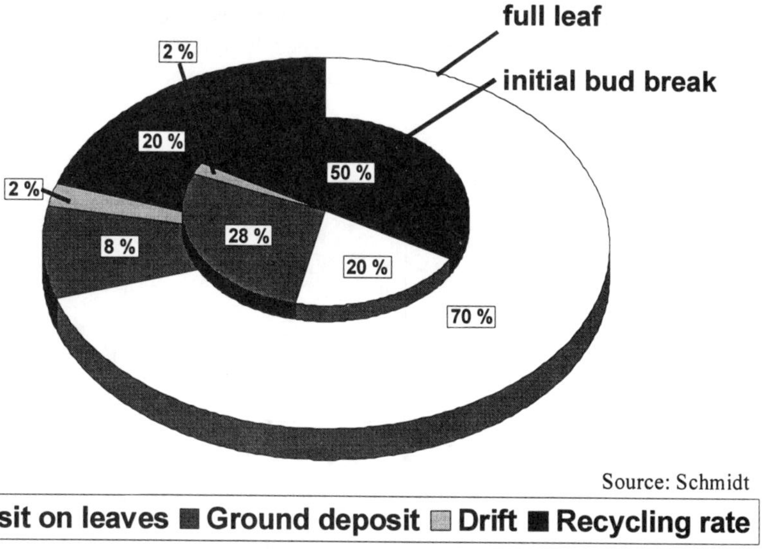

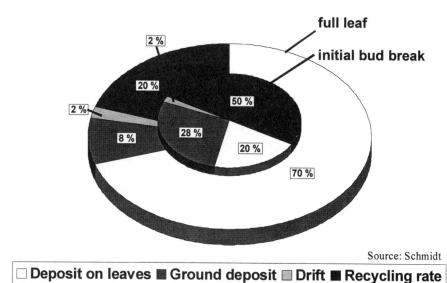

Figure 13 *Quantification of plant protection products applied by recycling sprayers with regard to target (leaves), non-target areas (ground deposit, drift) and recycling rate in fruit crops*

4.5 Field Sprayers with Air Assistance

Field sprayers with air assistance are believed to offer many advantages with regard to

- savings in plant protection products
- crop penetration and spray deposit
- biological efficacy
- performance per area unit
- observance of the most suitable time for treatment
- avoiding drift of plant protection products.

What has been said about this in lectures and publications is sometimes contradictory. This may be due to different surrounding conditions or starting points of investigations. One can certainly say that this technique can bring advantages in the cited fields; however, these advantages do not come all together, but only one by one. For instance, high-performance by liquid-saving application cannot at the same time economise on plant protection products and reduce drift. Equally, treatment in wind-exposed areas cannot at the same time offer labour-economic advantages or economies on plant protection products. Yet, the big advantage of this technique lies in its variability, which allows growers to use its advantages according to the situation. The high cost of acquisition is certainly one reason why such sprayers are preferably used on large farms or in special crops. Which air stream performance and speed is actually needed in practice will be determined by the local conditions first of all. But one has to note that more air is not necessarily good, but can also have negative effects. In the early stages of vegetation, for instance, the air stream must be well-dosed in order not to produce excessive drift.

4.6 Injector Nozzles

These nozzles suck in ambient air through lateral holes during spraying. The mixture of air and spray liquid which comes out at the nozzle outlet forms larger drops and fewer small droplets than if no air was fed in. This atomisation into larger drops can considerably reduce drift (Fig. 14).

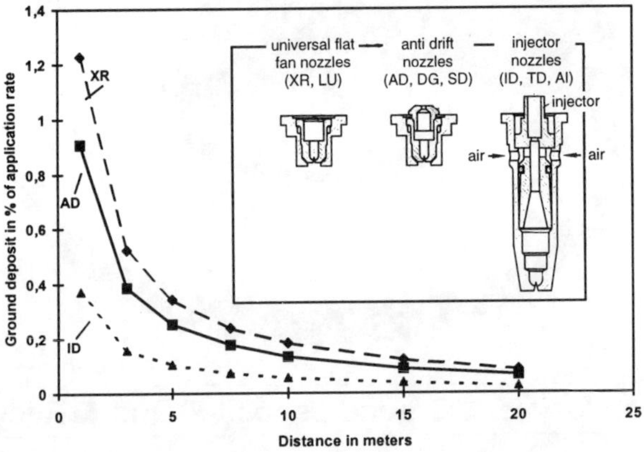

Figure 14 *Drift reduction (ground sediment) by modern flat fan nozzles (AD, DG, SD, ID, TD, AI) in field crops. This will allow to reduce buffer zones, for instance to surface waters*

The figure shows that drift from an XR and LU nozzle at a distance of 20 m is the same as from an ID, TD or AI nozzle at a distance of only 5 m. With regard to biological efficacy, however, no differences have been found to the other flat fan nozzles. The great choice in nozzles allows growers to carry out sprayings with high efficiency but low drift. Injector nozzles are meanwhile offered by all important nozzle manufacturers and have been tested to good effect by the BBA.

Still, a faultless application technique with exact dosing and distribution of the spray remains the prerequisite for good biological efficacy of the products. This includes to adjust the travelling speed to the target area (normally 6 to 8 km/h) and to spray crops only if the wind speed is not much more than 3 m/s.

4.7 Twin Fluid Nozzles

A few years ago an automatic drift control system with twin fluid nozzles was introduced into the market (Fig. 15). Pressurised air is used in the twin fluid nozzles for atomising the liquid and controlling the droplet size. In addition to the well known electronic control system for the liquid flow another control system for adjusting the air pressure in relation to the liquid pressure is part of this drift control system. The proportion between the air and liquid pressure is adjustable by either a wind sensor or by a dial plate or by both in combination.

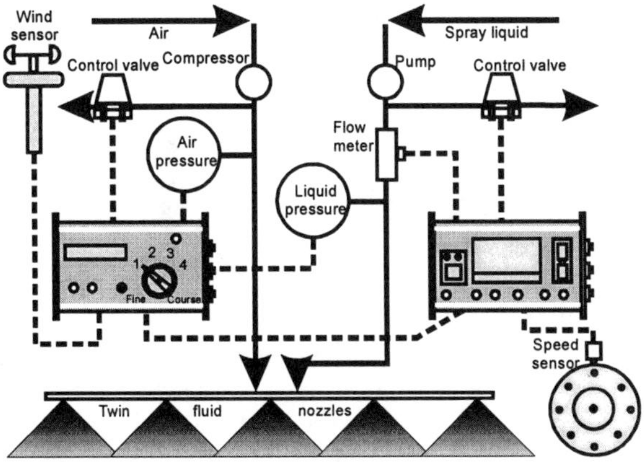

Figure 15 *Automatic drift control with twin fluid nozzles by adjusting the droplet size in regard to the field and the wind conditions*

The benefits of this system are:

- nearly constant droplet size over a wide range of application rate and forward speed
- adjustment of droplet size in regard to changing wind situations or in order to prevent contamination of surface water or neighbouring cultures and
- reduction of water amount per area unit which results in a higher performance.

4.8 Electrostatic Charging

Electrostatic charge is used in many ways in industry to good effect. In crop protection techniques, however, its advantages have not yet been proven. Electrostatic force obviously effects better spray deposit. Yet, it is not proven that this will bring economies in plant protection products. The author made research into electrostatic charging in crop protection equipment for several years and is confining his statements to the use of authorised products and of currently available sprayers re-equipped to work with electrostatic charging (contact charging)[26,27]. More efficiency could certainly be expected from products and equipment especially designed to work with electrostatic charging. Such projects have so far been impeded by the high cost. As we expect advantages from electrostatic charging, in particular with air assisted application in tall growing crops, we have compared orchard sprayers on the basis of a biological assessment. The results are shown in Fig. 16. To let the differences between the sprayers (rotary atomizer, electrostatic charging, conventional sprayers) stand out more clearly, we have reduced the application rate of the plant protection product in two steps from 100 % over 75 % to 50 %. Electrostatic charging brought only little advantage in the control of apple scab. The trials did not show any noticeable economies in plant protection products.

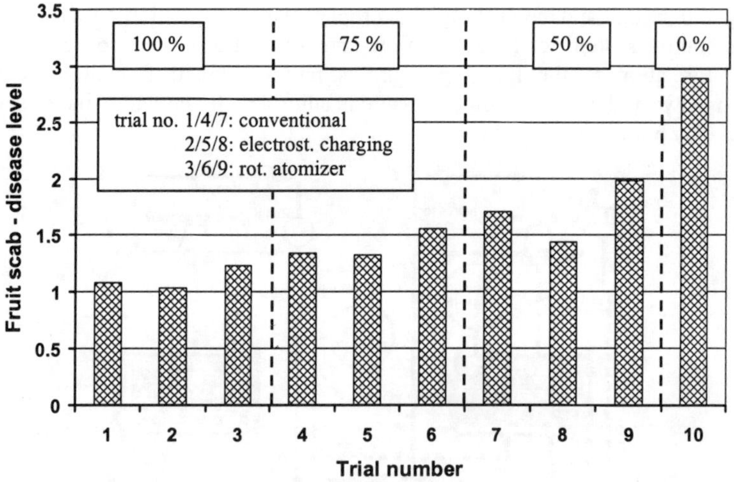

Figure 16 *Comparison of the biological efficacy of apple scab controlled by conventional and new air assisted sprayers (electrostatic charging, rotary atomizer)*

Using the same application rates and the same order of treatments, the sprayer with rotary atomizers fell behind conventional technique in the biological assessment.

More recent investigations on electrostatics hint at drift reduction, but the predicted advantages should be verified under practical conditions[28].

5 HOW NECESSARY ARE SPRAYER TESTS?

Plant protection equipment must comply with current standards of technique and function accurately, reliably and durably in practice. To ensure this, Germany has introduced legal regulations under which new plant protection equipment must be tested and used equipment regularly inspected[3,29,30].

The BBA publishes a list of all new equipment which meets the requirements and may be distributed. Fig. 1 which was shown in the beginning lists the current offer of field sprayers which have been declared by manufacturers and accepted by the BBA. Other EU Members have not (yet) introduced comparable testing procedures for new equipment, but are currently working on the definition of technical requirements and testing procedures for plant protection equipment in the framework of international and European standardisation.

Inspection of used plant protection equipment is intended to maintain its function in practical use. As sprayers are aging and wearing in practical use, their accuracy of dosage and distribution suffers. Germany has organised inspections of sprayers in use for more than thirty years now. It has made these inspections mandatory since the middle of 1993. After a transitional period, an annual 70 000 or so field sprayers will have to be inspected.

An inquiry in 24 European countries shows that 13 of 18 countries which answered are inspecting field sprayers and 11 of them are offering to inspect air assisted sprayers[30].

This means that most European countries consider inspections of used sprayers necessary for environmental reasons.

Table 1 provides answers to questions regarding the organisation and implementation of inspection measures for field sprayers.

Table 1 *Management and execution of the inspection of field sprayers in different European countries (AP: official stations; aS: approved stations; pW: private workshops; iP: integrated production organisations; Ber: advisory board)*

Field sprayers	Inspection since	Obligatory participation	Who inspects	Number of sprayers	Number of inspections per year	Official approved stations
	1	2	3	4	5	6
Austria	yes 1975	yes regional	p W	-	a few	yes
Belgium	yes 1995	yes	a P	28 000	10 000	no
Bulgaria	no	-	-	-	-	-
Croatia	(yes) (1984)	yes	a P	20 000	275	yes
Denmark	yes 1994	yes	a P	40 000	500	no
England (UK)	yes 1996	no	pW	-	-	yes
Finland	yes 1995	yes	Ber	40 000	6 500	yes
France	no	-	-	-	-	-
Germany	yes 1968	yes 1993	a P pW	167 000	70 000	yes
Hungaria	no	-	-	-	-	-
Ireland	no	-	-	-	-	-
Italy	yes 1985	no	a S	130 000	250	no
Netherlands	yes 1988	yes 1997	p W	25 000	3 000	yes
Norway	yes 1991	no	diverse	20 000	1 300	yes
Slovak Rep.	no	-	-	-	-	-
Slovenia	yes 1971	yes	a P	-	1 200	yes
Spain	yes 1990	no	a P p W	7 000	~ 50	no
Sweden	yes 1988	no	p W	22 500	1 700	yes
Swiss	yes 1989	yes only i P	a P	20 000	3 000	yes

Next to the name of the country, column 1 indicates whether or not and from which date field sprayers have been subjected to inspection. The following columns specify whether or not inspection is mandatory, who makes it, and how large the estimated number of field sprayers in the individual countries is. Additionally, these columns state the average number of inspected field sprayers per year and whether or not inspections are conducted by officially approved test stations (repair shops).

In Germany, the high number of inspections is surely due to their mandatory character. However, appropriate inspection equipment, which is adapted to practical requirements, as well as standardised regulations also contribute to the wide acceptance that the inspection of field sprayers has gained among German farmers.

6 WHAT IS THE CURRENT SITUATION WITH THE EUROPEAN STANDARDISATION?

Member States regard the basic requirements laid down in EU directives as fulfilled when products are manufactured in line with certain harmonised standards. Legislation and standardisation are therefore closely connected. Although there is no legislation in the field of environmental protection at the European level, standardisation work started to be co-ordinated in the middle of 1992[31]. There is a three-part document on sprayers which is to be published as a draft standard shortly. In essence, it regulates the following things: spray tanks, spray booms, fans, nozzles and filters, operative measuring equipment, controls, distribution accuracy, drift avoidance, cleaning of equipment and pesticide containers, band spraying facilities, etc.

A few important parameters -- classification of nozzles, sprayer boom movements -- have not yet been included in the aforementioned standard because relevant requirements and test methods must still be worked out by competent institutions in the Member States.

The BCPC (British Crop Protection Council) is heading a working group developing a classification system for assessing nozzles with regard to their drift potential[32,33]. The system shall also be suited to classify other technical facilities. For this purpose, the participating institutes carried out additional drift measurements in wind tunnels and under standardised conditions in the field. Fig. 17 shows a preliminary result of nozzle classification. The drift produced by the nozzles is described by the drift volume and the height of the point of concentration of the drift cloud. The DIX constant lines describe the ranges from high drift (DIX = 10) to low drift (DIX = 0.3). Some nozzles widely used in practice are entered as examples.

Boom oscillations of field sprayers, in particular of those with large working width, can cause considerable differences in the spray deposit. An EU research project currently works on testing procedures which would look into boom movements as well as unevenness of distribution caused by them (dynamic distribution quality) to include both as criteria of evaluation[5]. The results cannot always be compared because of the different technical outfit and experience of the institutes involved. Fig. 18 presents examples of the potential of improvement of usual field sprayers with regard to the dynamic distribution quality. For instance, better harmonisation of the components of a series field sprayer was already sufficient to stabilise the boom and considerably improved the quality of spray deposit.

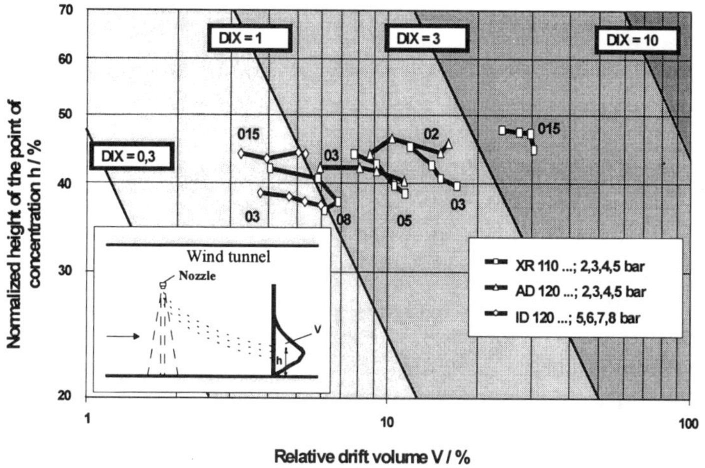

Figure 17 *Drift classification of nozzles. Drift cloud depending on the height of its point of concentration and its volume for some usual flat fan nozzles. The lines DIX = constant mean that there is constant absolute drift. Some DIX = constant lines entered as examples define the ranges from high (DIX = 10) to low (DIX = 0.3) drift.*

The technical requirements in the field of labour protection as to the safety of sprayers have been laid down in the European standard EN 907 in line with the EC Council Directive 89/392/EEC on machines.

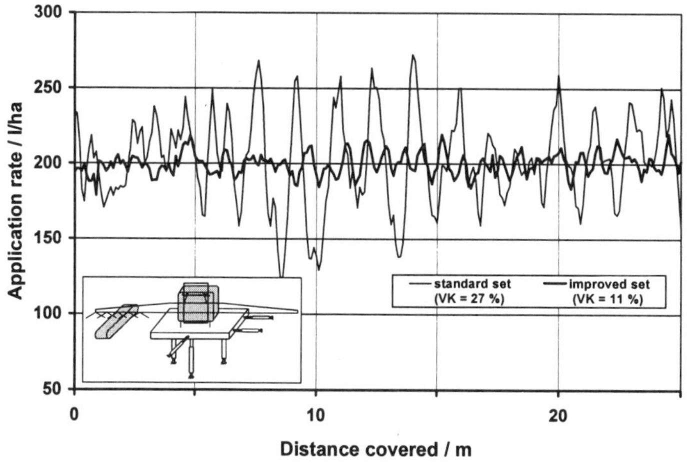

Figure 18 *Lengthwise distribution of a field sprayer under field conditions. Stabilising horizontal boom oscillations contribute decisively to equalising the spray deposit*

7 CONCLUSIONS

By far the largest share of plant protection products is applied by field and fruit crop sprayers. The performance of these sprayers, as far as they are up to the current standard of technique, is presented by the following criteria: dosage, spray distribution, spray deposit, deposition on the ground, and drift. Techniques which have been used for decades

1. crop sprayers which move a horizontal spray boom smoothly over the crop, always at about the same distance to the crop, and
2. crop sprayers which pass row crops and spray them with air-assistance

have stood the test of time and have reached a high technical level. Yet, there are a number of improvements in detail, such as of nozzles, air assistance, mobile farm electronics, and of completely new developments, such as site specific application or recycling sprayers, which allow to make the use of plant protection products more economical and efficient and reduce losses.

This paper presents methods to test crop sprayers and relevant European and international activities to standardise techniques. The growing number of common activities of different institutes of EU Member States is reflected in two research subjects.

Bibliographic References

[1] F. M. Brouwer, I. J. Terluin and F. E. Godeschalk, 'Pesticides in the EC', Agricultural Economics Research Institute (LEI-DLO), The Hague, 1994, p. 157

[2] C. Barnard, S. Daberkow, M. Padgitt, M. E. Smith and N. D. Uri, *The Science of the Total Environment,* 1997, **203,** 229

[3] H. Ganzelmeier, 'Yearbook Agricultural Engineering', H. J. Matthies, F. Meier, Frankfurt, No 10, 1998, Chapter 19, 215

[4] S. Rietz, B. Pályi, H. Ganzelmeier and A. László, *J. Agric. Engng. Res.,* 1997, **68,** 399

[5] C. Sinfort and A. Herbst, 'Bulletin in OEPP/EPPO Bulletin', 1996, **26,** 27

[6] F.-O. Ripke, *DLG-Mitteilungen,* 1998, H. 4, 36

[7] H. Ganzelmeier, 'Mitt. Biol. Bundesanst. Land-Forstwirtsch. Berlin-Dahlem', No 328, 1997, 115

[8] H. Ganzelmeier, D. Rautmann, R. Spangenberg and M. Streloke, 'Mitt. Biol. Bundesanst. Land-Forstwirtsch. Berlin-Dahlem', No 305, 1995, p. 111

[9] K. Schmidt and H. Koch, *Nachrichtenbl. Deut. Pflanzenschutzd.,* 1995, **47,** 161

[10] P. Kaul, H. Ganzelmeier, H. Henning and H.-J. Wygoda, *Nachrichtenbl. Deut. Pflanzenschutzd.,* 1995, **47,** 109

[11] G. Jahns and S. Rietz, 'Bulletin OEPP/EPPO Bulletin', 1996, **26,** 17

[12] H. Auernhammer, 'Innovationen in Technik und Bauwesen für eine nachhaltige Landwirtschaft', Kuratorium für Technik und Bauwesen in der Landwirtschaft, Darmstadt, Arbeitsheft 254, 1998, 64

[13] G. Wartenberg, *Landtechnik,* 1996, **51,** 196

[14] G. Wartenberg, B. Pallut and A. Giebel, *Z. Pflkankh. PflSchutz,* Sonderh. XVI, 1998, 317

[15] A. Häusler, H. Nordmeyer, P. Niemann et al, 'Innovative Verfahren zur Unkrauterkennung', Kuratorium für Technik und Bauwesen in der Landwirtschaft, Darmstadt, Arbeitsheft 236, 1996, 101

[16] H. Ganzelmeier, Yearbook Agricultural Engineering', H. J. Matthies, F. Meier, Frankfurt, No 11, 1999, Chapter 6 (printing)

[17] R. H. Biller and C. Sommer, 'Proceedings of International Conference on Agricultural Engineering', Madrid, Spanien, 23. – 26. September 1996, 799

[18] R. H. Biller, A. Hollstein and C. Sommer, 'Proceedings of First European Conference on Precision Agriculture', Warwick University, UK, 7. – 10. September 1997, Vol. 2, Technology IT and Management, 451

[19] G. Bäcker, O. Westphal and H. Göhlich, *Das Deutsche Weinmagazin,* 1995, H. 15, 24

[20] H. Koch, P. Weißer and H.-G. Funke, *Obstbau,* 1997, **22,** H. 12, 602

[21] A. Hollstein and R. Biller, 'Bericht zur 55. Internationalen VDI/MEG-Tagung Landtechnik, Braunschweig', 16./17. Oktober 1997, VDI-Bericht, Vol. 1356, 1997, 233

[22] H. Ganzelmeier, D. Rautmann, P. Zwerger and J. Schiemann,'Verfahren zur optischen Detektion und Behandlung von Kulturpflanzen und nicht biologisch transformierte Kulturpflanzen', Patentschrift DE 196 42 439 C 1, Patenterteilung 20. November 1997

[23] J. R. Rosell, A. Nogués and S. Planas, 'Proceedings of International Conference on Agricultural Engineering', Madrid, Spanien, 23. – 26. September 1996

[24] H. Göhlich, H. Ganzelmeier and G. Bäcker, 'Bulletin OEPP/EPPO Bulletin', 1996, **26,** 53

[25] H. Ganzelmeier and H.-J. Osteroth, *Gesunde Pflanzen,* 1994, **46,** 225

[26] E. Moser, H. Ganzelmeier and K. Schmidt, *Nachrichtenbl. Deut. Pflanzenschd.,* 1982, **34,** 57

[27] H. Ganzelmeier, *Erwerbsobstbau,* 1987, **29,** 87

[28] N. M. Western and E. C. Hislop, 'Proceedings of meeting of the Association of Applied Biologist, Optimising presticide applications', Bristol, England, 6./7. January 1997, Aspects of Applied Biology 48, 1997, 209

[29] S. Rietz, 'Mitt. Biol. Bundesanst. Land-Forstwirtsch. Berlin-Dahlem', No 347, 1998, 31

[30] H. Ganzelmeier, H. Henning, A. Jeske et al., 'Mitt. Biol. Bundesanst. Land-Forstwirtsch. Berlin-Dahlem', No 347, 1998, 41

[31] H. Ganzelmeier, *Nachrichtenbl. Deut. Pflanzenschutzd.,* 1997, **49,** 51

[32] Chr. Helck, A. Herbst and H. Ganzelmeier, 'Spray and nozzle classification', meeting of the reference nozzle and driftability Working Groups – Silsoe, April 1997 (unpublished)

[33] E. S. E. Southcombe, P. C. H. Miller, H. Ganzelmeier., J. C. van de Zande, A. Miralles and A. J. Hewitt, 'Proceedings of BCPC-Sypmposium', Brighton-Conference, UK, 1997

The Formulator's Toolbox – Product Forms for Modern Agriculture

Thomas S. Woods

DUPONT AGRICULTURAL ENTERPRISE, WILMINGTON, DE 19880-0402, USA

1 INTRODUCTION

Agriculture has become a high technology industry over the years, and the time and energy of agricultural workers are very valuable commodities. Moreover, the global community has become much more conscious of the effect of human activity on the world we live in. Accordingly, in this environment, crop protection products (CPP s) must meet many requirements in addition to the obvious one, that of optimally managing the pest, disease or weed they are designed to control. First and foremost, they must be safe to use. All users of CPP s should be able to expect that if they use a product responsibly according to directions, they will not sustain injury or illness through that use. Second, these products also must cause minimal impact on nontarget organisms and the environment in general. Third, CPP s must be designed so that the biological efficacy of the active ingredient is optimized, in order to ensure most economical use of the product in terms of cost and environmental loading. And finally, CPP s must meet the increasingly stringent requisites of world regulatory agencies. The responsibility for designing the contemporary CPP to meet these requirements lies with the formulation scientist, who must employ all the tools at his or her disposal to incorporate these attributes into the product. In recent years, there has been significant shift in formulation technology toward safer, more environmentally friendly, and more effective formulation types. To accommodate these trends, the formulation science community has developed several new approaches to meet these ever-increasing demands of modern agriculture.

The state of the art in crop protection product formulations and their influence on their surroundings have been examined by several observers. For example, several specialists who have taken rather different approaches to this topic are the following. Some ten years ago, Corty[1] examined the history of crop protection product formulation science and speculated on possible future trends, and Becher[2] took a philosophical and theoretical approach to the topic of formulation science. More recently, Tsuji[3] has written extensively in documenting recent trends in crop protection formulations; whereas, Beestman[4] has concentrated on the newer technologies which are beginning to have influence on this industry. Finally, Sanchez-Rasero[5] has examined the connection of formulation type and residues of active ingredients in food crops and the environment.

This paper will first review the traditional product forms in which CPP s have been offered over the years and then will examine the new tools in the formulator's toolbox which are available for engineering the crop protection product of today. Finally, the paper will address the possible directions agricultural formulation science may take in the future.

2 "TRADITIONAL" PRODUCT FORMS

2.1 Dustable Powders

The crop protection product form which perhaps most deserves the descriptor "traditional" is the dustable powder (DP). This product form consists of an active ingredient and any required formulants deposited onto a mineral powder, which is then milled to a small particle size using a mechanical energy mill. The powdered product is then applied directly to the target, either by ground machine, by aircraft, or by hand. The clouds of dust generated during use of this product form, especially when applied from aircraft, are very familiar to those who lived in agricultural areas during the time of their most prevalent use, 40 or so years ago. These agricultural dusts, while problematic in many ways, in fact have many positive characteristics, especially considering the level of technology available at the time of their introduction. These products were and are

- Easily manufactured and readily available.
- Immediately usable without a required mixing step.
- Easily applied with simple equipment.
- Long-lasting in storage.

However, of course, there are many drawbacks to the use of dustable powders. The dust clouds generated during application carry significant amounts of active ingredient away from the target area. Another problem in the use of DP products is that using the application techniques employed leads to nonhomogeneous application of the product to the target area. There is also danger of exposure to operators and bystanders, which is especially worrisome when the product contains a toxic substance. And of course, air-borne dust may represent an explosion hazard when improperly managed. But with all their faults, dustable powders were invaluable products to agriculture in prior days, and in fact, still find uses today in many parts of the world. And in fact, the DP product form may soon enjoy a resurgence of popularity if its apparent applicability to the application of biological products proves viable.

2.2 Wettable Powders

A major advance in the crop protection product forms available to farmers was the introduction of the wettable powder (WP) in the 1950 s. While similar to dustable powders in appearance and in manufacturing technology, WP s are designed to be suspended, usually in water, and then applied to the target crop, normally by spraying. In addition to the carrier diluent and active ingredient, wettable powders typically contain surface-active agents and other formulants to ensure proper behavior when mixed with the application fluid. The entire composition is generally milled to a small particle size to ensure homogeneity of the product and maximum efficacy of the active. Advantages of wettable powders include the following:

- Homogeneous application of the active ingredient to the target.
- Easily manufactured and economical to produce and purchase.
- Applicable to a wide range of active ingredients.
- Good retention of physical properties during storage and use.

Disadvantages of the WP product form include difficulty in measurement of the desired quantity of material, with the consequence that weighing of the product is generally required to ensure accurate application. In addition, since this product form can be quite dusty, exposure of the operator to potential toxicants may be possible during mixing. And there is also the possibility of dust explosion during product transfer. But in spite of these problems, wettable powders remain a key option in the formulator's toolbox, and many products continue to be offered in this form in all parts of the world. Where cost and application technology permit, however, many producers prefer to offer their products in newer forms which overcome some of the problems of wettable powders mentioned above.

2.3 Liquid Solutions

Two product forms which have enjoyed wide use in agriculture for many years may be characterized as liquid solutions, namely, the emulsifiable concentrate (EC) and the soluble concentrate (SL). The former type of composition EC is generally applicable to active ingredients which are soluble in water-immiscible, organic solvents, and the latter type SL is useful for water soluble actives in water-miscible systems. These compositions are designed to be added to water and to be applied to the target crop by spraying after emulsification or solution, respectively. In addition to the active ingredient and solvent or water carrier, these products typically contain emulsifiers, surfactants, or other ingredients to ensure the formation of homogeneous, non-crystallizing emulsions or solutions on addition to water. Advantages of liquid solutions include:

- Homogeneous application of the active ingredient to the target.
- Easily manufactured in very simple equipment.
- Easily measured volumetrically.
- Usually inexpensive and convenient to use.
- Generally most efficacious product form for a given active ingredient.

The primary disadvantage of EC s and some SL s is the presence of an organic solvent carrier, and in fact, their use is limited to those active ingredients which have sufficient solubility in water or suitable solvents. Moreover, the solvents which are generally applicable are volatile and somewhat flammable, and as such, may represent a risk to the user and the environment. Of course, water-based SL s offer very attractive options when sole use of water as the carrier is possible. In any case, the positive attributes of EC s and SL s are many, and their favorable cost-benefit ratio and ease of use and manufacture will ensure that products will continue to be offered in these forms in the future.

2.4 Suspension Concentrates

Another traditional liquid product form which avoids many of the problems of the soluble liquids is the suspension concentrate (SC). These products are water-based compositions in which water-insoluble active ingredients and inert ingredients are reduced to very small particle size by fluid energy milling to give a heterogeneous mixture of fine solid particles suspended in an aqueous medium. In addition to active ingredient, SC formulations may contain suspending agents, dispersants, and other formulants to ensure the production of a shelf-stable, pourable product. Advantages of SC formulations are the following:

- Homogeneous application of the active ingredient to the target.
- Easily measured volumetrically.
- Convenient to use.
- Small particle size ensures optimum bioefficacy of active.

Disadvantages of the SC product form include the difficulty of producing a solid-in-liquid suspension, fundamentally a thermodynamically unstable system, with shelf life adequate to ensure utility over a normal product life-span. Because of the tendency of the particles of many SC products to settle on storage, shaking of SC containers before use is often required. This step is objectionable to many users, and in fact for larger containers may be physically difficult or manually impossible. Also, the compositions of SC s are typically more complex than those of other product forms, requiring additional ingredients, material handling, and maintenance of inventory for operations personnel. And finally, due to complexity of composition and manufacture, cost of SC s may be higher than comparable solid or solution compositions.

2.5 Dry-applied Granules

Whenever the mode or locus of action of an active ingredient permits, the long-used dry-applied granule (GR) is an attractive option for the formulator to pursue. The GR product is designed to be applied to the soil, either onto the surface or in the furrow during planting, to provide control of the target pest. The product is comprised primarily of the active ingredient and a solid carrier, either mineral or biological in origin, and may also contain other formulants to ensure bioavailability of the active. Positive attributes of GR products include the following:

- Simple application techniques using widely available equipment.
- Generally good storage stability.
- Simple manufacturing technology.
- Inexpensive packaging and low overall cost.

Thus, for active ingredients for which direct dry application is feasible, the disadvantages of the GR product form are few. Nevertheless, GR products may be dusty, and since many of the active ingredients applied in this way may be highly toxic, protection from exposure for the applicator is very important. Moreover, these products are generally dilute in terms of active ingredient content, and since the carrier granules are high in bulk density , there are significant costs associated with shipping these products.

2.6 Seed Treatment Product Forms

It has long been recognized that certain active ingredients are useful to protect crop seeds or the plants which germinate from them from insects or diseases, either in storage or after planting. Actives useful for these purposes operate either by preventative contact action on the seed or in the soil against the targeted pest, or their activity manifests itself after systemic movement within the crop plant following germination. In either case, the active ingredient is traditionally applied to the seed by tumbling the seed with a solid or liquid formulation designed to adhere to the seed and yet be nonphytotoxic to the seed or embryonic plants. Compositions are similar to those of comparable formulations described above but may also contain additional ingredients to facilitate adherence of the product to the seed coat. Alternatively, in a more modern approach described by Clayton, *et al.,*[6] the active ingredients may be applied to seeds in a film coating, with the result that there is less dust and less exposure potential for the operator.

3 "CURRENT OFFERINGS" PRODUCT FORMS

3.1 Water Dispersible Granules

During the 1970 s, it was recognized that a new generation solid product form was needed as a updated version of the wettable powder. During this decade, several companies introduced products which incorporated many of the positive attributes of the wettable powder, yet avoided many of their problematic aspects.[7] These products, the earliest work on which was done by Albert and Weed,[8] came to be known as water dispersible granules (WG) or "Dry Flowables", These pourable products have the measurability characteristics of a suspension concentrate or "flowable", but exhibit the shelf stability properties of a solid product. They have the appearance of free-flowing small granules; the shape of which ranges from composite, popcorn-shaped to nearly smooth spheres, depending on their mode of manufacture. They are designed to be suspended in water, whereupon the granules break up to form uniform suspensions, similar to those from a wettable powder. Alternatively, WG products may also be applied dry using specially designed pneumatic applicators.[9] These WG products have been made by many manufacturing techniques, as reviewed by Gerety,[10] first by fluid bed granulation, and later by pan granulation, spray drying, and high speed mixing technologies. More recently, the industry has moved toward extrusion as the manufacturing method of choice, usually by paste extrusion, using water to plasticize a powder premix, or more recently by heat extrusion, using a premix designed to soften at elevated temperatures. Advantages of WG products over wettable powders are as follows:

- Uniform bulk density; thus, measurable volumetrically.
- Comparatively nondusty.
- Excellent shelf-life characteristics.
- Applicable to wide-range of water-insoluble active ingredients.

Primary among the disadvantages of the WG is higher cost, compared to the traditional product forms. Since the powder agglomeration step in manufacturing the WG represents an additional step beyond preparation of a wettable powder, and since most agglomeration processes require an additional expensive step to remove processing water, WG products are usually more costly than the traditional forms. In fact, this agglomerated characteristic of this product form requires that the granules break up easily and rapidly on dispersion in water, an additional challenge for the formulation scientist. Another possible disadvantage in moving to a WG product form is that the granules of these products may be somewhat fragile and require higher technology packaging for protection. Since many global markets can not tolerate the additional added cost of these products or their packaging materials, WG s have not totally supplanted the wettable powder or suspension concentratc in many world markets. Still, the "high-tech" image of the WG compared to its predecessors makes it highly desirable to many consumers, and we are likely to see an even greater preponderance of WG products in the crop protection marketplace of the future.

3.2 Ultra-low-volume Liquids

A very attractive formulation option to adopt when solubility characteristics of the active ingredient permits is that of the ultra-low-volume liquid (UL). These products are designed to be sprayed without dilution, and thus must be applied very carefully with specialized equipment. In addition to the active ingredient and a solvent system in which the active is very soluble, UL products often contain surface-active agents and drift control substances. Advantages of UL products include:

- High concentration leads to low material-handling requirements.
- Easy to transport and to use; no dilution is required.
- Optimum biological activity enabled for active ingredient.
- Good penetration of foliage and into less accessible areas.

First among disadvantages of UL formulations is the requirement for specialized application equipment. These may not be universally available; accordingly, UL product forms may not satisfy the needs of all farmers. In addition, the high solvent content of UL s may result in higher cost. Also, because of the highly concentrated application vehicle, UL products require near perfect weather conditions to ensure on-target delivery. And finally, the high concentration of UL products may represent a special handling hazard for more toxic active ingredients.

3.3 Oil-in-water Emulsions

For oil-soluble active ingredients, a relatively new product type which may be adopted by the formulation scientist is the oil-in-water emulsion (EW).[11,12] These products are prepared by dissolving the active ingredient in oil, and the oil solution is then emulsified into a water carrier. In addition to the active and the oil/water combination, EW products will require emulsifier surface active agents and may typically contain other formulants to enhance emulsion stability and influence biological activity. Advantages of the EW products include the following:

- The oil component may enhance biological activity of the active ingredient.
- Water carrier lowers organic solvent content and volatility.
- Generally good shelf life and transportability.
- Permits mixtures of actives with widely varying physical properties.

Among the disadvantages of EW products is their difficulty of formulation. The proper choice of oil and emulsifier is extremely important to avoid de-emulsification of the product on standing. Also, methods designed to predict shelf life may be inadequate to ensure a robust product. Finally, because of the presence of the organic oil in the formulation, EW products may represent a flammability or a skin-absorption toxicity hazard.

3.4 Microencapsulated Products

The search for viable microencapsulation technology to produce useful crop protection products has been underway for the last two or three decades, as reviewed by Stern and Becher.[13] The goal of these research efforts has been to develop techniques through which the bioavailability and/or the physical behavior of active ingredients can be controlled or influenced via containment of an active ingredient within a polymer shell or matrix. The active ingredient is then released by some mechanism at the desired time and location. The products in general are manufactured by preparing a formulated liquid composition of the active ingredient, which is then encapsulated by polymerization of monomers contained within droplets of the composition. Considerable resources from many private companies, public laboratories and universities have been applied to these efforts. From these efforts have come many technical successes, but relatively few commercial ones. Among the desirable characteristics which may be achieved by microencapsulation are the following:

- Lengthened field efficacy period of the active ingredient.
- Protection of the bioactive substance from adverse environmental conditions.
- Amelioration of acute toxicological properties of the active ingredient.
- Reduced volatility of the active ingredient and liquid formulants.
- Attractive alternative to the use of organic solvents for organic liquid and low-melting, organic solid active ingredients.

Among the difficulties associated with this product type are the problems associated with achieving the desired release characteristics of the active ingredient, while maintaining its optimum ultimate bioefficacy. These products also may be difficult to analyze, due to the requirement to release the active quantitatively from its polymer cage. In some cases, these products have been suspected to have increased toxic effects on pollinating insects; this outcome has been attributed to the similarity of the sizes of the microcapsules to those of pollen grains.

3.5 Premeasured Single-dose Products

A major innovation in crop protection delivery systems in recent years has been the introduction of several types of products offered in premeasured single-dose form. Where the value-in-use of a given product can justify the added cost of providing this convenience to the user, these product types have increasingly made their presence felt in the crop protection marketplace. Advances in several unrelated technologies have been achieved to permit the production of these products. First is the availability of polymer film which is compatible with formulated crop protection products and which is rapidly soluble in cold water. These polymer films, usually comprised of polyvinyl alcohol along with additives to influence their physical properties, are readily available from several suppliers, and have enabled manufacturers to offer products in water soluble, unit area packs. These packets may be added directly to the feeder tanks of spray equipment, where the polymer film dissolves, releasing the product to disperse in the spray water. An enhancement of this approach involves a technique called "bags-in-bags" in which mutually incompatible products may be used in combination as a single multiple-product pack by separately enclosing each product in its own water-soluble film pouch. Another version of the premeasured single dose approach is an extension of an established technology from the pharmaceutical industry, that of tabletting. This technique, usually applicable only to low-dose active ingredients, has been utilized by several manufacturers to supply a convenient offering to the agricultural user. The agricultural tablet (TB) is usually designed to be added to the spray water in conventional application equipment, where the tablet dissolves, assisted either by the presence of swelling agents and disintegrants or by effervescence. The TB form however has not found major applicability in the industry, due to its relatively high cost of manufacture, as well as the difficulty in packaging relatively fragile, sometimes moisture-sensitive, tablets for use in the rough-and-tumble world of agriculture. A version of the TB product form, which has made major inroads in the rice culture practices of Japan, is called the "Jumbo". These products are designed to be hand or machine applied onto the surface of rice paddy water, whereupon they sink to the bottom and slowly dissolve, releasing the active ingredient, usually a herbicide, to do its work. Clearly the advent of premeasured single dose products, while still in its infancy, promises to provide major opportunity to make the application of crop protection products more convenient and safer to use.

4 "EMERGING TECHNOLOGY" PRODUCT FORMS

4.1 Microemulsions

Microemulsion products for crop protection, also known as particular microemulsions,[14] are thermodynamically stable, isotropic dispersions of active ingredients, hydrocarbons, surface active agents, and water.[15] The basic technology of microemulsion formation has been the subject of a review by Ogino and Abe.[16] Because these products require a high concentration of surfactants, the resultant microemulsions generally are low in active ingredient content and are fairly costly. These limitations have prevented this product form from becoming widely utilized in commercial agriculture. Nonetheless, microemulsion products are physically attractive, clear fluids, from which the suspended microcrystals of active ingredient remain nonsettling for an indefinite period and which instantly disperse in spray water.[17] Because of these positive attributes, refinements of the technology are being sought by many researchers, in expectation that the product form will become more applicable in the future.[18]

4.2 Suspoemulsions

Another emerging product form which enables the preparation of mixture products from active ingredients with widely varying physical properties is the suspoemulsion (SE). Mulqueen *et al.*[19] and Winkle[20] have provided valuable reviews of the increasingly important uses of this product form. Suspoemulsion products typically contain one or more solvent-soluble active ingredients in an emulsion phase, combined with one or more low solubility active ingredients in a continuous aqueous suspension phase. The formulations scientist chooses formulants to provide a shelf-stable, pourable premix product which enables the farmer to facilitate the application of multiple active ingredients, so increasingly important in modern agricultural practice. We can expect to see an increasing importance of SE products in the crop protection marketplace as the technology improves, and as the use of multiple active ingredient premix products becomes a more universal practice.

4.3 Gelatinized Fluids

Among newer formulation types to find utility in agriculture are ones which can be characterized as gelatinized fluids, which have been reviewed by Osada and Gong.[21] These formulations differ from other types of liquid formulations described above in having higher viscosity. Gel formulations can be water based (GW) or solvent/emulsion based (GL), and the latter type has been offered in premeasured doses contained in water soluble film. This product form, while fairly new to agriculture, has seen wide use in many other types of products, for example, as tooth pastes, detergents, personal care products, and pharmaceuticals. Among the advantages of this product form are the following:

- No possibility of suspended particles settling from the formulated product.
- Offers the opportunity for novel packaging approaches.
- Can easily be used in direct injection equipment.
- Easily measured by volume or length of expressed gel.

But while these advantages contribute to making gel products attractive options for formulations scientists to develop, there are the sometimes offsetting negatives of increased formulations development cost and time, more costly packaging requirements, and the necessity to change traditional product-handling practices. Thus the industry has not been overly quick to embrace gel products as popular product forms, and the actual experience in the marketplace has been limited.

4.4 Biological Products

While products containing living biological substances have been commercially available for agricultural use for many years, as described in a review by Rodgers,[22] there remain many challenges in their formulation and use. The most common biological product for agricultural use has been *Bacillus thuringiensis (BT)* for insect control. *BT* products have most often been formulated as traditional wettable powders after spray drying or lyophylization of the fermentation medium in which they are cultured. However, suspension concentrates of *BT* have also been offered by some manufacturers. In addition to bacteria, viruses and fungi have also been shown to be useful in crop protection programs. Wherever the goals of prevention of insect, plant disease, and weed damage can be met by a highly target-specific biological product, these approaches are highly desirable, with usually favorable effects on beneficial species and the environment. However, living substances present special challenges to the formulation scientist, because in addition to requirements of good physical properties and convenience in use, the biological product must also keep the biological control agent functional. This requirement is in direct opposition to an outcome normally sought by the formulator, specifically, to prevent growth of microorganisms in a formulated product. This means that the product must not experience negative influences from solar radiation, temperature extremes, presence or absence of excess moisture, oxidants, and environmental contaminants such as ozone or other air pollutants. Formulators have utilized many different approaches to overcome these effects, including utilization of formulants such as UV absorbers or ozone destroyers, and microencapsulation for protection of biological agents. But while there have been some successes in these areas, there is much to be learned before formulation of biologicals can be considered a fully understood and easily applied discipline.

5 THE FUTURE OF CROP PROTECTION PRODUCT FORMS

5.1 General Considerations

The number of product form choices available to farmers is likely to continue to increase, as agriculture continues on its path to modernization. There is a strong impetus on the part of responsible manufacturers of crop protection products to move toward safer and ever-more effective and convenient formulation types. In general there already is and will continue to be less reliance on the use of volatile organic solvents, as well as more utilization of water-based and other environmentally benign formulations. Accompanying these trends among liquid formulation types, there will likely be continuing movement toward dry formulations, especially those which are highly concentrated, nondusty and easily measured. In very general and simplistic terms, the product forms of the future will be those which enable the farmer to use these products without touching or smelling, or in fact, without contacting at all, the active ingredient or its formulants. Improvement in the currently available technologies in microencapsulation and injection coating of powders will likely play a role in achieving these goals. In addition, there will be continuing movement toward concepts which otherwise minimize the impact of crop protectants on the environment. Among these trends is the increasing practice of offering isomerically pure or purer active ingredients. While not a trend in the science of formulations itself, but rather in manufacturing chemistry, the move toward isomeric enhancement may offer significant challenges for the formulations scientist. First among these considerations is that the enriched or isolated isomer of an active ingredient may have different physical properties than the isomer mixture, requiring fundamentally different techniques for its formulation. Also, the formulation may require the inclusion of formulants which prevent or minimize isomerization of a stereoisomeric or geometric isomer. And in addition, the crop protection formulation scientist must accommodate the higher concentration product forms which are a natural consequence of eliminating biologically noncontributing isomers from the product. We can also expect that the sciences governing adjuvant choice and inclusion of adjuvants in product compositions will become better understood in the future. Choosing the proper surface active agent,[23] either as a component of a formulation or as a tank-mixed adjuvant, to optimize biological activity of a crop protectant enables the user to minimize its use rate and cost, and also to reduce its potential for development of resistance. Some specific developments with possible implications for crop protection practices of the future are described in the paragraphs which follow.

5.2 Crop Plants as Delivery Systems

Among the most exciting developments to impact agriculture for a long time is the advent of the principle of the crop itself acting as its own protectant. The technology required to make this outcome reality has arisen from the tremendous advances in plant biotechnology, where crop plants are genetically engineered to resist the onslaught of insects and diseases, as well as to tolerate the application of broad spectrum herbicides for control of weed infestations. Regarding these developments as an offshoot of the sciences of crop protection formulations perhaps requires a stretch of the imagination, since biotechnology requires very different skills and training for the practitioner than the usual physicochemical skills of the formulation scientist. However, the idea may seem more natural if one describes what the formulator does as the creation of systems for delivery of a crop protectant to the place where it can do its work. In this sense, then, the biotechnological approach is only the next step in a natural evolution of delivery system technology, perhaps evolving to its ultimate state in the "synthetic seed" approach reported by Teng, et al.[24] Professor Hammock has ably addressed this topic in detail in a companion plenary lecture in this symposium.[25]

5.3 Impregnated Biodegradable Carriers

The increasing constraints on the crop protection delivery system scientist are causing this specialist to look for innovative ways to deliver an active ingredient to its target with minimal or no movement to off-target sites. One of the most attractive possibilities still in very early stages of development is the use of active ingredient-impregnated polymers, which are shaped and designed to be directly applied to the locus of action. The polymers employed may be completely water soluble, where very rapid release of an active substance on onset of water application is desirable, or they may be less soluble in water, if slower release of the active is preferred. Using these concepts, one can imagine the offering of soil insecticides impregnated in polymer ropes or other shapes, which can be buried along with seeds at planting for subsequent controlled release of the active. One can further imagine water soluble polymer nets or sheets which can be dropped over crop producing trees to dissolve and deliver their load of impregnated active ingredient when water is applied, either by natural rain or by sprinkler irrigation. Or alternatively, such impregnated water soluble polymers could be offered in perforated sheets or rolls similar to paper towels. The required amount of dissolvable polymer would be torn off and then added to the spray tank for dissolution and spraying in traditional fashion. All of these approaches meet the requirement to keep the product from contacting the user in any way. Some of the inherent difficulties of these approaches involving timing and speed of dissolution of the polymer matrices will inevitably be solved in future research programs.

5.4 Novel Structured Fluids

It seems clear that the structured fluid product form has not been adequately exploited for use in agriculture. Whereas the gel formulations described above offer some advantages over traditional liquid and solid formulations, their limitation of fixed viscosity makes them relatively inflexible for some uses. For example, a highly viscous gel formulation offered in a "tooth paste" tube package might require high-shear mixing before being used in traditional spraying operations. Soon to be available technology in "reversible" gels would enable the user to gain the benefits of the structured gel formulation while the product is at rest, but which would easily convert to a low viscosity, pourable form with the application of low shear. The unused portion in the container would then recongeal on standing to prevent any settling or product segregation. Structured fluids with other desirable properties can be envisioned. For example, a reversible gel product could be designed for application to soil to retain an active ingredient, and perhaps water or nutrients, in the root zone of a target crop plant. The product would be pourable on low shear agitation and would then congeal in the soil after application to provide the desired retention effect.

5.5 Triggered Release Products

From the discussion of microencapsulated products above, one can see that the possibilities for controlled release products for agriculture have not been fully realized. A long-held goal of visionary delivery system scientists is that of developing "triggered release" technology. To achieve this end, the product would be required to withhold release of its active ingredient until certain environmental conditions are met. For example, a rice herbicide might be required to be retained until the temperature of paddy water reached the point when weed seeds would germinate. The product would then completely release the herbicide to do its work when this "trigger" temperature is reached. Other "triggered-release" products could be designed to release a fungicide when the presence of conditions for development of plant disease organisms, such as high humidity or temperature, are encountered. The property of triggered release would avoid the need for the farmer practitioner to make the decision on when to apply the crop protection product. He or she could apply the product when convenient or when applying some other material, confident that it would accomplish its goal at the appropriate time. This technology has not yet reached fruition, though some tantalizing successes have been obtained under laboratory conditions. One can expect that technologists seeking this goal will eventually be successful, and yet another tool will be added to the formulator's toolbox.

6 CONCLUSION

From the above sections, it can been seen that the product form offerings available to the farmer have changed only slowly over the years, due in part to the traditional nature of agriculture as an enterprise. But as agriculture has become more progressive, the acceptance of new product forms has become easier and more rapid. In fact, it can be expected that the farmers of the future will not only embrace, but demand, crop protection products which are safer and more convenient to use. Providing these products will depend on a new generation of formulations scientists continuing to develop new tools for the toolbox to make the new offerings become reality. Making this happen will not be an accident. The ancient characterization of the practice of crop protection formulations as an art rather than a science has long been inappropriate. In fact, the description of this discipline by Corty, one of its most distinguished practitioners, as "the science of the infinite number of variables"[26] is very near the truth. Such complexity requires immense ingenuity by its adherents, along with rigorous application of good science and understanding of the practical utility of crop protection products. A thorough knowledge of physical chemistry, surface chemistry and physics, material science, analytical chemistry, organic and inorganic chemistry, and chemical engineering, not to mention government regulations, are absolute requirements for the formulations scientist, now and even more so in the future. These "keepers of the toolbox" will continue to have significant influence in the crop protection marketplace of the future and will ensure that the farmers of the next millennium will have at their disposal the products needed to feed and clothe an ever-growing world.

7 ACKNOWLEDGEMENTS

The author is grateful to George B. Beestman, James P. Foster, Thomas Cosgrove, and Dennis S. Bloemer of DuPont Agricultural Products for their review of the manuscript and their excellent suggestions for addition and improvement.

REFERENCES

1. Corty, C., "Pesticide Formulations and Application Systems, ASTM STP 980," D. A. Hovde and G. B. Beestman, Eds., American Society for Testing and Materials, Philadelphia, 1988, Vol 8, pp. 5-12.
2. Becher, P., "Pesticide Formulations and Application Systems, ASTM STP 980," D. A. Hovde and G. B. Beestman, Eds., American Society for Testing and Materials, Philadelphia, 1988, Vol 8, pp. 13-21.
3. Tsuji, K., "Abstracts, Formulations Forum '97," Association of Formulation Chemists, Las Vegas, Nevada, September 3-5, 1997, p. 2.
4. Beestman, G. B., "Pesticide Formulations and Adjuvant Technology," C.L. Foy and D.W. Pritchard, Eds., 1996, pp. 43-69.
5. Sanchez-Rasero, F ., "Caracteristicas de las Formulaciones de Plaguicidas. "Influencia Sobre los Residuos en Alimentos o Sobre el Medio Ambiente?" Presented at the IV Seminario Internacional sobre Residuos de Plaguicidas, Almería, España, November 26, 1996.
6. Clayton, P. B., Presly, A. H., and Rutherford, S. R., "Some Aspects of Film Coating Agrochemicals Onto Seed," British Crop Protection Council, Monograph 39, 1987, pp. 229-235.
7. Drummond, J. N., "Pesticide Formulation and Adjuvant Technology," C. L. Foy and D. W. Pritchard, Eds., CRC Press, Boca Raton, 1995, pp., 69-92.
8. Albert, R. F. and Weed, G. B., United States Patent 3,920,442, November 18, 1975.
9. Gandrud, D. E. And Skoglund, J. H., "Pesticide Formulations and Application Systems, ASTM STP 968," G. B. Beestman and D. I. B. Vander Hooven, Eds., American Society for Testing and Materials, Vol. 7, 1987, pp. 177-182.

10. Gerety, P. J., "Pesticide Formulations and Application Systems, ASTM STP 1183," P. D. Berger, B. N. Devisetty and F. R. Hall, Eds., American Society for Testing and Materials, Philadelphia, Vol. 13, 1993, p. 371-380.
11. Narayanan, K. S., "Pesticide Formulation and Adjuvant Technology," C. L. Foy and D. W. Pritchard, Eds., CRC Press, Boca Raton, 1995, pp. 115-174.
12. Hässlin, H. W., United States Patent 5,674,514, October 7, 1997.
13. Stern, A. J. and Becher, D. Z., "Pesticide Formulation and Adjuvant Technology," C. L. Foy and D. W. Pritchard, Eds., CRC Press, Boca Raton, 1995, pp. 93-114.
14. Green, John, DuPont Agricultural Products, Wilmington, Delaware, U.S.A., Personal Communication.
15. Beestman, G. B., *vide supra.*
16. Ogino, K. And Abe, M., "Surface and Colloid Science," E. Matijevic, Ed., 1993, **15,** 85.
17. Narayanan, *vide supra.*
18. See, for example, Nielsen, E., United States Patent 5,246,912, September 21, 1993.
19. Mulqueen, P. J., Paterson, E. S., and Smith, G. W., *Pestic. Sci.,* 1990, **29,** 451
20. Winkle, J. R., "Pesticide Formulation and Adjuvant Technology," C. L. Foy and D. W. Pritchard, Eds., CRC Press, Boca Raton, 1995, pp. 175-186.
21. Osada, Y. and Gong, J., *Prog. Polym. Sci.,* 1993, **18**, 187-226.
22. Rodgers, P. B., *Pestic. Sci.,* 1993, **39**, 117-129.
23. Tann, R. S., "Pesticide Formulations and Application Systems, ASTM STP 1328," G. R. Goss, M. J. Hopkinson, and H. M. Collins, Eds., American Society of Testing and Materials, Vol. 17, 1997, pp. 187-195.
24. Teng, W.-L., Liu, Y.-J., Lin, C.-P., and Soong, T.-S., United States Patent 5,250,082, October 5, 1993.
25. Hammock, B., in "Proceedings of the Ninth International Congress of Pesticide Chemistry," sponsored by the International Union of Pure and Applied Chemistry and the Royal Society of Chemistry, London, United Kingdom, August 3, 1998.
26. C. Corty, Personal Communication.

Modelling Foliar Penetration: Its Role in Optimising Pesticide Delivery

Jorg Schönherr, Peter Baur and Anke Buchholz

INSTITUTE OF VEGETABLE AND FRUIT SCIENCE, THE PHYTODERMATOLOGY GROUP, UNIVERSITY OF HANNOVER, D-31157 SARSTEDT, GERMANY

Abstract: Rates of foliar uptake are often limited by slow cuticular penetration of active ingredients (ai). Rates of cuticular penetration have been modelled based on two processes, namely solute mobility in cuticles (k^*) and partitioning of ai or adjuvants between the formulation residue, and cuticular waxes (K_{wxfr}). The model distinguishes between effects of ai and adjuvants on k^* and on driving force of penetration, which is proportional to the product $K_{wxfr} \cdot C_{fr}$, the latter variable being the concentration of the ai in the formulation residue. Solute mobilities in cuticles varied greatly among species but for neutral ai it depended only on molar volumes and not on lipophilicity. Solute mobility decreased exponentially with increasing size of ai but size selectivity (β') was the same for cuticles from all plant species tested and also for extracted cuticles (polymer matrix). Differences in solute mobility among species were related to the term k^{*o}, which represents the mobility of a hypothetical molecule having zero molar volume. While size selectivity did not change on extraction of cuticular waxes k^{*o} increased by orders of magnitude depending on plant species. Evidence is presented showing that differences in solute mobility among species and higher solute mobility in extracted cuticles are related to the tortuosity factor. Crystalline waxes increase the lengths of the diffusion paths and decrease mobility in cuticles proportional to the tortuosity factor. Solute mobilities increased with increasing temperatures and it is argued that at 5 to 10 °C solute mobilities can become so low that significant penetration of ai is prevented.

Rates of cuticular penetration can be manipulated by using appropriate adjuvants. The model distinguishes two adjuvant effects. Adjuvants which remain on the surface of the cuticles can serve as solvents and they affect driving forces via their effects on K_{wxfr} and C_{fr}. If these passive adjuvants are hygroscopic their effect on driving forces depends greatly on humidity. Active adjuvants are called accelerators. They are solvents as well, but in addition they penetrate into cuticles and increase solute mobility. These effects depend on type of accelerator, temperature, size of solutes, and plant species. Accelerators decrease viscosity of waxes which results in reduced size selectivity. Accelerators also reduce differences in solute mobility among plant species and penetration rates at low temperatures are greatly increased. By selecting the appropriate adjuvants, rates of foliar uptake can be manipulated and uniformity of biological responses of systemic ai can be improved, even under adverse environmental conditions.

1 INTRODUCTION

Numerous systemic plant protection agents (ppa) are sprayed on the foliage and active ingredients (ai) must penetrate into the leaves before they can become biologically active. With some of them long distance translocation via xylem or phloem may also be required. When studying delivery of foliar applied ppa it is conveniently subdivided into discrete processes such as droplet formation at the spray nozzle, flight and impaction on the leaf surface, penetration across the cuticle, diffusion in the apoplast, uptake into cells, loading of sieve tubes or tracheae, long distance translocation in phloem and xylem tissues and finally interactions with subcellular structures, organelles, enzymes or other sites of action. This scenario shows penetration of cuticles to be an early and essential step in delivery of ppa to sites of action and often this step can be rate limiting.

Modelling of foliar penetration is useful for it helps to identify the most important factors which influence foliar penetration. The simplest models are purely descriptive and are based on curve fitting.[1] Other examples are attempts to relate rates of translocation of ai to their ionisation constants and to octanol/water partition coefficients.[2-4] Cuticular penetration has also been correlated to octanol/water or cuticle/water partition coefficients and to molar volumes of solutes.[5,6] Enhancement of foliar penetration by ethoxylated alcohols varying in ethylene oxide contents has also been modelled.[7] These descriptive models relate to a particular situation and they are difficult to generalise. They are not suitable for analysing quantitative structure property relations (QSPR). Mechanistic models are more useful because they reveal the mechanisms of the processes involved and they provide information as to how rates of foliar uptake might be optimised. Here we shall present an improved quantitative model relating rates of foliar penetration to properties of cuticles, ai, adjuvants and to environmental factors. The model accounts for differences among plants species and stage of development of leaves and it is analytical as well as predictive. It is not intended to review of the vast amount of data collected world wide on all aspects of foliar penetration of ppa and we shall quote these contributions only if they contain data relevant for our modelling approach. General reviews on foliar uptake and related subjects can be found elsewhere.[8-11]

Similarities exist in quantitatively describing passive transport of xenobiotics across cell membranes, plant tissues and cuticles. Initially, we should like to point out some of those similarities while the major part of this contribution will stress the differences and focus specifically on mechanistic aspects of transport across cuticles. In plant cells specific carriers for xenobiotics are rarely available and many of the processes mentioned are (at least in part) physical in nature and may be described quantitatively in terms of rates of diffusion, laminar flow, permeability and partition coefficients and concentration gradients. Uptake into cells[12] as well as translocation in phloem or xylem vessels[3,4,13-15] are examples of modelling which rely on permeability and partition coefficients as well as on concentration gradients and dissociation constants of ppa. These models are based on a general transport equation of the form

$$\textit{Flux} = \textit{proportionality coefficient} \times \textit{driving force} \qquad (1)$$

This equation has been popular because it requires a minimum of information. Fluxes (J) expressed as amount translocated per area and time are measured and they are divided by the concentration difference (ΔC) across the barrier, membrane or vessel. The proportionality coefficient derived in this fashion is the permeability coefficient (P) and it is a simple empirical parameter relating fluxes to concentration differences. P is a lumped quantity[16] which depends on many factors such as the nature of the barrier, transport mechanism, size, charge and lipophilicity of solute and temperature, and interpreting P necessitates a large number of additional experiments. It is this complexity inherent in P which renders it rather useless for mechanistic models. Starting from Fick's law it can be shown that permeability of a membrane is related to solute mobility in the membrane as expressed by the diffusion coefficient (D), the thickness of the membrane (Δx) and to a partition coefficient (K_{ms}) which accounts for differential solubility of the permeand in the membrane and the adjacent solutions as shown in eq. (2):

$$P = \frac{DK_{ms}}{\Delta x} \qquad (2)$$

Finding a suitable lipid phase with similar sorptive properties as the membrane under consideration is clearly important, especially when solubility in the membrane phase cannot be measured directly. In modelling permeability of cell membranes which are surrounded by aqueous solutions, octanol/water partition coefficients (K_{ow}) or olive oil/water partition coefficients[12] have been used. K_{ow} were also selected to model phloem mobility,[4] permeability of cuticles[5] and the human skin.[17] Octanol/water partition coefficients are popular, because they are available for many compounds. Partition coefficients for cuticle/water and polymer matrix/water have been determined and found to be similar to K_{ow}-values.[6,18,19] Solubility of xenobiotics in the polymer matrix (extracted cuticles) was somewhat higher than in cuticles[18] indicating a smaller sorptive capacity of cuticular waxes. In fact, it was later demonstrated that wax/water partition coefficients were smaller by a factor of 10 than K_{ow}-values.[20,21] This suggests that wax/water partition coefficients might be more appropriate for modelling permeability of cuticles because the transport limiting barrier of cuticles is a wax/cutin composite[19,22,23] and Briggs and Bromilow[24] took this into consideration by suggesting that $\Delta\log K$ be used in modelling foliar uptake, with $\Delta\log K$ being the difference between the logarithms of the octanol/water and the alkane/water partition coefficients.

Finding the appropriate lipid phase is an important step in modelling passive permeation across cell membranes which are in contact with water on both sides. The situation is more complicated with plant cuticles where water is the liquid phase only on one side of the cuticle (the epidermal cell wall), while the outer surface of the cuticle is covered by water only during rain, fog or dew. Besides, ppa are always formulated and the non-volatile constituents of the formulation together with the active ingredients (and possibly other adjuvants) form a residue on the cuticle, once the water used as carrier has evaporated. Partitioning of ai between the formulation residue and the waxy limiting skin cannot be modelled using partition coefficients of the type lipid/water such as octanol/water or alkane/water. Only the equilibrium distribution between the inner surface of the cuticle and the water of the epidermal cell wall can be approximated using octanol/water partition coefficients.[18] Thus, owing to the heterogeneity of cuticles (inner and outer surfaces differ in chemistry and structure) and to the fact that during foliar penetration these surfaces have contact with different liquids or solutions, two partition coefficients must be used in modelling. Models relying on a single partition coefficient of the type lipid/water are inappropriate.

Finding the proper partition coefficients is only one part of the problem, because the driving force is proportional to the product *KC* on either side of the membrane.[19,23] Hence, measured flows are also proportional to ΔC and if these concentrations are not known (in addition to the appropriate partition coefficients) differences among plant species and ai cannot be interpreted or modelled.

K is the ratio of the equilibrium concentrations in the membrane and the adjacent liquid and these concentrations are not independent. The octanol/water partition coefficient is often used as a criterion for lipophilicity but it should be kept in mind, that very high partition coefficients observed with lipophilic actives are not due to very high solubility in octanol, but rather to a very low solubility in water. In fact cuticle water partition coefficients (K_{cw}) are inversely proportional to water solubility (S_w) and can be estimated from the equation

$$\log K_{cw} = 1.18 - 0.596(\log S_w) \qquad (3)$$

with the slope of the straight line (0.596) being related to the partial molar entropy of fusion of the solute.[6] Solutes having very high water solubility have small partition coefficients while the reverse is true for lipophilic solutes. Both C_w and K_{cw} may vary with ai by more than 6 orders of magnitude, but the products vary much less, as can be seen by comparing *KC* for methylglucose and bifenox. Cuticle/water partition coefficients for bifenox and methylglucose are 27500 and 0.13, respectively.[25] Water solubility of bifenox at 25 °C is only $3{\cdot}10^{-4}$ g l^{-1} while it is 900 g l^{-1} for methylglucose. Thus *KC* amounts to 117 and 8.25 for methylglucose and bifenox, respectively, and the *KC* ratio is 14.1. When saturated aqueous solutions of the two compounds are present on the leaf the driving force (*KC*) in the donor is larger by a factor of 14.1 with the polar methylglucose than with the lipophilic bifenox. A model using only partition coefficients would have predicted much higher rates of penetration for bifenox because the ratio of the partition coefficients is 211,538.

The corollary to this is that changing from the solvent water to some other solvent (for instance ethylene glycols, glycerol, or liquid surfactants) solubility of lipophilic ai can be increased greatly[25] but at the same time the partition coefficient is reduced as well and the product *KC* may not change much.[25, 26]

The importance of the product *KC* in determining driving forces and rates of foliar penetration was not recognised in the model by Briggs and Bromilow who restricted attention to partition coefficients.[24] Stock et al.[7] related rates of foliar uptake of model compounds differing in lipophilicity (K_{ow}) to the degree of ethoxylation of fatty alcohols.[7] Neither surfactant/cuticle partition coefficients nor solubility of actives in formulation residues were estimated or considered quantitatively. Numerous publications have shown type and amounts of surfactants to influence rates of foliar uptake and/or often biological effectiveness of ai. Surfactants and other adjuvants invariably influence driving forces of penetration via their effects on solubilities and partition coefficients.[19] The reasons for differences in performance of ai as affected by type and amount surfactants and other adjuvants cannot be identified in quantitative terms if partition coefficients and solubilities are not known. This is one of the reasons why the bulk of the literature on foliar penetration of ppa cannot be utilised for modelling velocity of foliar penetration.

2 ESSENTIAL ELEMENTS OF THE MODEL

Cuticular penetration is a purely physical process and takes place by simple diffusion, as canals, carriers or other special transport structures are absent in the transport limiting layer of the cuticle which is composed of the polymer cutin and associated waxes. The starting point of our modelling approach is Fick's law in one dimension which states that rates of diffusion across a plane or a membrane are proportional to the mobility (diffusion coefficient) in the membrane and the concentration gradient across it. Thus, flow rates are proportional to solute mobility and to the driving force. The driving force is the gradient in chemical potential across the barrier, but for convenience concentration gradients or differences are generally used. If donor, membrane and receiver constitute different phases (as is the case with cuticular penetration) two or more partition coefficients are needed to account for the fact that chemical potential not only depends on solute concentration but also on the properties of the solvents.[19]

We shall model diffusion of active ingredients from the formulation residue on the surface of the cuticle through the cuticle into the aqueous epidermal cell wall as depicted schematically in Figure 1. More specifically we shall focus on solute mobilities in cuticles as affected by plant species, properties of solutes, adjuvants and temperature. Mass flow into open stomata[27] which might be induced by wetters which decrease surface tension below 30 mN m^{-1} will not be considered even so it might be important with silicon surfactants and low viscosity oils.[28] Preferential penetration into glandular trichomes and cuticular ledges[29,30] will also be ignored.

The formulation residue represents the material remaining when the water used as carrier has evaporated. Depending on ambient humidity and polarities of ai as well as adjuvants (any compound in the formulation which is not ai) some water will be retained in the formulation residue. In Figure 1 the formulation residue is in the liquid state and contains an effective wetter such that it was drawn between the surface wax crystallites by capillary action and its thickness is larger over anticlinal walls. The cuticular membrane

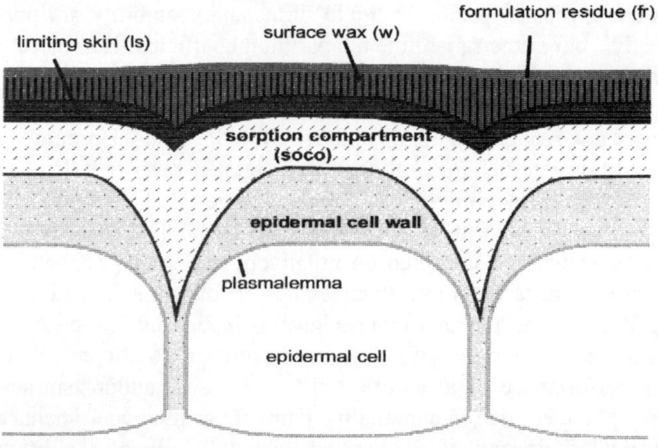

Figure 1 *Schematic drawing of a plant epidermis showing essential elements incorporated in the model.*

(CM) is composed of surface wax, the limiting skin and the sorption compartment. The limiting skin is a thin wax cutin composite[19,22,31] in which sorption capacity and solute mobilities are much lower than in the sorption compartment. The sorption compartment has sorption properties similar to wax free CM. If CM have pronounced cuticular pegs as in many species[32] they may extend between anticlinal walls of the epidermal cells and the distance between CM and plasmalemma can be much shorter than over the periclinal walls. Structural details and variability of cuticular morphology have been discussed in great detail in the recent review by Jeffree.[32]

Rates of cuticular penetration (J) will be modelled in terms of solute mobilities in the limiting skin and driving forces across CM as derived by Schönherr and Baur:[19,23]

$$J = \frac{M}{At} = k^* l_{ls}(K_{wxfr}C_{fr} - K_{mxw}C_w) \tag{4}$$

This steady state equation shows that the amount (M) of ai penetrating per unit area (A) and time (t) is the rate of penetration (J). In many cases the area covered by the formulation residue may decrease with time when liquid residues recede into the valleys formed by the network of the anticlinal walls (Figure 1). Annulus formation may also lead to unequal distribution of ai and adjuvants during droplet drying.[33] Both factors will tend to decrease penetration rates with time even under steady state conditions. The mobility of the ai (k^*) is related to the diffusion coefficient (D) in the limiting skin and can be determined using the UDOS (unilateral desorption from the outer surface of the CM) method.[34,35] The thickness of the limiting skin (l_{ls}) may be taken as one tenth of the thickness of the CM.[19,22] It is constant for CM of a given species but as we will explain below l_{ls} is the actual length of the diffusion path in the limiting skin and it is larger by orders of magnitude than the thickness of the limiting skin or the CM. The term in parentheses is the driving force across the CM. K_{wxfr} is the partition coefficient wax/formulation residue and C_{fr} is the concentration of the active in the formulation residue. The concentration of the ai in the water of the epidermal wall (C_w) and the polymer matrix/water partition coefficient (K_{mxw}) account for the driving force at the CM/cell wall interface. If ai are taken up rapidly into cells and are translocated C_w will be negligible making the term $K_{mxw}C_w$ zero. Under this condition the driving force is simply the product $K_{wxfr}C_{fr}$, and rates of penetration can be modelled using only 5 measurable quantities, namely contact area between formulation residue and CM, path lengths in the CM, mobility of ai in the limiting skin, the wax/formulation residue partition coefficient and the concentration of the ai in the formulation residue. The same equation may be used to model penetration of surfactants and other adjuvants and for this reason we prefer the general term solute.

3 SOLUTE MOBILITIES IN CUTICULAR MEMBRANES

3.1 Range of Solute Mobilities in CM from Selected Plant Species

Mobilities of bifenox in astomatous cuticular membranes ranged from 2×10^{-5} s^{-1} (peach leaf) to 1.5×10^{-8} s^{-1} (*Ilex paraguariensis*) which means that the highest and lowest mobilities differed by more than 3 orders of magnitude (Figure 2). If these mobilities are converted to diffusion coefficients as shown by Schönherr and Baur[34] a range of $1 \cdot 10^{-17}$ to

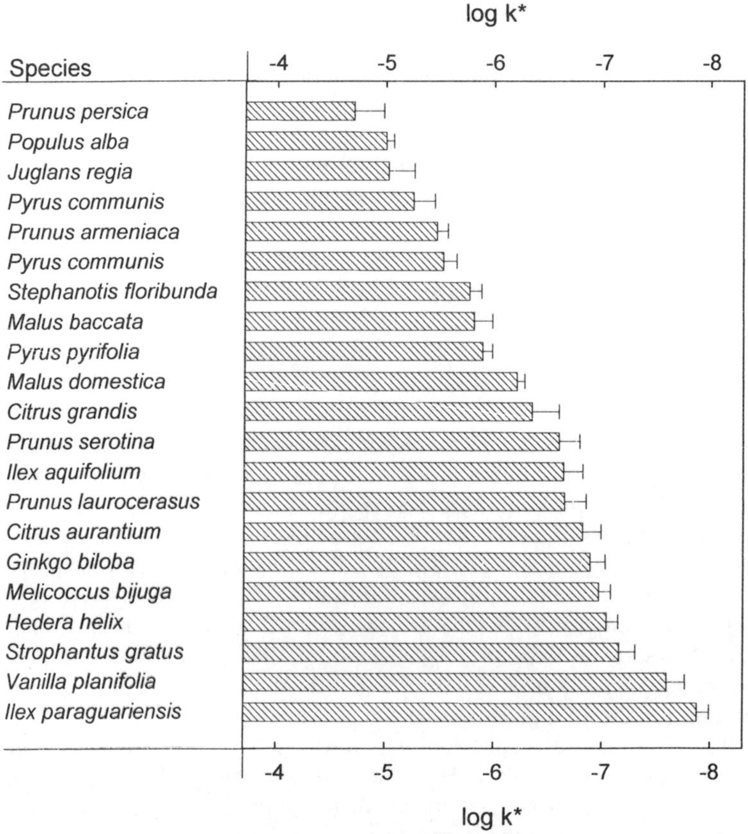

Figure 2 *Mobility (k*) of bifenox in astomatous cuticular membranes isolated from fully expanded leaves of 20 plant species. Two lots of pear leaf CM (c.v. 'Conference') are shown which were isolated in different years.*[42]

$5{\cdot}10^{-21}$ m^2s^{-1} is obtained. These figures are extremely low but they are of the same order and range as the diffusion coefficients measured directly using reconstituted waxes from various species.[21,31,36,37] This indicates that cuticular waxes are responsible for low solute mobilities in and barrier properties of cuticular membranes. Since these estimates are based on thickness of wax barriers rather than on actual lengths of the diffusion path, the "real" local solute mobilities in cuticular wax and in CM are larger by up to 3 orders of magnitude, as shall be shown later.

3.2 Solute Mobilities as Affected by Solute Properties

Mobilities (k^*) of ppa in cuticular membranes vary among plant species by orders of magnitude (Figure 2) but for a given species and at constant temperature they only depend on molar volume (V_x) of the penetrant. Plotting log(mobility) vs. molar volume resulted in linear graphs with all species and this is shown in Figure 3 for CM from apple and ivy

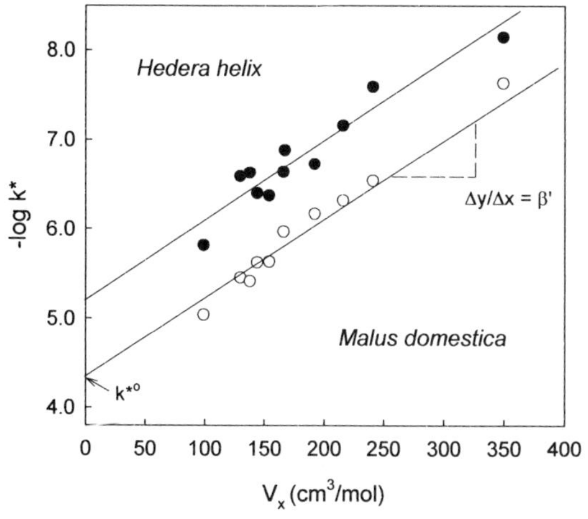

Figure 3 *The effect of solute size (V_x) on mobility (k*) in ivy and apple leaf cuticular membranes. The slope of the lines (β') is the size selectivity.*[38]

leaves. The slopes (*β'*) of the graphs are a measure of size selectivity and the y-intercept (k^{*o}) represents the mobility of hypothetical molecule having zero molar volume. Size selectivity was $(9.5\pm0.2)\cdot10^{-3}$ mol cm^{-3} for the CM of all plant species tested.[19,38,40] Within the range of molar volumes of ai (100 to 400 cm^3 mole^{-1}) solute mobilities decreased by a factor of 700. Thus, given equal driving forces rates of penetration decrease very rapidly with increasing size of ppa. Size selectivity (*β'*) did not depend on solubility of ai or on K_{ow}.

For the selection of plant species the y-intercept k^{*o} varied by more than 3 orders of magnitude from $4.68\cdot10^{-3}$ s^{-1} (*Populus alba*) to $5.37\cdot10^{-6}$ s^{-1} (*Ilex paraguariensis*) and accounted for all species related differences in solute mobility as might be caused by thickness of cuticles, amounts or composition of waxes, or structure of waxes. For a quantitative description of solute mobility in plant cuticles only two variables fully account for solute mobility as affected by properties of cuticles and solutes. No other descriptors are needed as long as solute mobility in cuticles does not depend on its concentration in the cuticle, as is the case with accelerator adjuvants (see below) and certain ai such as chlorfenvinphos.[62] All species related differences in solute mobility are incorporated in k^{*o} and once k^{*o} has been determined for a given species, mobility of any other ai can be predicted from the following simple linear model

$$\log k^* = \log k^{*o} - 0.0095 \pm 0.002 V_x \qquad (5)$$

3.3 Effect of Cuticular Waxes on Solute Mobility in CM

Extracting cuticular waxes increased permeability of cuticles by orders of magnitude[6,22,39] but exactly how waxes affected permeability remained unclear at that time. The puzzle was solved by analysing data on the basis of eq. (4) which separates wax and solute effects on mobility and partition coefficients. Extracting waxes increased solute mobility by orders of magnitude and this effect did not depend on solute size or polarity.[40-42] That is, only the y-intercept (k^{*o}) decreased on extraction while size selectivity (*β'*) remained unaffected (Figure 4). With *Citrus grandis* log k^{*o} decreased from -3.68 to -2.25 while with *Ilex paraguariensis* the corresponding figures were -5.26 and -2.38, respectively. Thus, extraction of waxes caused a reduction in k^{*o} by factors of 27 and 759, for *Citrus* and *Ilex* respectively. For pear leaf CM a reduction in k^{*o} by a factor of 182 has been reported[40] and for 5 other plant species Baur et al. found that k^{*o} decreased by factors ranging from 28 to 759.[41] In all of these cases size selectivity β' was not affected by extraction of waxes. Since size selectivity in CM and polymer matrix membranes (MX) was the same it follows that the effect of wax extraction on k^* is numerically identical to the effect on k^{*o} for all plant species and for all solutes, irrespective of their cuticle/water or wax/water partition coefficients.

It is quite remarkable that size selectivities in cutin and in the limiting skin composed of waxes and cutin were the same. This indicates that cutin in the limiting skin has the same viscosity as amorphous cuticular waxes. With all plant species k^* increased by extraction of waxes by more than 30 fold and this again shows that waxes associated with cutin are responsible for the barrier properties of cuticles. With some species (i.e. *Vanilla planifolia*), however, k^* for bifenox was extremely small ($2.14 \cdot 10^{-8}$ s^{-1}), yet extracting waxes increased mobility only 5-fold.[42] Clearly, cutin by itself can be a very efficient barrier in some plants. Since size selectivities in cutin and in CM were the same, yet mobilities increased on wax extraction by up to 3 orders of magnitude we are left with a new enigma. How can solute mobilities differ in two media having comparable viscosity? This problem was solved by studying temperature effects on solute mobility.

3.4 Effect of Temperature on Solute Mobility in CM

Rates of cuticular penetration generally increase with temperature and this represents a temperature effect on solute mobility[43] because sorption (partition coefficients) decreases with increasing temperature and solute concentrations in the CM.[44] With CM from some

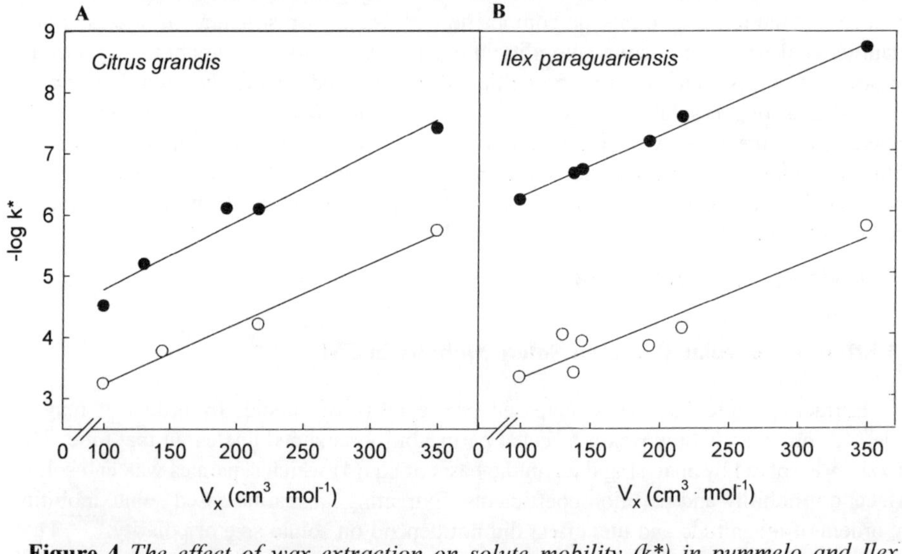

Figure 4 *The effect of wax extraction on solute mobility (k*) in pummelo and Ilex cuticular membranes. Extraction had no effect on size selectivity by reduced the y-intercept greatly.*[41]

species plotting log k* vs. 1/T resulted in linear Arrhenius graphs up to 78 °C while with other species slopes tended to decrease at temperatures >40°C (Figure 5A). At physiological temperatures up to 40 °C temperature effects on solute mobility were always reversible and together with the fact that activation energies were constant at ambient temperatures this shows that phase transitions in CM either did not occur or they had no effect on the structure of the limiting skin of the CM.[45] Activation energies depended on plant species and size of solutes and they ranged from 75 kJ mol^{-1} (apricot/IAA) to 189 kJ mol^{-1}(ivy/cholesterol) which corresponds to a range in Q_{10}-values from 3 to 14 at 20°C. With increasing temperature size selectivity as well as variability in *k** within and among plant species decreased (Figure 5A). The large (30-fold) difference in solute mobilities of tebuconazole and IAA observed at 25 °C decreased with temperature and at 60 °C it had vanished (Figure 5B).

Increasing solute mobilities causing higher rates of foliar uptake at high temperatures are desirable with ai. However, mobilities can be extremely low at temperatures around 5 to 10 °C which often occur in the field in temperate climates (dotted lines in Figure 5B). Compared to 25 °C solute mobilities are expected to be lower by factors of 26, 161 and 167 at 5 °C, for IAA, bifenox and tebuconazole, respectively. This is likely to result in reduced effects of growth regulators, weed killers or systemic fungicides when temperatures are low, especially with plant/ai combinations for which activation energies are high. Another undesirable effect of low temperatures is the fact that variability between species increases with decreasing temperatures (Figure 5B). Solute mobilities in cuticles follow a lognormal distribution and variability within a plant population increases with decreasing temperatures.[45,46] Both effects cause larger variabilities in rates of penetration of ppa at low temperatures and probably in larger variability in biological responses both within and between populations composed of different plant species.

Temperature effects on solute mobility can be modelled using the Arrhenius equation

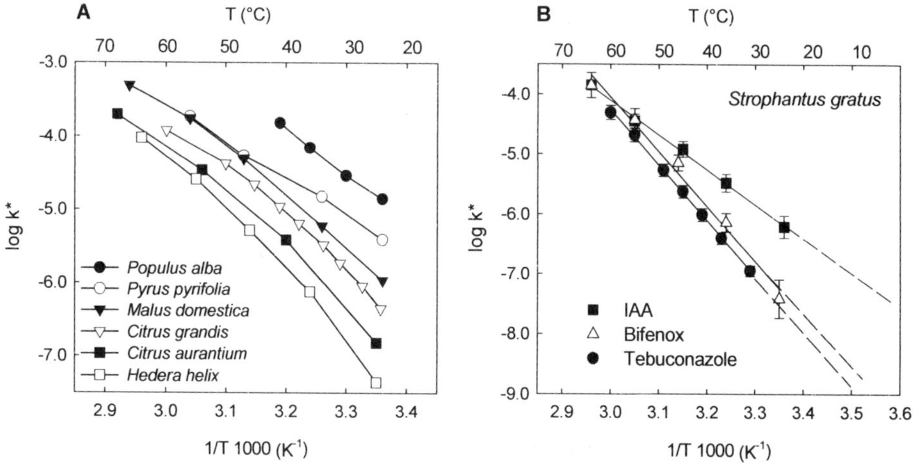

Figure 5 *Arrhenius graphs showing the effect of temperature (T) on mobility (k*) of bifenox in cuticular membranes from various plant species (A). Dependence on temperature of solute mobility (k*) increased in the order indoleacetic acid (IAA)<bifenox<tebuconazole, which is the order of increasing solute size (B).*[42]

$$\ln k^* = \ln k_o^* - \frac{E_D}{RT} \quad (6)$$

where R is the gas constant, T the Kelvin temperature, and k_o^* the y-intercept of eq. (6). (The intercepts of the Arrhenius plots (log k_o^*) must not be confused with the mobility of hypothetical compound having zero molar volume (k^{*o}) as defined by eq. (5)). Since E_D depends on species and molar volume of solute, eq. (6) is not very useful as a prognostic tool because k^* can be calculated only for a given species and solute. A more general and useful model can be derived when Arrhenius plots for a given species and solutes differing in molar volumes are available. Plotting log k^* vs. molar volume (V_x) gives straight lines for each temperature (Figure 6). The slopes of the lines (Figure 6A) represent the size selectivities at the various temperatures. These size selectivities increase with decreasing temperature. At 5 °C a compound having a molar volume of 100 cm^3 mol^{-1} has a 4000-fold higher mobility than one with 300 cm^3 mol^{-1}, but at 45 °C this difference is reduced to a factor of 3. Plotting the slopes of Figure 6A against temperature again resulted in a straight line (Figure 6B) which permits calculating β' at different temperatures:

$$\beta' = 0.02 - 3.9 \cdot 10^{-4}\, T \quad (7)$$

Plotting $\log k_o^*$ vs. E_D/R resulted in linear graphs[42,45] as shown in Figure 7 for cuticular membranes from 11 species and 5 different solvents. From the transition state theory of rate processes is known that k_o^* is proportional to entropy while E_D is related to enthalpy of the activation of diffusion.[47-49] A linear relationship between an entropy and an enthalpy term indicates that the free energy of the system was the same in CM of all species and the solutes differing greatly in size and lipophilicity experienced the same micro-environment along their diffusion paths. This most likely is an environment composed of CH_2-groups donated by the amorphous fraction of cuticular waxes[31] and by cutin.[42,45] This thermodynamic argument is in perfect agreement with the observation, that

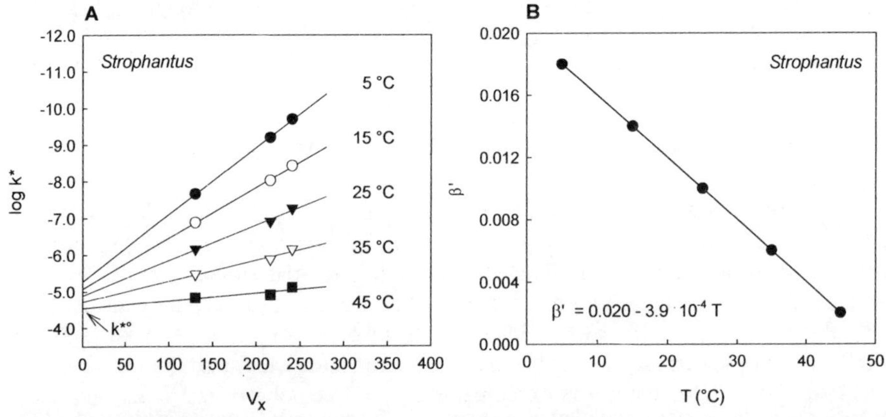

Figure 6 *The effects of molar volumes (V_x) and temperature on solute mobility (k^*) in Strophantus cuticular membranes (A). When slopes of the lines representing size selectivity at various temperatures (β') were plotted vs. temperature a linear dependence evolved (B).*[41]

size selectivity was the same in CM from all plant species (Section 3.2). Since wax

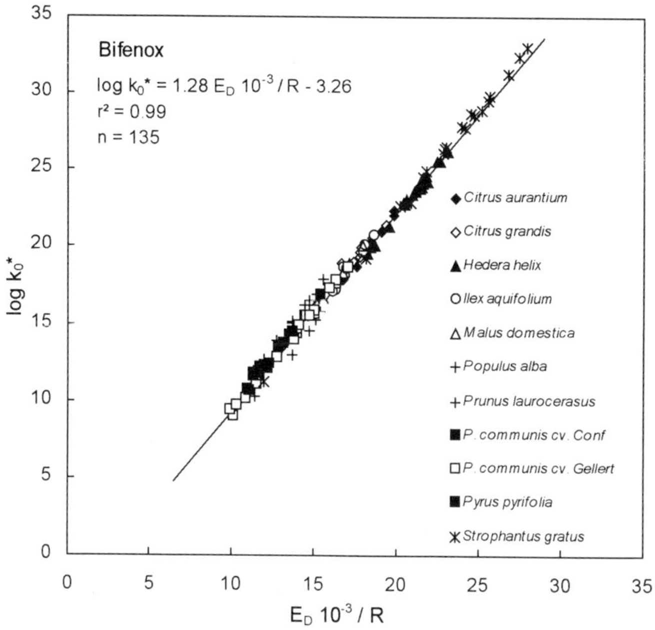

Figure 7 *Correlation between Arrhenius slopes (E_D/R) and y-intercepts of Arrhenius graphs (k^*_o). Data for 135 individual cuticular membranes from 11 plant species were plotted.*[42,45]

amounts, wax compositions and cuticle thickness differed greatly among species this is a very significant finding. Wax amounts and chemistry vary greatly depending on plant species and stage of development of leaves[50,51] but this has no consequences regarding viscosity and chemistry of the diffusion paths in plant cuticles which is an amorphous CH_2-phase in all plant species.

3.5 Effect of Temperature on Solute Mobility in Polymer Matrix Membranes

Extracting cuticular waxes from CM increases permeability[6,52] and mobility in CM[19,23,31] by up to 3 orders of magnitude. Plotting data for cuticular membranes and polymer matrix membranes (extracted cuticular membranes) resulted in two separate but parallel straight lines of identical slope (Figure 8). From the parallel displacement (indicated by the double arrow) the tortuosity factor was estimated.[42] It ranged from 50 (pear) to 3 300 (*Schefflera*) which means that the path length of diffusion in cuticular membranes is longer by factors of 50 to 3 300 compared to the paths in wax free polymer matrix membranes. The data also imply that the sole effect of cuticular waxes on solute mobility in cuticles is their effect on path length as suggested earlier by Baur et al. [45] Presumably, path lengths are increased by the blocking action of wax crystallites embedded in amorphous waxes.[31] Diffusion of solutes takes place in the amorphous wax fraction and/or in the cutin polymer and in both matrices solutes encounter a CH_2-group environment of comparable viscosity. It follows from this scenario that wax chemistry affects path lengths in CM via its effect on crystallinity and orientation of wax crystallites. Variability of solute mobility in CM both within and between plant species mainly reflects differences in paths lengths and tortuosity.

Above we had asked why solute mobilities in CM and MX-membranes can differ even though size selectivities and viscosities in MX and CM are identical. The thermodynamic argument based on the parallel displacement of the graphs for CM and MX as shown in Figure 8 now solves the riddle. While viscosities in CM and MX are the same, the length of the diffusion path in CM is much larger due to the blocking action of crystalline wax domains in the limiting skin. Cuticular waxes decrease permeabilities of cuticles by increasing the length of the diffusion path. Tortuosity of the diffusion path depends on crystallinity of waxes, particularly on size and orientation of crystalline wax plates in the limiting skin as argued by Riederer and Schreiber.[31] It follows, that the distance term (l_{ls}) in eq. (4), is not the thickness of the limiting skin but the path lengths in the limiting skin which can be estimated using the method shown in Figure 8 or by comparing solute mobilities in CM and MX as shown in Figure 4. The latter method is applicable only when size selectivities (β') in CM and MX are identical.

Amounts and composition of waxes change during leaf development[51] and it is probable that wax structure and crystallinity change during leaf ontogeny. Permeability of cuticles decreased during leaf expansion[51,52] and we suggest that permeabilities decrease, because tortuosity increases during leaf development until leaves are fully expanded.

3.6 Solute Mobility as Affected by Accelerator Adjuvants

Biological effectiveness of ppa and their rates of foliar penetration can be enhanced by a variety of adjuvants such as surfactants, emulsifiers, oils and solvents. This is often referred to as activation of foliar penetration and these adjuvants have been termed

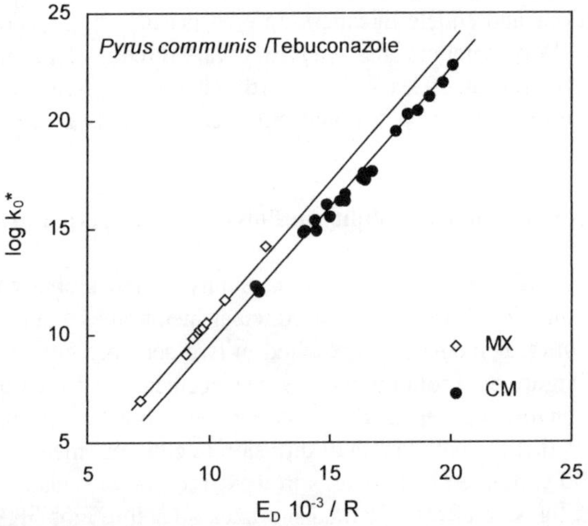

Figure 8 *Correlation between Arrhenius slopes (E_D/R) and y-intercepts of Arrhenius graphs (k_o^*). Data for individual cuticular membranes (CM) and polymer matrix membranes (MX) from pear leaves were plotted.*[42]

activators.[53-56] Rates of cuticular penetration are proportional to solute mobility in cuticles and to driving forces (eq. 4) and adjuvants invariably influence driving forces, while some may also increase mobility of solutes in the CM.[57,58] Thus, activators of foliar penetration may affect mobility and driving forces. These effects act independently and they cannot be separated when only rates of penetration are measured.[7]

In our model (eq. 4) effects of adjuvants on solute mobility and driving forces are considered separately and adjuvants that increase mobility of solutes in CM are called accelerators or accelerator adjuvants to distinguish them from activators having unknown mechanisms of action. Accelerators increase solute mobility by decreasing viscosity along the diffusion path.[59] Many compounds have been shown to have accelerator properties among them fatty alcohols,[58] ethoxylated alcohols[21,57-61] alkyl esters of dicarboxylic[61] and phosphoric acid esters.[23] This shows that surface activity is not necessary for accelerating activity and even active ingredients can have accelerating properties as shown with chlorfenvinphos.[62] Accelerators have been shown to increase solute mobility by more than 100-fold[21,23] and their effect was proportional to accelerator concentration in the CM.[21,62,63] Hence, penetration of accelerators into the CM is a prerequisite for their activity and to be useful they must penetrate faster than the ppa in absence of accelerators.[19] So far there is only limited information on rates of penetration of accelerators.[64] Octaethyleneglycol monododecylether ($C_{12}E_8$) penetrated very rapidly, but rate constants decreased with time and large differences between species existed (Figure 9).[64] Unfortunately, $C_{12}E_8$ concentration in the cuticle decreases with time as the surfactant penetrates into the leaf tissue and $C_{12}E_8$ is a good accelerator only for ppa which penetrate fast and can follow the accelarator front.

Unavailability of ^{14}C-labelled accelerators has till now prevented us from studying their mobilities in CM as affected by plant species and concentration and for this reason we cannot present a model relating accelerating activity to properties of accelerators and

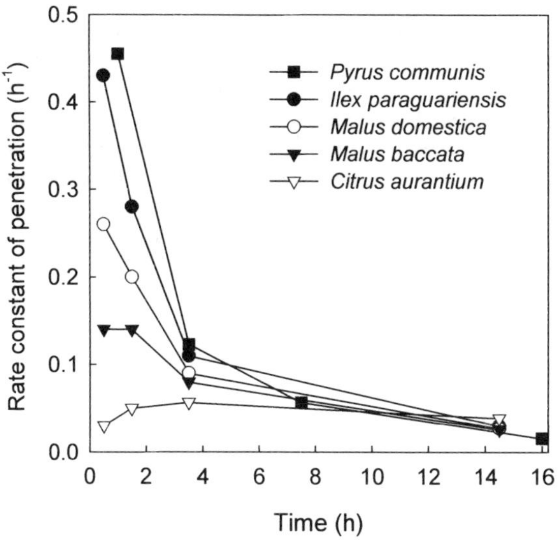

Figure 9 *Rate constants of penetration of octaethyleneglycol monododecylether ($C_{12}E_8$) at a concentration of $40 \cdot 10^{-3}$ g litre^{-1} as a function of time measured using cuticular membranes from 5 different plant species.*[64]

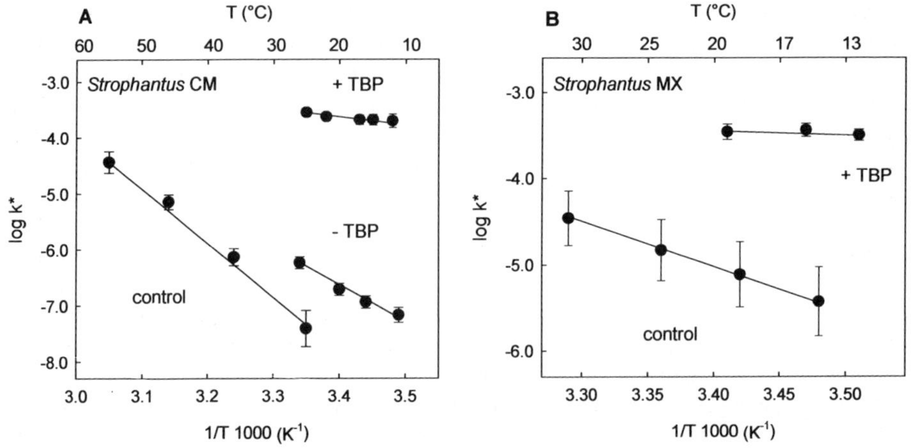

Figure 10 *Effects of temperature (T) and tributylphosphate (TBP) of mobility of tebuconazole in Strophantus leaf cuticular membranes (CM) and polymer matrix membranes (MX). Means and 95% confidence intervals were plotted as Arrhenius graphs.*[42]

cuticular membranes. Which type of accelerator at which concentration would increase rates of penetration best for a given species/ppa combination must therefore be tested experimentally and cannot be predicted at this time. However, the usefulness of accelerators is beyond question because they can drastically reduce size selectivity of CM[60,65,66] and they increase rates of penetration at low temperatures by increasing solute mobilities as demonstrated in Figure 10. Activation energy for diffusion of tebuconazole across CM was 195 kJ mol^{-1} and equilibrating the CM with 25 mmolar tributylphosphate decreased E_D to 29 kJ mol^{-1} (Figure 10A). The TBP concentration in the CM is not known but it is clear that TBP increased rate constants by more than 2 orders of magnitude and it almost eliminated the temperature effect on solute mobility. Clearly, accelerators are powerful tools for increasing rates of penetration of large ppa at low temperatures.

TBP increased k^* and decreased E_D with polymer matrix membranes as well, but the magnitude of the effect is somewhat smaller. The energy of activation was reduced by TBP from 114 to 18 kJ mol^{-1}. This shows that this accelerator increased fluidity of both polymer matrix and cuticular waxes and the effects of accelerators observed with CM may possibly be the sum of effects on wax and cutin. If this were true for all species and accelerators it would mean that studying accelerator effects using reconstituted waxes[21] might result in underestimates. Effects of TBP on tebuconazole mobility in CM were largely reversible as desorbing TBP decreased k^* and increased E_D to 125 kJ mol^{-1}. The original figure of 195 kJ mol^{-1} was not obtained and this might have been due to failure to desorb TBP quantitatively and/or to irreversible changes in cutin structure.

4 DRIVING FORCES OF CUTICULAR PENETRATION

We have shown above that mobilities of ppa vary greatly depending on their size, on temperature, and on plant species. Accelerators can increase mobilities greatly but no matter how high mobilities might be, rates of penetration will be zero if driving forces

(ΔKC) are zero. Accumulation of ppa in the epidermal cell wall can make KC in the receiver as large as in the donor (surface residue) and foliar penetration stops. With extremely lipophilic solutes this is a probable event and this is the mechanism by which amorphous constituents of cuticular waxes are locked in the cuticle. Penetration of ppa also ceases when they form solids on the surface of the CM. Water from the surrounding atmosphere or water penetrating the cuticle could serve as solvent, especially with polar and hygroscopic actives and/or adjuvants. Glycols, polyethyleneglycols, glycerol and ethoxylated surfactants sorb large amounts of water and if humidity is sufficiently high the spray residue on the cuticle can be in the liquid state even when the temperature is below the melting point of the adjuvants and actives.[25,26] Oil-based adjuvants could also serve as solvents[67] and their solvency would depend only on temperature but not on humidity. These adjuvants can be used to prevent solidification of ai and KC will be maximum when the solution is saturated.

Liquid adjuvants are always solvents but some adjuvants are accelerators as well and they increase mobility of ai in the cuticle. Penetration of accelerators is a prerequisite for their activity and this has consequences. On the surface of the CM accelerators invariably serve as solvents and if they penetrate faster than the ai a solid residue might form on the cuticle and penetration will cease. Penetration of accelerators causes a transient concentration in the CM[64] because a finite dose is deposited. Initially the concentration in the limiting skin increases, it reaches a maximum after some time and finally it decreases when the accelerator disappears in the tissue of the leaf. It follows from this scenario that rates of penetration of ai and accelerator must be well matched, when maximum rates of penetration of ai are aimed at. Mobilities of adjuvants and accelerators have not yet been studied in detail for lack of suitable radiolabelled accelerators. Work is in progress and once accelerator penetration has been modelled based on eq. (4) suitable accelerators for a particular ai can be predicted. Effects of adjuvants and humidity on driving forces of penetration are treated in more detail elsewhere.[68]

5 REFERENCES

1. P. J. McCall, *Aspects Appl. Biology*, 1988, **21**, 185.
2. G. G. Briggs, R. H. Bromilow, A. A. Evans, and M. Williams, *Pestic. Sci.*, 1983, **14**, 492.
3. D. A. Kleier, *Plant Physiol.* 1988, **86**, 803.
4. F. C. Hsu and D. A. Kleier, *J. Exp. Bot.*, 1996, **47**, 1265.
5. F. Kerler and J. Schönherr, *Arch. Environ. Contam. Toxicol.* 1988, **17**, 7.
6. J. Schönherr and M. Riederer, *Rev. En viron. Contam. Toxicol.* 1989, **108**, 1.
7. D. Stock, P. H. Holloway, B. T. Grayson and P. Whitehouse, *Pestic. Sci.*, 1993, **37**, 233.
8. D. F. Cutler, K. L.Alvin and C.E. Price (eds.), 'The Plant Cuticle' Academic Press, London, 1982.
9. A.Chamel, *Physiol. Veg.*, 1986, **24**, 491.
10. P. J. Holloway, R. T. Rees and D. Stock (eds.), 'Interactions Between Adjuvants, Agrochemicals and Target Organisms', Springer Verlag, Berlin, 1994.
11. G. Kerstiens (ed), 'Plant Cuticles - An Integrated Functional Approach', BIOS Scientific Publ. Oxford, 1996.

12. W. R. Lieb and W. D. Stein, In: 'Current topics in membrane transport' In: F. Bonner and A. Kleinzeller (eds.), Academic Press, New York, 1972, p. 1
13. B. T. Grayson and D. A. Kleier, *Pestic. Sci.,* 1990, **30**, 67.
14. G. G. Briggs, R. L. O. Rigitano, R. H. Bromilow, *Pestic. Sci.* 1987, **19**, 101.
15. K. Chamberlain, G. G. Briggs, R. H. Bromilow, A. A. Evans and Q.F. Chen, *Aspects Appl. Biol.* 1987, **14**, 293.
16. E.L. Cussler, 'Diffusion, Mass Transfer in Fluid Systems', Cambridge University Press, Cambridge, 1984.
17. R. O. Potts and R. H. Guy, Predicting skin permeability, *Pharmaceutical Res.*, 1992. **9**, 663.
18. F. Kerler and J. Schönherr, *Arch. Environ. Contam. Toxicol.*, 1988 **17**, 1.
19. J. Schönherr and P. Baur, *Pestic. Sci.*, 1994, **42**, 185.
20. L. Schreiber and J. Schönherr, *Pestic. Sci.,* 1993, **38**, 353.
21. M. Burghardt, L. Schreiber and M. Riederer, *J. Agric. Food Chem.,* 1998, **46**, 1593.
22. J. Schönherr and M. Riederer, *Arch. Environ. Contam. Toxicol.,* 1988, **17**, 13.
23. J. Schönherr and P. Baur, In: G. Kerstiens (ed), 'Plant Cuticles - An Integrated Functional Approach', BIOS Scientific Publ. Oxford, 1996, p. 134.
24. G. G. Briggs and R. H. Bromilow, In: 'Interactions Between Adjuvants, Agrochemicals and Target Organisms', P. J. Holloway, R. T. Rees and D. Stock (eds.), Springer Verlag, Berlin, 1994, p. 1.
25. P. Baur, H. Marzouk, J. Schönherr and B. T. Grayson, *J. Agric. Food Chem.,* 1997, **45**, 3659.
26. H. Marzouk, P. Baur and J. Schönherr, *Pestic. Sci.,* 1998, in press.
27. J. Schönherr and M. J. Bukovac, *Plant Physiol.,* 1972, **49**, 813.
28. M. Knoche, *Weed Res.,* 1994, **34**, 221.
29. J. Schönherr and M. J. Bukovac, *Planta,* 1970, **92**, 202.
30. J. Schönherr and M. J. Bukovac, *Physiol. Plant.,* 1978, **42**, 243.
31. M. Riederer, L. Schreiber, In: 'Waxes: Chemistry, Molecular Biology and Functions'. R. J. Hamilton (ed.), Oily Press, West Ferry, Scotland, 1995 p. 130.
32. C.E. Jeffree, In: 'Plant Cuticles - An Integrated Functional Approach', G. Kerstiens (ed.), BIOS Scientific Publ. Oxford, 1996, Chapter 2, p. 33.
33. R. F. Falk, In: 'Interactions Between Adjuvants, Agrochemicals and Target Organisms', P. J. Holloway, R. T. Rees and D. Stock (eds.), Springer Verlag, Berlin, 1994, Chapter 3, p. 53.
34. J. Schönherr and H. Bauer, In: 'Adjuvants and Agrichemicals' C. L. Foy, (ed.), CRC Press, Boca Raton, Fla. USA, 1992, Vol. 2, p.17.
35. H. Bauer and J. Schönherr, *Pestic. Sci.*, 1992, **35**, 1.
36. L. Schreiber, T. Kirsch and M. Riederer, *Planta*, 1996, **198**, 104.
37. L. Schreiber, M. Riederer and K. Schorn, *Pestic. Sci.*, 1996, **48**, 117.
38. A. Buchholz, P. Baur and J. Schönherr, *Planta*, 1998, in press.
39. M. Riederer and J. Schönherr, *Ecotox. Environ. Safety,* 1985, **9**, p.196.
40. P. Baur, H. Marzouk, J. Schönherr and H. Bauer, *Planta*, 1996, **199,** 404.
41. P. Baur, H. Marzouk and J. Schönherr, *Plant Cell Environ.* (submitted)
42. A. Buchholz, Doct. Diss., 1998, Universität Kiel, Germany.
43. P. Baur and J. Schönherr, *Chemosphere*, 1995, **30**, 1331.
44. M. Riederer and J. Schönherr, *Planta*, 1986, **169**, 69.
45. P. Baur, A. Buchholz and J. Schönherr, *Plant, Cell Environ.*, 1997, **20**, 982.

46. P. Baur, *Plant, Cell Environ.*, 1997, **20**, 167.
47. J. Crank and G.S. Park, 'Diffusion in Polymers', Academic Press, London, 1968.
48. S. Glasstone, K. Laidler and H. Eyring, 'The Theory of Rate Processes', McGraw Hill, New York, 1941
49. W. R. Vieth, "Diffusion in and Through Polymers", Carl Hanser Verlag, München, 1991.
50. E. A. Baker, In: 'The Plant Cuticle', D. F. Cutler, K. L.Alvin and C.E. Price, (eds.), Academic Press, London, 1982, p.139.
51. V. Hauke and L. Schreiber, Seasonal development of wax composition and cuticular transpiration of ivy (*Hedera helix* L.) sun and shade leaves. *Planta,*(in press)
52. J. Schönherr, *Planta,* 1976, **131**, 159.
53. D. Stock, B. M. Edgerton, R. E. Gaskin, P. J. and Holloway, *Pestic. Sci.*, 1992, **34,** 233.
54. M. Knoche and M. J. Bukovac, *J. Am. Soc. Hort. Sci.*, 1992, *117*, 80.
55. R. C. Kirkwood, *Pestic. Sci.*, 1993, **38**, 93.
56. D. Stock and P. J. Holloway, *Pestic. Sci.,* 1993, **38**, 165.
57. J. Schönherr, *Pestic. Sci.*, 1993, **38**, 155.
58. J. Schönherr, *Pestic. Sci.*, 1993, **39**, 213.
59. L. Schreiber, M. Riederer and K. Schorn, *Pestic. Sci.*, 1996, **48**, 117.
60. P. Baur, B. T. Grayson and J. Schönherr, *Pestic. Sci.*, 1997, **51**, 131.
61. P. Baur, Doct. Diss., Technische Universität München, 1993.
62. P. Baur, B. T. Grayson and J. Schönherr, *Pestic. Sci.*, 1996, **47,** 171.
63. M. Riederer, M. Burkhardt, S. Meyer, H. Obermeier and J. Schönherr, *J. Agric. Food. Chem.*, 1995, **43**, 1067.
64. P. Baur, J. Schönherr, *Z. Pfl. Krankh.*,1997, **104**, 380.
65. J. Schönherr and P. Baur, *Z. Pfl. Krankh.*, 1997, **104**, 246.
66. P. Baur and J. Schönherr, *Z. Pfl. Krankh.*,**105**, 84.
67. C. Gauvrit, In: 'Interactions Between Adjuvants, Agrochemicals and Target Organisms', P. J. Holloway, R. T. Rees and D. Stock (eds.), Springer Verlag, Berlin, 1994, Chapter 9, p. 171.
68. P. Baur, In: 'Recent Research Developments in Agricultural and Food Chemistry' (in press) 1999.

Natural Products

Chemistry and Insecticidal Activity of the Spinosyns

Gary D. Crouse[1], T.C. Sparks[1], C.V. DeAmicis[1], H.A. Kirst[2], J.G. Martynow[1], L.C. Creemer[2], T.V. Worden[1] and P.B. Anzeveno[1]

[1] DOW AGROSCIENCES, LLC, 9330 ZIONSVILLE ROAD, INDIANAPOLIS, IN 46268, USA
[2] ELANCO ANIMAL HEALTH RESEARCH AND DEVELOPMENT, 2001 W. MAIN STREET, GREENFIELD, IN 46140-0708, USA

1 INTRODUCTION

Twenty years ago, one of the main topics of the Fourth International IUPAC Congress of Pesticide Chemistry was natural products with insect regulatory and insecticidal activity.[1] In retrospect, the Congress described a science in its infancy. Although pyrethroids had begun to have a significant impact on the agricultural industry, the remainder of what were considered "biological" insect control materials were clearly not competitive with synthetic compounds. In the late 1970s, the focus of natural products in agricultural research was more on their behavioral or growth regulatory aspects, due largely to the emphasis on higher plants as sources of active secondary metabolites.

The 1978 proceedings provide an excellent contrast to today's view of natural product research as an important complement to the traditional synthetic approach. Advances in screening techniques, more rapid isolation and structural elucidation, and an expansion of the kinds of organisms screened, have resulted in the discovery of a great number of new biologically active molecules. Many comprehensive reviews and monographs related to the recent identification and characterization of natural products as pest control agents have appeared recently.[2-8] In particular, one monograph consists of a detailed description of insecticidal natural products; owing to the sheer number of active materials, the authors were unable to include growth regulators, repellents, or antifeedants.[9]

In 1978, nearly every major commercial pest control agent was derived from conventional synthesis programs. Even today, few natural products exist that can be considered as having major commercial agricultural importance. However, of the new pest control agents being developed today, perhaps even the majority can trace their origin, either directly or indirectly, to a natural product.

The most successful example of a synthetic class of pest control agents having been derived from a natural source is clearly the pyrethrins (Figure 1). Although the value of synthetic pyrethroids in modern pest control is undisputed, their potential utility was not always obvious. Structural complexity, low intrinsic activity and high photolability led some to doubt their commercial potential.[10] Through a combination of efforts, including trial and error, molecular modelling, and attribute-directed synthesis, over 200 patents and more than 30 commercial products related to pyrethrin I have been generated. Improved characteristics include photostability, increased intrinsic activity, decreased volatility, and improved selectivity toward beneficial insects. The most recent generation of products, exemplified by MTI-800 (1), bear little resemblance to the original pyrethrin I (2).

Many efforts to duplicate the pyrethroid success story with other classes of natural products have been brought to light recently. Researchers at Cyanamid were able to develop a new class of insecticides represented by the 2-arylpyrrole chlorfenapyr (3). The fungal metabolite dioxapyrrolomycin (4), itself too toxic and photo-reactive for commercial consideration, was the starting point for a structure modification program leading to (3).[11] Several companies have successfully developed synthetic analogs based on the terpenoid structure juvenile hormone I (6). The synthetic analogs such as (5) are more photostable and affect a broader range of insect pests, resulting in commercial levels of control against lepidopteran and other important insect classes.

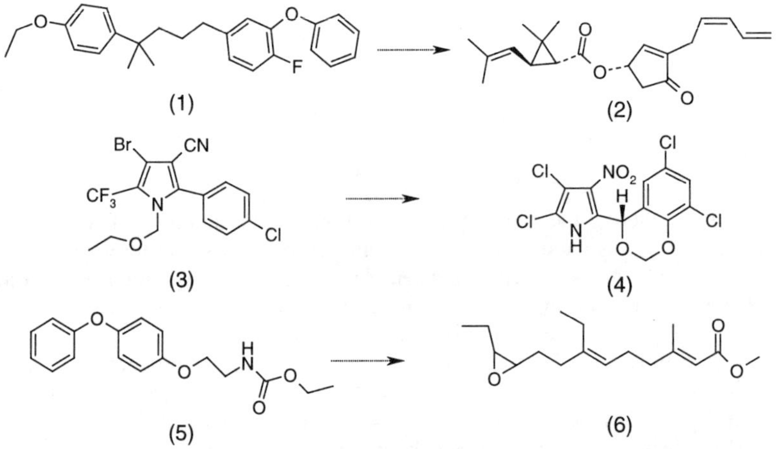

Figure 1

Examples of recent commercial products and the natural products from which they were derived

In many cases, the correlation between an active entity and a natural product to which it is related (mechanistically or structurally) is only discovered in retrospect. The dibenzoylhydrazides (*e.g.* tebufenozide, 7), first prepared as undesired byproducts in an unrelated area of research, were found to mimic the action of 20-hydroxyecdysone (8, Figure 2), resulting in incomplete molting and eventual death of many kinds of insect larvae. Subsequent modelling has determined how these structurally unrelated molecules overlap;[12] even so, this exercise offers few clues relative to how this knowledge might have been useful in selecting a diacylhydrazide as a steroid mimic, or how it might direct unrelated natural product modification efforts. Also, imidacloprid (9) is clearly related to nicotine (10), both functionally and structurally, but its discovery derived from a study of nitromethylene heterocycles (11), which themselves proved too environmentally unstable for commercial use.[13] However, it is an excellent lesson in the ability of synthetic chemists to generate modified structures that offer significant advantages over the natural material: while nicotine is still used for insect control, its volatility, disagreeable odor and high mammalian toxicity limit broader usage. The reduced basicity and volatility of imidacloprid relative to nicotine results in greater mobility within the plant, longer residuality, and less mammalian toxicity.

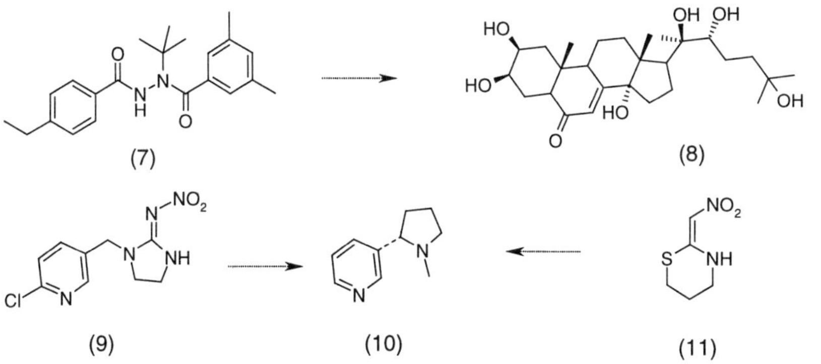

Figure 2

Examples of some commercial insecticides that function at the same site as a natural product, however which were not found through directed synthetic modification of the natural product

The potential of synthetic modification to improve upon a natural product is clearly demonstrated with avermectin analogs. In addition to its utility in human and veterinary health, abamectin (12) is currently marketed as an agricultural miticide. The 4''-epi-methylamino derivative (emamectin, 13), one of over 2000 synthetic modifications to avermectins and milbemycins prepared during the past 20 years, was found to exhibit an extraordinarily different spectrum of activity relative to its parent structure. While less active against mites by a factor of 10, it is over 1000 times more effective against southern armyworm (Table 1).[14]

Table 1 *Activity comparison of avermectins against mites and insects*

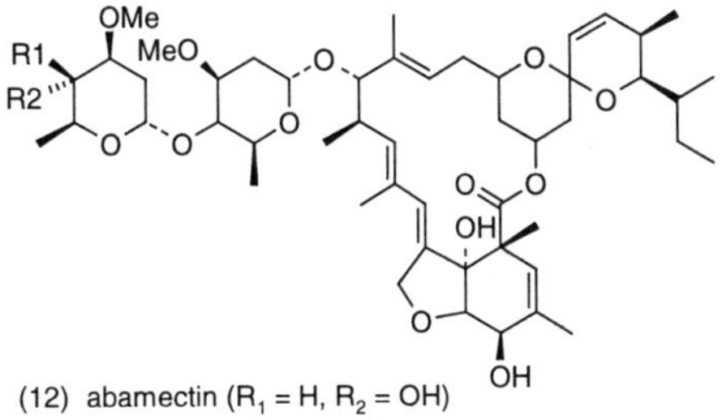

(12) abamectin (R_1 = H, R_2 = OH)
(13) emamectin, benzoate salt (R_1 = $NHCH_3$, R_2 = H)

	TSSM[a] LC_{90} (ppm)	SAW[b] LC_{90} (ppm)
(12)	0.03	8.00
(13)	0.25	0.004

[a] Two-spotted spider mite; from Ref. 14.
[b] southern armyworm

Similar modification efforts directed at herbicidal and fungicidal natural products have also led to important new fungicides and herbicides, making natural products one of the most important current sources both of new agricultural products and of leads for new semi-synthetic or entirely synthetic products.[7] Structural modification efforts are a routine component in the evaluation of new compounds that demonstrate pesticidal properties. Although each molecule requires a unique approach to the development of structure–activity information, the process usually proceeds by way of a series of logical progressions. With complex natural products, initial structure modifications are often based on what *can* be done. These progress to more quantitative SAR or attribute-based synthetic efforts as the particular chemistry of a material is better understood. Eventually, when the mode or mechanism of action and spatial and electronic factors are better understood, such advanced approaches as *de novo* design can be envisioned.

The spinosyns are a very recent example of a natural product with potential utility in agriculture (*vide infra*). Extensive field evaluations of the spinosyns have shown that this natural material has adequate efficacy, stability and environmental suitability to enable its use without chemical modification. Nevertheless, it was important to learn as much as possible about the chemistry of this complex new family of natural products. The remainder of this chapter will describe a variety of chemical transformations of spinosyns, and the effect of these structural modifications on insecticidal activity.

2 THE SPINOSYNS – ISOLATION AND DEVELOPMENT

The spinosyns (Figure 3) were discovered through screening of a culture broth of a soil organism obtained a Caribbean island in the early 1980s. The initial insecticidal activity was identified using a mosquito larvicide assay. The producing organism is a novel actinomycete named *Saccharospora spinosa*.[15] Subsequent isolation and structural elucidation work led to the identification of a new family of macrocyclic lactones called the spinosyns. The major factors are spinosyn A and spinosyn D, which are present in approximately 85:15 ratio and which together comprise over 90% of the active material (Figure 3).[16] Currently, over 24 natural spinosyn factors have been isolated and characterized. They differ predominantly in the methylation patterns of the amine and hydroxyl groups, and to a lesser extent in the C-methylation of the nucleus. These natural factors have been evaluated and their relative activities reported.[17]

When tested against a wide range of insect pests, the spinosyns were found to be extremely effective at controlling primarily members of the lepidopteran family of insects, which are major agronomic pests in such crops as cotton and vegetables. Just as importantly, most other characteristics necessary for effective pest control, such as environmental compatibility, speed of action, and low mammalian toxicity, were also close to ideal. Commercial launch of the first pest control products based on spinosyns occurred in 1997 and early 1998.

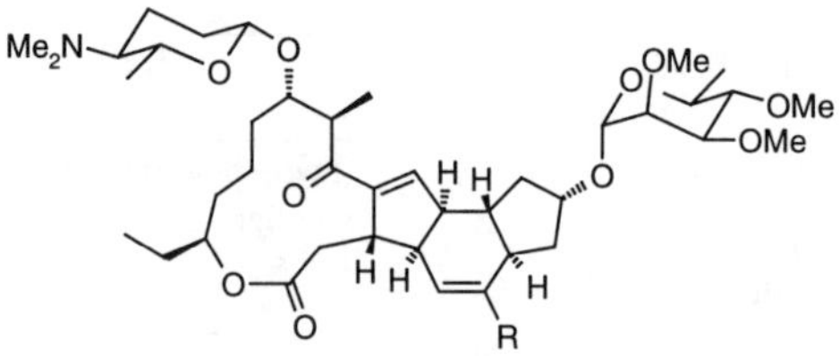

Figure 3

The two major components in spinosad are spinosyn A (14, R = H) and spinosyn D (15, R = CH_3)

The insecticidal spectrum exhibited by the spinosyns is quite broad; control of mites, aphids, lepidoptera, diptera, cockroaches, termites, and many other species has been observed. However, the greatest activity is against lepidopteran pests such as the tobacco budworm. The spinosyns appear to exert their effect on insects through alteration of nicotinic acetylcholine receptor function, although the mechanism is clearly different from other nicotinic agents such as imidacloprid. Some effects on the GABA-gated chloride channel system are also observed; however, their contribution to symptoms has not been established.[18]

3 CHEMICAL MODIFICATIONS OF SPINOSYNS

The spinosyns contain a number of reactive functional groups, and one of the first objectives was to learn whether these groups could be selectively modified. The general structure (Figure 4) shows the degree of complexity of the molecule, and also suggests opportunities where potential modification could occur. This review is intended as a general survey of the chemistry of this new class of macrolides, and the effect of particular modifications on insecticidal activity. In most cases, although additional examples have been made, only one specific example of any particular structural modification will be described.

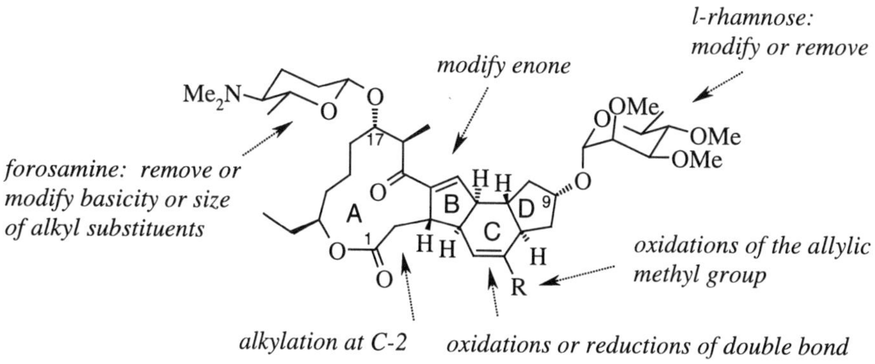

(14, R = H)
(15, R = CH_3)

Figure 4
Potential sites of modification of spinosyns. The letters indicate the four rings comprising the tetracyclic nucleus of spinosyns

3.1 Amine Modifications

N-dealkylation of spinosyn A (14) to spinosyn B (16) was accomplished using standard reagents (I_2/NaOAc, O_2/Pd_0, Scheme 1). Further dealkylation to spinosyn C (17) was considerably more difficult, requiring a stronger base (NaOMe). Yields were lower due to competing deamination to form the corresponding ketone (18). Acylation of the secondary amine (16) proceeded as expected to give, for example, the corresponding amide (19) or urea (20). Alternatively, realkylation with other alkyl halides generated mixed alkyl derivatives such as (21).

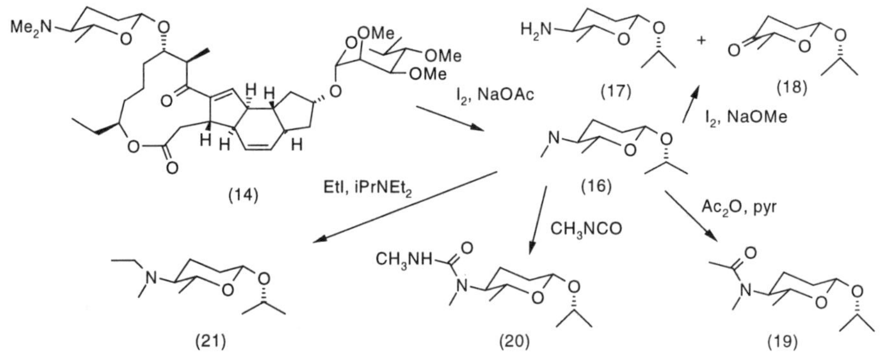

Scheme 1 *Amine modifications of spinosyns*

3.2 5,6-Double Bond Reactions

The lone non-conjugated double bond was easily hydrogenated (Equation 1) to the 5,6-dihydro spinosyn A derivative (22). Reduction of the double bond in spinosyn D was much more difficult, leading in most cases to over-reduction to the 5,6,13,14-tetrahydro derivative. However, use of trifluoroacetic acid as a co-solvent minimized the overreduction. A single isomer was formed; X-ray analysis confirmed an initial NMR assignment that the methyl group stereochemistry was as shown (23, Scheme 2).

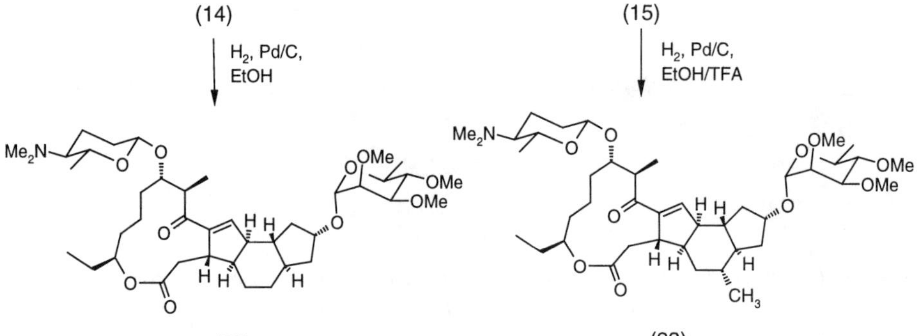

Scheme 2 *Hydrogenation of the non-conjugated double bond of spinosyn A and D*

Oxidation of the 6-methyl group on spinosyn D was considered an important objective; the position was considerably removed from the sugars, and could provide a useful handle for further modifications. Reaction of spinosyn D with selenium dioxide under standard conditions was quite slow and non-selective. Use of an acetic acid/dioxane solvent system, however, greatly facilitated the oxidation (Scheme 3). Under these conditions, rapid conversion to the 5-hydroxy derivative (24), along with two unsaturated derivatives (25 and 26), occurred at 0 °C. In addition to accelerating the rate of allylic oxidation, as reported by Shibuya,[19] the use of acetic acid also diminished the reactivity of the amine to this oxidant. Extended reaction with SeO_2/AcOH led to a low yield of a derivative in which the C-ring was aromatized (27). The indene-like structure of (27) also undergoes partial equilibration of the 13,14-double bond to the more substituted 3,14-position.

Despite other attempts using a variety of oxidation conditions, no products arising from allylic oxidation of the C-6 methyl group have yet been found. Selenium dioxide oxidation of spinosyn A, under the same conditions described above, was considerably slower, and led to highly complex mixtures.

Epoxidations of this double bond were complicated by the presence of the amine group, which is the region of highest electron density. However, instead of avoiding N-oxide formation, it was simpler to use an excess of oxidant, and to subsequently regenerate the free amine by stirring the product with a bisulfite solution. Thus, treatment of (15) with MCPBA in dichloromethane, followed by washing with an aqueous bisulfite solution, furnished predominantly the β-epoxide (28).

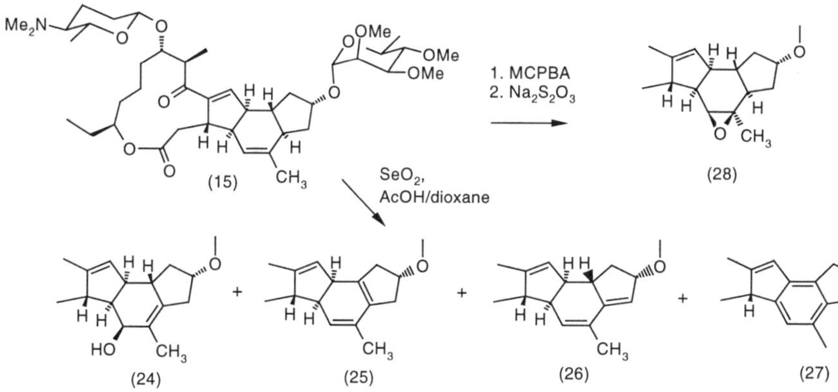

Scheme 3 *Oxidation of the 5,6-double bond of spinosyn D*

3.3 13,14-Double bond Reactions

Spinosyns degrade rapidly in sunlight, and one of the early areas of interest was to identify analogs that had greater photostability. As the enone is a likely site of photodegradation, considerable attention was spent investigating the chemistry of this portion of the molecule. A report detailing the chemistry on the aglycone of spinosyn A was published recently,[20] detailing some of the unusual reactivity of this 12-5-6-5 ring system. This chemistry has now been extended to the intact molecule, with similar results. Michael additions occur exclusively from the α face, due to steric crowding of the β face. The ketone carbonyl group is also hindered, leading to almost exclusive nucleophilic additions at C-13. Most nucleophilic reducing agents result in selective 1,4-reduction (Scheme 4). Choice of reaction conditions determines the stereochemistry of reprotonation at C-14. Under protic conditions, such as sodium borohydride in an alcoholic solvent, the 14-β isomer (29) predominates. Aprotic hydride reductions such as lithium tri-tert-butoxy aluminum hydride, yield predominantly the α isomer (30).

Treatment of spinosyn A with hydrogen peroxide under basic conditions furnished the corresponding epoxide (31). Assignment of stereochemistry was again based on NMR data, the structure has been assigned as the α-epoxide. Addition of hydroxylamine, rather than producing the expected 15-oxime, resulted instead in the 13-hydroxylamino derivative (32). Normal conditions for generating hydrazones at C-15 (dinitrophenyl-hydrazine in ethanol or acetic acid, for example) were unsuccessful; under more forcing conditions, loss of forosamine resulted.

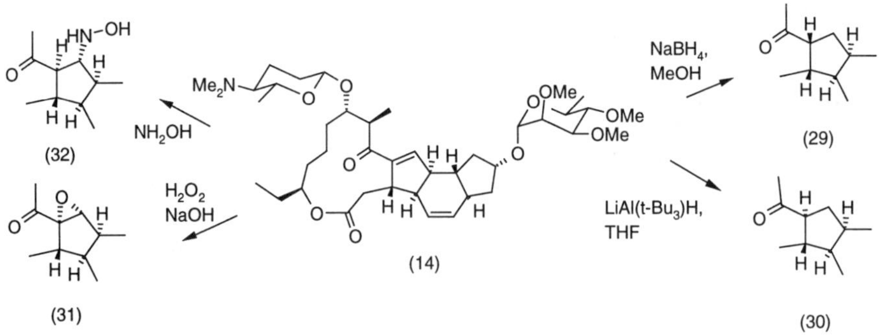

Scheme 4 *Reactions involving addition to the enone of spinosyn A*

3.4 Other Chemistry involving the Macrolide Ring

Treatment of spinosyn A with LDA in THF/HMPA solution at –78 °C led to kinetic deprotonation at C-2 (Scheme 5). Addition of electrophiles such as methyl iodide led cleanly to the 2-methyl product (33). This is further indication of the degree of steric complexity around the C-16 proton. The stereochemistry of addition is uncertain; while models suggest that approach from the β face (as drawn) is more sterically accessible, the dihedral angle for C2α-C3 and C2β-C3 are very similar, thus complicating NMR-based structural assignment of the incoming group.

Conditions for selectively opening the lactone were also not found. This reflects not only the competing lability of the forosamine, but the high level of steric hindrance around both carbonyl groups.

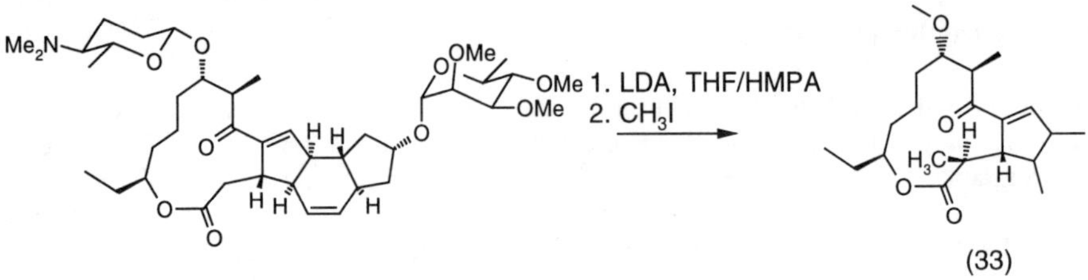

Scheme 5 *Deprotonation and alkylation of spinosyn A occurs at C-2*

3.5 Rhamnose Modifications

Because the parent (wild-type) strain of *S. spinosa* produces spinosad in only very minute quantities, an extensive strain selection program was conducted in order to increase the yield. An offshoot of this strain selection process was the identification of mutant strains that produced other spinosyns. Several strains were identified that possessed non-functional 2'- or 3'-, or 4'-*O*-methyltransferases. These strains provided the three mono-desmethyl spinosyns H, J, and K, respectively. Fermentation of these mutant strains successfully generated multi-gram quantities of these factors. These were critical starting materials for the further selective elaboration of the rhamnose moiety. Thus, alkylation under phase-transfer conditions led in high yield to the corresponding alkylated derivatives (35, 37, and 39, Scheme 6).

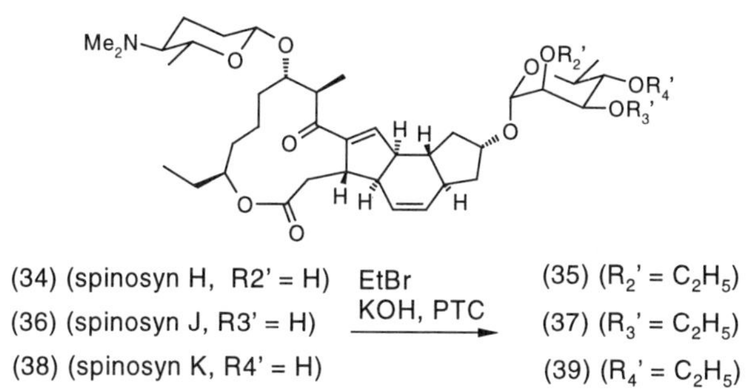

Scheme 6 *Alkylation of spinosyns H, J, and K*

3.5 Removal of Forosamine and/or Rhamnose

In order to study the effects of different sugars or other groups in place of the rhamnose and forosamine, conditions for selective removal and reattachment were necessary. Selective removal of the forosamine sugar, using mild acidic hydrolysis, was accomplished easily under mild acidic conditions (Scheme 7). More vigorous acidic hydrolysis was needed to accomplish removal of the rhamnose sugar to form the aglycone (41).[21] Conditions for selective hydrolytic removal of the rhamnose sugar from spinosyn A have not been found. The 3'-*O*-desmethylrhamnosyl (spinosyn J) derivative (36), however, offered a convenient starting material for generation of the C-9-pseudoaglycone. Swern oxidation of (36) to the 3'-ketone (42) and subsequent treatment with base (K_2CO_3) resulted in β-elimination of the monosaccharide to form (43).

The monosaccharides (40 and 43) were important precursors for the generation and evaluation of novel sugar derivatives of spinosyn A. Conditions for reattachment of forosamine have been described as part of Evans' total synthesis.[22] Alternatively, treatment of (43) with the 1-trichloroacetimidate of tri-*O*-ethylrhamnose generated the corresponding tri-*O*-ethylrhamnosyl analog (44).

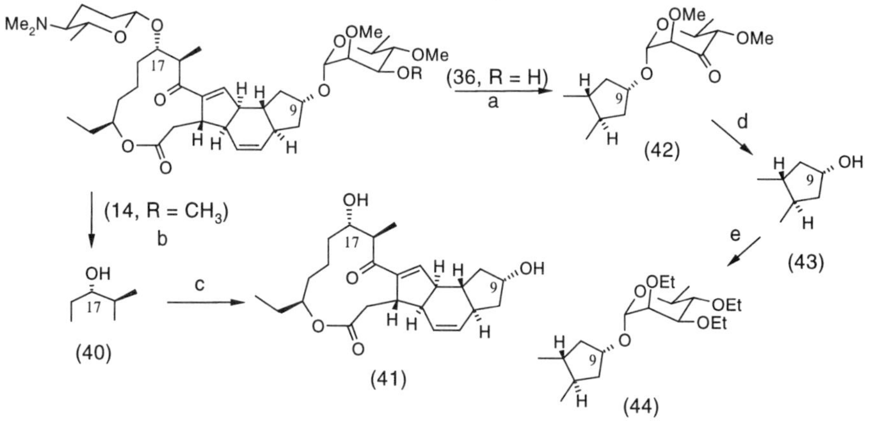

Scheme 7 *Removal and reattachment of forosamine and rhamnose from spinosyn A. Reagents: (a) NCS, Et_2S, Et_3N, CH_2Cl_2; (b) H_2SO_4, 30 °C; (c) H_2SO_4, 100 °C; (d) K_2CO_3, MeOH; (e) 2,3,4-tri-O-ethyl-1-trichloroacetimidoyl-L-rhamnose, PPTS, CH_2Cl_2*

4 INSECTICIDAL ACTIVITY

The number of synthetic and natural spinosyn analogs that have been characterized now numbers over 400. Standard evaluation of relative insecticidal activity of spinosyn analogs includes the tobacco budworm (TBW) neonate drench assay. The activities of the derivatives discussed above in the TBW neonate drench assay are shown in Table 2. As is frequently observed with natural product modification efforts, the majority of the derivatives or analogs show less insecticidal activity than the two major natural spinosyns, A and D. The two sugars each play a crucial role in insecticidal activity; removal of one or both results in nearly complete loss of activity. Steric changes about the amine group had little effect on insecticidal activity. Modifications that disturbed the amine basicity, on the other hand, were significantly more deleterious.

Modifications to the conjugated enone system were also detrimental to insecticidal activity; reduction or oxidation resulted in up to a 100-fold diminution in activity. Reduction of the 5,6-double bond, or incorporation of an additional double bond, had relatively little effect on activity. More significant changes, such as the 5-hydroxy (24) or indenyl (27) analogs, resulted in complete loss of activity.

Table 2 *Comparison of Insecticidal Activity of Selected Natural and Synthetic Spinosyns*

#	substitution or modification	parent factor[a]	TBW LC_{50}, ppm	#	substitution or modification	parent factor	TBW LC_{50}, ppm
14	**spinosyn A**	**A**	**0.3**	29	13,14-dihydro (14-β)	A	20
15	**spinosyn D**	**D**	**0.5**	30	13,14-dihydro (14-α)	A	4.7
16	**4''-NHCH$_3$**	**B**	**0.4**	31	13,14-epoxy	A	1.4
17	**4''-NH$_2$**	**C**	**0.8**	32	13-NHOH	A	5.6
18	4''-ketone	A	3.4	33	2-methyl	A	4.5
19	N-acetyl	B	5.7	**34**	**2'-OH**	**H**	**3.2**
20	N-CONHCH$_3$	B	46	35	2'-OEt	H	0.11
21	N-ethyl	B	0.4	**36**	**3'-OH**	**J**	**>64**
22	5-6-dihydro	A	0.5	37	3'-OC$_2$H$_5$	J	0.03
23	5,6-Dihydro	D	0.31	**38**	**4'-OH**	**K**	**0.9**
24	5-OH	D	>64	39	4'-OEt	K	0.34
25	7,11-dehydro	D	0.2	40	17-pseudoaglycone	A	>64
26	7,8-dehydro	D	0.6	41	aglycone	A	>64
27	indenyl	D	>64	42	3'- =O	J	7.6
28	5,6-β-epoxy	D	10	43	9-pseudoaglycone	A	>64
				44	2',3',4'-tri-*O*-Et	A	0.02

[a]Bold items indicates natural factors.

During the earliest phases of the synthetic program surrounding the spinosyns, most of the derivatives prepared were much less active than the parent spinosyns. Numerous attempts were made to gain insight into the reasons for this. Investigation of the quantitative structure–activity relationships (QSAR) were carried out using conventional multiple regression analysis of whole molecule properties and/or substituents, and comparative molecular field analysis (CoMFA). Although these early particular QSAR efforts were unsuccessful, a somewhat unconventional approach to QSAR, in the form of artificial neural networks (ANN),[23] proved to be far more successful. ANN analysis applied to spinosyn/spinosoid QSAR identified a number of interesting synthetic directions. An example of one such synthetic modification predicted by ANN-based

QSAR involved extending the alkoxy groups of the rhamnose moiety from *O*-methyl to *O*-ethyl. The resulting tri-*O*-ethylrhamnose analog (44) was significantly more active than spinosyn A (Table 2). Further synthetic investigation also revealed that the 3'-position, as predicted by ANN-based QSAR, had the greatest effect on biological activity. These studies clearly demonstrate that improvements in the biological activity of the spinosyns toward tobacco budworm are possible.

5 SUMMARY

The discovery of the spinosyns has created an invaluable tool for the control of agricultural insect pests, as well as for improving our understanding of the biochemistry of insect nervous systems, and for providing exciting challenges for synthesis chemists, molecular modellers and spectroscopists. Chemical modification of spinosyns has led to a better understanding of the relative importance of functional groups on activity; although most modifications result in diminished activity, the synthetically-derived alkylated rhamnose analogs are more efficacious than the natural factors. The molecule's uniqueness and complexity have challenged academic researchers as well; two total syntheses of spinosyn A have also been reported recently,[22,24] and other synthetic approaches have been reported as well.[25]

The great diversity of organisms in the world is still only beginning to be tapped for useful secondary metabolites. It is estimated that well over 90% of the world's fungi and bacteria may not have been taxonomically characterized, and many of the known organisms have not been adequately evaluated for potentially useful activity.[26] Industry and academia have made great strides in developing technology that will help to discover and exploit natural products for the control of pest in agriculture. The pesticide conference in 2018 will undoubtedly cover many more revolutionary advancements in the science of pest management.

References

1. K. Nakanishi, in: Advances in Pesticide Science - Part 2, H. Geissbuhler, G.T. Brooks and P.C. Kearney, eds., Pergamon Press, Oxford, 1978, p. 283.
2. J.P. Pachlatko, *Chimia*, 1998, **52**, 29.
3. M.J. Rice, M. Legg, K. A. Powell, *Pestic. Sci.*, 1998, **52**, 184.
4. G.A. Miana, Atta-Ur-Rahman, M. Iqbal Choudry, G. Jiliani, and H. Bibi, *Crit. Rep. Ecol. Chem.*, 1996, **35**, 241.
5. Z.-Z. Shang, *Crit. Rep. Ecol. Chem.,* 1996, **35**, 230.
6. G. Blunden, *Pestic. Sci.*, 1997, **51**, 483.
7. L.G. Copping, ed., 'Crop Protection Agents from Nature: Natural Products and Analogues', Royal Society of Chemistry, Cambridge, 1996.
8. C.R.A. Godfrey, ed., 'Agrochemicals from Natural Products', Marcel Dekker, New York, 1995.
9. S. Dev and O. Koul, 'Insecticides of Natural Origin', Harwood Academic Pub., Amsterdam, 1997.
10. K. Naumann, *Pestic. Sci.*, 1998, **52**, 3.
11. D.A. Hunt, *Pestic. Sci.* 1997, **50**, 201.

12. A.D. Mohammed-Ali, T.H. Chan, A.W. Thomas, G.M. Strunz, and B. Jewett, *Can. J. Chem.*, 1995, **73**, 550.
13. S. Kagabu and T. Akagi, *Nippon Noyaku Gakkaishi,* 1997, **22**, 84.
14. M.H. Fisher, in: 'Phytochemicals for Pest Control', P.A. Hedin, R.M. Hollingworth, E.P. Masler, J. Miyamoto, and D.G. Thompson, eds., ASC Symposium Series 658, American Chemical Society, Washington, D.C., 1995, p. 220.
15. M.M. Hoehn, K.H. Michel and R.C. Yao, *European Patent Appl. 398,588*, **1991.**
16. H.A. Kirst, K.H. Michel, J.W. Martin, L.C Creemer, E.H Chio, R.C Yao, W.M Nakatsukasa, L.D. Boeck, J.L Occolowitz, J.W Pashal, J.B. Deeter, N.D Jones and G.D. Thompson, *Tetrahedron Lett.,* 1991, **32**, 4839.
17. V.L. Salgado, G.B. Watson and J.J. Sheets, in: 'Proceedings of the 1997 Beltwide Cotton Production Conference', National Cotton Council, Memphis, TN, 1997, p. 1082.
18. C.V. DeAmicis, J.E. Dripps, C.J. Hatton, L.L. Karr, in: 'Phytochemicals for Pest Control', P.A. Hedin, R.M. Hollingworth, E.P. Masler, J. Miyamoto, and D.G. Thompson, eds., ACS Symposium Series 658, American Chemical Society, Washington, D.C., 1997, p.144.
19. T. Shibuya, *Syn. Comm.*, 1994, **24**, 2923.
20. J.G. Martynow and H.A. Kirst, *J. Org. Chem.*, 1994, **59**, 1548.
21. L.C. Creemer, H.A. Kirst, and J.W. Paschal, *J. Antibiotics*, 1998, submitted for publication.
22. D.A. Evans, W.C. Black, *J. Am. Chem. Soc.*, 1993, **115**, 4497.
23. J. Devillers, ed., 'Neural Networks in QSAR and Drug Design,' Academic Press, London, 1996.
24. L.A. Paquette, I. Collado, and M. Purdie, *J. Am. Chem. Soc.*, 1998, **120**, 2553.
25. W.R. Roush and A.B. Works, *Tetr. Lett.*, 1996, **37**, 8065.
26. N. Porter and M. Fox, *Pestic. Sci.*, 1993, **39**, 161.

Total Synthesis of Enzyme Inhibitors based on Carbohydrate Synthons

Bernd Giese[1] and A. O'Sullivan[2]

[1] DEPARTMENT OF CHEMISTRY, UNIVERSITY OF BASEL, CH-4056 BASEL, SWITZERLAND
[2] RESEARCH & DEVELOPMENT, NOVARTIS CROP PROTECTION AG, CH-4002 BASEL, SWITZERLAND

1 INTRODUCTION

After a natural product has been isolated, its pesticidal activity recognized, and its structure elucidated,[1,2] it becomes a potential focus of semisynthetic or total synthetic activity. The total synthesis of a pesticidally active natural product can be undertaken for a variety of reasons. Firstly to confirm the structure, if it were proposed solely on the basis of spectroscopy. Secondly to prepare the compound should it be difficult to obtain from its natural source. This can provide additional samples for initial testing or can produce large amounts for field trials, safety investigations or commercial use. Once a route to the natural compound is established, it is possible to modify it for the combinatorial preparation of analogs. Obviously a commercial synthesis is viable only for smaller molecules that can be prepared cheaply. Finally but most often motivating in academic institutes is the challenge involved in the planning and implementation of sophisticated synthetic strategies, and the hope that methodology acquired during such an enterprise would prove of lasting or general value.[3,4]

There are so few natural products on the market as agrochemicals, most of them of different biosynthetic origins, that generalizing remarks about their total synthesis would be specious.[1] Only selected insecticides, herbicides and fungicides which have shown a level of activity relevant enough to have generated industrial activity will be summarized if they relate to carbohydrate synthons. Avermectins and milbemycins have been irresistible targets for synthetic chemists. Carbohydrate chemistry has played a major role in many synthetic strategies described in more than 400 publications.[5–7]

Hydantocidin (**1**) is a pseudonucleoside showing a new mode of action with a total herbicidal activity about twice the level of glyphosate and matching the level of glufosinate against many weeds.[8] It provoked much industrial interest, although the cost of production has thus far prevented its commercialization.[9] Many syntheses have been described using a variety of strategies. Starting from an achiral allylic alcohol, Sharpless oxidation has been used to introduce chirality.[10] Tartrate and D-isoascorbic acid have been used from the chiral pool as starting materials,[10–12] and sugars have served as a basis for many syntheses: D-ribose four times,[13–16] and D-fructose three times.[17–19] Diacetonketogulonic acid (**2**) is obtained as an intermediate in the manufacture of vitamin C from D-glucose.[20]

Of the naturally occurring fungicides which are marketed, blasticidin has never been prepared, although a partial synthesis has been reported.[21] The synthesis of validamycin has been described once,[22] and similarly the synthesis of kasugamycin has been the subject of a single publication albeit in racemic form.[23] However polyoxin B (**3**) has captured the imagination of synthetic chemists throughout the world, and has been the subject of much synthetic work, together with the nikkomycins,[24] which have related structures and a similar mode of action. Initially the synthesis was aimed at the confirmation of the structure and the preparation of new analogs, but the novel polyhydroxylated aminoacid side chain and the unusual nucleotide core have stimulated its use as a model molecule to exemplify new specific methodology. This is evident from the various routes which have been described. Starting from racemic starting materials polyoxin has been prepared in racemic form or in optically active form through the use of chiral auxiliaries either in a stoichiometric or catalytic manner.[25–28] From the chiral pool a number of starting materials have been used. The following amino acids have served this purpose: serine,[29,30] hydroxyphenyl glycine,[31] and glutamic acid.[32] Sugars are clearly suitable starting materials. D-allose,[33] D-ribose,[34,35] D-glucose,[36] L-threitol,[37] D-mannitol,[38] L-arabinose[39] have all been utilized, and other chiral pool compounds such as tartate,[34,40–42] uridine,[43,44] and myoinositol,[45] have proved useful.

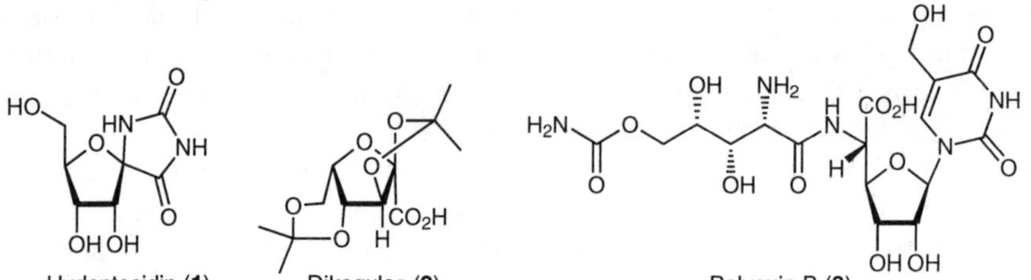

Hydantocidin (**1**) total herbicide | Dikegulac (**2**) plant growth regulator | Polyoxin B (**3**) fungicide

Figure 1 *Carbohydrate based inhibitor and regulators*

The cited syntheses describe the usefulness of carbohydrates from nature's chiral biodiversity for preparing agrochemical enzyme inhibitors. Two additional examples, the total syntheses of soraphen $A_{1\alpha}$ and trehazolin will be discussed in Sections 2 and 3.

2 TOTAL SYNTHESIS OF SORAPHEN $A_{1\alpha}$ FROM GLUCOSE AND MANNOSE

2.1 Retrosynthesis

In 1986 a new class of 18-membered macrolides called the soraphens was isolated by H. Reichenbach, G. Höfle *et al.* as metabolites of the microorganisms *Sorangium cellulosum.*[46] The parent compound, soraphen $A_{1\alpha}$ (**4**), contains ten stereogenic centers, with the methyl, methoxy, and hydroxy groups of the tetrahydropyran ring occupying axial positions.[47] Owing to this chemical structure as well as its antifungal activity, the novel natural product soraphen $A_{1\alpha}$ is an interesting target for synthesis.[48,49] It inhibits acetyl CoA carboxylase (ACC), the enzyme which catalyses the conversion of acetyl CoA into malonyl CoA in the lipid biosynthesis.[50] Inspection of the X-ray of soraphen (**4**) shows that the methyl group at C–2 is in the sterically favored position.[51] Therefore, as it appeared to be the thermodynamically more stable isomer, we decided to introduce the methyl group at C-2 as the last step of the synthesis. A retrosynthetic analysis of the macrocycle led to compounds (**5**) and (**6a**) which were synthesized starting from glucose and mannose, respectively.

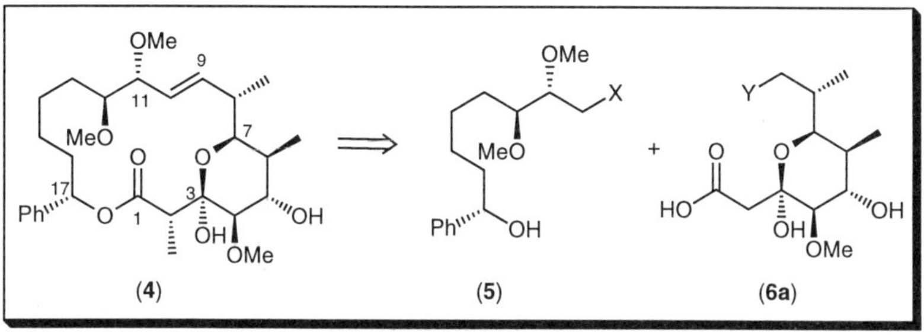

Figure 2 *Retrosynthesis of soraphen* $A_{1\alpha}$

Compound (**5**) was synthesized from the 2,3-dideoxysugar (**7a**) using an *umpolung* reaction with styrene epoxide (**8**).[14] The precursor of the dideoxysugar is glucose.

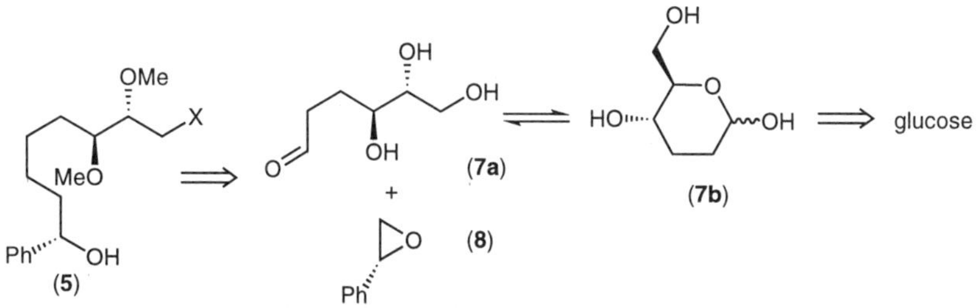

Figure 3 *Umpolung reaction with styrene epoxide*

The six-membered ring of (**6a**) has the framework of a L-ketose. Its derivative (**6b**), however, can tautomerize *via* the open-chain isomer (**6c**) into the D-aldose (**6d**), which can be obtained by alkylation at the aldehyde function of (**9**).

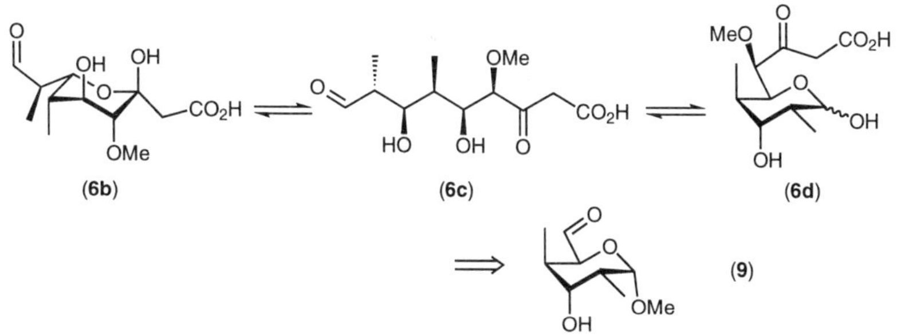

Figure 4 *Strategy to obtain aldehyde (6d)*

2.2 Synthesis of Building Block (9)

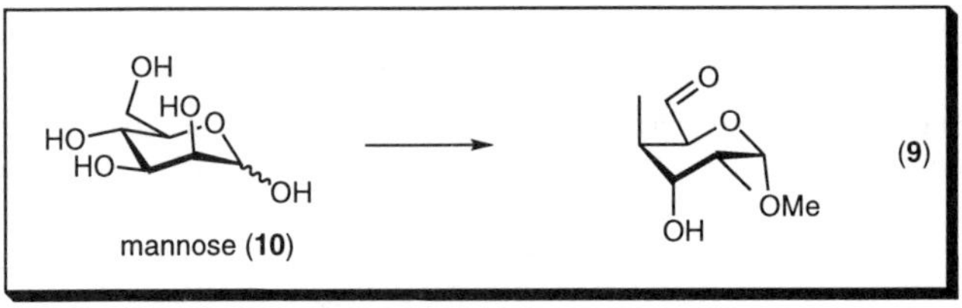

Figure 5 *Aldehyde (9) starting from mannose*

Hanessian *et al.* obtained aldehyde (**9**) from D-glucose.[52] We decided to synthesize (**9**) from ketone (**11**) which offers a suitable reactivity a) for the stereoselective introduction of a methyl group at C–2 with MeI, b) for the DIBAL reduction of the ketone function at C–3 into an axial OH group, c) for the introduction of the axial methyl group at C–4 by selective epoxide ring opening with MeMgX, and d) the oxidation at C–6 to an aldehyde function.

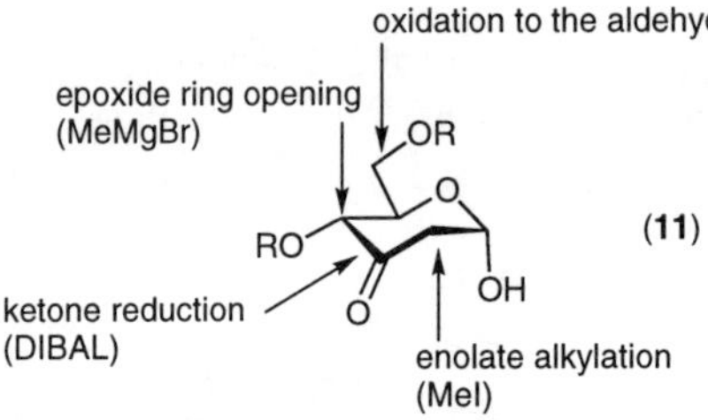

Figure 6 *Reactions at (11)*

The synthesis of (**11**) started with D-mannose (**10**) which, after protection, underwent *via* anion (**12**) a Klemer-Horton rearrangement to enolate (**13**).[53–56] Methylation and reduction led to (**14**) that was transformed into epoxide (**15**) using Viehe's salt.[57] Subsequent epoxide ring opening and oxidation at C–6 yielded aldehyde (**9**).

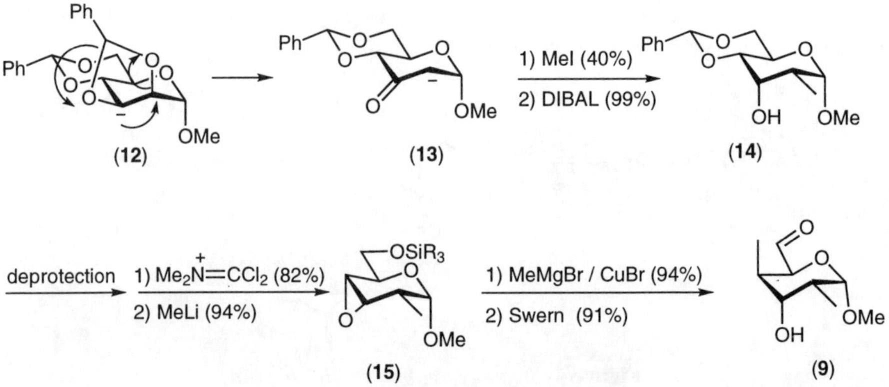

Scheme 1 *Synthesis of aldehyde (9)*

2.3 Selective C,C-Bond Formation at the Aldehyde Function of (9)

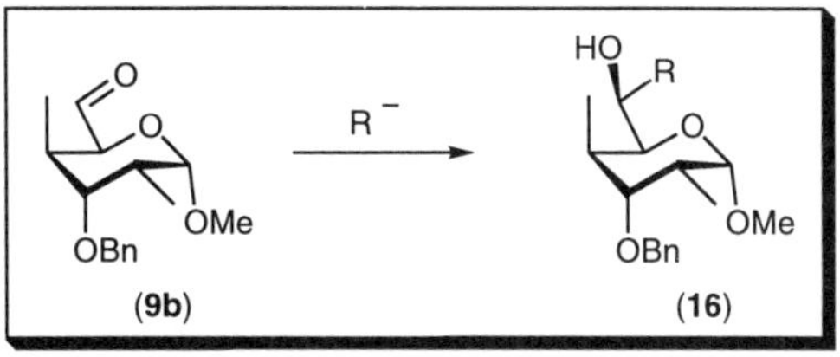

Figure 7 *Formation of (16)*

In order to arrive at the *R* configuration at C–4 of soraphen A, the *threo* configurated product (**16**) was required from the C,C-bond formation at the aldehyde function of (**9b**).

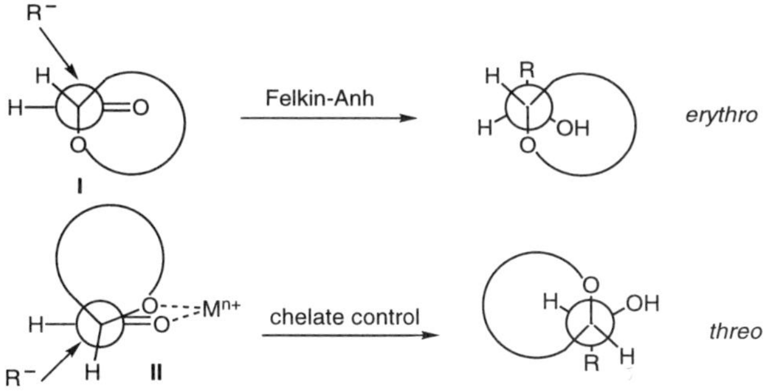

Figure 8 *Reactions of conformations I and II of (9)*

According to the Felkin-Anh rule[58,59] aldehyde (**9b**) would react *via* conformation **I** yielding the *erythro* configuration. However, under the conditions of chelate control,[60] conformation **II** of (**9b**) should be the reactive intermediate leading to the desired *threo* configurated product. In the event, the alkynylation of (**9b**) in the presence of $MgBr_2$ afforded stereoselectively (>25:1) compound (**17**). In the absence of Mg^{2+} salt the stereoselectivity vanished.

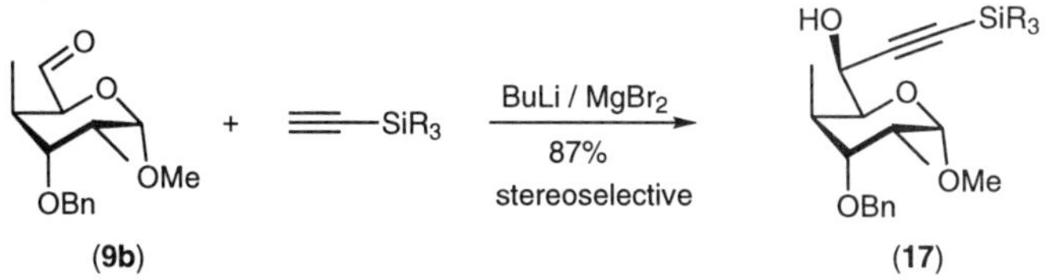

Scheme 2 *Alkylation of (9b)*

2.4 Wittig or Julia Olefination?

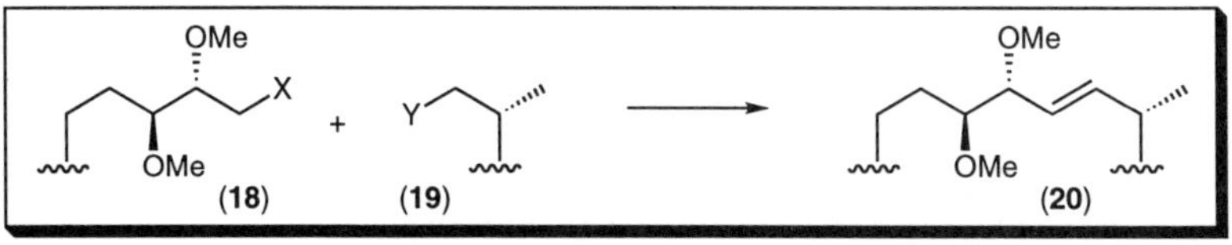

Figure 9 *Olefination strategy*

Ring opening and modification of the protecting groups of the sugar derivative (**17**) yielded aldehyde (**22**). Four combinations for the Julia or Wittig olefination reactions can be envisaged. As the Wittig and Julia reagents synthesized from (**5**) were very prone to elimination of methoxide, this building block (**5**) was used as aldehyde (**5a**) and its reaction partners (**21**) and (**23**) were prepared from (**22**).

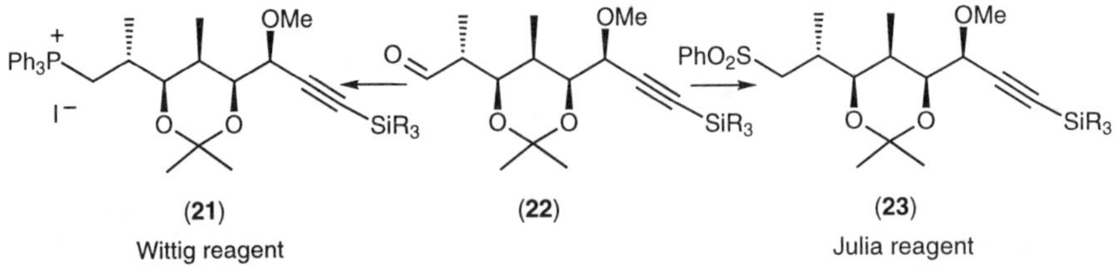

Scheme 3 *Wittig and Julia reagents*

It turned out that the aldehyde (**5a**) was also rather unstable under basic conditions due to the methoxy groups in β-position. Therefore, only the Julia olefination between (**5a**) and (**23**) afforded the C,C-coupled product (**24**) in acceptable yield.

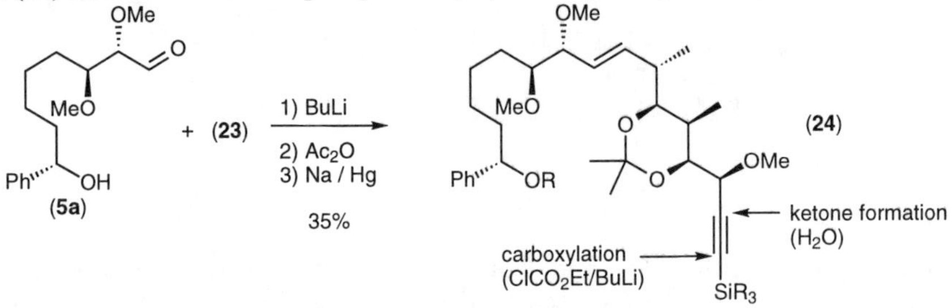

Scheme 4 *Preparation of (24)*

The ester group was introduced after desilylation and carboxylation with $ClCO_2Et$ and the acetylene was transformed into a ketone *via* enamine formation and subsequent hydrolysis. Manipulation of the protecting groups led to (**28**), the substrate for macrocyclization shown below.

2.5 Macrocyclization

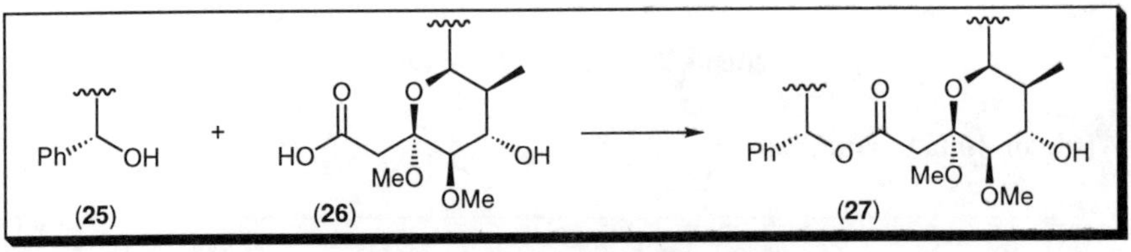

Figure 10 *Strategy of macrocyclization*

The formation of the macrocycle (**27**) turned out to be rather difficult. It was not possible to carry out this lactonization *via* the traditional acyl activation sequence.[61,62] Only the inverse reaction involving activation of the alcohol function yielded the product. Thus, the benzyl alcohol (**28**) was converted into the bromide with the Ghosez α-bromoenamine[63] and this bromide was cyclized in a substitution reaction with the Cs^+-carboxylate leading to the macrolactone (**29**) after deprotection. Both steps occurred with inversion of the configuration at the benzylic carbon atom, so that the overall stereochemistry was retention.

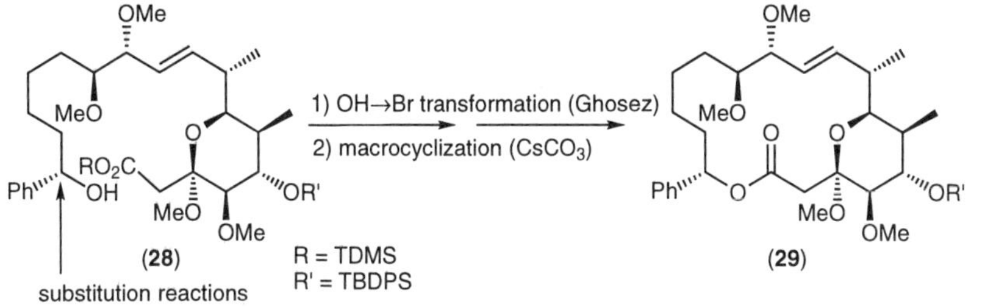

Scheme 5 *Macrocyclization to (29)*

2.6 Introduction of the Methyl Group

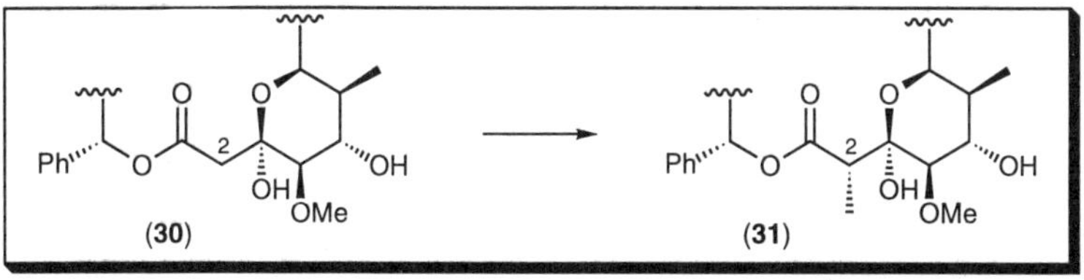

Figure 11 *Introduction of the C–2 methyl group*

The last step of the synthesis was the introduction of a methyl group at C–2 of the norsoraphen. The pK_a of the base was crucial for success. In the presence of strong bases, methylation occurred at both OH groups of the sugar residue. Only with bases of pK_a values below 30 did the equilibrium lead to the open chain enolate anion (**32c**) which was alkylated at the C-atom.

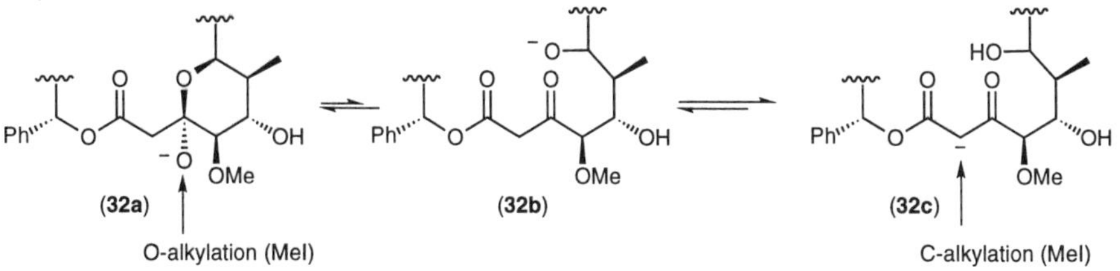

Scheme 6 *Deprotonation of norsoraphen (33)*

Potassium 2,6-di-tert-butyl phenoxylate was found to be a suitable base for the successful transformation of norsoraphen (**33**) into secosoraphen (**34**). After equilibration for 12 hours, the methylation was carried out at 0 °C. Quenching the reaction mixture after 30 min prevented dimethylation. The subsequent cyclization of (**34**) to soraphen $A_{1\alpha}$ (**4**) occurred spontaneously.

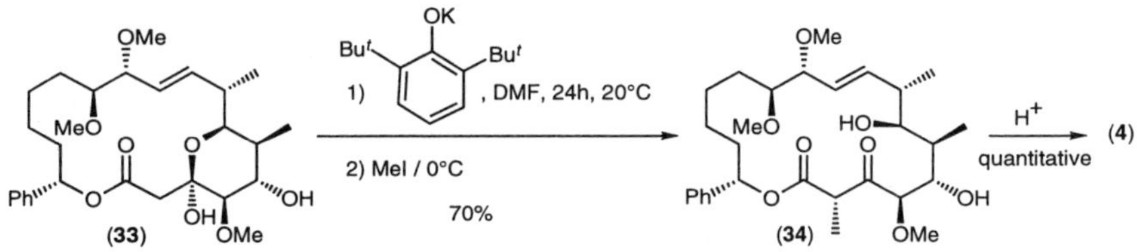

(33) (34)

Scheme 7 *Soraphen* $A_{1\alpha}$ *from norsoraphen*

3 TOTAL SYNTHESIS OF TREHAZOLIN FROM GLUCOSE

Trehazolin (**35**) is a specific inhibitor of trehalase, an enzyme that cleaves trehalose (**36**), the reserve carbohydrate of many insects and fungi. Thus, trehazolin or its derivatives and analogues may find application as insecticides and fungicides. Trehazolin (**35**) was first isolated in 1991 by O. Ando *et al.*[64] from the culture broth of *Micromonospora* strain SANK 62390 and several total syntheses were carried out in recent years.[65–69]

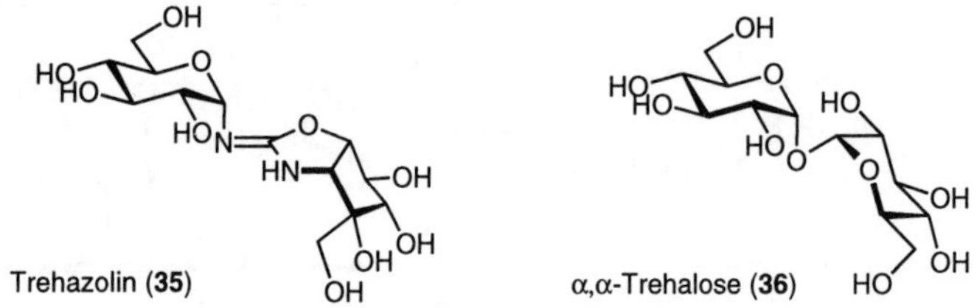

Figure 12 *Trehazolin and α,α-trehalose*

Our total synthesis started from glucose. The decisive step is the cyclization of the ketoxime (**38**) to trehazolamine (**37**) which is part of the aglycon of trehazolin (**35**). It is interesting to speculate that the biosynthesis of trehalozin may involve a similar cyclization of an intermediate arising from glucose.

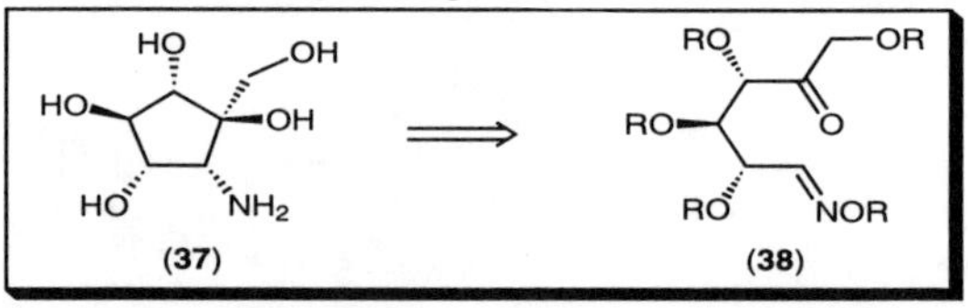

Figure 13 *Biosynthetic strategy for trehazolin*

It turned out that the radical induced cyclization of (**39**) generated a carbocycle (**40**) with the newly formed stereogenic centers having configurations opposite to those in trehazolin.

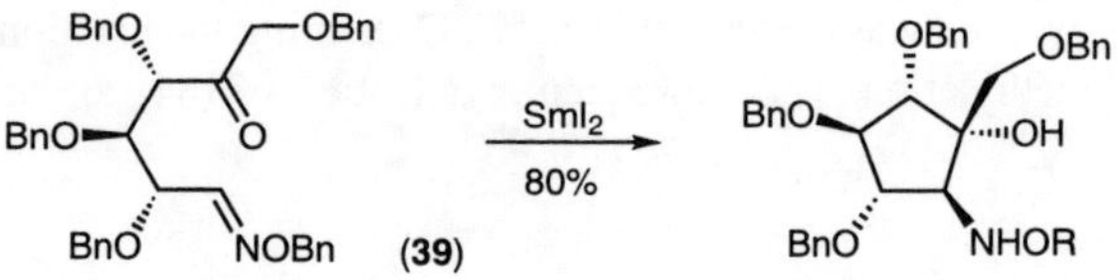

(39) (40)

Scheme 8 *Radical induced cyclization of (39)*

However binding the C–4 and C–6 hydroxy groups in a cyclic acetal function forced a *cis* fusion of the bicycle resulting from the radical cyclization, which provided the tertiary alcohol group with the correct stereochemistry (**41**→**42**). A cyclic acetyl protection of the C–1 oxime and C–2 hydroxy groups acted in a similar manner to direct the newly formed amino group into the desired configuration (**43**→**44**).

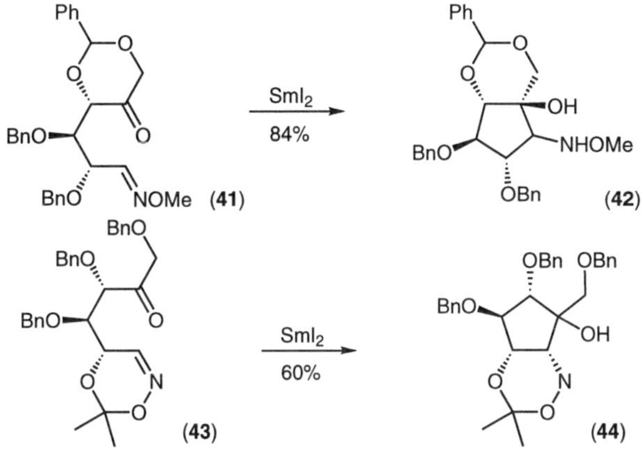

Scheme 9 *Radical induced cyclizations of (41) and (43)*

Using this synthetic trick, trehazolamine (**37**) was synthesized and coupled with isocyanate (**45**) via (**46**) to trehazolin (**35**).

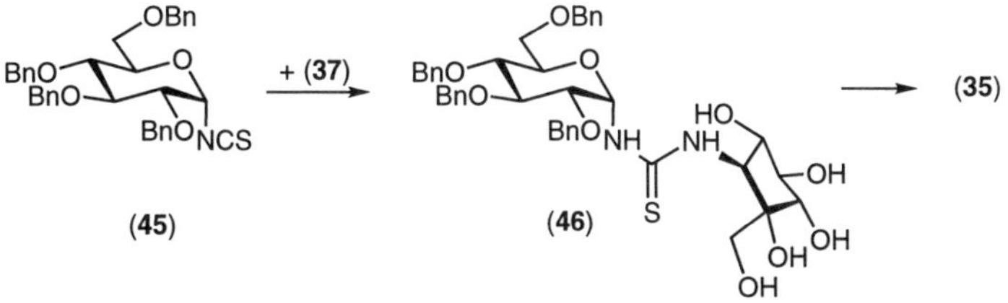

Scheme 10 *Final steps towards trehazolin (35)*

4 SUMMARY

Based on innovative synthetic strategies the carbohydrates glucose and mannose served as cheap synthons to prepare the enzyme inhibitors soraphen $A_{1\alpha}$ (**4**) and trehazolin (**35**).

References and Notes

1. J. P. Pachlatko, *Chimia*, 1998, **52**, 29.
2. 'Crop Protection Agents from Nature: Natural Products and Analogues', L.G. Copping, ed., The Royal Society of Chemistry, Cambridge, 1996.
3. S. J. Danishefsky, *Tetrahedron*, 1997, **53**, 8689.

4. P. A. Wender, S. Handy and D. L. Wright, *Chem. & Ind.* 1997, 6th October, p. 765.
5. G. I. Kornis in 'Agrochemicals from Natural Products', C. R. A. Godfrey, ed., Marcel Dekker Inc., New York 1995, p. 215.
6. H.-P. Fischer in 'Organic Synthesis Highlights II', H. Waldmann, ed., VCH, Weinheim 1995, p.261.
7. S. Hanessian, A. Ugolini, P. S. Hodges, P. Beaulieu, D. Dubé, C. André, *Pure Appl. Chem.* 1987, *59*, 299.
8. Glyphosate [$(HO)_2P(O)CH_2NHCOOH$] is a best selling contact herbicide. Glufosinate is a total herbicide, which has been synthesized and sold as a racemate for some years. As it contains only one chiral center the manufacture of the natural herbicidally active [S] isomer phosphinotricin [$CH_3P(O)(OH)$CH2CH2CH(NH)COOH] could become viable.[6]
9. S. Mio and H. Sano, *Annu. Rep. Sankyo Res. Lab.*, 1997, **49**, 91.
10. S. Mio, R. Ichinose, K. Goto, S. Sugai, and S. Sato, *Tetrahedron*, 1991, **47**, 2111.
11. N. Nakajima, M. Kirihara, M. Matsumoto, M. Hashimoto, T. Katoh, and S. Terashima, *Heterocycles*, 1996, **42**, 503.
12. N. Nakajima, M. Matsumoto, M Kirihara, M. Hashimoto, T Katoh, and S. Terashima, *Tetrahedron*, 1996, **52**, 1177.
13. A. J. Fairbanks and G. W. J. Fleet, *Tetrahedron*, 1995, **51**, 3881.
14. B.-T. Gröbel and D. Seebach, *Synthesis*, 1977, 357.
15. P. M. Harrington and M. E. Jung, *Tetrahedron Lett.* 1994, **35**, 5145.
16. S. Mio, Y. Kumagawa, and S. Sugai, *Tetrahedron*, 1991, **47**, 2133.
17. P. Chemla *Tetrahedron Lett.*, 1993, **34**, 7391.
18. M. Matsumoto, M. Kirihara, T. Yoshino, T. Katoh, and S. Terashima, *Tetrahedron Lett.*, 1993, **34**, 6289.
19. S. Mio, Y. Kumagawa, and S. Sugai, *Tetrahedron*, 1991, **47**, 2133.
20. H. Röper in 'Carbohydrates as Organic Raw Materials', F. W. Lichtenthaler, ed., VCH, Weinheim 1991, p. 277.
21. K. A. Watanabe and J. J. Fox, *Chem. Pharm. Bull.*, 1973, **21**, 2213.
22. S. Ogawa, T. Ogawa, T. Nose, T. Toyokuni, Y. Iwasawa, and T. Suami, *Chem. Lett.*, 1983, 921.
23. Y. Suhara, F. Sasaki, K. Maeda, H. Umezawa, and M. Ohno, *J. Am. Chem. Soc.*, 1968, **90**, 6559.
24. H.-P. Fiedler, in 'Sekundär Metabolismen bei Mikroorganismen', W. Kuhn and H.-P. Fiedler, eds., Attempto Verlag, Tübingen 1995, p. 79.
25. D. M. Gethin and N. S. Simpkins, *Tetrahedron*, 1997, **53**, 14417.
26. A. G. M. Barrett and S. A. Lebold, *J. Org. Chem.*, 1991, **56**, 4875.
27. Y. Auberson and P. Vogel, *Tetrahedron Lett.*, 1990, **46**, 7019.
28. B. M. Trost and Z. Shi, *J. Am. Chem. Soc.*, 1996, **118**, 3037.
29. P. Garner and J. M. Park, *J. Org. Chem.*, 1988, **53**, 2979.
30. P. Garner and J. M. Park, *J. Org. Chem.*, 1990, **55**, 3772.
31. F. Matsuura, Y. Hamada, and T. Shiori, *Tetrahedron Lett,* 1994, **35**, 733.
32. M. M. Paz and F. Javier, *J. Org. Chem.*, 1993, **58**, 6990.
33. H. Ohrui, H. Kuzuhara, and S. Emoto, *Tetrahedron Lett.,* 1973, **50**, 5055.
34. A. Dondoni, S. Franco, F. Junquera, F. L. Merchan, P. Merino and T. Tejero, *J. Org. Chem.*, 1997, **62**, 5497.

35. H. Akita, K. Uchida, and C. Y. Chen, *Heterocycles*, 1997, **46**, 87.
36. A. Chen, I. Savage, E. J. Thomas and P. D. White *Tetrahedron Lett.*, 1993, **34**, 6769.
37. R. F. W. Jackson, N. J. Palmer, M. J. Wythes, W. Clegg and M. R. J. Elsegood, *J. Org. Chem.*, 1995, **60**, 6431.
38. B. K. Banik, M. S. Manhas, and A. K. Bose, *J. Org. Chem.*, 1993, **58**, 307.
39. A. Duréault, F. Carreaux, and J. C. Depezay, *Tetrahedron Lett.*, 1989, **30**, 4527.
40. T. Mukaiyama, K. Suzuki, T. Yamada, and F. Tabusa, *Tetrahedron Lett.,* 1990, **46**, 265.
41. I. Savage and E. J. Thomas *J. Chem. Soc., Chem.Commun.*, 1989, 717.
42. A. K. Saksena, R. G. Lovey, V. M. Girijavallabhan, and A. K. Ganguly, *J. Org. Chem.,* 1986, **51**, 5024.
43. N. Damodaran, G. Jones, and J. Moffatt, *J. Am. Chem. Soc.*, 1971, **93**, 3812.
44. C. M. Evina and G. Guillerm *Tetrahedron Lett.* 1996, **37**, 163.
45. N. Chida, K. Koizumi, Y. Kitada, C. Yokoyama, and S. Ogawa, *J. Chem. Soc., Chem. Commun.,* 1994, 111.
46. H. Reichenbach, G. Höfle, H. Augustiniak, N. Bedorf, E. Forche, K. Gerth, H. Irschik, R. Jansen, B. Kurze, F. Sasse, H. Steinmetz, and W. Trowitzsch-Kienast, EP 282455; *Chem. Abstr.,* 1988, **111**, 132587.
47. N. Bedorf, D. Schomburg, K. Gerth, H. Reichenbach, and G. Höfle, *Liebigs Ann. Chem.,* 1993, 1017.
48. J. Rohr, *Nachr. Chem. Tech. Lab.,* 1993, **41**, 889.
49. S. Abel, D. Faber, O. Hüter, and B. Giese, *Angew. Chem. Int. Ed. Engl.,* 1994, **33**, 2460.
50. Gesellschaft für Biotechnologische Forschung mbH, Braunschweig-Stöckheim, 'Wissenschaflicher Ergebnisbericht', 1994.
51. Upon treatment of soraphen $A_{1\alpha}$ with base (thermodynamic conditions) no change of configuration occurred at the methylated C-atom α to the lactone. The crystal structure of **1** shows[47] that the methoxy group of the tetrahydropyran ring and the methyl group α to the lactone are in a sterically favorable position relative to each other. In the corresponding *epi* form, however, these methyl and methoxy groups are subject to 1,3-diaxial interactions. This might also be the reason why a possible dimethylation of norsoraphen **21** can be avoided under appropriate reaction conditions.
52. S. Hanessian, J.-R. Pougny, and I. K. Boessenkool, *Tetrahedron,* 1984, **40**, 1289.
53. A. Klemer and G. Rodemeyer, *Chem. Ber.,* 1974, **107**, 2612.
54. D. Horton and W. Weckerle, *Carbohydr. Res.,* 1975, **44**, 227.
55. See also: Y. Chapleur, *J. Chem. Soc., Chem. Commun.,* 1983, 141.
56. R. Tsang and B. Fraser-Reid, *J. Chem. Soc., Chem. Commun.,* 1984, 60.
57. U. Sunay, D. Mootoo, B. Molino, and B. Fraser-Reid, *Tetrahedron Lett.,* 1986, **27**, 4697.
58. M. Chérest, H. Felkin and N. Prudent, *Tetrahedron Lett.,* 1968, 2199.
59. N. T. Anh and O. Eisenstein, *Nouv. J. Chim.* 1977, **1**, 61.
60. M.T. Reetz, *Angew. Chem. Int. Ed. Engl.* 1984, **23**, 556. When using the magnesium salt of acetylene only the desired *erythro* isomer was detected. The formation of the *threo* isomer, however, was favored in a ratio of 8 : 1 when using the triisopropyloxytitaniumsalt.

61. D. Schummer, T. Jahn, and G. Höfle, *Liebigs Ann.,* 1994, 803.
62. D. Schummer, B. Böhlendorf, M. Kiffe, and G. Höfle in 'Antibiotics and Antiviral Compounds', K. Krohn, H. A. Krist, H. Maag, Eds., VCH, Weinheim, 1993, p.133.
63. F. Muneyamana, A.-M. Frisque-Hesbain, A. Devos, and L. Ghosez, *Tetrahedron Lett.,* 1989, **39**, 3077.
64. O. Ando, H. Satake, K. Itoi, A. Sato, M. Nakajima, S. Takahashi, H. Haruyama, Y. Ohkuma, T. Kinoshita, R. Enokita, *J. Antibiot.,* 1991, **44**, 1165.
65. S. Ogawa and C. Uchida, *Chem. Lett.,* 1993, 173.
66. C. Uchida, T. Yamagishi, S. Ogawa, *J. Chem. Soc., Perkin Trans. 1,* 1994, 589.
67. B. E. Ledford, E. M. Carreira, *J. Am. Chem. Soc.,* 1995, **117**, 11811.
68. Y. Kobayashi, H. Miyazaki, M. Shiozaki, *J. Org. Chem.,* 1994, **59**, 813.
69. S. Ogawa, C. Uchida, *J. Chem. Soc., Perkin Trans. 1,* 1992, 1939.

Natural Products with Antimicrobial Activity from *Pseudomonas* Biocontrol Bacteria

James M. Ligon[1], D.S. Hill[1], P. Hammer[1], N. Torkewitz[1], D. Hofmann[2], H.-J. Kempf[2] and K.-H. van Pée[3]

[1] NOVARTIS AGRIBUSINESS BIOTECHNOLOGY RESEARCH INC., RESEARCH TRIANGLE PARK, NC 27709, USA
[2] NOVARTIS CROP PROTECTION AG, CH-4002 BASEL, SWITZERLAND
[3] INSTITUTE OF BIOCHEMISTRY, TU-DRESDEN, GERMANY

1 INTRODUCTION

The biological control of soil-borne fungal pathogens of plants by naturally occurring microbes is now a well-known phenomenon. Microbes, including bacteria and fungi, that demonstrate this ability are taxonomically diverse. The characteristics that they share that are important determinants of this ability include the aggressive colonization of the plant rhizosphere and the production of antifungal metabolites. Many active bio-control microbes are bacteria of the genera *Pseudomonas* and *Burkholderia* (formerly classified as *Pseudomonas*). This group of bacteria is known to produce a diverse array of antifungal compounds.[1] In many cases the production of these compounds has been directly correlated with biocontrol activity (Table 1).[2]

Table 1 *Antifungal metabolites produced by* Pseudomonas *strains that have been implicated in the control of crop diseases*

Metabolite	Disease	Pathogen	Ref.
Phenazines	Take-all (wheat)	*Gaeumannomyces graminis* var. *tritici*	3
Pyoluteorin	Damping-off: cotton	*Pythium* spp.	4
	sugar beet	*Pythium ultimum*	4
2,4-Diacetyl-phloroglucinol	Take-all (wheat)	*G. graminis* var. *tritici*	5,6
	Damping-off (sugar beet)	*Pythium ultimum*	7
	Black root-rot (tobacco)	*Thielabiopsis basicola*	5
Pyrrolnitrin	Damping-off (Cotton, cucumber)	*Rhizoctonia solani*	8, 9
	Tan spot (wheat)	*Pyrenophora triticirepentis*	10
	Storage molds (pome fruit)	*Botrytis cinerea,*	11
		Penicillium expansum	11
	Seedling disease (Cotton)	*Thielaviopsis basicola*	12
		Alternaria spp.,Verticillium dahliae	12
	Dry rot (Potato)	*Fusarium sambucinum*	13
	Sclerotinia wilt (Sunflower)	*Sclerotinia sclerotiorum*	14
Oomycin A	Damping-off (cotton)	*Pythium* spp.	15
Agrocin 84	Crown gall (fruit trees)	*Agrobacterium tumefaciens*	16
Hydrogen cyanide	Black root-rot (tobacco)	*Thielabiopsis basicola*	17
Pseudobactin B10	Flax wilt	*Fusarium oxysporum*	18

The biological role of metabolite production by these bacteria appears to be in providing a competitive advantage in the colonization of the rhizosphere,[19] an environment that is rich in plant-exuded nutrients. Examples **(1)–(6)** of low molecular weight metabolites in this category with antifungal activity are shown in Figure 1. As a group, bacteria of the genus *Pseudomonas* and the closely related genus *Burkholderia*, represent a rich source of interesting compounds that are potential lead compounds for the development of agricultural chemicals. For example, the new fungicides, Fenpiclonil **(7)** and Fludioxonil **(8)** have recently been brought to the market (Figure 2).[20]

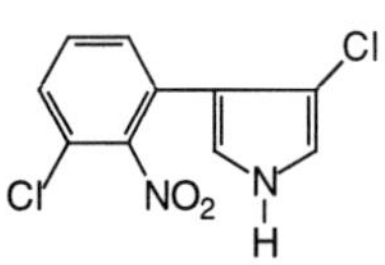

Pyrrolnitrin **(1)** PRN

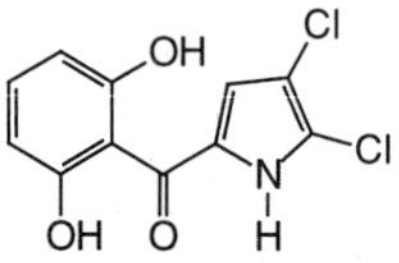

Pyoluteorin **(2)**

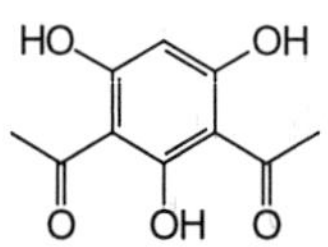

2,4-Diacetylphloroglucinol **(3)**

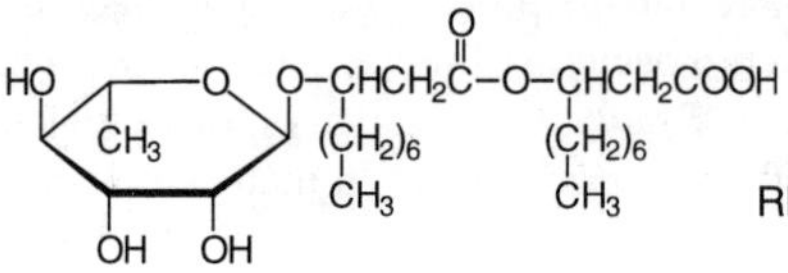

Rhamnolipid **(4)**

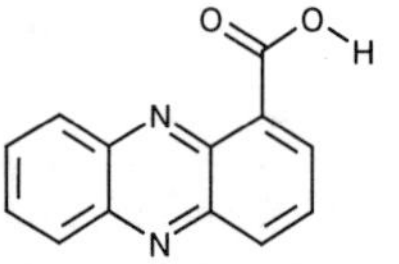

Phenazine-1-carboxylic acid **(5)**

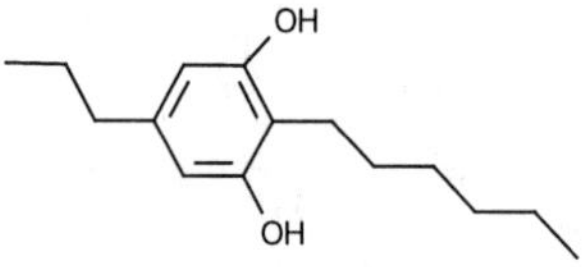

2-Hexyl-5-propylresorcinol **(6)**

Figure 1 *Low molecular weight metabolites with antifungal activities whose production by* Pseudomonas *strains has been correlated with biocontrol activity*

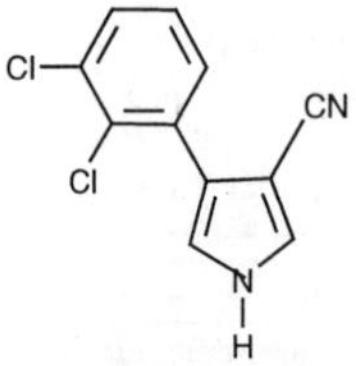

Fenpiclonil **(7)**

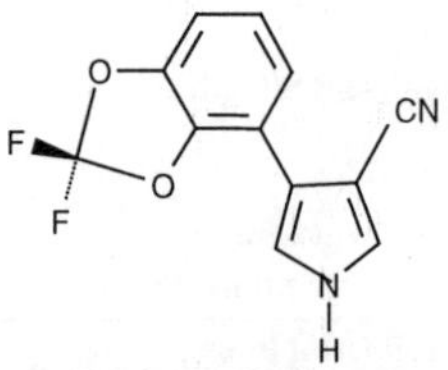

Fludioxonil **(8)**

Figure 2 *Synthetic analogs of pyrrolnitrin (1) that have been developed as commercial agricultural fungicides*

In the past few years, the genes that are responsible for the production of many antifungal metabolites have been isolated from the producing strains. These include the genes involved in the synthesis of phenazine-1-carboxylic acid (**5**),[21 22] pyoluteorin (**2**),[23] 2,4-diacetylphloroglucinol (**3**),[24 25] and pyrrolnitrin (**1**).[26] The study of these genes and the enzymes that they encode is resulting in a significant increase in the understanding of the biology and biochemistry of these interesting compounds. For example, recent studies of the genes involved in the synthesis of pyrrolnitrin (**1**) and the enzymes they encode, have led to the elucidation of the biochemical pathway by which this compound is synthesized[27] and the discovery of a new class of halogenases.[28] In addition, the application of modern principles of genetic engineering has the potential to result in the development of genetically modified biocontrol strains that produce higher amounts of the antifungal compounds and that are far superior to their wild-type counterparts in overall biocontrol activity. The case of pyrrolnitrin (**1**) will be used to illustrate the success of this approach.

2 PYRROLNITRIN: NATURAL METABOLITE AND LEAD STRUCTURE

Pyrrolnitrin (**1**), PRN (Figure 1), and its production by *Pseudomonas pyrrocinia* was first discovered in 1964 by Arima.[29] Since that time, this secondary metabolite has been found to be produced by numerous *Pseudomonas* and *Burkholderia* species and one *Myxococcus fulvus* isolate.[30] PRN is active against a wide range of *Deuteromycete, Ascomycete*, and *Basidiomycete* fungi, was developed as a topical antimycotic for human use,[31] and served as a lead structure for pharmaceutical research.[32] However, pyrrolnitrin was found to be unsuitable for use as an agricultural fungicide, since it is highly labile in sunlight.[33] Chemists at the former Ciba-Geigy discovered that substitution of the 3-chloro-pyrrole group with a 3-cyano group increased the stability of the compound under light and the overall and antifungal activity.[34] Further improvements of the activity were found with the PRN derivatives 4-(2,3-dichlorophenyl)pyrrole-3-carbonitrile (**7**) and 4-(2,2-difluoro-1,3-dioxol-4-yl)pyrrole-3-carbonitrile (**8**). These two compounds, called fenpiclonil and fludioxonil, respectively (Figure 2), were developed by Ciba-Geigy as agricultural fungicides for seed dressing and leaf applications. Both have good applicator, consumer, and environmental safety.

3 GENETICS OF METABOLITE BIOSYNTHESIS AND BIOLOGICAL DISEASE CONTROL

We have studied a wild-type strain of *P. fluorescens*, strain BL915, that is an active biocontrol agent for the control of *Rhizoctonia solani* infections of plants.[8] This strain produces pyrrolnitrin, as well as other antifungal metabolites, including 2-hexyl-5-propylresorcinol (**6**) and HCN, and enzymes such as chitinase and gelatinase. It was discovered that the production of all known antifungal activities by this strain is coordinately regulated by a 2-component regulatory system consisting of a sensor kinase and a response regulator.[35] Typically, two-component regulatory systems are activated upon the receipt of an environmental or internal physiological signal that activates the sensor kinase protein. This results in the phosphorylation of the response regulator that causes this protein to become an activator of gene transcription. Mutants with an inactivated *gacA* gene that encodes the response regulator protein, produce no antifungal metabolites or enzymes and have no biocontrol activity.[8] The *gacA* gene[35] and the *lemA* gene (unpublished results) that encode the sensor kinase protein have been cloned from *P. fluorescens* strain BL915 and sequenced.

Recently, the genes encoding the enzymes responsible for the synthesis of pyrrolnitrin were cloned from *P. fluorescens* strain BL915 and characterized (Figure 3).[26] It was found that four genes organized in a single transcriptional unit are required for PRN synthesis. Deletion mutations were constructed individually in each of the four PRN genes (*prn*) and each mutation resulted in a PRN-non-producing phenotype. In addition, the native GacA-regulated promoter preceding the coding region of the *prnA* gene was removed and replaced with the strong, constitutive P_{tac} promoter from *E. coli* such that the expression of the *prnABCD* gene cluster is under the control of the P_{tac} promoter. When the P_{tac}/*prnABCD* fusion was introduced into *E. coli,* PRN was produced by this bacterium.[26] Similarly, the P_{tac}/*prnABCD* genes were introduced into other bacterial strains known to be PRN-non-producing strains, including strains of *Pseudomonas* and *Enterobacter*, and these were also shown to produce pyrrolnitrin after receiving the modified *prn* genes. These results indicate that the *prnABCD* gene cluster is sufficient for the production of **(1)** in bacteria.

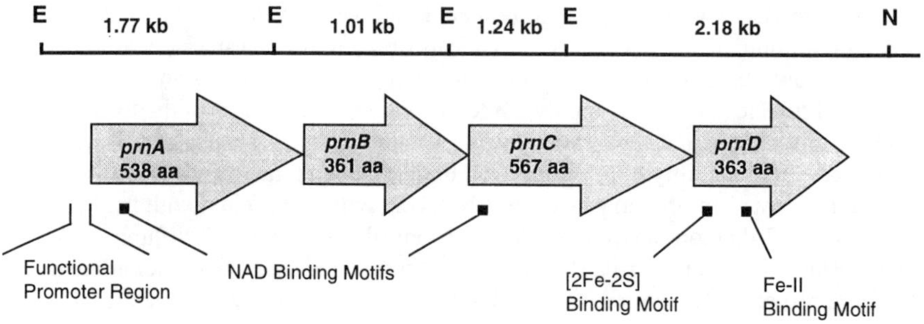

Figure 3 *The genetic organization of the prnABCD gene cluster from* P. fluorescens *strain BL915*

A restriction map showing the position of *EcoRI* sites and one *NotI* site and the distance between them appears above open boxes that depict the coding region of each gene. Special features identified in the nucleotide sequence of the individual genes are shown above.

Based on the feeding of cultures of PRN-producing strains with isotopically labelled and substituted tryptophan, this amino acid has been identified as the precursor for PRN biosynthesis.[36–40] Biosynthetic pathways have been proposed as early as 1967 by Hamill *et al.*[37] and have been further refined on the basis of tracer studies and the isolation of intermediates.[36 38–41,42] The *prn* genes and *Pseudomonas* strains with mutations in these genes were used recently as tools to clearly define the biosynthetic pathway leading to pyrrolnitrin synthesis (Figure 4, pathway **A**).[27] Analysis of the production of PRN intermediates by mutants blocked at one of the four enzymatic steps in the biosynthesis of PRN provided important information. For example, a mutation in the *prnB* gene resulted in the accumulation of 7-chloro-tryptophan (**10**), 7-CT, and no production of any other intermediates or PRN. Likewise, mutations in the *prnC* gene resulted in the accumulation of monodechloro-aminopyrrolnitrin (MDA) and no production of aminopyrrolnitrn (**11**), APRN or PRN. Mutations in the *prnD* gene resulted in the accumulation of APRN with no production of PRN. To further clarify which biosynthetic step was blocked in each deletion mutant, intermediate feeding experiments were conducted. Biosynthetic intermediates were added to cultures of the *prn* gene mutants and after a period of growth, the cultures were analysed to detect the production of PRN (**1**). The *prnA* mutant was unable to produce PRN except when it was supplied with exogenously added 7-CT (**10**). Similarly, the *prnB* and *prnC* mutants were able to produce PRN only if supplied with MDA (**11**) or APRN (**12**), respectively.

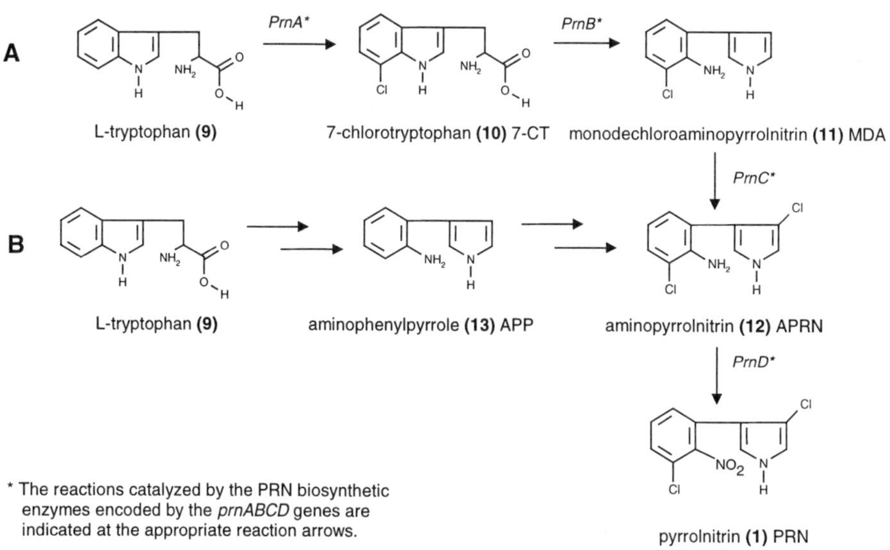

Figure 4 *Biosynthetic pathways **A** for pyrrolnitrin as proposed by van Pée et al.*[42] *and pathway **B** suggested by Chang et al.*[39]

Together, these results support the biosynthetic pathway **A** shown in Figure 4 and indicate that the *prnA* gene encodes a tryptophan halogenase that catalyses the chlorination of tryptophan to form 7-CT (**10**); *prnB* encodes the enzyme that catalyses the complex rearrangement of 7-CT (**10**) from the indole to the phenylpyrrole structure and the decarboxylation to form MDA (**11**); *prnC* encodes an MDA halogenase that catalyses a second chlorination in the 3 position of the pyrrole ring to form APRN (12); and *prnD* encodes an enzyme that oxidizes the amino group of APRN(**12**) to a nitro group to form pyrrolnitrin (**1**). These conclusions were verified by using a *P. fluorescens* BL915 mutant that lacked the entire *prnABCD* gene cluster due to a large deletion of these genes from the chromosome. When the *prnA* gene alone was expressed in the mutant lacking all other *prn* genes, the production of 7-CT **(10)** was detected. Similarly, when the *prnB* gene was expressed in the mutant lacking all other *prn* genes, the production of MDA **(11)** was detected when 7-CT **(10)** was supplied to the culture. Similar experiments with the *prnC* and *prnD* genes demonstrated that the expression of these genes in the deletion mutant caused the production of APRN **(12)** and PRN (1) if MDA **(11)** and APRN **(12)**, respectively, were supplied exogenously. This study has clearly demonstrated that pyrrolnitrin is synthesized in *P. fluorscens* strain BL915 by the pathway **A** shown in Figure 4.

Chang *et al.* reported the conversion of exogenously supplied aminophenylpyrrole, APP **(13)** by *Pseudomonas* to PRN.[39] Similarly, Zhou *et al.* reported the conversion of APP **(13)** to APRN **(12)** in a cell-free system.[40] These workers concluded that APP **(13)** is an intermediate in the pyrrolnitrin biosynthesis and that ring rearrangement precedes chlorination (Path **B** in Figure 4). However, the study by Kirner *et al.* demonstrated that APP accumulated only in strains in which the *prnB* gene is over-expressed and APP was not detected in cultures of the *prnA* mutant strain, as would be expected if the ring rearrangement (catalysed by the *prnB* gene product) occurs prior to the first chlorination step (catalysed by the *prnA* gene product).[27] Therefore, it is now clear that the pathway **A** shown in Figure 4, first proposed by van Pée *et al.*, is the correct pathway for the biosynthesis of pyrrolnitrin in *P. fluorescens* strain BL915.[42]

Comparison of the *prnABCD* gene sequences with the DNA databases using BLAST revealed some interesting features of the genes and the enzymes they encode. First, the PrnA and PrnC enzymes that encode halogenating enzymes have no homology to one another except that each contains a consensus NAD-binding domain near its NH_2-terminus (Figure 3). This is consistent with the finding that these enzymes require NADH as a cofactor for activity.[28] The PrnC enzyme was found to be highly similar to a chlorinating enzyme that is involved in the biosynthesis of the antibiotic tetracycline by *Streptomyces aureofaciens*.[43] The PrnB enzyme had no matches in the databases. PrnD was found to contain two domains that are commonly found in bacterial dioxygenases (Figure 3).[26] One is the motif of Rieske iron–sulfur proteins that is involved in the binding of [2Fe-2S] clusters and the other is the proposed binding site domain for the mononuclear nonheme Fe(II) that is thought to be involved in binding molecular oxygen. These features are consistent with the function that has been proposed for this enzyme, the oxidation of the amino group of APRN **(12)**.

Since PRN contains two chlorine atoms, it has been used for many years as a model for the study of organic halogenations by van Pée.[44] The detection of chloroperoxidase from the fungus *Caldariomyces fumago* and the development of a simple spectrophotometric assay for the detection of halogenating enzymes based on the synthetic compound monochlorodimedone as organic substrate, resulted in the subsequent isolation of a number of haloperoxidases from different organisms.[45,46] All such enzymes produce hypohalogenic acid which is the actual halogenating agent and therefore, halogenation catalysed by haloperoxidases lacks substrate and regiospecificity.[44,47] Recently, genetic investigations demonstrated that haloperoxidase-type enzymes are definitely not involved in the biosynthesis of chlorotetracycline[43] or pyrrolnitrin.[48] Comparison of the nucleotide sequences of a gene encoding a chloroperoxidase cloned from *Pseudomonas pyrrocinia* and thought to have a role in the synthesis of pyrrolnitrin[49] with the genes from *P. fluorescens* strain BL915 encoding the PrnA and PrnC enzymes that have been shown to be involved in this process did not reveal any similarities.[26] This has led to the conclusion that the chloroperoxidases do not have a role in pyrrolnitrin synthesis and that the PrnA and PrnC enzymes represent a novel class of halogenases that do not require peroxide for activity.

4 GENETIC MODIFICATION OF *PSEUDOMONAS FLUORESCENS* FOR INCREASED METABOLITE PRODUCTION AND BIOCONTROL ACTIVITY

In the past few years, we have attempted to utilize the recently acquired knowledge about the regulation, genetics, and biochemistry of the synthesis of antifungal metabolites by *P. fluorescens* strain BL915 to generate genetically modified versions of the strain with increased metabolite production for the purpose of improving the biocontrol activity of the strain. Some modifications have been directed at the two-component system that regulates metabolite synthesis, while others were specifically targeted at improving PRN production. One example of genetic modifications affecting the regulatory system is one in which we cloned an 11 kilobase DNA fragment from strain BL915 that contains the native *gacA* gene and introduced a plasmid containing this fragment into the same strain. This created a condition in which, in addition to the single copy of the *gacA* gene present in the chromosome, there are additional plasmid-borne copies of the gene in each cell. This modified strain, designated BL915(pPRN-E11), was shown to produce about 2.5-fold more PRN than the parent strain (Figure 5, column *4*). Subsequently, other modifications of the *gacA* gene were constructed and tested. The native *gacA* gene has an unusual TTG translation initiation codon which is known to result in reduced translational efficiency of genes compared to the more common ATG initiation codon. Using polymerase chain reaction (PCR) technology, we changed the first base in the coding sequence of the chromosomal *gacA* gene in *P. fluorescens* strain BL915 from a thymidine (T) to an adenine (A), thereby changing the translation initiation codon of the *gacA* gene from the less efficient TTG codon to the more efficient ATG codon. The resulting strain, BL915-ATG/*gacA*, produced about twice as much PRN as the parent strain (Figure 5, column *3*). In another modification of the *gacA* gene we sought to increase the level of transcription of the gene by replacing the native promoter with the strong P_{tac} promoter from *E. coli*. The resulting modified strain, BL915-P_{tac}/*gacA*, also produced about twice as much PRN as the parent strain (Figure 5, column *2*).

Another class of genetically modified *P. fluorescens* strain was constructed by modification of the *prnABCD* gene cluster derived from this strain. In a similar approach as was used to construct the modified strain BL915-P_{tac}/*gacA* described above, we replaced the native GacA-regulated promoter of the *prnABCD* gene cluster with the more active P_{tac} promoter. A DNA fragment containing the modified P_{tac}/*prnABCD* genes was cloned into a broad host-range plasmid and this was introduced into strain BL915. The resulting modified strain, BL915(P_{tac}/*prnABCD*),

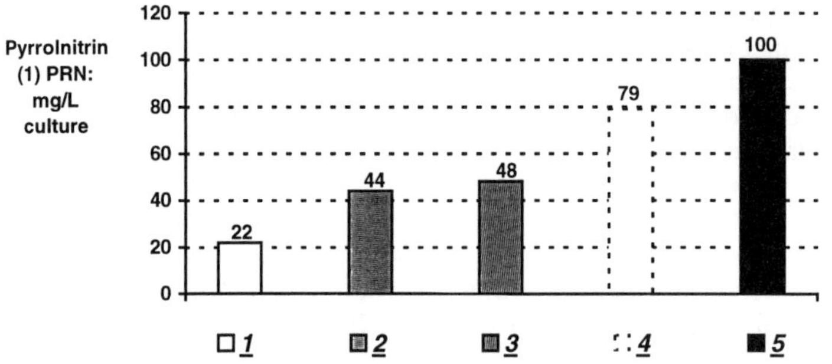

Figure 5 *PRN production by* P. fluorescens *strain BL915 and derivatives. The values shown represent the amount of PRN produced in culture by strains BL915 [column 1], BL915-P_{tac}/gacA [column 2], BL915-ATG/gacA [column 3], BL915(pPRN-E11) [column 4], and BL915(P_{tac}/prnABCD) [column 5].*

contains multiple plasmid-borne copies of the P_{tac}/*prnABCD* genes. It was shown to produce approximately four times as much PRN as the parent strain (Figure 5, column 5).

The genetically modified strains described above were tested for biocontrol activity in two pathosystems, including cucumber and impatiens, with *Rhizoctonia solani* as the pathogen. Since strain BL915 is very effective when used at the rate of 2×10^8 cells/g soil and we were interested in determining whether the modified BL915 strains have increased biocontrol activity, we tested all modified strains at a rate one-tenth that of the normal rate (2×10^7 cells/g soil). The results shown in Table 2 demonstrate that in most cases, the modified strains exhibited increased biocontrol activity relative to parent strain BL915. Strain BL915(P_{tac}/*prnABCD*) had nearly as much biocontrol activity at the ten-fold lower rate in both pathosystems as the parent strain at the normal rate (Table 2). Strain BL915(pPRN-E11) showed similar improved activity in the impatiens trials, but was not improved over the parent strain in the cucumber trials. At the low rate, strains BL915-P_{tac}/*gacA* and BL915-ATG/*gacA* each had higher activity than the parent strain in both pathosystems, but not as much activity as the parent strain applied at the high rate. These results demonstrate that the genetic modifications of strain BL915 described herein have resulted in increased production of PRN and in a concomitant increase in biocontrol activity.[50]

5 SUMMARY

Nature has proven to be a rich source of interesting compounds with commercial potential. Bacteria belonging to the genera *Pseudomonas* and *Burkholderia* are known to be the source of a number of these and will undoubtedly be the basis for new compounds yet to be discovered. Only by taking an interdisciplinary approach to the study and development of these compounds, can mankind derive the maximum benefit from this resource. PRN is an excellent example of this principle.

Table 2 *Biocontrol activity of modified BL915 strains in two pathosystems against* Rhizoctonia solani

Treatment	**Stand** (14 Days after Planting)		**Treatment**	**Stand** (14 Days after Planting)	
Test 1:	**Cucumber**	**Impatiens**	**Test 2:**	**Cucumber**	**Impatiens**
Healthy Control	113 (120)	120 (120)	Healthy Control	54 (60)	119 (120)
Pathogen Control	86	92	Pathogen Control	31	78
BL915-Hi	97	111	BL915-Hi	54	106
BL915-Lo	80	91	BL915-Lo	41	73
BL915(pPRN-E11)	75	103	BL915-P_{tac}/*gacA*	46	80
BL915(*prnABCD*)	101	105	BL915-ATG/*gacA*	44	98

Cucumber seeds or impatiens transplants were planted in infested soil except in the case of the healthy control plants which were planted in uninfested soil. All strains were applied as a drench at 2 x 10^7 cells/g soil, except for the high rate of BL915 which was applied at a 10-fold higher rate. Since the strains were tested in separate trials (Test 1 and Test 2), results for the controls are shown with each test group. The number of plants planted for each treatment is shown in parentheses with the results for the healthy controls.

PRN has been used as a lead structure to develop important new commercial fungicides. Wild-type *Pseudomonas* strains that produce PRN have potential to be developed as biocontrol products for the control of soil-borne plant pathogens. A thorough study of the genetics and biochemistry of PRN synthesis has made it possible to create genetically modified biocontrol bacteria with improved metabolite production and biocontrol activity. Genetically modified biocontrol strains with improved activity will likely be the biocontrol products in the future. Finally, and completing the circle from chemistry to biochemistry and genetics and back, the enzymes that participate in the biosynthesis of natural metabolites could be used in a biocatalysis strategy in the modification or synthesis of new chemical structures. Therefore, the field of natural products offers a great opportunity for chemists, biochemists, biologists and geneticists and will undoubtedly stimulate many fruitful interactions among scientists from these disciplines.

References

1. T. Leisinger and R. Margraff, *Microbiol. Rev.* 1979, **43**, 422.
2. D.N. Dowling and F. O'Gara, *Trends in Biotechnol.* 1994, **12**, 133.
3. L.S. Thomashow and D.M. Weller, *J. Bacteriol.* 1988, **170**, 3499.
4. C.R. Howell and R.D. Stipanovic, *Phytopathology* 1980, **70**, 712.
5. C. Keel et al., *Mol. Plant-Microbe Interact.* 1992, **5**, 4.
6. M.N. Vincent et al. *Appl. Environ. Microbiol.* 1991, **57**, 2928.

7. P. Shanahan, D.J. O'Sullian, P. Simpson, J.D. Glennon, and F. O'Gara, *Appl. Environ. Microbiol.* 1992, **58**, 353.
8. D.S. Hill, D.S., J.I. Stein, N.R. Torkewitz, A.M. Morse, C.R. Howell, J.P. Pachlatko, J.O. Becker, and J.M. Ligon, *Appl. Environ. Microbiol.* 1994, **60**, 78.
9. D.S. Hill, D.S., P.E. Hammer, and J.M. Ligon, in 'Current Topics in Plant Molecular Biology', P.M. Gresshoff (ed.), CRC Press, Inc., Boca Raton, 1997, pp. 41-50.
10. W.F. Pfender, J. Kraus, and J.E. Loper, *Phytopathology*, 1993, **83**,1223.
11. W. Janisiewicz, et al. *Plant Dis.* 1991, **75**, 490.
12. C.R. Howell and R.D. Stipanovic, *Phytopathology,* 1979. **69**, 480.
13. K.D. Burkhead et al. *Appl. Environ. Microbiol.* 1994, **60**, 2031.
14. T.J., McLoughlin et al. *Appl. Environ. Microbiol.* 1992, **58**, 1763.
15. N. Gutterson, *Crit. Rev. in Biotechnol.* 1990, **10**, 69.
16. A. Kerr, *Agri. Sci.* 1989, **2**, 41.
17. C. Voisard, C. Keel, D. Haas, and G. Défago, *EMBO J.* 1989, **8**, 351.
18. J.W., Kloepper, J. Leong, T. Teintze, and M.N. Schroth, *Nature,* 1980, **286**, 885.
19. M. Mazzola, R.J. Cook, L.S. Thomashow, D.M. Weller, and L.S. Pierson III, *Appl. Environ. Microbiol.* 1992, **58**, 2616.
20. K. Gehmann, R. Nyfeler, A. Leadbeater, D. Nevill, and D. Sozzi, *Proc. Brighton Crop Prot. Conf. Pest Dis.* 1990, **2**, 399.
21. L.S. Pierson III and L.S. Thomashow, 'Cloning and heterologous expression of the phenazine biosynthetic locus from *Pseudomonas aureofaciens*' pp. 30-84. In 'Molec. Plant-Microbe Interact.', 1992, **5**, 330.
22. L.S. Pierson III, T. Gaffney, S. Lam, and F. Gong, *FEMS Microbiol. Lett.* 1995, **134**, 299.
23. B. Nowak-Thompson, S.J. Gould, and J.E. Loper, *Gene*, 1997, **204**,17.
24. M.G. Bangera and L.S. Thomashow, *Molec. Plant-Microbe Interact.* 1996, **9**, 83.
25. L.S. Thomashow and M.G. Bangera, *GenBank Accession* 1997, No. U41818.
26. P. Hammer, D.S. Hill, S. Lam, K.-H. van Pée, and J. Ligon. *Appl. Environ. Microbiol.* 1997, **63**, 2147.
27. S. Kirner, P. Hammer, D.S. Hill, A. Altmann, I. Fischer, L. Weislo, M. Lanahan, K.- H. van Pée, and J. Ligon, *J. Bacteriol.* 1998, **180**, 1939.
28. K. Hohaus, A. Altmann, W. Burd, I. Fischer, P. Hammer, D.S. Hill, J. Ligon and K.-H. van Pée, *Angew. Chem. Int. Ed. Engl.* 1997, **36**, 2012.
29. K. Arima, H. Imanaka, M. Kousaka, A. Fukuda, and C. Tamura, *Agric. Biol.* Chem. 1964, **28**, 575.
30. K. Gerth, W. Trowitzsch, V. Wray, G. Höfle, H. Irschik, and H. Reichenbach, *J. Antibiotics*, 1982, **35**, 1101.
31. S. Tawara, S. Matsumoto, T. Hirose, Y. Matsumoto, S. Nakamoto, M. Mitsuno, and T. Kamimura, *Jpn. J. Med. Mycol.* 1989, **30**, 202.
32. U. Suminori, K. Kazuo, K. Tanaka, and H. Nakamura, *Chem. Phar. Bulletin (Tokyo),* 1969, **17**, 559.
33. A. Ueda, H. Nagasaki, Y. Takakura, H. Nishikawa, and A. Nakada, European Patent Application No. 198292:890.
34. R. Nyfeler and P. Ackermann, in 'Synthesis and Chemistry of Agrochemicals, D.R. Baker, J.G. Fenyes, And J.J. Steffens (eds), American Chemical Society, Washington, 1992, pp.395-404.

35. T. Gaffney, S. Lam, J. Ligon, K. Gates, A. Frazelle, J. Di Maio, S. Hill, S. Goodwin, N. Torkewitz, A. Allshouse, H.-J. Kempf, and J.O. Becker, Molec. *Plant-Microbe Interact.* 1994, **7**, 455.
36. D.H. Lively, M. Gorman, M.E. Haney, and J.A. Mabe, *Antimicrob. Agents Chemother.* 1966, **1966**, 462.
37. R. Hamill, R. Elander, M. Mabe, and M. Gorman, *Applied Microbiol.* 1967, **19**, 721.
38. R. Hamill, R. Elander, M. Mabe, and M. Gorman, *Antimicrob. Agents Chemother.* 1970, **1967**, 388.
39. C.J. Chang, H.G. Floss, D.J. Hook, J.A. Mabe, P.E. Manni, L.L. Martin, K. Schroder, and T.L. Shieh, *J. Antibiotics*, 1981, **24**, 555.
40. P. Zhou, U. Mocek, B. Seisel, and H.G. Floss, *J. Basic Microbiol.* 1992, **32**, 209.
41. L.L. Martin, C.-J. Chang, and H.G. Floss, *J. Am. Chem. Soc.* 1972, **94**, 8942.
42. K.-H. van Pée, O. Salcher, and F. Lingens, *Angew. Chem. Int. Ed. Engl.* 1980, **19**, 828.
43. T. Dairi, T. Nakano, K. Aisaka, R. Katsumata, and M. Hasegawa, *Biosci. Biotechnol. Biochem.* 1995, **59**, 1099.
44. K.-H. van Pée, *Annu. Rev. Microbiol.* 1996, **50**, 375-399.
45. P.D. Shaw and L.P. Hager, *J. Am. Chem. Soc.* 1959, **81**, 6527.
46. L.P. Hager, D.R. Morris, F.S. Brown, and H. Eberwein, *J. Biol. Chem.* 1966, **241**, 1769.
47. M.C.R. Franssen, *Biocatalysis* ,1994, **10**, 87.
48. S. Kirner, S. Krauss, G. Sury, S.T. Lam, J.M. Ligon, and K.-H. van Pée. *Microbiology* ,1996, **142**, 2129.
49. C. Wolfframm, K.-H. van Pée, and F. Lingens, *FEBS Lett.* 1988, **238**, 325.
50. J. Ligon, S. Lam, T. Gaffney, D.S. Hill, P. Hammer, and N. Torkewitz, in 'Biology of Plant-Microbe Interactions' (G. Stacey, B. Mullin, and P. Gresshoff, eds.), International Society for Molecular Plant-Microbe Interactions, St. Paul, MN, USA, 199

Modification of Plant Secondary Metabolism by Foreign Phytoalexin Genes

Rüdiger Hain

CROP PROTECTION BUSINESS GROUP, RESEARCH, MOLECULAR TARGET RESEARCH & BIOTECHNOLOGY, BAYER AG, D-51368 LEVERKUSEN, GERMANY

1 INTRODUCTION

A variety of defense mechanisms, either constitutive or inducible, contribute to the resistance of plants to pathogens. Genetic engineering of plants has raised the possibility of improving the resistance of crop plants to microbial pathogens by novel means and strategies[1,2]. In particular induced defense reactions have been proposed as playing an important role in disease resistance. Among these defense responses, the accumulation of antimicrobial compounds called phytoalexins has been studied extensively[3]. Phytoalexins are low molecular weight, antimicrobial secondary plant products whose synthesis is rapidly induced in plants in response to microbial infection, stress or treatment of plant tissues with a wide range of naturally occurring or synthetic compounds (biotic or abiotic elicitors). In recent years direct evidence for the involvement of phytoalexins on disease resistance has been shown by different approaches. Besides *in vitro* data on the antimicrobial activity of phytoalexins, one approach made use of phytoalexin detoxifying genes found in certain pathogens[4]. Further support for this notion was provided by transgenic plants that showed increased disease resistance to fungi based on the expression of a novel phytoalexin (**5**).

This paper describes the modification of secondary metabolism using key genes in phytoalexin biosynthesis (**Figure 1**). As examples genes for stilbene synthase [STS], pinosylvin synthase [PSS] and bibenzyl synthase [BBS] from grape, pine and orchids respectively were used.

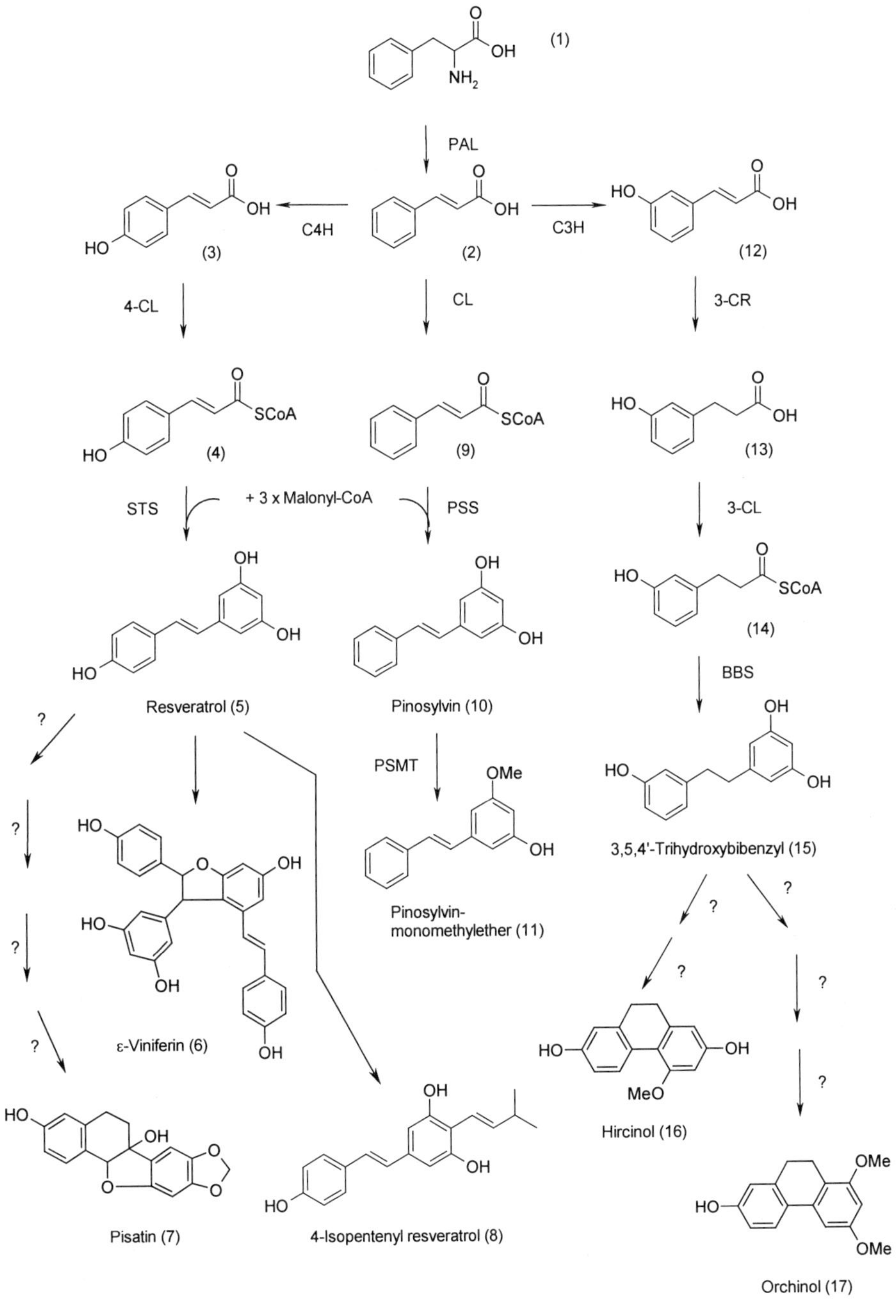

Figure 1 *Phytoalexin biosynthesis in peanut, pea, grape, pine and orchid*

2 DISEASE RESISTANCE RESULTS FROM FOREIGN PHYTOALEXIN EXPRESSION IN PLANTS

The strategy of engineering pathogen resistance by the introduction of novel phytoalexins is to date limited by the availability of only a few biosynthetic genes[7]. In most cases such genes are members of small gene families[6,8,9] whose individual transfer may not lead to the level of expression required for enhanced disease resistance in the foreign plant. This is particularly so because, due to association with their original promoters, expression patterns differ among gene familiy members with respect to developmental control, tissue-specificity, inducibility and promoter strength[8]. So far, stilbene synthase genes have been studied most intensively. STS have been isolated from several unrelated species such as peanut, grapevine and pine[6,10,11]. In most cases the biosynthesis of stilbenes with phytoalexin properties is inducible by pathogenic fungi, ozone, UV light and other such environmental stress conditions[12-15]. Stilbene formation specifically requires the sole presence of the key enzyme STS which can be further classified according to its substrate preference. The STS enzymes from peanut and grapevine use malonyl-CoA and 4-coumaroyl-CoA (**4**) as substrates to form the phytoalexin trihydroxystilbene resveratrol[5,16,17](**5, Figure 1**). There is no resveratrol synthase in tobacco, though the precursor molecules, which are also used by chalcone synthase, the key enzyme of flavonoid biosynthesis, are present throughout the plant kingdom. The inducible formation of resveratrol (**5**) as a *de novo* produced phytoalexin was found to occur in transgenic tobacco plants carrying a STS gene from peanut[5]. Moreover, by transfer of two stilbene synthase genes from grapevine and their expression under the control of their natural promoters, an increase in disease resistance in transgenic tobacco plants was achieved[6]. However, the amount of STS mRNA in tobacco which accumulated after induction was no more than 5% of the amount found in grapevine.

Due to the fact that transfer of a whole STS gene family to a foreign plant is impracticable, we tried to optimize the expression of single STS genes by using heterologous promoters and enhancer elements. Our first approach was to increase the expression of a grapevine STS gene in tobacco by the insertion of transcriptional enhancer elements (**Figure 2**), yet maintaining the natural regulation defined by the STS promoter. It is known that upstream sequences of the 35S RNA promoter from Cauliflower Mosaic Virus (CaMV) are able to stimulate transcription from heterologous promoters[18,19], although their properties do not strictly meet the definiton of enhancers as being orientation-independently active at distances greater than 1 kb 5´ or 3´ to the transcription start site[20,21]. In our approach we used a tetramer of the -343 to -90 region which was successfully employed in T-DNA tagging[22]. In additon, an almost linear correlation between the enhancing activity of the -209 to -46 fragment and its multimerization of up to four copies had been previously described[23]. Insertion of the tetrameric enhancer 1.5 kb upstream of the transcription start site (VE5+) led to a significant increase in STS mRNA accumulation without alteration of the kinetics of induction. This enhancement was striking in transgenic plants derived from *Agrobacterium tumefaciens*-infected leaves (about 14-fold) and in shoots regenerated from protoplasts after direct gene transfer (about 4-fold). The enhancing property was lower in the inverse orientation, corresponding to earlier findings[19].

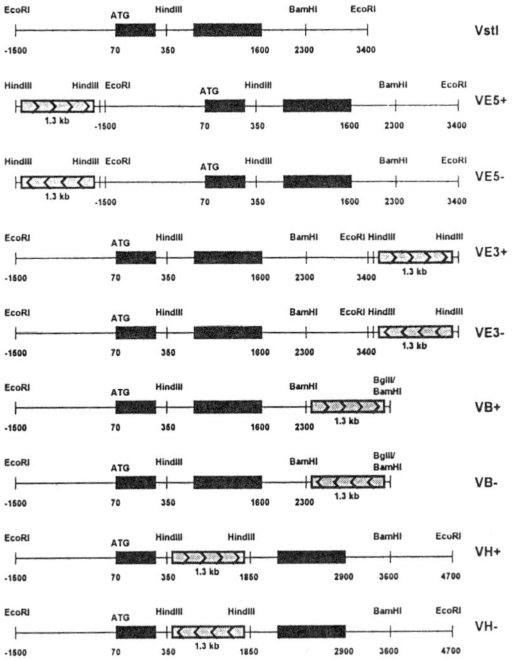

Figure 2 *Schematic representation of the STS/enhancer plasmid constructions pVst1, pVE5+/–, pVE3+/–, pVB+/– and pVH+/–.*
All nine *Vst1*/enhancer constructs were cloned into a binary vector and transferred to tobacco. The multiple enhancer elements (bars with arrows) were orientated at all positions in the original forward orientation as found in the 35S RNA promoter (+-orientation) as well as in the inverse orientation (–-orientation). Black bars are exon regions.

A loss of orientation independence was also found when the enhancer fragment was located 3' of the STS coding region, at a distance of 1.8 kb from the stop codon. Moreover, only a 5-fold stimulation was attained in the 3' position compared to a much higher activity at the 5' position confirming previous observations[19,23]. Insertion of the 1.3 kb enhancer fragment into the intron of the STS gene, turned out to have a negative impact on transcription efficiency, presumably by interference with RNA processing. In biological trials in the greenhouse VE5+ enhancer plants were reproducibly less susceptible towards *Botritis cinerea* than control plants. In comparison to line Vst1 expressing the unmodified grapevine STS gene, line VE5+ exhibited an additional attenuation of disease incidents of approximately 20%. Thus, the level of disease resistance based on the inducible synthesis of an additional phytoalexin can be influenced positively by modulation of the corresponding promoter strength.

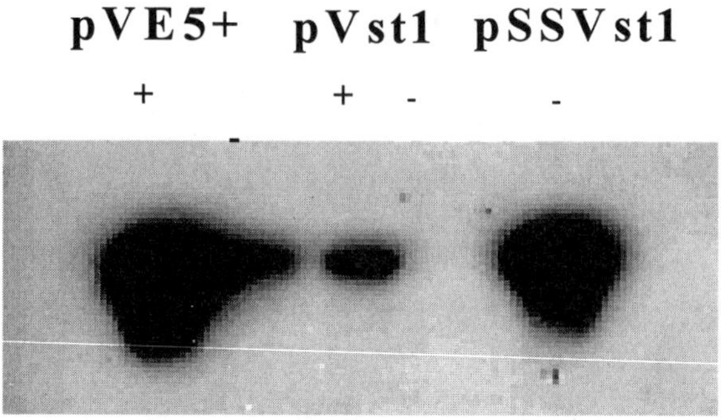

Figure 3 *Northern blot of RNA isolated from transgenic tobacco*
(-, not induced, +, induced). Middle lanes: pVst1, STS gene under natural promoter, left lanes, pVE5+, STS gene under natural promoter with enhancer elements, right lanes, pSSVst1, STS gene under control of a double CaMV35S RNA promoter

	kan^r	Vst1	VE5+
Trial A	7.3	19.4	63.7
Trial B	5.0	21.9	42.2
Trial C	2.9	37.1	32.3
Trial D	2.1	47.3	60.8
Trial E	4.3	44.7	62.2
Trial F	10.2	38.3	55.3
Trial G	3.4		36.0

Table 1 *Influence of STS gene expression on the resistance of N. tabacum towards grey mould*
Average disease reduction (%) compared to SR1 wild-type tobacco is presented. All experiments were done with F1 generations, plants being hemizygous (67 %) or homozygous (33 %) for the respective STS gene. After inoculation with *B. cinerea* spores (2 x 10^5 spores/ml) and subsequent incubation for 4 to 6 days, infection density and symptom development were evaluated using six leaves. In every trial 30-40 plants for each line were tested. All trials (A-G) represent data of biological tests with transgenic tobacco plants expressing the inducible grapevine STS gene *Vst1*.

3 APPLICATION OF THE PHYTOALEXIN TECHNOLOGY IN PLANT PROTECTION AND PLANT BREEDING

STS genes have been transferred to a number of plants including tobacco, tomato, potato, oil seed rape, rice as well as in grapevine plants. Besides the synthesis of the foreign phytoalexin, enhanced disease resistance against *Botrytis cinerea* has been demonstrated in tobacco[5]. Experiments in transgenic tomato plants, expressing STS genes from grapevine revealed an increase in resistance against *Phytophthora infestans*[24]. Data with transgenic potato plants indicate enhanced resistance to *P. infestans* and *Fusarium sulfureum* (Stahl pers. com). Furthermore, STS expression in rice indicate an enhanced disease resistance to rice blast (*P. oryzae*)[25]. In the meantime STS has also been expressed in wheat resulting in induced resveratrol synthesis and first phytopathological experiments have shown promising results[26]. Very interestingly the transfer of a grape STS under control of a heterologues inducible promoter back into grape resulted in very high expression of resveratrol (**5**) and preliminary results showed enhanced disease resistance of grape plants to the grey mould (*B. cinerea*)[27].

Taken together, these results imply a more general relevance of the STS system described. For enhanced disease resistance, the particular pathogen has to trigger foreign phytoalexin biosynthesis in the host plant. Molecular analysis of the expression of STS genes show that these genes can be induced in heterologous plants at least by perthotrophic fungi[5]. *B. cinerea* or *A. longipees* trigger resveratrol synthesis in tobacco as well as *P. infestans* in tomato and potato. In terms of agronomical relevance it might be of importance, whether the foreign phytoalexin biosynthesis is also induced by biotrophic fungal pathogens (e.g. *Erysiphe graminis, Bremia lactuca, Plasmopara viticola, Puccinia spp.*) in the respective target plant.

The regulatory pattern of STS expression which has evolved is of general relevance for disease resistance reactions of plants. The antimicrobial compound is only transiently induced at the site of necessity. This reflects one of the most sophisticated means in plant protection generated by nature. Therefore, in our opinion strategies for enhancing disease resistance should make use of the regulatory pattern of such inducible defense genes rather than expression additional defense mechanisms constitutively.

4 MODIFICATION OF STILBENE SYNTHASE EXPRESSION PATTERN CAUSES FLOWER COLOUR CHANGES AND MALE STERILITY IN PLANTS

In another strategy, the effects of constitutive stilbene synthesis were studied. A rather common approach to enhance the expression of a foreign gene makes use of the 35S RNA promoter which produces constitutively high rates of transcription. In the case of phytoalexin genes we considered that the permanent synthesis of the antimicrobial compound was possibly undesirable and could have negative effects on the physiology and the metabolic profiles in the transgenic plant. To study the effects of permanent STS expression in plants, the *Vst1* gene from grapevine was expressed under control of a 35S RNA promoter strengthened by the duplication of its enhancer region (see **Figure 3**). The effects of high foreign phytoalexin gene expression were analysed in tobacco and petunia as models. Interestingly, very high constitutive STS expression has a dramatic influence on flower colour and pollen development in these plants. This new observation resulted from the constitutive overexpression in tobacco plants. A very high constitutive level of STS expression caused white and male sterile flowers, while the remaining organs did not reveal any changes[28].

However, in petunia we observed effects on physiology and phenotype of the plant. Petunia plants expressing STS constitutively to a very high level showed smaller and curly leaves, shorter internodes as well as smaller flowers and earlier senescens[29].

Analysis of several transgenic lines demonstrated that the regulation and the level of STS expression directly determined pollen development and flower colour. Diminished coloration as well as deficient microsporogenesis seem to result from a marked reduction of flavonoid biosynthesis. In plants expressing the STS gene under control of the tapetum-specific promotor of *A. majus* only pollen development was impaired while the flower colour as well as the phenotype remained unchanged[28,29].

Nuclear male sterility has been genetically engineered by different strategies[30-35]. Here, we have shown a new strategy for the engineering flower colours and male sterilility in plants exploiting the direct competition of STS and CHS for their precursors (**Figure 4**). This means that the level of constant STS expression determines the amounts of malonyl-CoA (**18**) and 4-coumaroyl-CoA (**4**) which are permanently depleted from the metabolic pool and hence are not available for the synthesis of flavonoids needed for flower colour and pollen development. Other approaches are also connected with an alteration of flower pigmentation[36, 37]. Usually, the inhibition of flavonoid biosynthetic genes (e.g. chalcone synthase or dihydroflavonol 4-reductase) is accomplished by antisense-inhibition or cosuppression. Taken together these experiments support the notion that flavonoids not only determine flower colour, but also play an essential role in the development of the male gametophyte.

Preliminary biological tests in the greenhouse have shown that symptom density after inoculation with grey mould was slightly lower on male sterile plants expressing STS constitutively high in comparison to control plants. Nevertheless, disease resistance was not as pronounced as in plants expressing the inducible grapevine STS gene. Taking into account that following fungal attack the viral 35S RNA promoter used in these experiments was downregulated at the level of transcription, this observation was not surprising. The 35S RNA promoter was negatively influenced by fungal infection in several independent experiments[38]. This phenomenon could be explained by a silencing of viral promoters through pathogen-induced defense mechanisms. In regard to *trans*-factors, the 35S RNA promoter could be in direct competition to the endogenous plant's defence gene promoters as was shown by competition experiments and insertion of additional *as-1* elements into the 35S RNA promoter[39,40]. In this respect, it seems likely that the induction of endogenous resistance genes following pathogen attack results in a downregulation of 35S RNA promoter. Another explanation could be a negative influence of the depletion of malonyl-CoA (**18**) and 4-coumaroyl-CoA (**14**) needed for other plant owned defence mechanism such as natural phenolic compound synthesis.

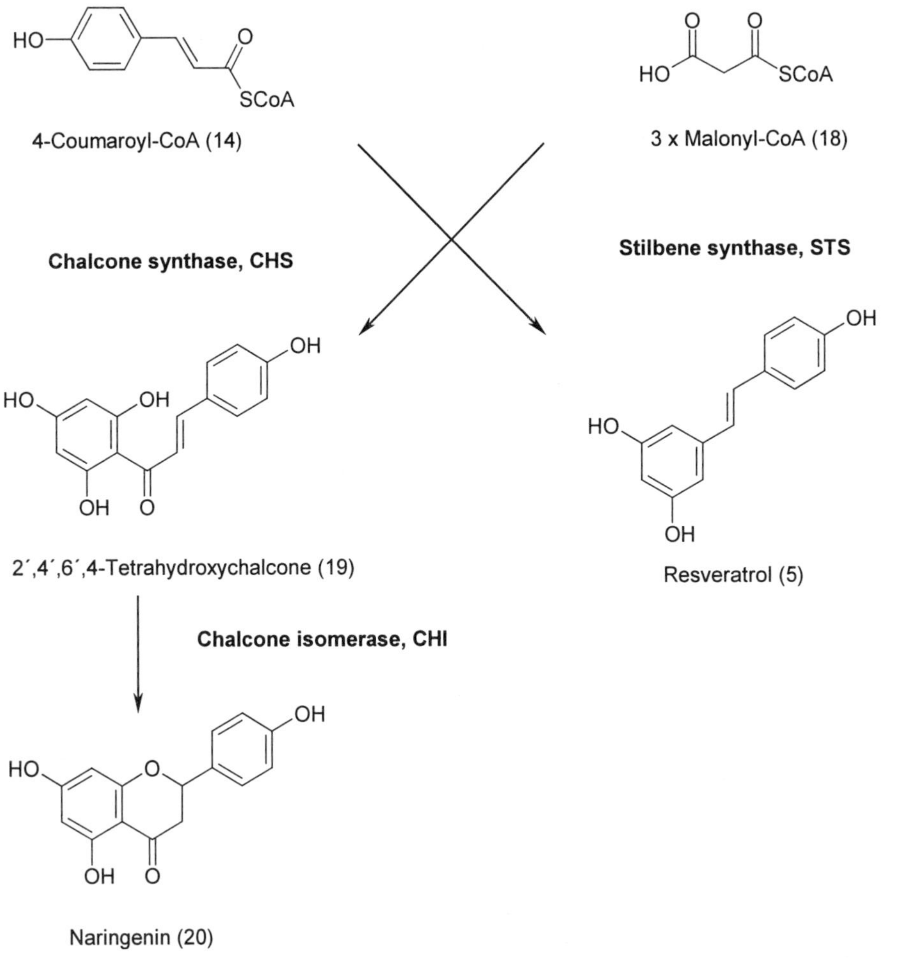

Figure 4 *Common substrates used by chalcone- and stilbene synthase*

So far, our observations led to the conclusion that in terms of engineering disease resistant plants, the enhancement of the natural STS expression pattern is more favourable than the constitutive STS expression. Perhaps the transient expression of phytoalexin genes could be further optimized by multimerization of corresponding pathogen-responsive *cis*-elements or similar improvements of signal transduction. Moreover, it remains to be clarified whether the systemic impairment of flavonoid biosynthesis in our male sterile plants may influence their tolerance against environmental stresses (e.g. UV-B). Such possible disadvantages can be avoided by tissue- and/or developmentally restricted STS expression since we have demonstrated that a tapetum-specific STS expression was sufficient for male sterility in tobacco.

These findings open possibilities to exploit stilbene synthase genes for disease resistance and male sterility - in the same transgenic crop. Future experiments will be needed and focussed on restoring fertility in our male sterile plants for the development of a novel hybrid seed system using STS genes.

5 FUTURE ASPECTS OF ENGINEERING SECONDARY PATHWAYS IN PLANTS

5.1 **Disease Resistance**

There are several genes involved in phytoalexin biosynthetic pathways whose regulation and inducibility have been investigated intensively in the plant of origin and which might provide potential in engineering improved disease resistance (**Figure 4**). One proposed means of increasing the plant´s tolerance spectrum is the modification of phytoalexins so that a certain pathogen might be unable to degrade the new compound[41]. A feasable approach may be the alteration of stereochemistry. Some enzymes, associated with stereoisomerism in plants, have been studied intensively[42, 43]. Pisatin (**7**), the major phytoalexin in pea, differs from most known pterocarpanoid phytoalexins by its (+) optical stereochemistry. The enzyme responsible for this step, an isoflavone oxidoreductase, has been isolated and characterized[44]. Since it is known that some pathogens are more sensitive to the (-) enantiomer than to the (+) enantiomer of a phytoalexin, or vice versa[45], transfer of genes responsible for stereospecificity from one plant to another could broaden the spectrum of fungal resistance. The corresponding isoflavone oxidoreductase from chickpea has already been described and a cDNA clone has been isolated[46]. This is the first step in synthesizing more potent fungitoxic compounds in plants which cannot be degraded by the pathogen.

Pinosylvinsynthase [PSS], a stilbene synthase in *Pinus sylvestris*, forms the 3,5-dihydroxystilbene, pinosylvin (**10**), which is more hydrophobic than resveratrol[47]. A cDNA clone for PSS has been isolated and intergeneric gene transfer has been performed from scots pine to tobacco. PSS has been expressed under control of the 35S RNA promoter of CaMV and a STS promotor from grapevine. Pinosylvinsynthase activity has been demonstrated in crude extracts of tobacco. The antimicrobial activity of the pine phytoalexin expressed in tobacco to different pathogens has to be evaluated. However, in the case of PSS it might be necessary to provide higher amounts of the precursor cinnamic acid (**2**). This could possibly be achieved by downregulation of endogenous cinnamate hydroxylases. The same result might be obtained by overexpressing cinnamate-CoA-ligase (**CL, Figure 1**)

This development opens the possibility to synthesize a second foreign phytoalexin besides resveratrol and might increase disease resistance and broaden the spectrum of antimicrobial activity of such novel plants.

The fungicidal bioactivity of phytoalexins may also be enhanced by means of their chemical modification. There is evidence that the fungitoxic properties of stilbenes is related to their hydrophobicity[48,49]. Therefore, modifications, e.g. methylation, might influence their biological activity. The transfer of genes like resveratrol methyltransferase or pinosylvin monomethyltransferase [PSMT] to new host plants could result in the production of the more fungitoxic pterocarpans or pinosylvin monomethylether (**10**), respectively (**see Figure 1**).

Throughout, the family of Orchidaceae bibenzyls, 9,10-dihydrophenanthrens and phenanthrens occur as phytoalexins. Very recently, bibenzyl synthase [BBS], catalyzing the commited step in the biosynthesis of these phenolic constituents (**see Figure 1**), has been purified from the orchid *Bletilla striata*[50]. The corresponding cDNA has been isolated. With such genes the synthesis of dihydrophenanthrene phytoalexins such as hircinol (**16**) and/or orchinol (**17**) is feasible in foreign plants provided that the substrates of BBS are present or inducible in the target plant. So far, enzymatically active BBS has been expressed in tobacco. However, the product of bibenzyl synthase was not yet detectable in crude extracts of transgenic tissue or plant material most likely the precursor molecules are not present in tobacco in sufficient amounts and thus it can be envisioned that 3C cinnamate hydroxylase [3CH], the 3C reductase [3CR] as well as the 3C CoA-ligase [3CL] gene have to be transferred in addition to provide the right precursors. Moreover, it is not clear whether 3,5,4, trihydroxy bibenzyl (**15**) can be converted into hircinol (**16**) in tobacco without the addition of further genes. This is an example for a more complex situation and shows that metabolic engineering might have limitations. Nevertheless, there are other enzymes like naringenin-7-O-methyltransferase [NOMT] which has been identified in rice, and which is reponsible for the synthesis of the rice phytoalexin sakuranetin[51] (**21**) biotransformed from naringenin (**20**), (**Figure 5**). The gene has not yet been cloned but it can be envisioned that this gene could be used in many plants for experiments to enhance disease resistance.

5.2 Functional Food by Modification of Secondary Plant Metabolism

5.2.1 *Definition:*

Functional food is defined as unprocessed or processed (plant) nutrition, which provides additional health benefits to consumers or animals. Nutriceuticals are natural bioactive ingredients or supplements to overcome nutritional deficiencies or to prevent disease. The markets for medical nutrition and health food will certainly increase in the future.

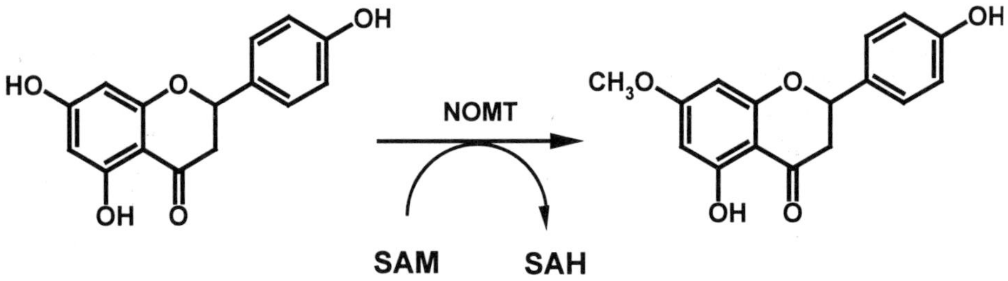

Naringenin (20) **Sakuranetin** (21)

SAM: S-adenosyl-L-methionine
SAH: S-adenosyl-L-homocysteine
NOMT: naringenin-7-O-methyltransferase

Figure 5 *Biosynthesis of sakuranetin in rice*

5.2.2 ***Application:***

The expression of STS genes could be promising tools interfering with the plant secondary metabolism with respect to

- the induction of disease resistance in plants, thus preventing the formation of toxic metabolites from phytopathogens;
- the flower colour changes in ornamentals;
- the development of a new hybrid seed systems;
- the production of functional food and forage by metabolic engineering of beneficial natural nutriceuticals and phytopharmaceuticals or by eliminating toxic (allergic) plant constituents[52].

5.2.3 ***Biology:***

Resveratrol (**5**), the product of STS expression, was recently described as a potential phytopharmaceutical[53]. It is a component in red wine and seem to be responsible for the so called French paradox, describing the observation that french people have a relatively low incidence of heart attack. In addition *in vitro* and animal experiments have shown that resveratrol possesses additional biological activities:
Protection against artheriosclerosis by its antioxidant activity, inhibition of platelet aggregation, modulation of hepaticapolipoprotein and lipidsynthesis and the production of pro atherogenic eicosanoids by human platelets[53].

Pharmascience is a leader in the production pharmaceutical grade resveratrol under the trade name of Resverin™. This company provides grant programs for research on (**5**) and states that resveratrol is a powerful antioxidant in wine and grapes and has been demonstrated to have important activity on the cardiovascular system, on cancer chemoprevention and neuroprotection.

5.2.4 ***Biotechnology:***
Thus, future uses of STS genes to enrich functional food with the phytopharmaceutical resveratrol (**5**) can be envisioned. On the one hand resveratrol could be isolated from plants overexpressing STS and be applied as a natural food supplements in the same way as synthetic material, or the STS gene could be inserted into the crop plant leading to induce the biosynthesis of (**5**) prior to processing the crop to a health food product.

6 SUMMARY AND CONCLUSIONS

Considerable progress in enhancing resistance of plants to fungal pathogens by transfer of foreign genes has been made in model systems. However, none of the strategies seem to have resulted in acceptable resistance under field conditions. Further investigations with existing tools and/or the combination of defense mechanisms might become necessary[54]. The intriguing strategy of engineering pathogen resistance by the modification of phytoalexin biosynthetic genes is to date limited by the availability of only a few biosynthetic genes. Nevertheless, the expresssion of foreign antimicrobial natural compounds may become a successful strategy in molecular plant breeding in the future. Increasing knowledge of the molecular basis of host/pathogen interactions as well as on genes being identified in gene discovery efforts via emerging plant genome projects, functional genetics, metabolic engineering and metabolic profile investigations will certainly lead to further opportunities for the modification of plant secondary metabolism[52]. We have shown that depending on the regulation of foreign phytoalexin genes biosynthetic profiles can be modified in ways which even lead to changes of plant phenotypes. The engineering of secondary metabolism will certainly not only open further possibilities to synthesize antimicrobial metabolites but also to express flavours, phytopharmaceuticals, chemical intermediates, nutriceuticals or even natural crop protection agents in plant cells or in cell culture bioreactors. The example of STS gene transfer may encourage such endeavours leading to novel and beneficial agricultural practices.

References

1. C.J. Lamb, J.A. Ryals, E.R. Ward and R.A. Dixon, *Biotechnology*, 1992, **10**, 1436.
2. I. Chet, Biotechnology in Disease Control, New-York, Wiley-Liss, 1993.
3. A.J. Anderson: In Mycotoxins and Phytoalexins, 3rd Edition, Eds. Sharma & Shalunkhe, Boca Rotan, Florida, CRC Press, 1991, 569.
4. W. Schäfer, D. Straney, L. Ciuffetti, H.D. VanEtten and O.C. Yoder, *Science*, 1989, **246**, 247.
5. R. Hain, B. Bieseler, H. Kindl G. Schröder and R. Stöcker, *Plant.Mol. Biol.*, 1990, **15**, 325.

6. R. Hain, H.J. Reif, E. Krause, R. Langebartels, H. Kindl, B. Vornam, W., E. Schmelzer,P.H. Schreier, R. Stöcker and K. Stenzel, *Nature*, 1993, **361**,153.
7. R. Fischer and R. Hain, *Current Opinion in Biotechnology* 1994**,** **5**, 125.
8. T. Lanz, G. Schröder and J. Schröder, *Planta*, 1990*,* **181**, 169.
9. W. Wiese, B. Vornam, E. Krause and H. Kindl, *Plant.Mol. Biol*, **26**, 667.
10. G. Schröder, J.W.S. Brown and J. Schröder, *Eur. J. Biochem.,* 1988, **172**, 161.
11. A. Schwekendiek, G. Pfeffer and H. Kindl, *FEBS Letters* 1992, **301**, 41.
12. A. Schöppner and H. Kindl, *J. Biol. Chem.*, 1984, **259**, 6806.
13. F. Melchior and H. Kindl, *Arch. Biochem. Biophys.*, 1991, **288**, 552.
14. J. Kangasjärvi, J. Talvinen, M. Utriainen, and R. Karjalainen, *Plant, Cell and Environment*, 1994*,* **17**, 783.
15. B. Grimmig, R. Schubert, R. Fischer, R. Hain, p.H. Schreier, C. Betz, C. Langebartels, D. Ernst, H. Sandermann Jr, *Acta Physiologiae Plantarum*, 1997, **4**, 467.
16. J.L. Ingham, *Phytochemistry*, 1976*,* **15**, 1791.
17. P. Langcake, *Physiol. Plant. Path.*, 1976, **9**, 77.
18. R. Kay, A. Chan, M. Daly and J. McPherson, *Science*, 1987, **236**, 1299.
19. J.T. Odell, F. Nagy, N.H. Chua, *Nature*, 1985*,* **313**, 810.
20. Y. Gluzman and T. Shenk, 1983, *Current Communication in Molecular Biology,* Cold Spring Harbor, New York
21. G. Khoury and P. Gruss, *Cell*, 1983, **33**, 313.
22. H. Hayashi, I. Czaja,H. Lubenov, J. Schell and R. Walden, *Science*, **258**, 1350.
23. R.X. Fang, F. Nagy, S. Sivasubraminiam and N.H. Chua, *The Plant Cell,* 1989*,* **1**, 141.
24. J. E. Thomzik, K. Stenzel, R. Stöcker, P.H. Schreier, R. Hain and D.J. Stahl, Physiol.Mol. Plant Pathol., 1997, **51,** 265**.**
25. P. Stark-Lorenzen, B. Nelke, G. Hänßler, H.P. Mühlbach, J.E. Thomzik, *Plant Cell Reports*, 1997, **16**, 668.
26. G. Leckband and H. Lörz, *Theoret Appl Genet*, 1998, in press
27. P. Coutos-Thevenot , M.C. Mauro, C. Breda, D. Buffard, R. Esnault, R. Hain and M. Boulay, VII. Symposium international sur la genetique et l'amelioration de la vigne, Montpellier (France), 1998.
28. R. Fischer, I. Budde and R. Hain*,* *Plant Journal*, 1997, **11**(3), 489
29. U. Teuschel and R. Hain, in prep.
30. W.K.F. Nacken, P. Huijser, J.P. Beltram, H. Saedler and H. Sommer*, Mol. Gen.Genet.*, 1991, **229**, 129.
31. C. Mariani, M. De Beuckeleer, M. de Block, and R.B. Goldberg, *Nature*, 1990, **347**, 737.
32. C. Napoli, C. Lemieux and R. Jorgensen, *Plant Cell*, 1990, **2**, 279.
33. A.R. van der Krol, P. J. Lenting, J. G. Veenstra, I.M. van der Meer, R.E. Koes, A.G. Gerats, J.N. Mol and A.R. Stuitje, *Nature*, 1988, **333**, 866.
34. I.M. van der Meer, M.E. Stam, A.J. van Tunen, J.N. Mol and A.R. Stuitje, *Plant Cell*, 1992, **4**, 253.
35. F. Vedel, M. Pla, V, Vitart, S. Gutierres, P. Chetrit and R. De Pape, *Plant Physiol. Biochem.*, 1994, **32**, 601.
36. G. Forkmann, *Plant Breeding*, 1991, **106**, 1.
37. G. Forkmann, *Current Opinion in Biotechnology*, 1993, **4**, 159.

38. E. Lam, P.N. Benfey, P.M. Gilmartin, R.X. Fang and N.H. Chua, *Proc. Natl. Acad. Sci.* USA, 1989, **86**, 7890.
39. R. Hain and R. Fischer, unpublished
40. G. Neuhaus, G. Neuhaus-Url, F. Katagiri, K. Seipel and N.H. Chua, *The Plant Cell,* 1994, **6**, 827.
41. H.D. Van Etten, D.E. Matthews and P.S. Matthews, *Annu.Rev. Phytopathol.*, 1989, **27**, 143.
42. N.L. Paiva, R. Edwards, Y. Sun, G. Hrazdina and R.A. Dixon, *Plant.Mol. Biol.*, 1991, **17**, 653.
43. S. Daniel, K. Tiemann, U. Wittkampf, W. Bless, W. Hinderer and W. Barz, *Planta*, 1990, **182**, 270.
44. Y. Sun, Q. Wu, H.D. Van Etten and G. Hrazdina, *Arch. Biochem. Biophys.*, 1991, **284**,167.
45. L.M. Delserone, D.E. Matthews and H.D. Van Etten, *Phytochemistry*, 1992, **31**, 3813.
46. K. Tiemann, D. Inze, M. Van Montagu and W. Barz, *Eur. J. Biochem.*, 1991, **200**, 751.
47. R. Gehlert, A. Schöppner and H. Kindl, *Molec. Plant Microbe Inter.*, 1990, **3**, 444
48. T.P. Schultz, T.F. Hubbard, J. Le Hong, T.H. Fisher and D.D. Nicholas, *Phytochemistry*, 1990, **29**, 1501.
49. V. Pont and R. Pezet, *J. Phytopathol.*, 1990, **130**, 1.
50. T. Reinecke and H. Kindl, *Phytochemistry*, 1994, **35**, 63
51. R. Rakwal, M. Hasegawa and O. Kodama, *Biochem. Biophys. Res. Com.*, 1996, **222**, 732.
52. R. A. Dixon, C.J. Lamb, S. Masoud, V.J.H. Sewalt and N.L. Paiva, *Gene*, 1996, **179**, 61.
53. G.J. Soleas, E.P. Diamandis and D.M. Goldberg, *Clinical Biochemistry*, 1997, **30**, 91.
54. R.A. Dixon and N.L. Paiva, *The Plant Cell*, 1995, **7**,1085.

Mode of Action and Resistance

A Prognosis for Discovering New Herbicide Sites of Action

Leonard L. Saari

DUPONT AGRICULTURAL PRODUCTS, STINE-HASKELL RESEARCH CENTER, PO BOX 30, NEWARK, DE 19714, USA

1 INTRODUCTION

Herbicides have been in general agricultural use since the 1940s, and their use has contributed significantly to large increases in the food supply. The search for herbicides that affect new, different sites of action (SOAs) is inspired by many reasons. Among them are the need for new resistant-management tools, increased weed spectrum, new patent and market areas, and a possibly better environmental fit. The continued efficacy of herbicides, however, is threatened by the emergence of herbicide-resistant weeds, which result from the selection pressure applied by herbicide use. The first case of triazine resistance was reported in 1970[1] and introduced a new dimension to weed control. The true impact of herbicide resistance was better appreciated as a result of the increased reporting of resistance to other herbicide classes and weeds species during the late 1980s (reviewed in 2 and 3). In part, the increase in resistance reflects an increased dependence on single-SOA herbicides as well as providing the time required for resistant weeds to become visibly present.

Similar to herbicides, human antibiotics have been in use since the early 1940s. Resistance of bacteria to the action of antibiotics has become increasingly serious. Methicillin-resistant strains of *Staphylococcus aureus* (MRSA) appeared over 20 years ago; as a result, the use of the glycopeptide antibiotic, vancomycin, increased to control this microbe and consequently, vancomycin-resistant strains of *enterococci* emerged.[4] The probable appearance of MRSA also resistant to vancomycin is of great concern for public health. Already, partial vancomycin resistance has been observed in MRSA in a clinical setting,[5] illustrating the potential consequence of evolving resistance.

In addition to the other recommendations for managing herbicide resistance,[6] an effective resistance management tool for both herbicides and antibiotics is to alternate products with different SOAs. However, a major concern in human health is that no new chemical classes of antibiotics have been introduced into clinical practice in over 30 years.[7] The main areas where antibiotics exert their effects have remained constant: protein synthesis, DNA/RNA synthesis, cell wall synthesis, and folate metabolism.[8]

Herbicides with different sites or modes of action have been discovered continuously since 1942.[9] Figure 1 shows the introduction of several herbicides based on their announcement in the literature (documentation of introduction to the literature was more accurate and obtainable than introduction into agricultural use). The rate of discovery appears to be declining slightly with time, and no new herbicide classes have been added since the mid-1980s. Is the situation for discovering herbicides with new SOAs approaching that of antibiotics, or is it more favorable? To evaluate this query, the following questions will be addressed in this paper: How prevalent have resistant weeds become? What is the current status of commercial herbicides with regard to SOA, sales, resistance, and chemical classes? What is the frequency between resistance and (i) site or mode of action, (ii) resistance mechanism, and (iii) type of inhibitor? What is the potential for discovering herbicides with new SOAs?

2 SITES AND MODES OF ACTION

Differentiating between the SOA and the mode of action (MOA) of a herbicide is important. As used in several technical circles, the SOA refers to the molecular event that is disrupted or inhibited by the binding of a xenobiotic molecule to a biologically-relevant macromolecule. For example, sulfonylurea and imidazolinone herbicides inhibit acetolactate synthase (ALS; also known as acetohyroxyacid synthase or AHAS) (reviewed in 10). Similarly, aryloxyphenoxypropionates inhibit acetyl-CoA carboxylase (ACCase) (reviewed in 11) and trifluralin disrupts microtubule formation (reviewed in 12). In contrast, the MOA refers to the process or processes, which cause(s) the weed to succumb. While the primary SOA is often known, the MOA for a herbicide is not as easily understood because of the pleiotropic effects (i.e., secondary and tertiary effects) of the herbicide. In addition to the SOA, herbicide classification is also performed by similar MOA. For instance, Sherman[13] categorized photosystem I (PSI) diverters such as paraquat and protoporphyrinogen oxidase (PPO) inhibitors such as acifluorfen together in a category called "photooxidative stress generators". The SOAs are different: paraquat affects PSI whereas acifluorfen inhibits PPO. The action of PSI diverters produces superoxide and hydroxyl radicals whereas the result of PPO inhibition creates singlet oxygen. Both herbicides, however, require light for action and

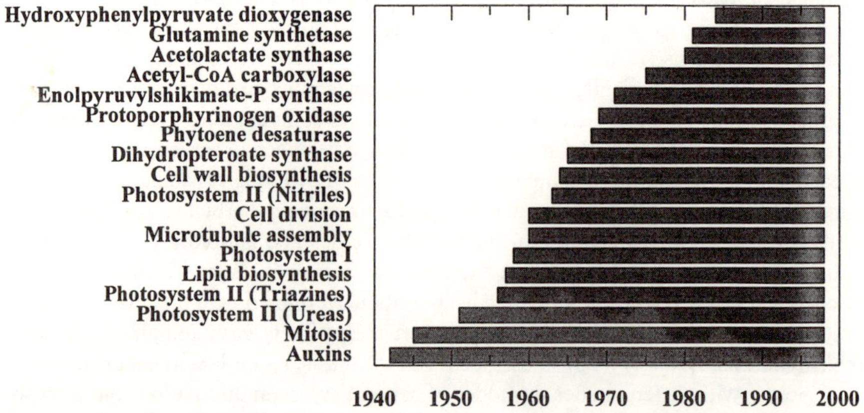

Figure 1 *Year of the Announcement for Herbicides Affecting the Above Processes (Introduction into Literature)*

produce highly-reactive and tissue- damaging oxygen species. Even though the SOAs are different, these herbicides are classified together, in this case, because of a similar MOA.

2.1 Herbicide Categories

Among the most recent and successful attempts to categorize herbicides by SOA/MOA has been performed by Retzinger and Mallory-Smith[14] and the Herbicide Resistance Action Committee[15] (Table 1). Three herbicide SOAs relate to amino acid biosynthesis and include (i) inhibitors of branched-chain amino acid synthesis (ALS inhibitors),[10] (ii) inhibitors of aromatic amino acid synthesis, namely inhibitors of enolpyruvylshikimate phosphate synthase (EPSPS) (reviewed in 16), and (iii) inhibitors of glutamine synthesis, that is, inhibitors of the enzyme, glutamine synthetase (GS) (reviewed in 17). The ALS inhibitors include sulfonylureas, imidazolinones, triazolopyrimidines, and pyrimidinyl benzoates. Only one EPSPS inhibitor exists, glyphosate. Glufosinate is the sole example of a GS inhibitor herbicide. PPO is the SOA for herbicides such as acifluorfen and oxadizon (reviewed in 18). These herbicides impede the biosynthesis of tetrapyrroles, which are required in chlorophyll and heme synthesis. Herbicides such as the dinitroanilines interfere with cell division by binding to the monomer involved in microtubule assembly, tubulin.[12] Norflurazon is an example of a herbicide that inhibits phytoene desaturase (PDS), an enzyme involved in the biosynthesis of carotenoids (reviewed in 19). Carotenoids are pigments that protect chlorophyll from bleaching. A molecule that is required in the carotenoid biosynthetic pathway is plastoquinone whose precursors are produced by p-hydroxyphenylpyruvate dioxygenase (HPPD),[20] the SOA of isoxaflutole and sulcotrione. ACCase is the SOA of the aryloxyphenoxypropionate and cyclohexanedione herbicides.[11] Inhibition of this enzyme results in the cessation of lipid synthesis. Finally, photosynthesis itself has two sites where disruption can occur: PSI (reviewed in 21) and photosystem II (PSII) (reviewed in 22). Triazines, bromoxynil, substituted ureas, and several other herbicide classes bind the D1 protein of PSII and interrupt photosynthetic election flow. In PSI, electrons are diverted by bipyridyl herbicides such as paraquat. Dihydropteroate

Table 1 *Sites and Modes of Action of Herbicides*

- Photosystem I electron (PSI)
- Photosystem II (PSII)
- Protoporphyrinogen oxidase (PPO)
- 4-OH Phenylpyruvate dioxygenase (4-HPPD)
- Phytoene desaturase (PDS)
- Acetolactate synthase (ALS)
- Enolpyruvulshikimate phosphate synthase (EPSPS)
- Glutamine synthetase (GS)
- Acetyl-CoA carboxylase (ACCase)
- Microtubule assembly
- Dihydropteroate synthase (DHPS)
- Uncoupling
- Unknown molecular sites
  - Cell division
  - Lipid synthesis (not ACCase)
  - Synthetic auxins
  - Auxin transport
  - Mitosis
  - Cellulose biosynthesis
  - Carotenoid biosynthesis (not PDS)

synthase (DHPS) is implicated as the SOA of asulam because plants transformed with a resistant form of bacterial DHPS are resistant to the herbicide.[23] Some of the herbicide SOAs are specific to plants and/or microbes. Those that are also found in mammals include PPO, 4-HPPD, GS, ACCase, membrane disruption, microtubule assembly, and DHPS. Of course, if the SOA is unknown, it is also unknown whether there is commonality between plants and animals. The importance of whether plants and animals share a SOA has to do with the increased possibility that acute and chronic toxicity might be avoided in mammals if the target was absent in that organism. For some herbicides, which act on plant but not mammalian targets, such as the ALS inhibitors that inhibit branched-chain amino acid synthesis, the acute toxicity levels are extremely favorable.[24]

There are other herbicide classes that are commercially important for which specific SOA are not identified (Table 1). For example, auxin herbicides, which include 2,4-D and dicamba, have an unknown SOA although an auxin-binding protein has been implicated.[25] Chloroacetamides, categorized as cell division inhibitors,[15] comprise a very important commercial class of herbicides worldwide, but the SOA for this class is also unknown. There are several other herbicides whose SOAs are undiscovered, but auxins and chloroacetamides are among the more prominent ones.

2.2 Metabolic Processes, Chemical Families, and Herbicide Sales

At least eight different metabolic processes are affected by commercial herbicides (Table 2). Of approximately 243 total herbicide active ingredients produced, 20%, 17%, and 13% of the active ingredients interfere with photosynthesis, amino acid biosynthesis, and lipid biosynthesis, respectively. The largest number of herbicides associated with the inhibition of photosynthesis and amino acid biosynthesis is from the triazine and sulfonylurea herbicide classes, respectively. The 243 commercial herbicides comprise three groups: 147 herbicides interfere with 11 known SOA (ACCase, ALS, PSI, PSII, PPO, PDS, 4-HPPD, EPSPS, GS, DHPS, and microtubules); 63 herbicides affect eight well-described MOA but are of unknown SOA [bleaching (not

Table 2 *Metabolic Processes Affected by Herbicides*

	Number of Active Ingredients[a]
Photosynthesis (I and II)	49
Amino acid biosynthesis (ALS + EPSPS + GS)	41
Lipid biosynthesis (ACCase + others)	32
Cell division	19
Tetrapyrrole synthesis	18
Auxin function	16
Microtubule assembly	13
Carotenoid biosynthesis	12
Other processes	10
Unknown	33
Total	243

[a] For herbicides found in The Pesticide Manual, 11th edition

PDS), mitosis, cell division, cell wall synthesis, uncoupling, lipid synthesis (not ACCase), auxin function, and auxin transport]; 33 herbicides have an unknown SOA and are not associated with any particular MOA.

Another way to appreciate SOA/MOA is to compare the associated total sales (Table 3).[26] Total worldwide herbicide sales were approximately US $14.5 billion in 1996. PSII inhibitors comprise the greatest proportion of the sales at 16%. The ALS inhibitors, a new class relative to PSII herbicides, share the second spot with EPSPS inhibitors (glyphosate) at $1.9 billion in sales. Sales of herbicides whose SOA affects the biosynthesis of amino acids is the fastest growing class. Following in order of worldwide sales are cell division inhibitors, tetrapyrrole synthesis inhibitors, the auxins, ACCase inhibitors, microtubule assembly inhibitors, and carotenoid biosynthesis inhibitors. Seventy-six herbicide products comprise US $11.7 billion or 81% of the total sales. Approximately 19% of herbicide sales are less than $50 million, and are not included in the various herbicide categories. In addition to some of the commercially-important herbicide classes, there are newer SOA such as the 4-HPPD whose inhibitors will undoubtedly increase in sales with time.

Of the eleven, known SOA for commercial herbicides, the number of chemical families found to inhibit each SOA varies from one to ten (Table 4).[15] For example, four chemical families (sulfonylureas, imidazolinones, triazolopyrimidines, and pyrimidinyl benzoates) inhibit ALS. On average, about three commercial chemical classes have been found for each SOA, suggesting that opportunities are limited after the first few chemical classes are discovered. An exception appears to be the PSII inhibitors, which comprise ten classes. However, even this example results in about three chemical classes if one separates the PSII inhibitors into the different sub-classes

Table 3 *1996 Worldwide Herbicide Sales for Different Sites/Modes of Action*

SOA/MOA	Sales (US$, Millions)	Number of Active Ingredients Included
PSII	2,260	16
EPSPS	1,950	1
ALS	1,921	15
Cell division	1,075	7
Auxins	1,010	9
ACCase	895	7
Microtubule	710	3
PSI	655	2
Lipid (-ACCase)	410	5
Carotenoid (-HPPD)	305	4
PPO	250	4
GS	130	1
4-HPPD	105	2
< $50 million	2,800[a]	-
Total	14,476	

[a] Total of products with sales of less than $50 million each

of triazines, ureas, and nitriles. This sub-classification was done in part because of different cross-resistance patterns for herbicide-resistant mutants.[14] Recalculating the number for the PSII subclasses results in five, two, and three chemical families.

3 HERBICIDE RESISTANCE IN WEEDS

3.1 Relation to Site of Action

Most major economic classes of herbicide have resistance with which to contend. Herbicide use must be carefully managed in order to preserve these valuable agronomic tools. In spite of resistance, herbicides remain useful in most parts of the world. A total of 127 weed species have evolved resistance to one or more herbicide active ingredients.[27,28] The PSII inhibitors, ALS inhibitors, PSI diverters, ACCase inhibitors, and auxins have 67, 37, 27, 14, and 13 resistant weed species, respectively, associated with them. While long residual properties are common to many of the herbicides that have selected for resistance, such as some of the PSII and ALS inhibitors, it is more likely that sequential applications played the key role for the selection pressure imposed by PSI diverters. For some herbicides such as the lipid biosynthesis inhibitors, sufficient time for selection has occurred to finally yield resistant biotypes. Of significance is the relatively recent selection of a glyphosate-resistant biotype of *Lolium rigidum*.[27-29]

Six weed species, *Alopecurus myosuroides*, *Avena fatua*, *Echinochloa crus-galli*, *Eleusine indica*, *L. rigidum*, and *Poa annua*, are associated with resistance to at least four herbicide classes.[28] Several other species have developed resistance to at least three classes. All of the above six species are grasses, and it is interesting to speculate that grasses may have an increased ability to adapt to herbicide selection pressure relative to broadleaves.

Table 4 *Number of Commercial Chemical Families Found for Each Site of Action*

SOA	Number of Commercial Chemical Families
Acetyl-CoA Carboxylase	2
Acetolactate synthase	4
Dihydropteroate synthase	1
Enolpyruvylshikimate-P synthase	1
Glutamine synthetase	1
Phytoene desaturase	3
Protoporphyrinogen oxidase	5
Photosystem I	2
Photosystem II	10 (5 + 2 + 3)[a]
Microtubule assembly	4
4-OH Phenylpyruvate dioxygenase	3
Average	~3

[a] Numbers in parenthesis are for PSII inhibitor subclasses (HRAC Groups C1, C2, and C3, respectively)[15]

Once resistance to a herbicide has been documented in *L. rigidum*, the time required for resistance to appear for subsequently-applied herbicides apparently decreases.[30] Is this phenomenon generally true for all herbicides? Figure 2 plots the time of active ingredient introduction (into the literature, not agronomic use) and the time of appearance of resistance for those herbicides associated with resistance as a function of SOA/MOA. Potential errors include the unknown variation in time between literature and agronomic introduction. Secondly, the way in which various herbicides were used may highly affect the selection pressure exerted. For example, PSI diverters may have never selected resistant weeds if sequential applications had not been made. Thirdly, whether resistance is observed at all is dependent on the observational interest in finding it for a particular herbicide. The trend based on ACCase- and ALS- inhibitor resistance (Figure 2) suggests that less time is required for resistance to appear to more newly-developed herbicides, similar to the phenomenon described for *L. rigidum*.[30] The development of herbicides with increasing selection pressure, resistant crops and the trend to rely on single-SOA herbicides will favor resistance occurring more quickly.

Several herbicide classes are not yet affected by resistance in weeds. These include inhibitors of PPO, PDS, 4-HPPD, GS, DHPS, mitosis, cell wall synthesis, and auxin transport.[28] Lack of selection pressure[31] for these herbicides may in part explain the apparent lack of resistance. Selection pressure may be reduced because of alternating herbicides with different SOAs, little continuous herbicide use, and/or using herbicides with low residual activity. Also contributing to the non-appearance of resistance is the probable low initial frequency of resistant biotypes that may be due to multiple sites being inhibited or resistance mechanisms not being readily available. For some SOA/MOA, reduced fitness of resistant biotypes and slow resistant-seed turnover also contribute to the lack of resistance. Finally, insufficient interest or effort in finding resistant weeds undoubtedly has contributed to not observing resistance for some herbicide classes.

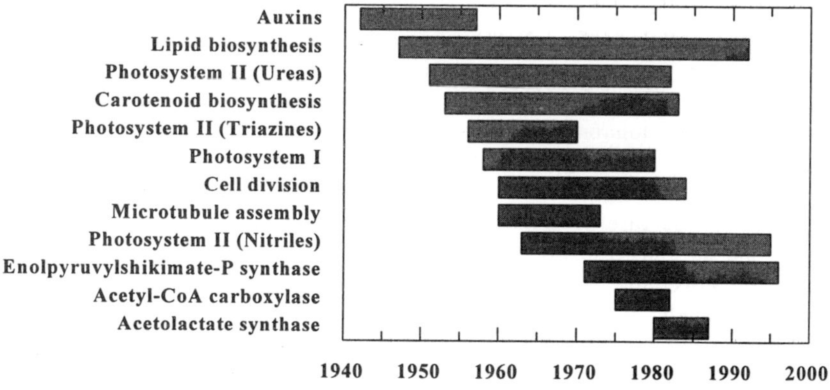

Figure 2 *Time of Herbicide Introduction (into Literature) (left edge of bar) and Appearance of Resistance (right edge of bar).*

3.2 Relation to Mechanism of Evolved Herbicide Resistance

There are several evolved mechanisms of herbicide resistance in weeds, the type of resistance that occurs because of herbicide selection. Evolved resistance does not imply any type of mechanism. The mechanisms available include target-site alteration, metabolic detoxification, reduced metabolism of pro-pesticides, and changes in uptake, translocation, or sequestration.

3.2.1 Target-Site Resistance. Reduced target-site sensitivity as a resistance mechanism is known for inhibitors of PSII,[22] ALS,[10] ACCase,[11] and microtubule assembly.[12] For those herbicides with an undefined SOA such as the chloroacetamides, it is unknown whether any of the observed resistance is due to decreased target-site sensitivity, although other mechanisms such as metabolism can be identified.

3.2.2 Target-Site Overexpression. The second type of target-site alteration that could confer resistance is overexpression. That is, the target site (e.g., an enzyme) has more copies produced *in vivo* as a result of herbicide selection. Herbicide resistance in weeds due to this mechanism is unknown. There are, however, cases where target amplification has resulted in laboratory plant cells being resistant to herbicides such as in *Daucus carota*[32] and *Nicotiana tabacum*.[33] The resistance observed in these species is actually due to two mechanisms; one is target-site insensitivity as evidenced by an increase in the ALS I_{50} values, and the other is gene amplification. To obtain the resistant lines, selection using ever-higher step-wise concentrations of the herbicide was required. Also, the gene amplification was found in cell lines, not whole plants. It is not clear whether this phenomenon will be found in whole plants; currently, it is only found in plant cell lines and other microorganisms. Central metabolism may be much less adaptable in eukaryotes relative to prokaryotes. If step-wise selection is a condition for obtaining resistance via this mechanism, then the appearance of resistant weeds with an overexpressed target in the field is unlikely.

3.2.3 Metabolism-Based Resistance. An increased rate of metabolic detoxification of a xenobiotic by a weed defines metabolism-based resistance, of which several examples have been previously cited (see 2). This mechanism is more affected by environmental conditions such as temperature since a chemical reaction is required for the metabolism. Examples include *L. rigidum* biotypes that have become resistant to chlorsulfuron, simazine and chlorotoluron due to increased metabolism of these herbicides[34] and resistant *A. myosuroides* biotypes that degrade chlorotoluron.[35]

3.2.4 Herbicide Availability. The resistance mechanism of sequestration is exemplified by the case of paraquat resistance in *Conyza* and *Erigeron* spp. The evidence includes many observations; first, photosynthesis is functional in both resistant and susceptible plants.[36] Second, herbicide sequestration was implied when paraquat apparently reached the active site in sensitive but not resistant plants.[37] Furthermore, paraquat was shown to accumulate preferentially in vacuoles of resistant plants relative to sensitive plants.[38] An alternate hypothesis for the observed paraquat resistance was that increased enzyme activity capable of degrading toxic oxygen species produced by paraquat action (e.g., superoxide dismutase) was responsible. However, the correlation between resistance and the activity of these degrading enzymes was not established. Thus, the overall body of evidence supports sequestration of the herbicide thereby making it unavailable for contacting the target site.

3.2.5 Reduced Activation of Pro-Herbicides. The reduced activation of pro-herbicides had not been documented as a mechanism of resistance in weeds until recently. *A. fatua* was found to be resistant to triallate due to the decreased metabolism of the pro-herbicide, triallate, to the more active, metabolism product, triallate sulfoxide.[39] Both sensitive and resistant biotypes are equally sensitive to the active sulfoxide metabolite. Triallate conversion to the sulfoxide in resistant plants was demonstrated to be one twelfth the rate observed in susceptible plants.

3.2.6 Multiple Resistance. A biotype of *L. rigidum*, VLR69, was shown to be resistant to nine classes of herbicides after 21 years of exposure to five herbicide classes.[40-42] Specifically, this biotype was resistant to diuron, chlorotoluron, chlorsulfuron, triasulfuron, imazaquin, atrazine, simazine, metribuzin, ametryn, ACCase inhibitors, and metolachlor but not resistant to oxyfluorfen or dinitroaniline herbicides. While there may be several resistance mechanisms at work in this weed biotype, two mechanisms, decreased target-site sensitivity and increased metabolism, were shown to be present. In other words, this was a case of multiple resistance.

Several resistance mechanisms accumulate in resistant *L. rigidum* after herbicide selection because this weed is an obligate out-crosser.[34] Upon herbicide application, the majority of the susceptible biotypes are removed from the population leaving the biotypes with various resistance mechanisms to exchange their genomes. Eventually, these survivors would accumulate several resistance mechanisms since the only other weeds with which to cross-pollinate are also resistant.

3.2.7 Relative Frequency of Resistance Mechanisms. The frequency with which the various herbicide resistance mechanisms are observed (Table 5) is a result of both scientific and social phenomena. Ideally, the initial frequency of different resistance mechanisms combined with the fitness costs[31] associated with a mechanism would be reflected. However, many of the same factors that influence the observed rate of observed resistance in weeds, as discussed previously, also affect the rate of seeing various resistance mechanisms. For example, the high number of resistance mechanisms associated with decreased binding at the target site is in part due to the survey work done by DuPont- and ALS Inhibitor Resistance Working Group (AIRWG)-sponsored research on resistant weeds in North America. Also, if more of the numerous resistant *L. rigidum* sites in Australia were characterized with regard to the resistance mechanism, the number of other mechanisms, especially metabolism-based, would undoubtedly increase. Given the current data, it is apparent that the decreased binding of the herbicide at the target site is the most prevalent mechanism followed by herbicide metabolism. Changes in uptake, translocation, or sequestration and reduced pro-herbicide metabolism have been reported infrequently. While reduced metabolism has only been reported once, relatively few herbicides are pro-herbicides, suggesting that the

Table 5 *Frequency of Documented Herbicide-Resistance Mechanisms in Weeds*

Resistance Mechanism	Number of Occurrences
Decreased binding at target site	>1,000 (ref. 43)
Overexpression of target site	0?
Metabolic detoxification	Dozens
Changes in uptake, translocation, or sequestration	<5
Reduced metabolism of pro-herbicides	1

frequency of this mechanism is greater than comparisons with other frequencies would indicate.

3.3 Relation to Type of Inhibitor

Most herbicides that inhibit known enzymic sites can be classified as either extraneous-site or reaction-intermediate inhibitors.[44] Extraneous-site inhibitors bind to a site extraneous to the active site. Since the extraneous site is not directly important to catalysis, the amino acid residues that bind the inhibitor are not as vital to the survival of the organism, and consequently, there is a higher potential for genetic mutations causing amino acid substitutions conferring resistance. In contrast, reaction-intermediate analogs are stable analogs of reaction intermediates and can bind to the enzyme very tightly. Since the catalysis is directly dependent on residues that would bind a reaction intermediate or analog, it is assumed that amino acid changes conferring resistance would be less likely. Other mechanisms of resistance, however, may not be affected with regard to the inhibitor type. Reaction-intermediate analogs, however, probably are generally simpler molecules, and may not present as many metabolic handles as more complex molecules for deactivation via metabolism. Among the eight, defined enzymic SOAs (ACCase, ALS, PPO, PDS, 4-HPPD, EPSPS, GS, and DHPS), only GS has a reaction-intermediate analog, phosphinothricin, that has been commercialized as a herbicide.[44]

The discovery of reaction-intermediate analogs as herbicides has been more difficult than that of extraneous-site inhibitors. Even if true reaction-intermediate analogs were found as initial lead discovery compounds, further attempts to produce analogs with better characteristics would probably result in some of the binding energy between the enzyme and inhibitor being derived from extraneous-site amino acids. From a resistance point of view, these hybrid inhibitors (part reaction-intermediate analog and part extraneous-site inhibitor) would be more prone to target-site resistance than true reaction-intermediate analogs. These hybrid inhibitors may even bind in a mutually-exclusive manner with true reaction-intermediate analogs because partially-overlapping binding sites exist.

4 POTENTIAL FOR DISCOVERING NEW HERBICIDE SITES OF ACTION

Finding new herbicide SOAs is an important endeavor, and statistical and empirical evidence suggests several novel SOAs are yet to be discovered. Also, new technologies such as genomics, combinatorial chemistry, high throughput screening, and informatics will add to our ability to explore and find new targets.

4.1 Number of Additional Herbicide Targets Available

It is generally acknowledged that there are about 40,000 to 60,000 genes in a typical plant. If 0.1, 1.0, or 10% of these genes are valid herbicide targets, then a plant with 50,000 genes would have about 50, 500, and 5,000 targets, respectively. Currently, only 11 commercial herbicide targets (SOA) have been identified, leaving several to be discovered.

A potential source of new targets is the 33 herbicides with unknown SOA (Table 2). In addition, several of the 63 herbicides that are classified with regard to MOA would potentially have new SOAs. Although some of the herbicides that affect different MOAs may have similar SOAs, there potentially is a total of 96 novel SOAs (i.e., 33 + 63) remaining to be identified. Some chemical optimization of inhibitors for these targets has occurred in spite of not knowing the molecular target. However, knowledge of the actual SOA might allow for both mechanistic and structural studies resulting in new chemistries that inhibit these targets.

4.2 Potential New Commercial Sites of Action

There exist some recently reported SOAs that may yield commercial inhibitors upon further study. Adenylosuccinate synthetase and AMP deaminase, enzymes involved with purine metabolism, are apparently inhibited by the *in vivo* phosphorylation products of hydantocidin[45,46] and carbocyclic coformycin,[47] respectively. If phosphorylation *in situ* is required for activity on plants, then both of these compounds are pro-herbicides. The inhibition of auxin transport is a newly-revisited herbicide SOA[48] and is reported as the SOA of a new maize herbicide known as diflufenzopyr.[49] Finally, an enzyme-directed approach has identified potent inhibitors of an enzyme involved in histidine biosynthesis, imidazoleglycerolphosphate dehydratase, which also inhibit plant cell growth and are reversible by histidine.[50]

4.3 Impact of New Technologies

New technologies will have a large impact on the discovery of new herbicide SOAs. Genomics, combinatorial chemistry, and informatics help take advantage of serendipity through the sequencing of huge numbers of genes or the synthesis of large numbers of chemical compounds. That is, there are approximately 10^{30} to 10^{50} possible molecules in molecular space[51] of which only a fraction have been synthesized. Combining this potential with having access to 50,000 plant genes in the future elevates the probability of discovering new herbicidal SOAs.

Genomics allows greater access to these genes and is defined as the mapping, sequencing, and analyzing of genomes.[52] Genomics is further divided into structural and functional genomics. Structural genomics is the study of the structure of the genome through genetic and physical methods. Currently, there is much effort to further define the structural genome of *Arabidopsis* by sequencing expressed sequence tags (ESTs) and the entire genome (to be completed in year 2000). ESTs are 450-550 base pair DNA fragments produced by isolating the mRNA from an organism of interest, producing cDNA from the mRNA, cloning the cDNA into a plasmid, and sequencing the cDNA insert. Each unique EST identifies a gene.

Functional genomics is the analysis of gene function. New SOAs are likely to be revealed through the study and characterization of genes of unknown function as well as those of known function. Two powerful tools are used: the creation and characterization of mutants deficient in a specific gene product and the capability to monitor changes in gene expression of genes simultaneously. The ability to monitor the expression of large numbers of genes should identify changes in gene expression specifically associated with a particular herbicide. That is, the inhibition of an enzyme may result in a change in the expression level for that gene or other genes in the pathway. These changes, when visualized, are a fingerprint of the particular enzyme inhibited. Both *in vivo* and *in vitro* DNA array technologies exist such as the deposition of pre-made gene segments on glass slides[53] or *in situ* oligonucleotide synthesis.[54] For example, cDNA from plants could be arrayed and attached to a glass slide and tested for hybridization with fluorescent-labeled cDNA prepared from mRNA isolated from plants plus and minus herbicide treatment. The visualization of the expression differences should identify genes involved with herbicide action, and the structural information should help identify the gene itself.

5 SUMMARY

Using several herbicides with different SOAs is sound herbicide resistance management. The current situation for herbicides, with regard to resistance, is not as serious as for antibiotics, but a faster rate of discovering new SOAs would greatly aid resistance management. There appear to be several potential new herbicide targets based on statistical and empirical evidence. Only eleven SOAs have been defined at the molecular level whereas perhaps hundreds to thousands of enzyme targets remain to be discovered. Herbicide resistance has been selected for in at least 127 weed species and at least four different resistance mechanisms identified. New technologies such as genomics, bioinformatics, high throughput screening, and combinatorial chemistry will aid in the search for new, effective herbicide SOA.

References

1. G. F. Ryan, *Weed Sci.*, 1970, **18**, 614.
2. 'Herbicide Resistance in Plants, Biology and Biochemistry', S. B. Powles and J. A. M. Holtum, eds., CRC Press, Inc., Boca Raton, FL, 1994, 353 pp.
3. 'Weed and Crop Resistance to Herbicides', R. DePrado, J. Jorrin and L. Garcia-Torres, eds., Kluwer Academic Publishers, London, 1997, 340 pp.
4. P. S. Barie, *World J. Surg.*, 1998, **22**, 118.
5. H. Hanaki and K. Hiramatsu, *Jpn. J. Antibiot.*, 1997, **50**, 794.
6. Herbicide Resistance Action Committee, 'Guideline to Management of Herbicide Resistance', HRAC Publicity Office, 4002 Basel, Switzerland.
7. B. B. Finlay and P. Cossart, *Science*, 1997, **276**, 718.
8. D. T. W. Chu, J.J. Plattner and L. Katz, *J. Med. Chem.*, 1996, **39**, 3853.
9. 'The Pesticide Manual, Eleventh Edition', C. D. S. Tomlin, ed., The British Crop Protection Council, Surrey, UK, 1997, 1606 pp.
10. L. L. Saari, J. C. Cotterman and D. C. Thill, in 'Herbicide Resistance in Plants, Biology and Biochemistry', S. B. Powles and J. A. M. Holtum, eds., CRC Press, Inc., Boca Raton, FL, 1994, Chapter 4, p. 83.
11. B. J. Incledon and J.C. Hall, *Pestic. Biochem. Physiol.* 1997, **57**, 255.
12. R. J. Smeda and K. C. Vaughn, in 'Weed and Crop Resistance to Herbicides', R. DePrado, J. Jorrin and L. Garcia-Torres, eds., Kluwer Academic Publishers, London, 1997, Chapter 9, p. 89.

13. T. D. Sherman, K. C. Vaughn and S. O. Duke, in 'Herbicide Resistant Crops', S. O. Duke, ed., CRC Press, Inc., Boca Raton, FL, 1996, Chapter 2, p. 13.
14. E. J. Retzinger, Jr. and C. Mallory-Smith, *Weed Technol.* 1997, **11**, 384.
15. Herbicide Resistance Action Committee, 'Classification of Herbicides according to the Mode of Action', HRAC Publicity Office, 4002 Basel, Switzerland.
16. W. E. Dyer, in 'Herbicide Resistance in Plants, Biology and Biochemistry', S. B. Powles and J. A. M. Holtum, eds., CRC Press, Inc., Boca Raton, FL, 1994, Chapter 8, p. 229.
17. M. D. Devine, S. O. Duke and C. Fedtke, 'Physiology of Herbicide Action', PTR Prentice Hall, Englewood Cliffs, New Jersey, 1993, Chapter 13, p. 274.
18. S. O. Duke, J. Lyndon, J. M. Becerril, T. D. Sherman, L. P. Lehnen, Jr. and H. Matsumoto, *Weed Sci.*, 1991, **39**, 465.
19. M. D. Devine, S.O. Duke and C. Fedtke, 'Physiology of Herbicide Action', PTR Prentice Hall, Englewood Cliffs, New Jersey, 1993, Chapter 8, p. 141.
20. D. L. Lee, M. P. Prisbylla, T. H. Cromartie, D. P. Dagarin and S. W. Howard, *Weed Sci.* 1997, **45**, 601.
21. C. Preston, in 'Herbicide Resistance in Plants, Biology and Biochemistry', S. B. Powles and J. A. M. Holtum, eds., CRC Press, Inc., Boca Raton, FL, 1994, Chapter 3, p. 61.
22. J. W. Gronwald, in 'Herbicide Resistance in Plants, Biology and Biochemistry', S. B. Powles and J. A. M. Holtum, eds., CRC Press, Inc., Boca Raton, FL, 1994, Chapter 2, p. 27.
23. F. Guerineau, L. Brooks, J. Meadows, A. Lucy, C. Robinson and P. Mullineaux, *Plant Mol. Biol.*, 1990, **15**, 127.
24. E. M. Beyer, M. J. Duffy, J. V. Hay and D. . Schlueter, in 'Herbicides: Chemistry, Degradation, and Mode of Action, P. C. Kearney and D. D. Kaufman, eds., Dekker, Inc., New York, 1988, Vol. 3, p. 117.
25. S. R. Webb and J. C. Hall, *Pestic. Biochem. Physiol.*, 1995, **52**, 137.
26. County NatWest WoodMac (Wood MacKenzie), The NatWest Investment Bank Group, London, 1997.
27. I. M. Heap, *Pestic. Sci.*, 1997, **51**, 235.
28. I. M. Heap, SurveyWeb
29. S. B. Powles, D. F. Lorraine-Colwill, J. J. Dellow and C. Preston, *Weed Sci.*, 1998, (in press).
30. S. B. Powles and J. M. Matthews, in 'Resistance '91: Achievements and Developments in Combating Pesticide Resistance', I. Denholm, A. L. Devonshire, and D. W. Hollomon, eds., Elsevier Science Publishers, London, 1992, p. 75.
31. J. Gressel and L. A. Segel, in 'Herbicide Resistance in Plants', H. M. LeBaron and J. Gressel, eds., John Wiley & Sons, Inc., New York, 1982, p. 325.
32. S. Caretto, M. C. Giardina and C. Nicolodi, *Theor. Appl. Genet.*, 1994, **88**, 520.
33. C. T. Harms, S. L. Armour, J .J. DiMaio, L. A. Middlesteadt, D. Murray, D. V. Negrotto, H. Thompson-Taylor, K. Weymann, A. L. Montoya, R. D. Shillito and G. C. Jen, *Mol. Gen. Genet.*, 1992, **233**, 427.
34. F. J. Tardif, C. Preston and S. B. Powles, in 'Weed and Crop Resistance to Herbicides', Kluwer Academic Publishers, Dordrecht, The Netherlands, 1997, Chapter 12, p. 117.

35. J. Menendez, J. Jorrin and R. De Prado, in 'Weed and Crop Resistance to Herbicides', Kluwer Academic Publishers, Dordrecht, The Netherlands, 1997, Chapter 18, p. 161.
36. C. Preston, in 'Herbicide Resistance in Plants, Biology and Biochemistry', S. B. Powles and J.A.M. Holtum, eds., CRC Press, Inc., Boca Raton, FL, 1994, Chapter 3, p. 61.
37. E. P. Fuerst, H. Y. Nakatani, A. D. Dodge, D. Penner and C. J. Arntzen, *Plant Physiol.*, 1985, **77**, 984.
38. M. M. Lasat, J. M. DiTomaso, J. J. Hart and L. V. Kochian, *Physiol. Plant.*, 1997, **99**, 255.
39. A. J. Kern, D. M. Peterson, E. K. Miller, C. C. Colliver and W. E. Dyer, *Pestic. Biochem. Physiol.*, 1996, **56**, 163.
40. M. W. M. Burnet, J. T. Christopher, J. A. M. Holtum and S. B. Powles, *Weed Sci.*, 1994, **42**, 468.
41. M. W. M. Burnet, Q. Hart, J. A. M. Holtum and S. B. Powles, *Weed Sci.*, 1994, **42**, 369.
42. C. Preston, F. J. Tardif, J. T. Christopher and S. B. Powles, *Pestic. Biochem. Physiol.*, 1996, **54**, 123.
43. H. M. Brown and J. C. Cotterman, *Chem. Plant Prot.*, 1994, **10**, 47.
44. L. M. Abell, *Weed Sci.*, 1996, **44**, 734.
45. D. L. Siehl, M. V. Subramanian, E. W. Walters, S-F. Lee, R. J. Anderson and A. G. Toschi, *Plant Physiol.*, 1996, **110**, 753.
46. D. R. Heim, C. Cseke, B. C. Gerwick, M. G. Murdoch and S. B. Green, *Pestic. Biochem. Physiol.*, 1995, **53**, 138.
47. J. E. Dancer, R. G. Hughes and S. D. Lindell, *Plant Physiol.*, 1997, **114**, 119.
48. M. V. Subramanian, S. A. Brunn, P. Bernasconi, B. C. Patel and J. D. Reagan, *Weed Sci.*, 1997, **45**, 621.
49. *Agrow*, 1998, **299**, 19.
50. D. Ohta, I. Mori and E. Ward, *Weed Sci.*, 1997, **45**, 610.
51. P. L. Myers, *Curr. Opin. Biotechnol.*, 1997, **8**, 701.
52. P. Hieter and M. Boguski, *Science*, 1997, **278**, 601.
53. M. Schena, D. Shalon, R. W. Davis, P. O. Brown, *Science*, 1995, **270**, 467.
54. D. J. Lockhart, H. Dong, M. C. Byrne, M. T. Follettie, M. V. Gallo, M. S. Chee, M. Mittmann, C. Wang, M. Kobayashi, H. Horton and E. L. Brown, *Nat. Biotechnol.* 1996, **14**, 1675.

ABC Transporters and their Impact on Pathogenesis and Drug Sensitivity

A.C. Andrade, L.-H. Zwiers and Maarten A. De Waard

LABORATORY OF PHYTOPATHOLOGY, WAGENINGEN AGRICULTURAL UNIVERSITY, PO BOX 8025, 6700 EE WAGENINGEN, THE NETHERLANDS

Abstract: This review presents an outline of the multifunctional properties of ABC transporters in different biological systems. A well known function of these transport proteins is protection of organisms against toxic compounds. This also applies to plant pathogens. We propose that ABC transporters can play an important role in plant pathogenesis and fungicide sensitivity and thus can be regarded as potential target sites for the discovery of new biologically active compounds.

1 INTRODUCTION

Transport is one of the most important and fascinating aspects of life and an essential requirement in all organisms. Unicellular organisms need to maintain their homeostatic balance with constant uptake and allocation of nutrients and the secretion of toxic (waste) products. They must also be able to sense changes in their biotic and abiotic environment. In addition, multicellular organisms need to transport metabolites and information to and from organs. Multicellular organisms even possess specialised organs (tissues) for transport functions, *e.g.*, the blood and nervous system in animals and the vascular tissue in plants.

The main barrier for any transport event is the plasma membrane. Compounds can passively cross this barrier by diffusion. Transport by diffusion is possible only down a concentration gradient and is limited to solutes able to partition in hydrophobic membranes. Therefore, transport of most compounds over membranes is mediated by membrane bound proteins with specialised transport functions. With the unravelling of the genomes from different organisms the importance of membrane transporters becomes obvious. For instance, the complete genomic sequence of the gram-positive bacterium *Bacillus subtilis* possesses 2379 protein encoding ORFs with a known function. Of these proteins, 381 are likely to be involved in transport.[1] This means that about 16% of the genes of this organism codes for membrane transporters. Several types of membrane transporter systems can be distinguished.

1.1 Ion Channels

Ion channels are membrane complexes mediating the movement of ions across plasma membranes as well as membranes of cell organelles. These channels form a pore allowing the passive flux of ions down its electrochemical gradient. The opening of these channels is generally gated. This means that the opening is regulated by changes in membrane potential, membrane stretching or binding of a ligand. Ion channels play a role in diverse functions such as osmoregulation, cell growth, development, and nutrient uptake.[2]

1.2 Facilitators

In contrast to ion channels, facilitators or carriers bind molecules to be transported and undergo a reversible change in conformation during transport. Based on the energy source driving the transport facilitators can be classified in primary and secondary active transport systems.

1.2.1 Primary Active Transport Systems. Transporters belonging to this system couple transport to ATP hydrolysis. This provides the energy to transport solutes against an electrochemical gradient. Besides proton translocating ATPases two other families of ATP utilising transporters are described. The P-type ATPases that make up a large superfamily of ATP-driven pumps involved in the transmembrane transport of charged substrates and the ATP binding cassette (ABC) transporters.[3]

1.2.2 Secondary Active Transport Systems. Transporters belonging to this system derive the energy needed for transport from an electrochemical gradient over the membrane. Facilitated diffusion, the transport of solutes down its own electrochemical gradient, is generally mediated by uniporters. When transport of solutes takes place against an electrochemical gradient, the energy to drive this process is supplied by the symport or antiport of H^+ or other ions down their electrochemical gradient. A well characterised group is the major facilitator (MF) superfamily of transporters. Members of this superfamily function as H^+-substrate antiporters which use the proton motive force to drive transport.[4]

This review describes ABC transporters and presents an overview of their structural diversity and multifunctional character in a variety of biological systems. Emphasis will be on ABC transporters of (filamentous) fungi.

2 ABC TRANSPORTERS

2.1 Significance

ABC transporters are members of a large superfamily of transporters. Generally, they are located in plasma membranes and intracellular membranes and include both influx and efflux systems. ABC transporters are present from archae bacteria to man but became especially known for their involvement in multidrug resistance (MDR) in tumour cells.[5] The phenomenon of MDR is accompanied by a massive overproduction of ABC transporters.[6] Besides MDR, they are also involved in various diseases such as cystic fibrosis, diabetes, adrenoleukodystrophy and the Zellweger syndrome. Furthermore, they play a role as peptide transporters in antigen presentation and in chloroquine resistance in the malarial parasite *Plasmodium falciparum*.[7-9]

2.2 Abundance of ABC Transporters

ABC transporter encoding genes are present in genomes of species representing all three domains of life *e.g.* archae, eubacteria and eukaryotae. In several of these classes of organisms, ABC transporters constitute the largest family of proteins (Table 1). Analysis of transport proteins in seven complete genomes of prokaryotic organisms shows that ABC-transporter and MF superfamilies account for an almost invariant fraction (0.38 to 0.53) of all transport systems per organism. The relative proportion of the two classes of transporters varies over a tenfold range, depending on the organism.[10]

In eukaryotes the number of ABC transporters reported in literature is steadily increasing. In *Saccharomyces cerevisiae*, to date the only eukaryotic organism with the complete genome sequenced, 29 ABC-transporter proteins have been identified. In ongoing genome sequencing projects on other eukaryotic species, sequences homologous to ABC transporters have been detected as well.

2.3 Molecular Architecture

ABC-transporter proteins are characterised by the presence of several highly conserved amino acid sequences in their ABC domain. Two of these motifs , the Walker A [G-$(X)_4$-G-K-(T)-$(X)_6$-I/V] and Walker B [R/K-$(X)_3$-G-$(X)_3$-L-$(\text{hydrophobic})_4$-D], are found in any ABC transporter and in many other proteins which bind and hydrolyse nucleotides.[11-14] The Walker motifs are separated by 120-170 amino acids including a motif characteristic for ABC transporters. This so-called ABC signature, [L-S-G-G-$(X)_3$-R-hydrophobic-X-hydrophobic-A] is highly conserved among ABC transporters only.[15] The presence of multiple membrane spanning regions is also characteristic for ABC transporters.

All members of the ABC-transporter superfamily have a modular architecture. The majority of the ABC transporters in higher organisms consists of two transmembrane domains (TMD), each with six predicted membrane spanning regions, and two

Table 1 *Number of ABC transporters in species representing different domains of life*

Domain	*Category*	*Species*	*Genome size (nt)*	*ABC transporters* [a]	*Ref.*
Archae	Euryarchaeotae	*Archaeoglobus fulgidus*	2,178,400	40	16
Eubacteria	Firmicutes	*Bacillus subtilis*	4,214,807	77	1
Eubacteria	Proteobacteria	*Escherichia coli*	4,639,221	79	17
Eukaryotae	Fungi	*Saccharomyces cerevisiae*	12,069,313	29	18

[a] The figures do not give the number of transport systems, since these can be assembled from different polypeptides.

intracellular located nucleotide binding folds (NBF) in a two times two-domain configuration. The nucleotide binding domain can be either located at the amino terminus or at the carboxy terminus of the polypeptide, yielding proteins with a $[TMD\text{-}NBF]_2$ or $[NBF\text{-}TMD]_2$ configuration. The best characterised examples of ABC transporters with the $[NBF\text{-}TMD_6]_2$ and $[TMD_6\text{-}NBF]_2$ configuration are the yeast multidrug transporter PDR5 and the human multidrug transporter P-glycoprotein (P-gp or MDR1), respectively.[19-22]

The domains can be formed as separate polypeptides or as a single polypeptide with one or more domains fused. Separate polypeptides subsequently aggregate to form functional transporters. In eukaryotic organisms the polypeptides are generally composed of at least two domains but usually contain all four domains.[23] The so-called "half-sized" transporters with a [TMD-NBF] or [NBF-TMD] configuration are likely to function as dimers.[24]

Multidrug resistance associated proteins (MRP) form a subfamily of ABC transporters with a $TMD_4\text{-}[TMD_6\text{-}NBF]_2$ topology. The main difference with other ABC transporters resides in the presence of an additional transmembrane spanning region at the protein amino terminus. An additional difference is the presence of the so-called R-region located between the two homologous halves. The R-region is involved in regulation of the protein. MRPs act as glutathione-S-conjugate carriers and have been identified in a broad variety of organisms. The best described example is the human MRP involved in broad spectrum drug resistance.[25-27]

2.4 Substrates

Substrates of ABC transporters range from 107 kDa proteins (*e.g.* haemolysin) to ions (*e.g.* Cl^-).[23, 28] Most of the mammalian MDR proteins are (by definition) able to transport a wide variety of compounds although substrate specific transporters also occur. For instance, the human ABC-transporter P-gp (MDR1) has 93 known substrates from various chemical classes either of natural or synthetic origin. The main denominator is their high hydrophobicity.[29] Recently, a screening of the structures of these 93 substrates for potential spatial relationships between structural elements responsible for interaction with P-gp revealed that the presence of two or three electron donor groups with a spatial separation of 2.5 or 4.6 Å could be correlated with interaction with P-gp.[30]

Eukaryotic organisms also contain ABC transporters with a specific substrate range. For example, STE6 from *S. cerevisiae* involved in secretion of the mating factor and TAP1 and TAP2 involved in human antigen presentation.[31, 32] Bacterial ABC transporters involved in drug resistance have a very specific substrate specificity and are known as specific **d**rug **r**esistance transporters (SDR). Only one bacterial ABC transporter involved in MDR, LmrA from *Lactococcus lactis*, has been detected.[33, 34]

Although ABC transporters are generally described as transporters some can also act as channels and regulators of channels. The cystic fibrosis transmembrane conductance regulator (CFTR) is an ABC transporter with channel function. The associated chloride channel is time and voltage independent and requires ATP hydrolysis for opening.[35, 36] The human P-gp seems to control an associated ATP-dependent volume regulated chloride channel activity.[37, 38]

3 P-GLYCOPROTEIN

The human P-gp (MDR1) is probably the best characterised ABC transporter involved in multidrug resistance. Detailed structure-function relationship studies have been performed and its structure has been determined to 2.5 nm resolution.[39] P-gp was first described in hamster cell lines in which the MDR phenomenon was correlated with the overexpression of a 170 kDa protein.[20] In human, two P-gp homologues, MDR1 and MDR3, have been identified. MDR1 is involved in broad-spectrum drug resistance and MDR3 in the translocation of phosphatidylcholine.[40] The overexpression of MDR1 in resistant cells with a low and high degree of resistance is due to elevated mRNA levels caused by regulatory mutations and gene amplification, respectively.[41]

3.1 Catalytic Sites

Biochemical evidence and amino acid sequence information suggest that P-gp has ATPase activity. Membrane bound or purified P-gp preparations show a basal ATPase activity which can be stimulated by several drugs.[42] Both nucleotide binding folds bind and hydrolyse ATP.[43] Synthetic half-sized P-gp molecules also display basal ATPase activity. However, interaction between both halves seems necessary for stimulation of ATPase activity by drugs.[44] This is also demonstrated by mutating either of the two nucleotide binding domains. Inactivation of NBF_1 results in a block of ATP hydrolysis in NBF_2 and abolishes the drug extrusion capacity of the cells, and *vice versa*.[45] Interaction between nucleotide binding sites was also demonstrated by vanadate-trapping experiments. This inhibitor of ATPase activity traps ADP in a catalytic site and trapping of ADP at only one site is sufficient to block ATPase activity of the entire protein.[46, 47]

These results have lead to a model for the catalysis mediated by P-gp in which both catalytic sites alternately undergo ATP hydrolysis. ATP binding at one site promotes ATP hydrolysis at the other. This induces a conformational change preventing the hydrolysis of the new-bound ATP. This new conformation has a high energetic state and relaxation of this conformation leads to the release of ADP and P_i, and transport of a substrate.[48]

3.2 Substrate Binding and Transport

Photoaffinity labelling and mutant analysis indicate that both membrane-bound halves of ABC transporters are involved in substrate binding. The substrate binding sites are located at the cytoplasmic site of the membrane, especially in transmembrane loops 4, 5, 6, 10, 11 and 12.[49-52]

The way ABC transporters expel their substrates is not completely understood. ABC transporters probably act as “hydrophobic membrane cleaners” by detecting drugs which partition in membranes because of their hydrophobic nature. The possibility that transport out of the cytosol also contributes to the efflux can not be excluded.[33, 53, 54] The result of both transport processes is reduced accumulation of toxic compounds at their intracellular target site. The recently determined structure of P-gp revealed a large central pore forming a chamber within the membrane.[39] Whether this pore is involved in the transport process or whether the transport occurs through conformational changes upon ATP binding and hydrolysis remains unclear.

4 PHYSIOLOGICAL FUNCTIONS

4.1 Prokaryotes

Bacterial ABC transporters can be functionally grouped in two major distinct subfamilies. The superfamily of importers is responsible for transport of nutritional substrates. These transporters are also called periplasmic permeases and have a multisubunit component system with a similar structural organisation.[55] The presence of a periplasmic binding protein and the synthesis of the import system subunits (NBF and TMD) as separate polypeptides are distinctive features to the eukaryotic ABC proteins. The histidine permease from *Salmonella typhimurium* is a well-characterised member of this subfamily. It is composed of the histidine-binding protein (HisJ) as the receptor, and the membrane-bound complex formed by two copies of HisP (NBF) plus the HisQ and HisM (TMD).[56]

The subfamily of ABC exporters is involved in secretion of proteins, peptides and non-protein compounds.[57] In general, a basic functional structure for ABC exporters is composed of dimeric molecules.[58] The ATP-binding motif of this subfamily shows a higher degree of similarity with the eukaryotic ABC proteins as compared to the above mentioned ABC importers. In addition, some ABC exporters have their NBF and TMD domains synthesised as a single polypeptide. In gram-negative bacteria, additional export proteins are required for transport to the extracellular medium. For instance, HlyD and TolC which are involved in the secretion of haemolysin in *E. coli*.[59] Other examples of prokaryotic ABC transporters are the export system of proteases A, B and C in the phytopathogenic bacterium *Erwinia chrysanthemi*,[60] the secretion machinery of peptide antibiotics (bacteriocins) from *Lactococcus lactis*,[61] and the β-1,2-glucans oligomers export systems of the plant pathogen *Agrobacterium tumefaciens* (CHVA)[62] and the symbiont *Rhizobium meliloti* (NDVA).[63] β-1,2-Glucans oligomers are involved in the attachment of the bacteria to plant cells. Therefore, CHVA can be regarded as a virulence and NDVA as a nodulation factor.

4.2 Eukaryotes

4.2.1 Yeasts. With the unravelling of the complete genome sequence of *S. cerevisiae* 29 ABC proteins were identified by sequence homology.[18] Only ten of these proteins have a known physiological function. The 29 encoded ABC polypeptide sequences could be divided in six subfamilies. The majority of the proteins have the tetra-domain modular architecture comprising nine proteins with the $[NBF\text{-}TMD]_2$ and seven with the reverse $[TMD\text{-}NBF]_2$ topology. Furthermore, "half-sized" [TMD-NBF] proteins, which likely function as dimers, were detected. For instance, the peroxisomal ABC transporters PXA1 and PXA2 form heterodimers and are involved in long-chain fatty acid transport and β-oxidation.[24] The yeast ABC proteins with a known physiological function, different from a role in MDR, are listed in Table 2. The MDR proteins of *S. cerevisiae* are discussed in section 5.1.

In the fission yeast *Schizosaccharomyces pombe* and the human pathogen *Candida albicans*, ABC proteins have also functionally been described. All of them have an orthologue in the genome of *S. cerevisae*.[64-68]

Table 2 *ABC transporters from* Saccharomyces cerevisiae *with an identified physiological function*

Gene	*GenBank number*	*Size (aa)*	*TMD*	*Topology*	*Knock-out*	*Function*	*Ref.*
STE6	Z28209	1290	12	$[TMD + NBF]_2$	Viable	**a**-pheromone export	31
ATM1	Z49212	690	5	TMD + NBF	Restricted growth	Mitochondrial DNA maintenance	69
PXA1	L38491	870	5	TMD + NBF	Viable	VLCFA beta-oxidation	70
PXA2	X74151	853	6	TMD + NBF	Viable	Interaction with PXA1	70
GCN20	D50617	752	0	$[NBF]_2$	Viable	Interactions with tRNA and GCN2	71
YEF3	U20865	1044	3	$[NBF]_2$	No growth	Aminoacyl-tRNA binding to ribosomes	72

4.2.2 Filamentous Fungi. Members of the ABC-transporter superfamily have been described for at least seven fungal species (Table 3). The saprophyte *Emericella nidulans*, the human-pathogens *Aspergillus flavus* and *Aspergillus fumigatus* and the plant pathogens *Magnaporthe grisea* (rice blast), *Botryotinia fuckeliana* (grey mould), *Mycosphaerella graminicola* (wheat leaf blotch) and *Penicillium digitatum* (citrus green mould).[73-77] In addition, many other members are expected to be revealed in ongoing fungal genome sequencing projects. We screened available expressed sequence tags (EST) data bases of *E. nidulans* and *N. crassa* for potential homologues of ABC transporters with the conserved motifs listed in Table 3.[78, 79] The search was performed with the BLAST program of sequence alignment and yielded seven homologous sequences from *E. nidulans* and two from *N. crassa* (Table 4).[80] EST clones identical to atrC and atrD, two previously characterised genes from *E. nidulans*, were also detected.[81]

The physiological relevance of ABC transporters in filamentous fungi is probably high.[82] For instance, a number of them may be involved in secretion of secondary metabolites, which in the case of fungitoxic compounds, can act as a self-protection mechanism. Similarly, ABC transporters may provide protection against toxic metabolites produced by other micro-organisms present in particular ecosystems. Plant pathogenic fungi have to cope with a variety of plant defence compounds and they may posses ABC transporters which function in protection against the toxic action of such compounds as well. These hypotheses are supported by the observation that a wide variety of natural compounds such as isoflavonoids, plant alkaloids and antibiotics can act as substrates of ABC transporters.[29, 30] In addition, specific ABC transporters of filamentous fungi may function in secretion of a mating factor as shown for several yeast species. Therefore, ABC transporters can mediate processes important for survival of fungi in nature and hence, may function as significant parameters in the population dynamics of these organisms.

4.2.3 Higher Eukaryotes. Basically, the majority of the ABC transporters characterised in higher eukaryotes have an orthologue in the *S. cerevisae* genome or at least a very close homologue with similar substrate specificity. However, due to evolutionary speciation, physiological needs may be different and account for differences in ABC-transport proteins. This is well illustrated by the high number of MRP-like transporters already characterised in

Table 3 *Multiple alignment of conserved sequences from reported ABC-transporter proteins from filamentous fungi*

Species	*Gene*	*GenBank number*	*Domain*		
				N terminal	
			Walker A	ABC signature	Walker B
E. nidulans	*atrA*	Z68904	LGRPGTGCSTFL	VSGGERKRVSIAE	AAWDNSSRGLD
E. nidulans	*atrB*	Z68905	LGRPGSGCTTLL	VSGGERKRVSIIE	FCWDNSTRGLD
B. fuckeliana	*pgp1*	Z68906	LGRPGSGCSTFL	VSGGERKRVSIAE	VSWDNSTRGLD
M. grisea	*abc1*	AF032443	LGPPGSGCSTFL	VSGGERKRVTIAE	QCWDNSTRGLD
A. fumigatus	*mdr1*	U62934	VGPSGSGKSTVV	LSGGQKQRIAIAR	LLLDEATSALD
A. flavus	*mdr1*	U62932	VGPSGSGKSTII	LSGGQKQRIAIAR	LLLDEATSALD
			* * * *	*** * *	* **
				C terminal	
			Walker A	ABC signature	Walker B
E. nidulans	*atrA*	Z68904	MGVSGAGKTTLL	LNVEQRKLLTIGV	LFLDEPTSGLD
E. nidulans	*atrB*	Z68905	MGSSGAGKTTLL	LSVEQRKRVTIGV	IFLDEPTSGLD
B. fuckeliana	*pgp1*	Z68906	MGASGAGKTTLL	LSVEQRKRVTIGV	LFLDEATSGLD
M. grisea	*abc1*	AF032443	MGVSGAGKTTLL	LNVEQRKRLTIGV	LFVDEPTSGLD
A. fumigatus	*mdr1*	U62934	VGPSGCGKSTTI	LSGGQKQRVAIAR	LLLDEATSALD
A. flavus	*mdr1*	U62932	VGASGSGKSTTI	LSGGQKQRIAIAR	LLLDEATSALD
A. fumigatus	*mdr2*	U62936	VGPSGGGKSTIA	LSGGQKQRIAIAR	LILDEATSALD
			* ** ** *	* * *	** ** **

Asterisks indicate identical amino acid residues.

the genome of *Arabidopsis thaliana*.[83] These transporters share with YCF1, the closest yeast homologue, glutathione-S-conjugate transport activity. In addition, plant MRP proteins have the property to transport chlorophyll catabolites.[84, 85] Other physiological functions of ABC transporters in higher eukaryotes have been described as well.[27, 28, 40, 86-88]

5 MULTIDRUG RESISTANCE

The use of cytotoxic compounds such as drugs in clinical medicine and disease control agents in agriculture is an essential component of human life. However, the widespread and sometimes excessive use of these compounds has resulted in a high selection pressure resulting in drug resistant populations. This phenomenon is of major concern to society.

Table 4 *Partial sequences of putative ABC transporters detected in the Expressed Sequence Tags (EST) databases from* Emericella nidulans *(E.n.) and* Neurospora crassa *(N.c.)*

Species	*EST clone*	*GenBank number*	*Walker A*	*ABC signature*	*Walker B*	*S. cerevisae homologue*	*BLAST score* [a]
E.n.	h8h04a1.r1	AA785885		NVEQRKRLTIGV	LFLDEPTSGLD	PDR10	e^{-47}
E.n.	c9e04a1.f1	AA783966		SGGQKQRLCIAR	LLLDEATSSLD	MDL1	e^{-17}
E.n.	o8f05a1.f1	AA787659			LLLDESTSALD	YCF1	e^{-09}
E.n.	e7d04a1.r1	AA784517	GPNGSGKTTLM			YEF3B	e^{-30}
E.n.	m7a02a1.r1	AA786886	GRNGAGKSTLM			YPL226	e^{-23}
E.n.	e4a06a1.r1	AA784449	GLNGCGKSTLI			YPL226	e^{-07}
E.n.	k5a05a1.f1	AA786673			SFLDEPTNTVD	YEF3B	e^{-44}
N.c.	NCM8C11T7	AA901957		SQGQRQLVGLGR	VIMDEATASID	YLL015	e^{-20}
N.c.	NCC3E5T7	AA901865		SDGQKSRIVFAL	LLLDEPTNGLD	YER036	e^{-65}

[a] Based on homology of the full EST clone.

In general, the major mechanisms underling the mechanism of resistance in prokaryotes and eukaryotes can be classified as follows: (a) enzymatic inactivation or degradation of drugs, (b) alterations of the drug target-site and, (c) decreased drug-accumulation caused by energy-dependent drug efflux. More than one mechanism may operate in concert and the sum of different alterations represents the final resistant phenotype.

In several cases the resistance mechanism not only conferred decreased sensitivity to a specific drug (and analogues) used during the selection process, but also to several structurally and functionally unrelated compounds. This phenomenon, termed **multidrug resistance** (MDR) has been described to operate in a broad range of organisms. It relates to decreased accumulation of drugs via energy-dependent drug efflux systems. The majority of the transport proteins involved in drug-extrusion as determinants of MDR belong either to the ABC transporter or the MF superfamilies.

5.1 ABC Transporters in MDR in Prokaryotes and Lower Eukaryotes

In prokaryotes most of the characterised efflux-systems involved in MDR utilise the proton motive force as energy source for transport and act via a drug/H^+ antiport mechanism. The first example of a prokaryotic ABC transporter involved in MDR is the LMRA protein from *Lactococcus lactis*.[89] The gene encodes a 590 aa membrane protein with the TMD_6-NBF topology. The protein probably functions as a homodimer. Functional studies performed in *E. coli* indicate that its substrate specificity comprises a wide range of hydrophobic cationic compounds, very similar to the pattern displayed by the human MDR1. Surprisingly, when expressed in human lung fibroblast cells, LMRA was targeted to the plasma membrane and also conferred typical multidrug resistance, confirming the evolutionary relation of these two proteins.[34]

Genes encoding ABC transporters in parasitic protozoa have been isolated and analysed from *Plasmodium, Leishmania*, and *Entamoeba spp*., and variation in the copy number and/or levels of expression have been implicated in drug resistance.[90]

From *S. cerevisiae*, at least four members of the ABC transporter superfamily are involved in MDR: PDR5, SNQ2, YCF1 and YOR1.[19, 26, 91, 92] PDR5 and SNQ2 have the $[NBF\text{-}TMD]_2$ topology and preferential substrate specificity for aromatic cationic compounds, whereas YCF1 and YOR1 have the $[TMD\text{-}NBF]_2$ orientation and substrate specificity for anionic compounds. Despite its inverted topology and low sequence similarity, PDR5 seems to be the yeast functional homologue of the human MDR1, if substrate specificity is considered.[93] The presence of several other ABC proteins from *S. cerevisiae* with high homology to the ones involved in MDR and with common regulatory mechanisms suggests that other ABC transporters may be involved in MDR of *S. cerevisiae* as well.[94] In yeast species such as *S. pombe* and *C. albicans* multidrug-efflux systems based on overproduction of ABC transporters have also been identified. Examples are CDR1 and CDR2 from *C. albicans* and PMD1 and BFR^+ from *S. pombe*.[95-98]

5.2 MDR in Filamentous Fungi

MDR in filamentous fungi has been reported for laboratory generated mutants of *E. nidulans* selected for resistance to azole fungicides. In genetically defined mutants, resistance to azoles is based on an energy-dependent efflux mechanism which results in decreased accumulation of the compounds in fungal mycelium.[99] This mechanism also operates in other species such as *P. italicum*, *B. fuckeliana, Nectria haematococca* and probably *M. graminicola*.[99, 100] In our laboratory, ABC transporter encoding genes from *E. nidulans*, *B. fuckeliana* and *M. graminicola*, have been isolated and are currently functionally characterised.[73, 75, 76, 81] The isolated genes display a high degree of homology with PDR5 and PMD1, yeast ABC transporters involved in MDR. AtrB from *E. nidulans* complements a PDR5 null mutant of *S. cerevisiae*, suggesting indeed a role in fungicide sensitivity and resistance.

Very recently, the involvement of an ABC transporter (PMR1) in azole resistance has been established for field isolates of the phytopathogenic fungus *P. digitatum*, the causal agent of citrus green mold.[77] Another example is MDR1 from the human pathogen *A. fumigatus* which confers decreased sensitivity to the antifungal compound cilofungin when overexpressed in yeast.

6 EVOLUTIONARY ASPECTS OF ABC TRANSPORTERS

The ubiquitous occurrence of ABC transporters throughout the living world indicates the ancient character of this superfamily of proteins. They are believed to date back in evolutionary time for more than 3 billion years.[101] Thus, the understanding of evolutionary relationships among these transporters might be helpful in elucidating the origins of multidrug efflux systems, their physiological functions, and more important, the nature of their substrate specificity.

Recently, two paradigms on the evolution of bacterial multidrug transporters have been proposed. The first one describes that the transporters have evolved to protect cells from structurally diverse environmental toxins. The second one states that the transporters initially functioned in transport of specific physiological compounds (or a group of structurally related natural compounds) with the ability to expel drugs being only a fortuitous side effect.[102] Experimental evidence has been proposed for both hypotheses, but it is unlikely that transport proteins have evolved numerous distinct binding sites for structurally dissimilar molecules and therefore, a physiological substrate is likely to exist.[103] Furthermore, the presence of accessory factors as determinants of substrate specificity, such as the periplasmic binding proteins of the prokaryotic uptake systems or the eukaryotic glutathione-S-conjugate export pumps, could explain, in part, the accommodation of structurally unrelated compounds by ABC proteins.

Comparison of multidrug transport systems from six complete genomes of bacteria (three pathogenic and three non-pathogenic), indicates that, with one exception (*Methanococcus jannaschii*), the number of multidrug efflux pumps is approximately proportional to the total number of encoded transport systems as well as the total genome size. Therefore, the similar numbers of chromosomally encoded multidrug efflux systems in pathogens and nonpathogens suggests that these transporters have not arisen recently in pathogenic isolates in response to antimicrobial chemotherapy.[101] However, during speciation, novel ABC transport proteins with modified substrate specificity might have evolved, as a result of fusions, intragenic splicings, duplications and deletions, in order to accomplish the different needs of organisms occurring in distinct environments. This can be illustrated by the occurrence of bacterial ABC transporters as separate subunits (*e.g.*, NBF and TMD) and by the inverted topology of domains observed in eukaryotic proteins. In addition, neither homologous proteins nor a characteristic motif of the so-called cluster I of yeast ABC transporters have, as yet, been found in prokaryotes.[18]

The considerations mentioned above and the experimental data available suggest that a MDR phenotype is not primarily caused by the appearance of a novel transport protein with a modified substrate profile but rather by an increased expression level of a pre-existent transport system as a result of alterations in regulation of such proteins (PDR in yeast) or gene amplification (MDR in mammalian).

7 PERSPECTIVES

Since the early 1980s the significance of ABC transporters for drug sensitivity and resistance has been recognised in the medical field. A similar interest in the role of ABC transporters in agriculture only started recently. Now, there is a growing awareness that ABC transporters can be involved in mechanisms of natural insensitivity and acquired resistance in a wide range of organisms. In this review, we provide evidence that this also holds true for (pathogenic) filamentous fungi.

In the treatment of MDR-cancer cells inhibitors of ABC-transporter activity are used as synergists of drugs to reduce the MDR phenotype. If MDR would be the main mechanism of resistance to azoles, similar inhibitors could be useful in mixtures with these fungicides to increase control of azole resistant populations of (plant) pathogenic fungi. If ABC transporters also play a role in protection against plant defence compounds and/or secretion of pathogenicity factors, inhibition of ABC transporter activity would result in enhanced host resistance and/or reduced virulence of the pathogen. Both processes would reduce disease development. As described for *S. cerevisiae* and *S. pombe* specific ABC transporters can be responsible for the transport of a mating factor.[64, 104]

Inhibition of the activity of such specific ABC transporters would prevent mating, reduce the genetic variation and retard the epidemiology of plant pathogenic fungi.

In *S. cerevisiae* and *C. albicans*, ABC transporters with an $[NBF]_2$ configuration have been described. These so-called cluster IV ABC transporters interact with tRNA and act as elongation factors. ABC proteins of this cluster are interesting target sites for antifungal compounds as they seem to be absent from mammals.[18, 71]
Although evidence is accumulating that fungal ABC transporters are involved in pathogenesis and (fungicide) resistance, more research is needed to assess the full significance of ABC transporters in these phenomena. Knock-out mutants and mutants overexpressing ABC transporters will help gaining insight in the physiological functions of ABC-transporters. Knock-out mutants lacking the natural insensitivity provided by ABC transporters can also be used as tools to screen for compounds with intrinsic fungitoxic activity.

Acknowledgements

The authors thank CAPES, Brazil (ACA) and Novartis Crop Protection AG, Switzerland (LHZ) for financial support.

References

1 F. Kunst, N. Ogasawara, I. Moszer, A. M. Albertini, G. Alloni, V. Azevedo, M. G. Bertero, P. Bessieres, A. Bolotin, S. Borchert, et al., *Nature*, 1997, **390**, 249.
2 A. Garrill, S. L. Jackson, R. R. Lew, and I. B. Heath, *Eur. J. Cell Biol.*, 1993, **60**, 358.
3 B. Andre, *Yeast*, 1995, **11**, 1575.
4 M. E. Fling, J. Kopf, A. Tamarkin, J. A. Gorman, H. A. Smith, and Y. Koltin, *Mol. Gen. Genet.*, 1991, **227**, 318.
5 C. F. Higgins, *Annu. Rev. Cell Biol.*, 1992, **8**, 67.
6 W. T. Beck, and M.K. Danks, in 'Characteristics of multidrug resistance in human tumour cells.', ed. I. B. Roninson, New York, 1991.
7 S. J. Foote, J. K. Thompson, A. F. Cowman, and D. J. Kemp, *Cell*, 1989, **57**, 921.
8 G. Lombard Platet, S. Savary, C. O. Sarde, J. L. Mandel, and G. Chimini, *Proc. Natl. Acad. Sci. USA.*, 1996, **93**, 1265.
9 J. J. Neefjes, F. Momburg, and G. J. Hammerling, *Science*, 1993, **261**, 769.
10 I. T. Paulsen, M. K. Sliwinski, and M. H. Saier, *J. Mol. Biol.*, 1998, **277**, 573.
11 G. F. Ames, C. S. Mimura, and V. Shyamala, *FEMS Microbiol. Rev.*, 1990, **6**, 429.
12 J. E. Walker, M. Saraste, M. J. Runswick, and N. J. Gay, *EMBO J.*, 1982, **1**, 945.
13 G. F. L. Ames, K. Nikaido, J. Groarke, and J. Petithory, *J. Biol. Chem.*, 1989, **264**, 3998.
14 L. Bishop, R. J. Agbayani, S. V. Ambudkar, P. C. Maloney, and G. F. L. Ames, *Proc. Natl. Acad. Sci. USA*, 1989, **86**, 6953.
15 J. M. Croop, *Cytotech.*, 1993, **12**, 1.
16 H. P. Klenk, R. A. Clayton, J. F. Tomb, O. White, K. E. Nelson, K. A. Ketchum, R. J. Dodson, M. Gwinn, E. K. Hickey, J. D. Peterson, et al., *Nature*, 1997, **390**, 364.
17 F. R. Blattner, G. Plunkett, 3rd, C. A. Bloch, N. T. Perna, V. Burland, M. Riley, J. Collado Vides, J. D. Glasner, C. K. Rode, G. F. Mayhew, et al., *Science*, 1997, **277**, 1453.

18 A. Decottignies and A. Goffeau, *Nat.Genet.*, 1997, **15**, 137.
19 E. Balzi, M. Wang, S. Leterme, L. Vandyck, and A. Goffeau, *J. Biol. Chem.*, 1994, **269**, 2206.
20 R. L. Juliano and V. Ling, *Biochim. Biophys. Acta*, 1976, **455**, 152.
21 M. M. Gottesman, C. A. Hrycyna, P. V. Schoenlein, U. A. Germann, and I. Pastan, *Annu. Rev. Genet.*, 1995, **29**, 607.
22 J. A. Endicott and V. Ling, *Annu. Rev. Biochem.*, 1989, **58**, 137.
23 M. A. Blight and I. B. Holland, *Mol. Microbiol.*, 1990, **4**, 873.
24 N. Shani and D. Valle, *Proc. Natl. Acad. Sci. USA.*, 1996, **93**, 11901.
25 S. P. Cole, G. Bhardwaj, J. H. Gerlach, J. E. Mackie, C. E. Grant, K. C. Almquist, A. J. Stewart, E. U. Kurz, A. M. Duncan, and R. G. Deeley, *Science*, 1992, **258**, 1650.
26 Z. F. Cui, D. Hirata, E. Tsuchiya, H. Osada, and T. Miyakawa, *J. Biol. Chem.*, 1996, **271**, 14712.
27 A. Broeks, B. Gerrard, R. Allikmets, M. Dean, and R. H. A. Plasterk, *EMBO J.*, 1996, **15**, 6132.
28 M. P. Anderson, R. J. Gregory, S. Thompson, D. W. Souza, S. Paul, R. C. Mulligan, A. E. Smith, and M. J. Welsh, *Science*, 1991, **253**, 202.
29 M. M. Gottesman and I. Pastan, *Annu. Rev. Biochem.*, 1993, **62**, 385.
30 A. Seelig, *Eur. J. Biochem.*, 1998, **251**, 252.
31 K. Kuchler, R. E. Sterne, and J. Thorner, *EMBO J.*, 1989, **8**, 3973.
32 J. C. Shepherd, T. N. Schumacher, P. G. Ashton Rickardt, S. Imaeda, H. L. Ploegh, C. A. Janeway, Jr., and S. Tonegawa, *Cell*, 1993, **74**, 577.
33 H. Bolhuis, H. W. Van Veen, D. Molenaar, B. Poolman, A. J. M. Driessen, and W. N. Konings, *EMBO J.*, 1996, **15**, 4239.
34 H. W. Van Veen, R. Callaghan, L. Soceneantu, A. Sardini, W. N. Konings, C. F. Higgins, K. Szabo, E. Bakos, E. Welker, M. Muller, et al., *Nature*, 1998, **391**, 291.
35 C. E. Bear, C. Li, N. Kartner, R. J. Bridges, T. J. Jensen, M. Ramjeesingh, and J. R. Riordan, *Cell*, 1992, **68**, 809.
36 J. R. Riordan, J. M. Rommens, B. Kerem, N. Alon, R. Rozmahel, Z. Grzelczak, J. Zielenski, S. Lok, N. Plavsic, J. L. Chou, et al., *Science*, 1989, **245**, 1066.
37 S. P. Hardy, H. R. Goodfellow, M. A. Valverde, D. R. Gill, F. V. Sepulveda, and C. F. Higgins, *EMBO J.*, 1995, **14**, 68.
38 M. A. Valverde, T. D. Bond, S. P. Hardy, J. C. Taylor, C. F. Higgins, J. Altamirano, and F. J. Alvarez Leefmans, *EMBO J.*, 1996, **15**, 4460.
39 M. F. Rosenberg, R. Callaghan, R. C. Ford, and C. F. Higgins, *J. Biol. Chem.*, 1997, **272**, 10685.
40 A. Van Helvoort, A. J. Smith, H. Sprong, I. Fritzsche, A. H. Schinkel, P. Borst, and G. Van Meer, *Cell*, 1996, **87**, 507.
41 A. V. Gudkov, O.B. Chernova and B.P. Kopnin, in 'Karyotype and amplicon evolution during stepwise development of multidrug resistance in Djungarian hamster cell lines.', ed. I. B. Roninson, New York, 1991.
42 M. K. al Shawi and A. E. Senior, *J. Biol. Chem.*, 1993, **268**, 4197.
43 M. K. al Shawi, I. L. Urbatsch, and A. E. Senior, *J. Biol. Chem.*, 1994, **269**, 8986.
44 T. W. Loo and D. M. Clarke, *J. Biol. Chem.*, 1994, **269**, 7750.
45 T. W. Loo and D. M. Clarke, *J. Biol. Chem.*, 1995, **270**, 22957.
46 I. L. Urbatsch, B. Sankaran, J. Weber, and A. E. Senior, *J. Biol. Chem.*, 1995, **270**, 19383.

47 I. L. Urbatsch, B. Sankaran, S. Bhagat, and A. E. Senior, *J. Biol. Chem.*, 1995, **270**, 26956.

48 A. E. Senior, M. K. Alshawi, and I. L. Urbatsch, *FEBS Lett.*, 1995, **377**, 285.

49 T. W. Loo and D. M. Clarke, *J. Biol. Chem.*, 1995, **270**, 843.

50 A. R. Safa, R. K. Stern, K. Choi, M. Agresti, I. Tamai, N. D. Mehta, and I. B. Roninson, *Proc. Natl. Acad. Sci. USA*, 1990, **87**, 7225.

51 L. M. Greenberger, *J. Biol. Chem.*, 1993, **268**, 11417.

52 X. P. Zhang, K. I. Collins, and L. M. Greenberger, *J. Biol. Chem.*, 1995, **270**, 5441.

53 H. Bolhuis, H. W. Van Veen, J. R. Brands, M. Putman, B. Poolman, A. J. M. Driessen, and W. N. Konings, *J. Biol. Chem.*, 1996, **271**, 24123.

54 M. M. Gottesman and I. Pastan, *J. Biol. Chem.*, 1988, **263**, 12163.

55 C. A. Doige and G. F. Ames, *Annu. Rev. Microbiol.*, 1993, **47**, 291.

56 R. E. Kerppola, V. K. Shyamala, P. Klebba, and G. F. Ames, *J. Biol. Chem.*, 1991, **266**, 9857.

57 M. J. Fath and R. Kolter, *Microbiol. Rev.*, 1993, **57**, 995.

58 C. Wandersman, *Res. Microbiol.*, 1998, **149**, 163.

59 C. Wandersman and P. Delepelaire, *Proc. Natl. Acad. Sci. USA.*, 1990, **87**, 4776.

60 S. Letoffe, P. Delepelaire, and C. Wandersman, *EMBO J.*, 1990, **9**, 1375.

61 G. W. Stoddard, J. P. Petzel, M. J. van Belkum, J. Kok, and L. L. McKay, *Appl. Environ. Microbiol.*, 1992, **58**, 1952.

62 G. A. Cangelosi, G. Martinetti, J. A. Leigh, C. C. Lee, C. Theines, and E. W. Nester, *J. Bacteriol.*, 1989, **171**, 1609.

63 S. W. Stanfield, L. Ielpi, O. B. D, D. R. Helinski, and G. S. Ditta, *J. Bacteriol.*, 1988, **170**, 3523.

64 P. U. Christensen, J. Davey, and O. Nielsen, *Mol. Gen. Genet.*, 1997, **255**, 226.

65 D. F. Ortiz, L. Kreppel, D. M. Speiser, G. Scheel, G. McDonald, and D. W. Ow, *EMBO J.*, 1992, **11**, 3491.

66 D. F. Ortiz, T. Ruscitti, K. F. McCue, and D. W. Ow, *J. Biol. Chem.*, 1995, **270**, 4721.

67 I. Balan, A. M. Alarco, and M. Raymond, *J. Bacteriol.*, 1997, **179**, 7210.

68 M. Raymond, D. Dignard, A. M. Alarco, N. Mainville, B. B. Magee, and D. Y. Thomas, *Mol. Microbiol.*, 1998, **27**, 587.

69 J. Leighton, *Methods Enzymol.*, 1995, **260**, 389.

70 E. H. Hettema, C. W. van Roermund, B. Distel, M. van den Berg, C. Vilela, C. Rodrigues Pousada, R. J. Wanders, and H. F. Tabak, *EMBO J.*, 1996, **15**, 3813.

71 C. R. Vazquez de Aldana, M. J. Marton, and A. G. Hinnebusch, *EMBO J.*, 1995, **14**, 3184.

72 M. G. Sandbaken, J. A. Lupisella, B. DiDomenico, and K. Chakraburtty, *J. Biol. Chem.*, 1990, **265**, 15838.

73 G. Del Sorbo, A. C. Andrade, J. G. Van Nistelrooy, J. A. Van Kan, E. Balzi, and M. A. De Waard, *Mol. Gen. Genet.*, 1997, **254**, 417.

74 M. B. Tobin, R. B. Peery, and P. L. Skatrud, *Gene*, 1997, **200**, 11.

75 H. Schoonbeek, and M.A. De Waard, 12th International Rheinhardsbrunn Symposium, Friedrichroda, 1998, p. in press.

76 L.-H. Zwiers, and M.A., De Waard, Modern Fungicides and Antifungal Compounds 12th International Rheinhardsbrunn Symposium, Friedrichroda, 1998, p. in press.

77 R. Nakaune, K. Adachi, O. Nawata, M. Tomiyama, K. Akutsu, and T. Hibi, *Appl. Environ. Microbiol.*, submitted.

78 B. A. Roe, S. Kupfer, S. Clifton, R. Prade, and J. Dunlap, in 'Aspergillus nidulans and Neurospora crassa cDNA sequencing project (Fifth data release - April 18, 1998)', 1998.

79 M. A. Nelson, S. Kang, E. L. Braun, M. E. Crawford, P. L. Dolan, P. M. Leonard, J. Mitchell, A. M. Armijo, L. Bean, E. Blueyes, et al., *Fungal Genet Biol*, 1997, **21**, 348.

80 S. F. Altschul, T. L. Madden, A. A. Schaffer, J. Zhang, Z. Zhang, W. Miller, and D. J. Lipman, *Nucl. Acids Res.*, 1997, **25**, 3389.

81 A. C. Andrade, X. Xuei, G. Del Sorbo, P. L. Skatrud, and M. A. De Waard, European Conference of Fungal Genetics (4th), Leon - Spain, 1998, p. 146.

82 M. A. De Waard, *Pestic. Science*, 1997, **51**, 271.

83 R. Tommasini, E. Vogt, J. Schmid, M. Fromentau, N. Amrhein, and E. Martinoia, *FEBS Lett.*, 1997, **411**, 206.

84 R. Tommasini, E. Vogt, M. Fromenteau, S. Hortensteiner, P. Matile, N. Amrhein, and E. Martinoia, *Plant J.*, 1998, **13 (6)**, 773.

85 Y. P. Lu, Z. S. Li, Y. M. Drozdowicz, S. Hortensteiner, E. Martinoia, P. A. Rea, Z. S. Li, M. Szczypka, Y. P. Lu, D. J. Thiele, et al., *Plant Cell*, 1998, **10**, 267.

86 S. Ruetz and P. Gros, *Cell*, 1994, **77**, 1071.

87 M. F. Luciani and G. Chimini, *EMBO J.*, 1996, **15**, 226.

88 C. C. Paulusma, P. J. Bosma, G. J. Zaman, C. T. Bakker, M. Otter, G. L. Scheffer, R. J. Scheper, P. Borst, and R. P. Oude Elferink, *Science*, 1996, **271**, 1126.

89 H. W. Van Veen, K. Venema, H. Bolhuis, I. Oussenko, J. Kok, B. Poolman, A. J. Driessen, and W. N. Konings, *Proc. Natl. Acad. Sci. USA.*, 1996, **93**, 10668.

90 B. Ullman, *J. Bioenerg. Biomembr.*, 1995, **27**, 77.

91 A. Decottignies, L. Lambert, P. Catty, H. Degand, E. A. Epping, W. S. Moye Rowley, E. Balzi, and A. Goffeau, *J. Biol. Chem.*, 1995, **270**, 18150.

92 Z. S. Li, Y. P. Lu, R. G. Zhen, M. Szczypka, D. J. Thiele, and P. A. Rea, *Proc. Natl. Acad. Sci. USA*, 1997, **94**, 42.

93 M. Kolaczkowski, M. van der Rest, A. Cybularz Kolaczkowska, J. P. Soumillion, W. N. Konings, and A. Goffeau, *J. Biol. Chem.*, 1996, **271**, 31543.

94 E. Balzi and A. Goffeau, *J. Bioenerg. Biomembr.*, 1995, **27**, 71.

95 D. Sanglard, F. Ischer, M. Monod, and J. Bille, *Antimicrob. Agents. Chemother.*, 1996, **40**, 2300.

96 D. Sanglard, F. Ischer, M. Monod, and J. Bille, *Microbiology U K*, 1997, **143**, 405.

97 K. Nishi, M. Yoshida, M. Nishimura, M. Nishikawa, M. Nishiyama, S. Horinouchi, and T. Beppu, *Mol. Microbiol.*, 1992, **6**, 761.

98 K. Nagao, Y. Taguchi, M. Arioka, H. Kadokura, A. Takatsuki, K. Yoda, and M. Yamasaki, *J. Bacteriol.*, 1995, **177**, 1536.

99 M. A. De Waard, J.G.M. Van Nistelrooy, C.R. Langeveld, J.A.L. Van Kan, and G. Del Sorbo, Modern Fungicides and Antifungal Compounds 11th International Rheinhardsbrunn Symposium, Friedrichroda, 1995, p. 293.

100 T. Joseph-horne, D. Hollomon, N. Manning, and S. L. Kelly, *Appl. Environ. Microbiol.*, 1996, **62**, 184.

101 M. H. Saier, I. T. Paulsen, M. K. Sliwinski, S. S. Pao, R. A. Skurray, and H. Nikaido, *Faseb J.*, 1998, **12**, 265.

102 A. A. Neyfakh, *Trends Microbiol.*, 1997, **5**, 309.

103 R. K. Poole, *Trends Microbiol.*, 1997, **5**, 340.

104 J. P. McGrath and A. Varshavsky, *Nature*, 1989, **340**, 400.

Resistant Target Sites and Insecticide Discovery

Vincent L. Salgado

RHONE-POULENC AG COMPANY, 2 T.W. ALEXANDER DRIVE, RESEARCH TRIANGLE PARK, NC 27709, USA

1 ABSTRACT

Until very recently, only a handful of insecticide target sites could deliver the fast kill, low use rates and high insect selectivity needed for commercial success, leading to the development of many competing products with identical modes of action. The resulting over-exploitation of these few target sites led to widespread occurrence of resistant forms, which have recently been characterized at the molecular level.

Dieldrin resistance accounts for 60% of all insecticide resistance cases. A single ala to ser substitution at position 302 of the GABA receptor of *D. melanogaster* was found to confer very high levels of resistance to dieldrin, and an homologous mutation has since been found to confer dieldrin resistance in many other species.

While *kdr* and *super-kdr* nerve-insensitivity resistance to DDT and pyrethroids has long been attributed to reduced sensitivity of nerve membrane sodium channels, the mechanism of resistance is only now being elucidated at the molecular level, and appears to be due to several distinct mutations in the voltage-dependent sodium channel.

Six residues around the active-site pocket of acetylcholinesterase have been identified where mutations impede access of organophosphate and carbamate insecticides, with little effect on acetylcholine hydrolysis. These mutations can accumulate in a single gene to give very high levels of resistance.

In recent years, the agrochemical industry has focused on the discovery of leads with novel modes of action and has developed a number of promising new products acting at novel target sites not currently affected by resistance. These include avermectins acting on an inhibitory glutamate receptor, imidacloprid, acting on the nicotinic acetylcholine receptor, spinosad, acting at a novel site on the nicotinic acetylcholine receptor, indoxacarb, which acts at a novel site on the voltage-dependent sodium channel, and the diacylhydrazines, which mimic the action of ecdysone at the ecdysone receptor. These successes have proven that novel target sites with great potential can be found. Also, novel compounds acting at resistant target sites can sometimes be effective. A case in point is fipronil, which, though acting at the GABA receptor, does not suffer high levels of cross-resistance with dieldrin.

While these innovations derived from families discovered with traditional screening methods, new developments in high throughput synthesis and screening promise to further accelerate the pace of discovery of novel insecticides.

2 INTRODUCTION

Since the 1970s, increasing concern over the environmental impact of organophosphate, carbamate and organochlorine insecticides, as well as the increasing prevalence of resistance to these and the newer pyrethroids, has stimulated the agrochemical industry to search for novel classes of insecticides. Because of the difficulty in finding molecules that simultaneously meet requirements for high insecticidal activity, low toxicity to mammals and other non-target species, adequate persistence in the field with low residues at harvest, and low cost, many of the first products to be developed from leads discovered over the last three decades are only now being commercialized. At the same time, bacterially derived insecticidal Bt toxins, which for 50 years have seen only limited use in sprays in organic farming, some Integrated Pest Management systems and in Canadian forestry, are now being genetically engineered into crops that last year in the U.S. alone were planted on nearly 9 million acres.

While insecticide classes have always been defined by chemical structure, target site and mode of action have always been recognized as the most fundamental attributes of those classes. Mode of action has become increasingly important to the discovery of new classes with the recognition that much insecticide resistance is target-site-based, and it will become even more important as it becomes generally accepted that target-site-based *in vitro* screens are the best if not the only way to take advantage of the opportunities to screen the microgram quantities of millions of different compounds that are becoming available through combinatorial chemistry.

3 TARGET SITE RESISTANCE

Insects have developed resistance to insecticides by five broad types of mechanisms: behavioral avoidance, enhanced metabolic degradation, decreased metabolic activation (of proinsecticides), decreased penetration and reduced target site sensitivity. This paper focuses on resistance due to reduced target site sensitivity and its relationship to mode of action and new product discovery.

Target site-based insecticide resistance is particularly important with the three most heavily exploited target sites: the GABA receptor, the voltage-dependent sodium channel and the acetylcholinesterase enzyme. It is also the most common mechanism of resistance to Bt toxins.

3.1 GABA Receptor Target Site Resistance

Dieldrin resistance is extremely common, at one time accounting for more than 60% of reported cases of insecticide resistance[1]. A single gene conferring 4000-fold resistance to dieldrin (*Rdl*) was cloned from field-isolated *Drosophila melanogaster* and found to code for a GABA-gated chloride channel subunit[2]. Dieldrin resistance was associated with a point mutation of ala to ser at position 302 in the second transmembrane domain, which is thought to line the pore of the chloride channel. The mutant and wild-type receptor isoforms were heterologously expressed and the mutation was shown to confer very high levels of resistance to the chloride channel blocking actions of dieldrin and picrotoxin[3]. An homologous mutation has also been shown or has been demonstrated to confer dieldrin resistance in the species shown in Table 1, among others. While resistance to dieldrin is greater than 1000-fold, resistance to other cyclodienes and other families of chloride channel blockers, including the new insecticide fipronil (see below) is much lower.

Table I *Dieldrin resistance confers very high levels of resistance to dieldrin and lower levels of resistance to other chloride channel blockers (na = information not available).*

		Resistance Ratio at the LD_{50}				
Species	Strain	dieldrin	endosulfan	lindane	fipronil	ref.
Blatella germanica	Cld-R	17,000	na	na	7.7	4
Musca domestica	OCR	3100	na	na	31	4
Musca domestica	na	na	na	na	20	5
Drosophila simulans	ser-302	1700	6	3	23	6
Drosophila melanogaster	ser-302	5100	9	5	73	6

3.2 Sodium Channel Target Site Resistance

The voltage-dependent sodium channel is a large (>240 kd) transmembrane protein composed of a single polypeptide with more than 2000 residues arranged in four homology domains (I-IV), each containing six transmembrane α-helical segments 1-6 (Figure 1). These 24 transmembrane helices and the interconnecting loops are arranged to form a voltage-gated sodium-selective pore with at least nine distinct target sites for natural toxins, insecticides and drugs. One of these sites is the target for insecticidal sodium channel modulators, including DDT, the natural pyrethrins and the pyrethroids, while another site appears to be shared by hydrophobic sodium channel blockers, including the dihydropyrazoles and indoxacarb (Section 4.5) as well as local anesthetics.

Target site resistance to DDT and pyrethroids, known as knockdown resistance or *kdr,* gives moderate levels (<100X) of resistance to DDT and pyrethroids in many insect species. *Super-kdr*, which confers higher levels (>100-fold) of resistance to type I pyrethroids and extremely high levels (>1000-fold) to type II or α-cyano pyrethroids, has been described in house fly, horn fly and diamondback moth. In house fly, German cockroach, diamondback moth, tobacco budworm, green peach aphid, yellow fever mosquito and Colorado potato beetle, *kdr* resistance has been associated with a mutation at a particular leu residue in transmembrane segment 6 of homology domain II (IIS6) of the voltage-dependent sodium channel (leu-his in tobacco budworm[7], leu-phe in others[8-13]. On the other hand, *kdr* resistance in one haplotype of *H. virescens* was associated with a val-met replacement in IS6 at position 421[7]. The S6 segments of each homology domain are thought to comprise part of the pore, so the mutations in the two different S6 regions may be close together in the pore and part of the same pyrethroid binding site. *Super-kdr* in house flies[8] and horn flies[10] was associated with the leu-phe mutation in IIS6, in combination with a met-thr replacement in the intracellular IIS4-S5 loop. On the other hand, a highly resistant strain of diamondback moth from Taiwan, which has the leu-phe mutation in IIS6, also had a thr-ile mutation just into IIS5, only 11 amino acids away from the *super-kdr* mutation in the flies and giving a similar highly resistant phenotype[14]. All of these mutations are summarized in Figure 1.

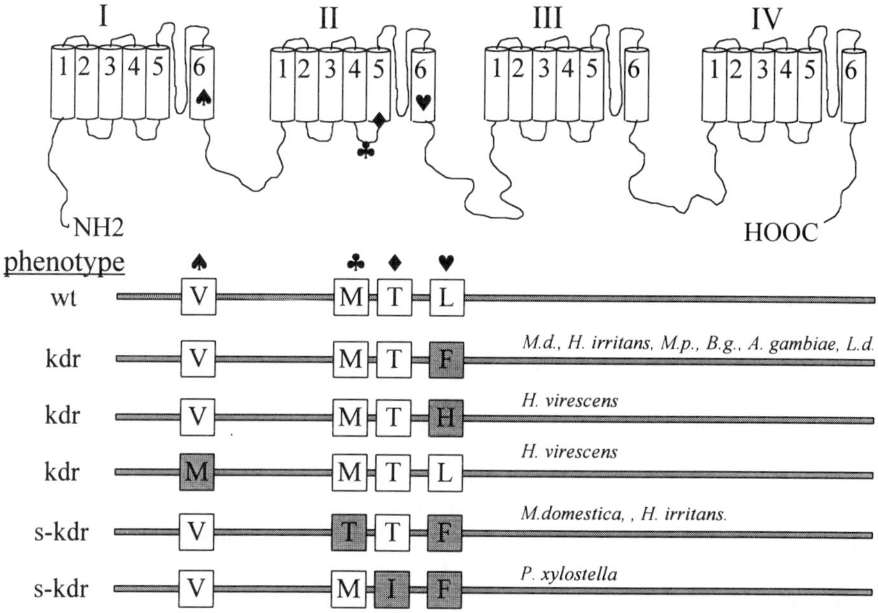

Figure 1 *Mutations in the voltage-dependent sodium channel of several insect species conferring the indicated pyrethroid resistant phenotypes. (Species:* Musca domestica[8], Myzus persicae[9], Haematobia irritans[10], Blatella germanica[11,12], Anopheles gambiae[13], Heliothis virescens[7], Plutella xylostella[14]*).*

Recent work on the effects of pyrethroids on heterologously expressed sodium channels has shown that both the *kdr* and *super-kdr* mutations greatly reduce the affinity of the sodium channels for pyrethroids, and also reduce the effectiveness of the pyrethroid molecules that do bind[15,16].

3.3 Acetylcholinesterase

Organophosphate and carbamate resistance due to insensitive acetylcholinesterase is a serious problem in some insect species. Sequence analysis of the *Ace* gene from several resistant field strains of *Drosophila melanogaster* led to the identification of point mutations at four residues that conferred reduced sensitivity to OPs and carbamates[17]. Homologous mutations at two of these residues, as well as two additional point mutations, were identified in house fly[18], bringing the total to six resistance-conferring mutations in this target site. Any one of these mutations confers only moderate resistance, but combinations of two or more of them can yield highly resistant acetylcholinesterases.

The crystal structure of the acetylcholinesterase of the electric fish *Torpedo* has been solved. The enzyme is a ball with a deep gorge on one side, on the bottom of which is the active site pocket, containing a group of amino acids known as the catalytic triad, which participate directly in the hydrolysis of acetylcholine[19]. Inhibitors are compounds that also fit into the site and interact with the catalytic triad.

The effects of the six resistance mutations can be understood from the location of corresponding residues in the *Torpedo* enzyme. All are positioned near the active site in such a way that they could impede access of the inhibitors with little effect on hydrolysis of the smaller acetylcholine molecule[17].

4 THE NEW INSECTICIDES

In recent years, a number of new products with novel or under-exploited modes of action and no known target site resistance have either been brought to market or are near commercialization. Despite their novel modes of action, these have all evolved from leads discovered by traditional whole insect screens rather than target-based screens.

4.1 Neonicotinoids

The neonicotinoids are based on the nitromethylene heterocycle insecticides discovered by Shell Development Company in the 1970s[20] and shown to interact with the agonist site on the insect nicotinic receptor. Though very active in the lab and very selective for insects, these compounds could not be developed until the photolabile nitromethylene group was replaced with the more photostable nitroimine to obtain imidacloprid[21]. Introduced by Bayer in 1991, imidacloprid has rapidly become one of the leading insecticides. Since nicotine has been used very little in recent decades, the neonicotinoid target site is essentially novel in insect control. At least seven newer neonicotinoids, some shown below with the part of the structure common to nicotine shown in bold, are currently under development by various companies. Nicotine bears a positive charge on the amine nitrogen at physiological pH and is active at both insect and vertebrate nicotinic receptors. Neonicotinoids, on the other hand, are not charged and are highly selective for the insect receptors, possibly because the tertiary amine nitrogen bears a partial positive charge that is sufficient to interact with the insect, but not the vertebrate, receptor[22]. Also, the nitro or cyano group of the neonicotinoids bears a partial negative charge, which is thought to enhance binding to the insect but not the vertebrate receptor[23].

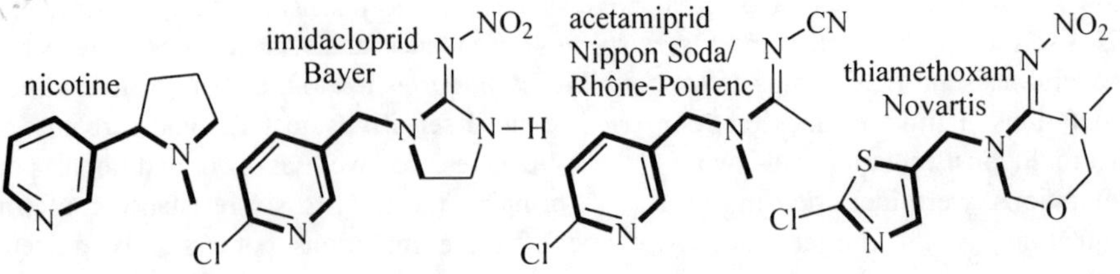

4.2 Spinosad

Spinosad is a new broad-spectrum insect control product, introduced by Dow AgroSciences in 1997, that is a naturally occurring mixture of 85 % spinosyn A and 15 % spinosyn D(see structure), produced by the soil actinomycete *Saccharopolyspora spinosa*[24].

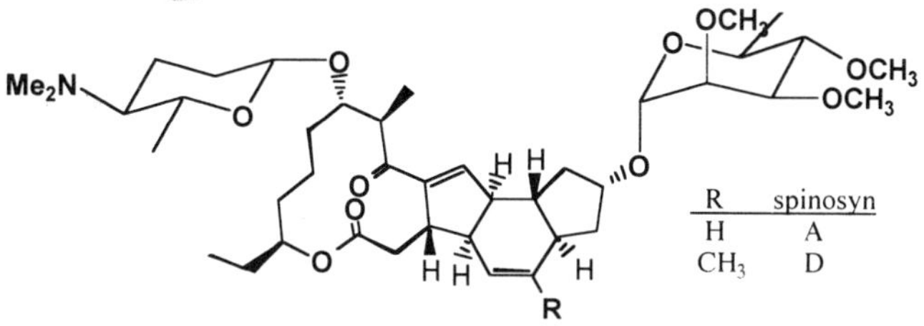

Spinosyns excite the insect nervous system by allosterically activating nicotinic receptors and prolonging the responses of those receptors to agonists such as nicotine and acetylcholine. The nicotinic receptors, whether activated by acetylcholine or spinosyns, are blocked by the nicotinic antagonist α-bungarotoxin with an I_{50} of 2 nM. While effects of spinosyns on nicotinic receptors are sufficient to account for the observed symptoms, spinosyns also antagonized GABA responses and activated a picrotoxin-sensitive chloride current in some experiments[25]. The effects of spinosyns on ligand-gated ion channels are analogous to those of avermectins, suggesting that these two groups of macrolides may act at homologous sites: whereas avermectins allosterically activate inhibitory glutamate receptors and enhance agonist responses (see below), spinosyns have analogous effects on nicotinic receptors. Furthermore, both avermectins and spinosyns modulate GABA receptors.

4.3 Fipronil

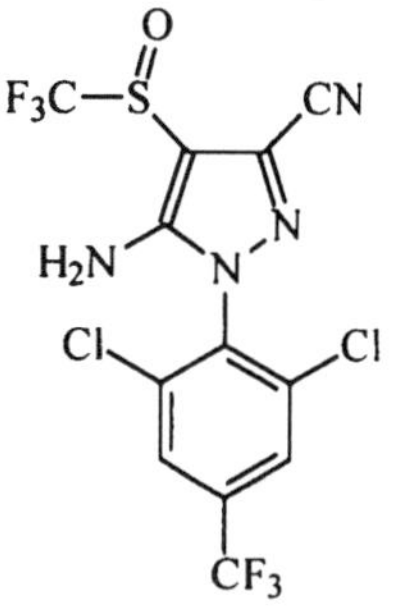

Fipronil is a novel broad-spectrum phenylpyrazole insecticide introduced by Rhône-Poulenc in 1993, with many uses in agriculture, veterinary medicine and urban pest control. Fipronil blocks GABA-gated chloride channels[26]. This was once one of the major insecticide modes of action, but its use has been steadily declining since the 1960s primarily because of the unwelcome persistence and toxicity of organochlorines in the environment. Fortunately, while dieldrin resistance alleles (Section 3.1) are thought to be prevalent in many insect populations, they have been found to confer only moderate levels of resistance to fipronil in some species and minimal resistance in others (Table 1). The level of cross-resistance to fipronil conferred by Rdl has not been determined for many insects that are controlled by fipronil, but after several years of use in more than 60 countries, the field performance of fipronil has not been affected by resistance.

4.4 Avermectins

The avermectins are derived from a family of macrocyclic lactones produced by the soil microorganism *Streptomyces avermitilis*. Abamectin (avermectin B1) was introduced in 1985 as a crop insecticide against mites and some insect species. The semisynthetic derivative ivermectin (22,23-dihydroavermectin B1) was introduced commercially in 1981 as an anthelmintic. The recently introduced semisynthetic emamectin benzoate has unprecedented lepidoptera activity, controlling all lepidopterous pests on various crops at rates of 8-16 g/ha. While abamectin resistance is known in some species, there is not yet convincing evidence for target site resistance, and there is no cross-resistance with emamectin benzoate[27].

Avermectins allosterically activate inhibitory glutamate receptors and potentiate the responses of those receptors to glutamate[28]. With the introduction of emamectin benzoate, the inhibitory glutamate receptor is essentially a new target site for most lepidopterous pests.

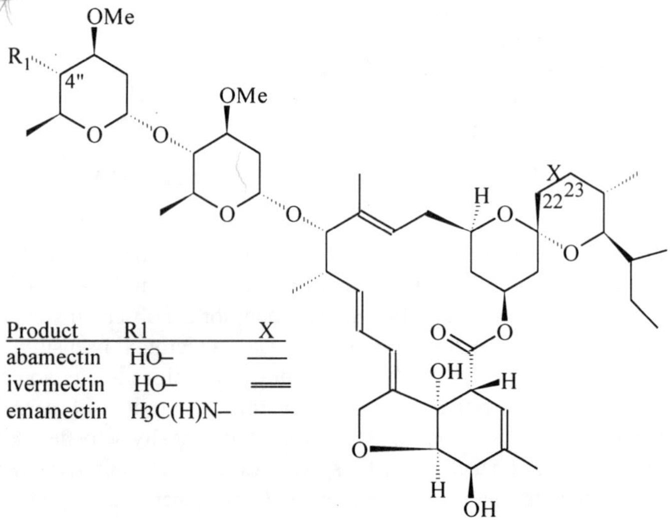

4.5 Indoxacarb /Dihydropyrazoles

DuPont is developing a novel lepidoptera insecticide, indoxacarb, derived from the dihydropyrazoles, a class first discovered at Philips-Duphar in the 1970s[29] and shown to block voltage-dependent sodium channels in a manner indistinguishable from the local anesthetics[30]. The proinsecticide indoxacarb is only weakly active against sodium channels, but it is N-decarbomethoxylated to decarbomethoxyindoxacarb, a potent sodium channel blocker. This transformation is catalyzed by an esterase/amidase-like enzyme that is inhibited by the esterase inhibitors DEF, DFP and paraoxon and thus may be antagonized by organophosphates[31]. Indoxacarb and the dihydropyrazoles act at a different target site from the pyrethroids, and *kdr* insects are not cross-resistant to them.

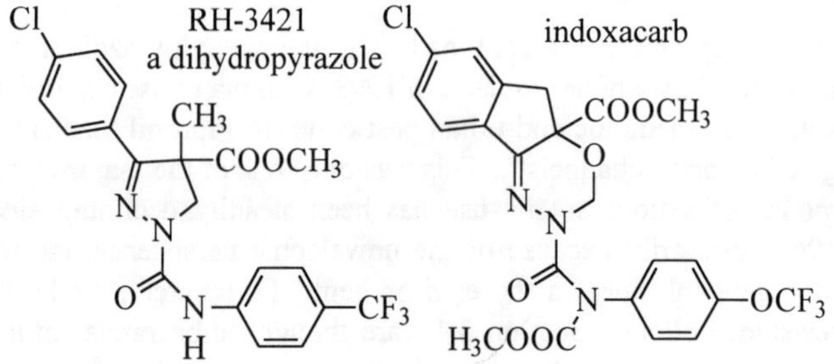

4.6 Diacylhydrazine Ecdysone Agonsists

While most of the new insecticides are neuroactive, the very important diacylhydrazine class of insect growth regulators was recently introduced by Rohm and Haas, with the launch of Confirm (tebufenozide) for control of lepidopterous pests at 100-300 g/ha. Intrepid (methoxyfenozide) is slated for introduction in 1999 and will offer a broader worm spectrum at 3-4X lower rates, and Mach 2 (halofenozide) is being introduced this year for control of soil-dwelling lepidoptera and coleoptera larvae in turf.

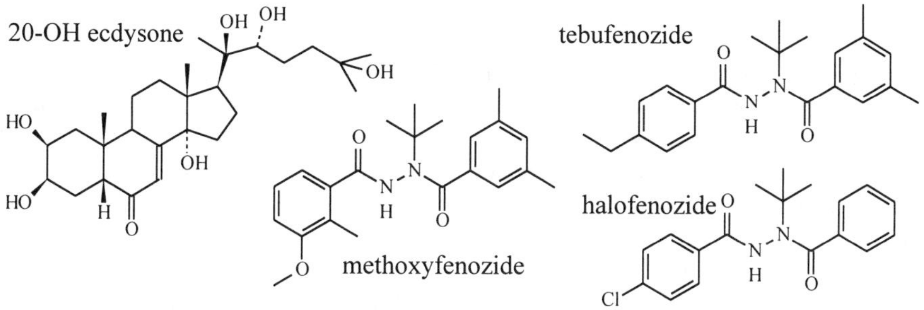

The diacylhydrazines are relatively fast-acting for insect growth regulators, as they induce larvae to begin a molt within a few hours of exposure, in contrast to most other insect growth regulators, which disrupt the progress of a naturally occurring molt. The diacylhydrazines induce a molt by mimicking the insect molting hormone 20-hydroxyecdysone. This hormone, which is normally released into the hemolymph to trigger a molt, enters the nucleus and binds to a receptor that is a heterodimer of the ecdysone receptor and the ultraspiracle protein. This hormone-protein complex then binds to ecdysone-responsive promoters and initiates gene transcription[32]. Diacylhydrazines bind to the ecdysone recognition site of the receptor heterodimer and mimic the action of the natural hormone, thereby inducing a molt that cannot be completed because of the persistence of the compound[33, 34].

4.7 Chlorfenapyr

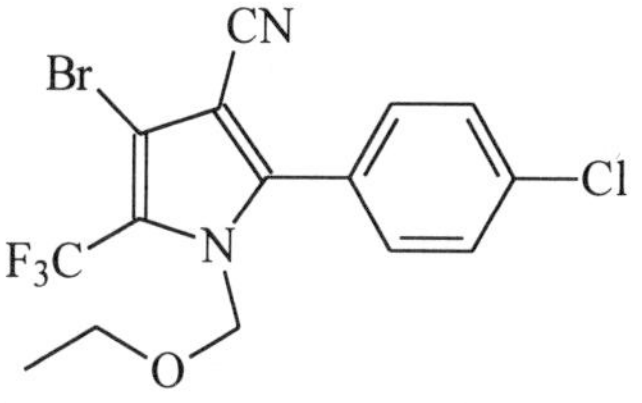

Chlorfenapyr, commercialized by American Cyanamid, is a proinsecticide that is converted in the insect to an uncoupler, by removal of the N-ethoxyethyl moiety[35]. As uncouplers have no target site *per se*, target site resistance should not be possible, and chlorfenapyr is therefore useful against many resistant insects. However, although safened by its pro-insecticide nature, there are concerns about toxicity to non-target organisms, because of the nonspecific mode of action of uncouplers.

4.8 Pymetrozine

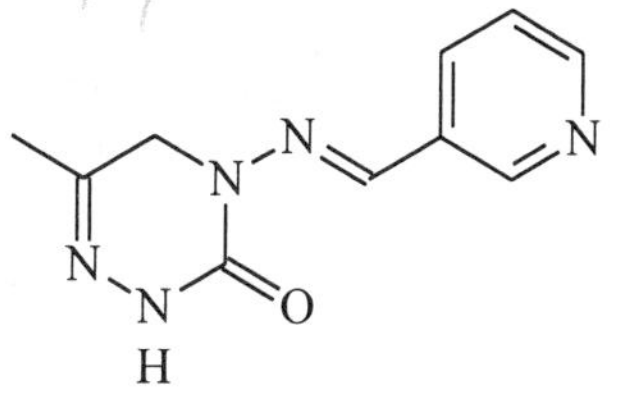

Chess (pymetrozine) was introduced in 1997 for systemic, contact and translaminar control of whiteflies, aphids, stinkbugs and hoppers. It is very selective for sucking insects and specifically inhibits stylet penetration behavior and cibarial pumping by an unknown mechanism, after uptake through feeding, topical application or injection[36].

5 NEW APPROACHES TO INSECTICIDE DISCOVERY

The discovery of new insecticides is particularly challenging because of the large number of hurdles that must be met to achieve registration and commercial success. From a biological standpoint, action at a target site is far from sufficient to kill an insect. The compound must not only have lethal action at the target site, but it also must be stable in the environment long enough to be taken up by the target insects. Then it must be able to enter the insect in large enough quantities, and resist metabolism and excretion long enough to exert its toxic effect.

In addition to biological activity, insecticides must pass many other hurdles set by the marketplace and by concerns for human and environmental safety. Safety can be particularly challenging for insecticides, because of the molecular similarity of insects to man and other animals. However, while certain classes of insecticides are inherently toxic to all animals, it has often been possible to obtain insect-selective analogs. While chlorfenapyr and many other insecticides have achieved insect selectivity through selective metabolism, target site selectivity can also play a role, as described in Section 4.1 for the neonicotinoids. In addition, there are some target sites, such as the ecdysone receptor described above, that do not occur in vertebrates.

The 1990s have seen the introduction of an unprecedented number of new classes of insecticides with novel or underexploited modes of action, discovered by traditional *in vivo* screening. Now, as we prepare to enter the next millenium, rapid advances in biotechnology, bioinformatics, robotics, high throughput screening (HTS), genomics, combinatorial chemistry, structure-based design and other developing technologies promise to revolutionize insecticide research by accelerating the discovery of new families with desirable modes of action.

5.1 *In vivo* High Throughput Screening

The traditional insecticide screen, previously carried out manually at the rate of approximately 10,000 compounds per year, has been miniaturized at many companies and is now being done at a rate of 100,000-500,000 compounds per year, using mictotiter plates and robotics[37]. However, increasing throughput alone is insufficient for the discovery of new modes of action. More hits are generated, but as with traditional screening, many of these hits are non-specific toxicants, and methods are needed for rapidly identifying the compounds with promising modes of action. Traditional mode of action determination through physiology and biochemistry may be needed if the target site is novel or the compound requires bioactivation, and this can be time-consuming. On the other hand, if the compound is intrinsically active, known target sites can be tested in a series of assays either in-house or by companies that provide this service.

5.2 *In vitro* High Throughput Screening

For many years now, pharmaceutical companies have relied on *in vitro* high throughput screening to identify leads with modes of action chosen for their proven or expected therapeutic value. On the other hand, a legacy of focusing on *in vivo* activity, and our ability to screen directly on target organisms, has slowed the widespread adoption of *in vitro* screening by the agrochemical industry. In fact, while most companies do appear to have implemented *in vitro* screening programs for agrochemical lead discovery to a greater or lesser degree, using various reporter gene, radioligand binding and enzyme assays, many of those attempts may have failed because of premature expectations of *in vivo* activity from the hits. For the *in vitro* approach to be successful -- indeed for it to add anything at all to parallel *in vivo* screening of the same compounds, it must be recognized that hits with high intrinsic activity on validated target sites, even if they have no *in vivo* activity, should be considered leads with bioavailability problems.

Current *in vivo* and *in vitro* microscreens in agrochemical companies use assays in plate densities up to 96 wells. Ultra high throughput screens are currently being developed and implemented in the pharmaceutical industry that will allow the screening of 100,000 or more compounds in one day in plates with 4- or 16-fold higher densities and correspondingly smaller assay volumes. These developments are driving the development of combinatorial chemistry methods to synthesize vast libraries of compounds in quantities of only a few hundred pmol[38]. Such quantities will unquestionably be too small to screen *in vivo*, so *in vitro* screening must be adopted in order to take advantage of these compounds.

One of the greatest challenges of high throughput *in vitro* screening for insecticide leads is the selection of appropriate target sites. Emphasis is currently placed on validated target sites - those with proven high insecticidal potential, but this approach may not yield the compounds with truly novel sites that are needed for resistance management. However, new approaches for identifying and validating novel target sites have been proposed. Mapping of insect genomes, coupled with other genetic techniques, permits the identification of genes whose inactivation or activation leads to death of the insect. On the other hand, it is not yet clear how many of these genes would contain receptor sites for small molecules and which of those would provide good insect control.

References

1. G.P. Georghiou, The management of the resistance problem, in 'Pesticide Resistance: Strategies and Tactics for Management', National Academy Press, Washington, 1986, p. 14.
2. R.H. ffrench-Constant, D.P. Mortlock, C.D. Shaffer, R.J. MacIntyre and R.T. Roush, *Proc. Natl. Acad. Sci.* , 1991, **88**, 7209.
3. R.H. ffrench-Constant, T.A. Rocheleau, J.C. Steichen and A.E. Chalmers, *Nature*, 1993, **363**, 449.
4. J.G. Scott and Z. Wen, *J. Econ. Entomol.*,1997, **90**, 1152.
5. L.M. Cole, R. A. Nicholson and J.E. Casida, *Pestic. Biochem. Physiol.*, 1993, **46**, 47.
6. L.M. Cole, R.T. Roush and J.E. Casida, *Life Sciences*, 1995, **56**, 757.
7. Y. Park, M.F.J. Taylor and R. Feyereisen, *Biochem. Biophys. Res. Comm.*, 1997, **239**, 688.
8. M.S. Williamson, D. Martinez-Torres, C.A. Hick and A.L. Devonshire, *Mol. Gen. Genet.*, 1996, **252**, 51.
9. D. Martinez-Torres, A.L. Devonshire and M.S. Williamson, *Pestic. Sci.*, 1997, **51**, 265.
10. F.D. Guerrero, R.C. Jamroz, D. Kammlah and S.E. Kunz, *Insect Biochem. Molec. Biol.*, 1997, **27**, 745.
11. K. Dong, *Insect Biochem. Molec. Biol.*, 1997, **27**, 93.
12. M. Miyazaki, K. Ohyama, D.Y. Dunlap and F. Matsumura, *Mol. Gen. Genet.*, 1996, **252**, 61.

13. D. Martinez-Torres, F. Chandre, M.S. Williamson, F. Darriet, J.B. Berge, A.L. Devonshire, P. Guillet, N. Pasteur and D. Pauron, *Insect Molec. Biol.*, 1997, **7**, 179.
14. T.H. Shuler, D. Martinez-Torres, A.J. Thompson, I. Denholm, A.L. Devonshire, I.R. Duce and M.S. Williamson, *Pestic. Biochem. Physiol.* 1998, **59**, 169.
15. D.M. Soderlund, T.J. Smith, S.H. Lee, P.J. Ingles and D.C. Knipple, in "Neurotox '98'", ed. D. Beadle, Royal Soc. Chem., Cambridge , in press.
16. C. Cohen, in "Neurotox '98'", ed. D. Beadle, Royal Soc. Chem., Cambridge , in press.
17. A. Mutero, M. Pralavorio, J.-M. Bride and D. Fournier, *Proc. Natl. Acad. Sci.*, 1994, **91**, 5922.
18. A.L. Devonshire, F.J. Byrne, G.D. Moores and M.S. Williamson, in 'Structure and Function of Cholinesterases and Related Proteins', eds. B.P. Doctor, D.M. Quinn, R. L. Rotundo, and P. Taylor., in press.
19. J.L. Sussman, M. Harel, F. Frolow, C. Oefner, A. Goldman, L. Toker and I. Silman, *Science*, 1991, **253**, 872.
20. S.B. Soloway, A.C. Henry, W.D. Kollmeyer, W.M. Padgett, J.E. Powell, S.A. Roman, C.H. Tieman, R.A. Corey and C.A. Horne, Nitromethylene insecticides, in 'Advances in Pesticide Science', eds. H. Geissbuhler, G.T. Brooks and P.C. Kearney, Pergamon Press, 1979, Oxford, vol. 2, p. 206.
21. K. Shiokawa, S.-I. Tsuboi, K. Iwaya and K. Moriya, *J. Pestic. Sci.*, 1994, **19**, 329.
22. I. Yamamoto, M. Tomizawa, T. Saito, T. Miyamoto, E.C. Walcott and K. Sumikawa, *Arch. Ins. Biochem. Physiol.*, 1998, **37**, 24.
23. A. Nakayama and M. Sukekawa, *Pestic. Sci.*, 1998, **52**, 104.
24. H.A. Kirst, K. H. Michel, J. S. Mynderse, E. H. Chao, R. C. Yao, W. M. Nakatsukasa, L. D. Boeck, J. Occlowitz, J. W. Paschel, J. B. Deeter and G. D. Thompson, in 'Synthesis and Chemistry of Agrochemicals III', eds. D. R. Baker, J. G. Fenyes and J. J. Steffens, Am. Chem. Soc., Washington, D. C. 1992, p 214.
25. V.L. Salgado, G.B. Watson and J.J. Sheets, *Proc. Beltwide Cotton Conf.*, 1997, 1082.
26. D.B. Gant, J.R. Bloomquist, H.M. Ayad, and A.E. Chalmers, *Pestic. Sci.*, 1990, **30**, 355.
27. S.M. White, D.M. Dunbar, R. Brown, B. Cartwright, D. Cox, R.K. Eckel, R.K. Jansson, P.K. Mookerjee, J.A. Norton, R.F. Peterson and V.R. Starner, *Proc. Beltwide Cotton Conf.*, 1997, 1078.
28. J.P. Arena, K.K. Liu, P.S. Paress, E.G. Frazier, D.F. Cully, H. Mrozik and J. Schaeffer, *J. Parasitol.*, 1995, **81**, 286.
29. R. Mulder, K. Wellinga and J.J. van Daalen, *Naturwissenschaften*, 1975, **62**, 531.
30. V.L. Salgado, *Molecular Pharmacology*, 1992, **41**, 120.
31. Wing, K.D., M.E. Schnee, M. Sacher and M. Connair, A novel oxadiazine insecticide is bioactivated in lepidopteran larvae, *Arch. Insect Biochem. Physiol.,* 1998, **37**, 91.
32. T.-P. Yao, B.M. Forman, Z. Jiang, L. Cherbas, J.-D. Chen, M. Mckeown, P. Cherbas and R.M. Evans, *Nature*, 1993, **366**, 476.
33. K.D. Wing, *Science*, 1987, **241**, 467.
34. T.S. Dhadialla, G.R. Carlson and D.P. Le, *Annu. Rev. Entomol.*, 1998, **43**, 545.
35. R.W. Addor, T.J. Babcock, B.C. Black, D.G. Brown, R.E. Diehl, J.A. Furch, V. Kameswaran and V.M. Kamhi, *ACS Symp. Ser.,* 1992, **504**, 283.
36. P. Harrewijn and H. Kayser, *Pestic. Sci.*, 1997, **49**, 130.
37. Anonymous, *Courier Agrochem.,* 1997, **1**, 4.
38. A.W. Czarnik, *Anal. Chem.*, 1998, **70**, 378A.

Molecular Approaches to Antifungal Molecule Discovery

Marie-Claire Grosjean-Cournoyer[1], Pierre-Henri Clergeot[2], Andrew Payne[3], Viviane Brozek[1], Derek W. Hollomon[3], Marie-Pascale Latorse[1] and Marc-Henri Lebrun[2]

[1] RHONE-POULENC AGRO, 14/20 RUE PIERRE BAIZET, 69009 LYON, FRANCE
[2] LABORATOIRE DE PHYSIOLOGIE VEGETALE UMR CNRS-RPA 41, 14/20 RUE PIERRE BAIZET, 69009 LYON, FRANCE
[3] IACR-LONG ASHTON RESEARCH STATION, CELL BIOLOGY DEPARTMENT, BRISTOL BS41 9AF, UK

1 INTRODUCTION

Improvement of crop quality and yields depends on the use of chemicals that protect plants against diseases. Plant protection needs the development of new pesticides including antifungal molecules with novel modes of action, high biological activity and specificity, low ecological impact and low risk of inducing resistance.

Compounds used in plant protection against fungal diseases can be classified in three groups depending on their mode of action:

- molecules acting directly on an essential function involved in fungal growth or reproduction
- molecules inhibiting an essential function for pathogenicity
- molecules stimulating plant defence mechanisms.

Most antifungal compounds currently used belong to the first group and disrupt essential metabolic functions inhibiting fungal growth. The second group of molecules, interfering specifically with the infection process, might not necessarily inhibit fungal growth. For example, tricyclazole has no fungitoxic activity *in vitro*[1] but inhibits melanin biosynthesis and has a good protective effect on whole plants. Melanin has a role in the rigidity of the appressorial wall of *Magnaporthe grisea* and *Colletotrichum lindemuthianum*,[2] a property essential for generating the osmotic pressure required for mechanical penetration of the plant cuticle.[3] The third group of compounds disrupt infection process by acting indirectly on the fungus via an induction of plant defences, e.g. Acibenzolar (benzo(1,2,3)thiadiazole-7-carbothioic acid S-methyl ester or BTH).[4,5]

The discovery of new molecules requires a high throughput programme of chemical synthesis and screening. An increase in our knowledge of vital functions in phytopathogenic fungi and their pathogenicity determinants can lead to the characterisation of new targets for antifungal molecules. Such targets should allow the design of biochemical or cellular high throughput screening (HTS) procedures.

In this paper, we describe how molecular approaches can lead to the discovery of novel antifungal molecules acting on the infection process of fungi. These strategies could also be used to design screening systems for antifungal molecules acting on vital targets or on plant defence mechanisms.

2 IDENTIFICATION OF PATHOGENICITY GENES

2.1 Background

Even if strategies used by fungi to attack plants are quite diverse, there are common steps in the infection process:

- attachment to the host
- penetration of the host
- colonisation of the host
- reproduction / dissemination of the pathogen.

All these steps of the infection cycle involve functions that are potential targets for new fungicides.

Recent developments in fungal molecular biology opened the way for a systematic analysis of genetic determinants involved in the interaction between the fungus and the plant. This systematic analysis relies on the discovery and characterisation of genes specifically expressed during the infection process, or mutants affected in their pathogenicity.

2.2 Differential screening

The discovery, by differential screening, of genes specifically expressed during the infection process was successful in *M. grisea*,[6] *Botrytis cinerea*,[7] *Uromyces fabae*,[8] *Phytophthora infestans*[9] and *Cladosporium fulvum*.[10] Isolation of genes specifically expressed during the infection process does not preclude their importance in pathogenicity. Inhibition of their expression by gene deletion[6] or by an antisense strategy[11,12] is required to assess their importance. Talbot et *al.*[6] identified the gene, MPG1, from the rice blast fungus *M. grisea* as a cDNA clone differentially expressed during infection. Sequence comparison suggested that the gene encodes a hydrophobin. Mutants in which this gene was deleted had a reduced ability to infect their host and had altered hydrophobin structures on their cell walls.

Differential gene expression provides powerful ways to analyse pathogen development. Gene expression can be analysed using small amounts of tissue including appressoria. But, once the pathogen entered the host tissue, its low biomass relative to host plant tissue makes such an analysis difficult. This limitation can be overcome by other approaches such as insertional mutagenesis.

2.3 Insertional mutagenesis

Mutants affected in the infection process can be obtained by insertional mutagenesis through random integration into the genome of a plasmid carrying a selectable marker. The effect of the insertion(s) on pathogenicity is studied *in planta* and strains showing an altered, or deficient infection phenotype are studied further. This strategy was used with success to identify pathogenicity-deficient mutants, and to characterise the corresponding genes.[13-17] Such insertional mutagenesis requires a good transformation protocol and conditions favouring single copy integration to simplify the characterisation of the mutation. In *M. grisea,* restriction enzyme mediated integration[18] (REMI), obtained by adding a restriction enzyme to the plasmid in the transformation protocol, increases the transformation rate[19] and the number of single copy integration. As shown in Table 1, the transformation rate was increased by 6-fold at the optimal restriction enzyme concentration (*Hind*III: 40 U/ml and *Bgl*II: 4 - 10 U/ml).

A library of *M. grisea* transformants obtained by REMI insertional mutagenesis was developed in our laboratory. The strategy used to obtain and analyse these transformants is presented in Figure 1. 2300 transformants were obtained and 1200 purified transformants were tested for pathogenicity defects on rice plants. Among these 1200 transformants, 1.2% presented deficiencies in pathogenicity, either total (n=5) or partial (n=10). This frequency is comparable to those already obtained for *Ustilago maydis,*[14] *C. lindemuthianum*[17] and *M. grisea.*[16] It is important to genetically link the integration event with the pathogenicity phenotype in order to rule out loss of pathogenicity due to spontaneous mutations. Six pathogenicity deficient transformants were analysed by crossing with a wild type strain and two of these transformants contained mutations that cosegregated with the selectable marker (hygromycin resistance). The function(s) of the mutated genes is currently analysed. In the case of a known function, biochemical or cellular tests could be designed to screen molecules inhibiting this function. For a gene with an unknown function, the screening of molecules could be designed on the basis of the inhibition of its expression.

Enzyme / Concentration (Units)	0	1	4	10	40
*Hind*III	24	15	27	42	150
*Bgl*II	24	30	156	152	9

Table 1 *Transformation rate (transformants /µg DNA) of M. grisea strain P1.2 with circular pAN7.1*[20] *and different concentrations of two restriction enzymes, HindIII and BglII.*

Transformation of the recipient strain
with a plasmid using REMI

↓

Purification of the transformants

↓

Screening for pathogenicity defects
Phenotypic characterisation

↓

Genetic analysis

↓

Recovery of the flanking sequences

↓

Gene characterisation and analysis

Figure 1 *Characterisation of genes involved in the infection process by insertional mutagenesis in M. grisea.*

3 USE OF REPORTER GENES TO SURVEY THE EXPRESSION OF POTENTIAL TARGETS *IN PLANTA* AND TO DESIGN HIGH TRHOUGHPUT SCREENING SYSTEMS

3.1 GFP as a vital marker

The green fluorescent protein isolated from the jellyfish *Aequorea victoria*[21] is a small protein of 238 amino-acids responsible for bioluminescence. The formation of the fluorescent chromophore is not species-dependent. Unlike other reporter genes, it can be used as an universal vital marker, particularly since its optimisation by mutagenesis which led to an increase in brightness in various organisms. GFP has been used successfully in filamentous fungi such as *U. maydis*,[22] *Aspergillus nidulans*,[23] *Mycosphaerella graminicola* or *M. grisea* (Figure 2).

3.2 Gene expression studies

The gene coding for the GFP has been used as a reporter gene to study the regulation of gene expression in *U. maydis*.[24] To evaluate the potential of GFP in *M. graminicola*, the fungus causing *Septoria* leaf blotch in wheat, for gene expression both *in vitro* or *in planta*, the fungal strain ST16 was cotransformed with pAN7-1[20] and either pICLGFP (gift of Dr P. Bowyer - Long Ashton Research Station) which contains the modified GFP gene, sGFP-TYG gene,[25] fused to the isocitrate lyase promoter of *Neurospora crassa*,[26] or pGPD-GFP which contains the sGFP-TYG gene under the control of the glyceraldehyde-3-phosphate dehydrogenase (GPD) promoter of *A. nidulans*. Analysis of the promoter region of the acetate-induced isocitrate lyase gene has been described in *A. nidulans*.[27] The isocitrate lyase gene was transcribed during growth on acetate but not during growth on glucose. *M. graminicola* hygromycin resistant pICL-GFP transformants were assessed for GFP expression *in vitro* in presence or in absence of glucose, or potassium acetate at 50 mM (Figure 3). As shown for *A. nidulans*,[27] the expression of the GFP in the pICL-GFP *M. graminicola* transformant was repressed by glucose and induced by potassium acetate (Figure 3). Therefore, GFP has a good potential to study gene expression in *M. graminicola*.

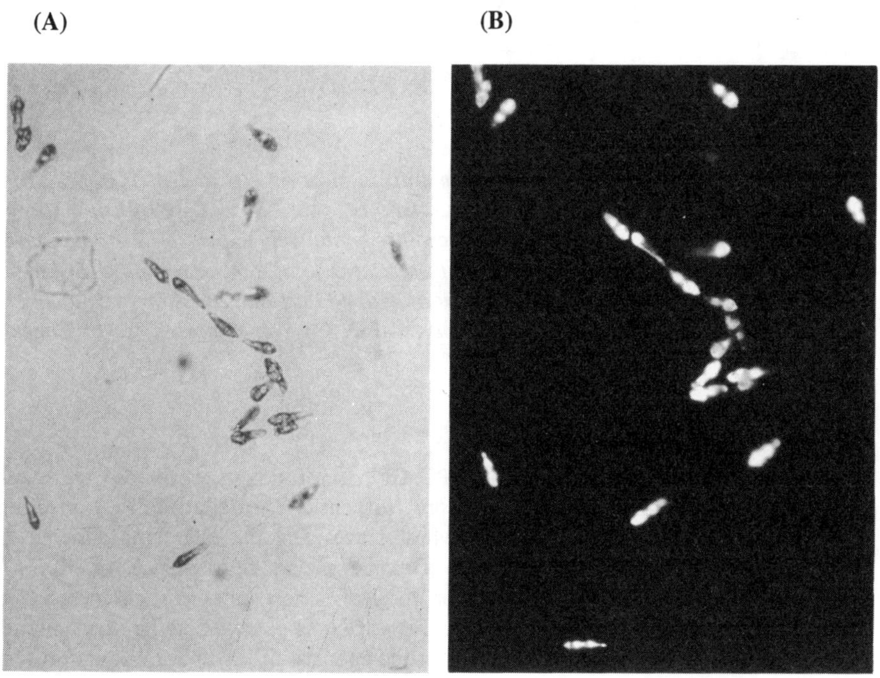

Figure 2 *GFP expression in M. grisea spores. M. grisea transformed with a constitutive GFP construct was analysed by microscopy using a Nikon optiphot without further manipulation of the object. A) Observation in light microscopy. B) Observation in fluorescence microscopy using the Nikon filter set BV-1A. Non transformed cells did not fluoresce under theses conditions. The bar represents 50 µm.*

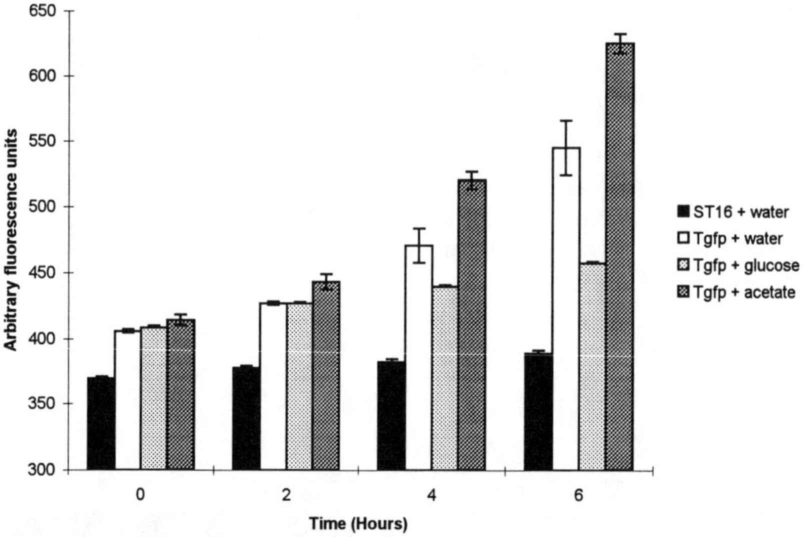

Figure 3 *GFP fluorescence of M. graminicola strain ST16 and of a pICL-GFP transformant (Tgfp) in the presence of different carbon sources. 100 μl of a spore suspension in water were added to microtitre plate well. Glucose and potassium acetate were added to the spore suspensions at a final concentration of 50 mM. GFP fluorescence was measured by excitation at 485 nm and emission at 530 nm at different time points using LS50B Luminescence Spectrophometer (Perkin Elmer, Seer Green, UK).*

M. graminicola pICL-GFP and pGPD-GFP transformants were used to follow the GFP expression *in planta*. Spore suspensions of the GFP expressing transformants were inoculated onto wheat leaves and the infection followed *in planta* by microscopic observations. Despite strong fluorescence of the *M. graminicola* pICL-GFP transformant on the leaf surface prior to penetration and following pycnidial development, no fluorescence was observed within plant tissue. *M. graminicola* is believed to grow within the apoplast of leaves and sugars present in the apoplast might be sufficient to repress the ICL promoter. By comparison, the transformant carrying the GFP gene under the control of the constitutive GPD promoter fluoresced inside the plant. This showed that GFP reporter gene constructs can be used to follow expression of a gene *in planta* and in living cells. Such information is very useful in characterising the role of a gene in pathogenicity.

3.3 HTS design

In addition to its role for studying gene expression during the infection process, GFP can be used as a reporter gene in screening systems. The general strategy to test molecules inhibiting the expression of a gene is presented on Figure 4. The promoter of a gene described as either a key gene in pathogenicity or in a pathway involved in pathogenicity is fused to the GFP reporter gene. The construct is used to transform the selected fungus. Genes expressed specifically during the infection process usually need to be induced for *in vitro* expression as shown for the cutinase gene of *Fusarium solani* f. sp. *pisi*.[28] The right conditions for the *in vitro* induction of the expression of the GFP promoter fusion have to be found. The inductive condition could be starvation as shown for the expression of MPG1.[6] Such a screen is a cellular test where the transformant strain is grown in presence of the inducer of the GFP expression and molecules to be tested. The molecules which inhibit the expression of the GFP could be easily detected using such conditions in microwell plates.

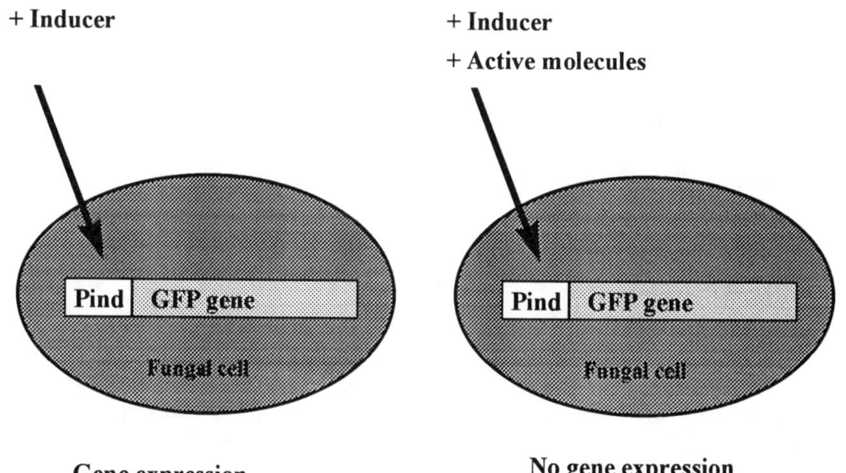

Figure 4 *General strategy for the detection of molecules inhibiting the regulation of the expression of key genes or key pathways in pathogenicity. The promoter of interest (Pind) is fused to the GFP reporter gene. The transformed strain is grown in presence of the GFP expression inducer added to the molecules to be tested. The active molecules are those which inhibit the GFP expression.*

4 CONCLUSIONS

The results from molecular and genetic studies of plant fungal pathogenicity clearly showed that a large number of genes and biochemical pathways are involved in the infection process. Our knowledge of infection process will increase significantly over the forthcoming years, giving access to new targets and strategies for the screening of antifungal molecules. Antifungal compounds inhibiting the development of the fungus on its host plant are very attractive. They should be highly selective as they will inhibit pathways that are specific to pathogenic fungi. They are likely to have a low toxicity as they inhibit pathways that are not essential for cell survival or growth. Occurrence of resistance might be reduced as the selection for resistant strains should only occur during attempted infection. For all these criteria, inhibitors of pathogenicity are likely to be good antifungal molecules.

Several new fungicides groups, including anilinopyrimidines, phenylpyrolles, quinolines and strobilurins, have recently been introduced as disease control agents. These products came from discoveries in the 1980s, and molecular technologies have contributed little to their success. Their modes of action are not all known, but together they emphasise the wide range of metabolic events that can be selectively inhibited to generate useful fungicides. Molecular technologies have added a new dimension to conventional approaches to fungicides discovery. They offer a more rational approach to target-site validation through better understanding of the interaction between plants and their fungal pathogens, provide tools to analyse in detail the molecular basis of inhibition, and provide a platform for High Throughput Screening essential to optimise the coupling of combinatorial chemistry and lead discovery.

Acknowledgements

P.-H. Clergeot was supported by a fellowship from the Bio Avenir program funded by Rhône-Poulenc with the participation of the Ministère de la Recherche et de l'Espace and of the Ministère de l'Industrie et du Commerce extérieur. We thank Drs. R. Beffa and B. Cournoyer for useful comments on the manuscript

References

1. C.P. Woloshuk, P. M. Wolkow and H. D. Sisler, *Pestic. Sci.*, 1981, **12**, 86
2. P.Vidhyasekaran, 'Fungal pathogenesis in plants and crops, Molecular biology and host defense mechanisms', Marcel Dekker, Inc, New York, 1997, Part 2, p. 106.
3. F.G. Chumley and B. Valent, *Mol. Plant-Microbe Interact.*, 1990, **3**, 135.
4. L. Friedrich, K. Lawton, W. Ruess, P. Masner, N. Specker, M. Gut Rella, B. Meier, S. Dincher, T. Staub, S. Uknes, J.-P. Métraux, H. Kessmann and J. Ryals, *Plant J.*, 1996, **10**, 61.

5. U. Schaffrath, E. Freydl and R. Dudler, *Mol. Plant-Microbe Interact.*, 1997, **10**, 779.
6. N.J. Talbot, D. J. Ebbole and J. Hamer, *Plant Cell*, 1993, **5**, 1575.
7. E. P. Benito, T. Prins, and J.A.L. van Kan, *Plant Mol. Biol.*, 1996, **32**,947.
8. M. Hahn and K. Mendgen, *Mol. Plant-Microbe Interact.*,1997, **10**, 427.
9. C. M. J. Pieterse, M.B.R. Riach, T. Bleker, G. C. M. van den Berg-Velthuis and F. Govers, *Physiol. Mol. Plant. Path.*, 1993, **43**, 69.
10. M. Coleman, B. Henricot, J. Arnau and R. P. Oliver, *Mol. Plant-Microbe Interact.*,1997, **10**, 1106.
11. H.J. Judelson, R. Dudler, C. M. J. Pieterse, S. E. Unkles and R. W. Michelmore, *Gene*, 1993, **133**, 63.
12. H. Prokisch, O. Yarden, M. Dieminger, M. Tropschug, and I. B. Barthelmess, *Mol. Gen. Genet.*, 1997, **256**, 104.
13. S. Lu, L. Lyngholm, G. Yang, C. Bronson, O. C. Yoder and B. G. Turgeon, *Proc. Natl. Acad. Sci.* USA, 1994, 91, 12649.
14. M. Bölker, H. U. Böhnert, K. H. Braun, J. Görl and R. Kahmann, *Mol. Gen. Genet.*, 1995, **248**, 547.
15. Z. Shi and H. Leung, *Mol. Plant-Microbe Interact.*, 1995, **8**, 949.
16. J. A. Sweigard, A. M. Carroll, L. Farrall, F. G. Chumley and B. Valent, *Mol. Plant-Microbe Interact., 1998, 11, 404.*
17. M. Dufresne, J. A. Bailey, M. Dron and T. Langin, *Mol. Plant-Microbe Interact.*,1998, **11**, 99.
18. R. H. Schiestl and T. D. Petes, *Proc. Natl. Acad. Sci.* USA, 1991, **88**, 7585.
19. Z. Shi, D. Christian and H. Leung, *Phytopathology*, 1995, **85**, 329.
20. P. J. Punt, R. P. Oliver, M. A. Dingemanse, P. H. Pouwels and C. A. M. J. J. van den Hondel, *Gene*, 1987, **56**, 117.
21. M. Chalfie, Y. Tu, G. Euskirchen, W. W. Ward, and D. C. Prasher, *Science*, 1994, **263**, 802.
22. T. Spelling, A. Bottin and R. Kahmann, *Mol. Gen. Genet.*, 1996, **252**, 503.
23. J. M. Fernandez-Abalos, H. Fox, C. Pitt, B. Wells and J.H. Doonan, *Mol. Microbiol*, 1998, **27**, 121.
24. A. Bottin, J. Kämper and R. Kahmann, *Mol. Gen. Genet.*, 1996, **253**, 342.
25. J. Sheen, S. Hwang, Y. Niwa, H. Kobayashi and D.W. Galbraith, *Plant J.*, 1995, **8**, 777.
26. L.D.S. Gainey, K. Kölble and I.F. Connerton, *Mol. Gen. Genet.*, 1991, **229**, 253.
27. P. Bowyer, J. R. De Lucas and G. Turner, *Mol. Gen. Genet.*, 1994, **242**, 484.
28. J.T. Kämper, U. Kämper, L. M. Rogers, P.E. Kolattukudy, *J. Biol. Chem.*, 1994, **269**, 9195.

Metabolism

Cytochrome P450s and other Xenobiotic Metabolizing Enzymes in Plants

H. Ohkawa, H. Imaishi, N. Shiota, T. Yamada and H. Inui

DEPARTMENT OF BIOLOGICAL AND ENVIRONMENTAL SCIENCE, FACULTY OF AGRICULTURE, KOBE UNIVERSITY, KOBE 657-8501, JAPAN

1 Introduction

Xenobiotics are transformed in higher plants. These biotransformation processes are grouped in three main phases known as Phase I or conversion, Phase II or conjugation, and Phase III or compartmentation. Phase I reactions include oxidations, reductions and hydrolyses, whereas Phase II reactions are conjugations with glutathione, sugars or amino acids. In Phase III, xenobiotic conjugates are converted to secondary conjugates or insoluble bound residues and are deposited in the vacuole or other compartments of plant cells.[1] The biotransformations of xenobiotics are dependent upon the plant species, giving rise to pesticide detoxification and activation or herbicide selectivity and resistance.

2 Plant Enzyme Systems Metabolizing Xenobiotics

In the metabolism of xenobiotics in plants, the Phase I reactions are mostly catalyzed by cytochrome P450 monooxygenases and esterases. Cytochrome P450 monooxygenases localized on the microsomes consist of a number of cytochrome P450 (P450 or CYP) species and a generic NADPH-cytochrome P450 oxidoreductase (P450 reductase), catalyzing oxidative reactions of endogenous and exogenous lipophilic compounds including pesticides. Currently, over 60 P450 sequences from plants are known, but the function for very few has been identified. Important P450 enzymes whose characterization remains elusive or poorly understood are involved in the biosynthesis of sterols, glucosinolates, phenylpropanoids/flavonoids, salicylic acid, jasmonic acid, gibberellic acids, brasinosteroids and alkaloids, and in the metabolism of xenobiotics including herbicides.[2] Herbicide chemicals such as chlortoluron, atrazine, bentazone and chlorsulfuron are known to be metabolized by P450 enzymes in plants.

The Phase II of xenobiotic metabolism was mediated by glutathione S-transferases (GST). GST isozymes of maize in xenobiotic detoxification are the most studied and GSTIV was found to play a major role in the tolerance of maize to the herbicide alachlor.[3] The role of glucosyltransferases associated with xenobiotic metabolism is also studied well.

The metabolism of chlortoluron in wheat is shown in Figure 1. Chlortoluron has been used for selective weed control in wheat, which can detoxify the herbicide via ring-methyl hydroxylation and N-demethylation. Both reactions appear to be mediated by P450 enzymes. Metabolism via ring-methyl hydroxylation is primarily responsible for chlortoluron resistance in wheat, whereas N-demethylation makes a relatively minor contribution toward chlortoluron resistance. The metabolism of atrazine in maize is shown in Figure 2. Maize is highly resistant to triazines. Rapid detoxification of atrazine via glutathione conjugation is the primary mechanism of resistance with three GST isozymes responsible for catalyzing the conjugation. Atrazine can also be metabolized via

N-dealkylation. Mono-N-dealkylation reduces the binding affinity of atrazine for the D1 protein which is the target of the herbicide action, and thereby rapidly reduces its phytotoxicity. However, removal of both N-alkyl groups is necessary for complete detoxification. As a mechanism for herbicide resistance, N-dealkylation is considerably less efficient than is glutathione conjugation, although P450 is thought to catalyze N-dealkylation. This has not been demonstrated *in vitro*.[4]

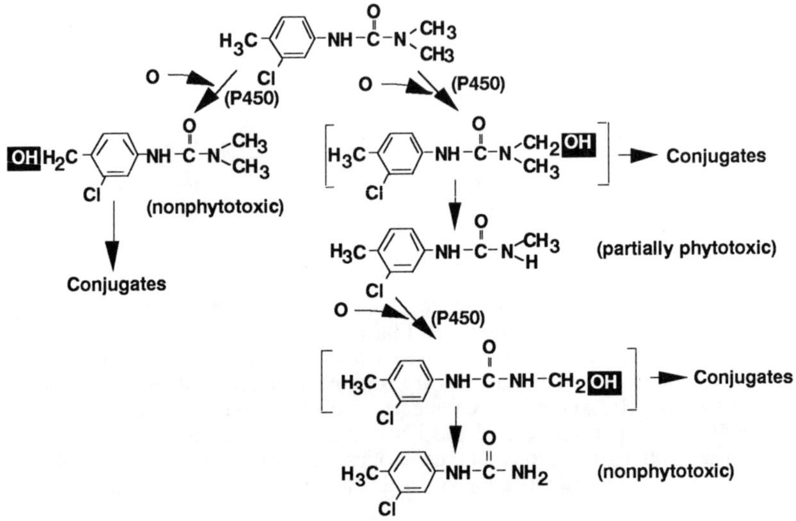

Figure 1 *Metabolism of chlortoluron in wheat*

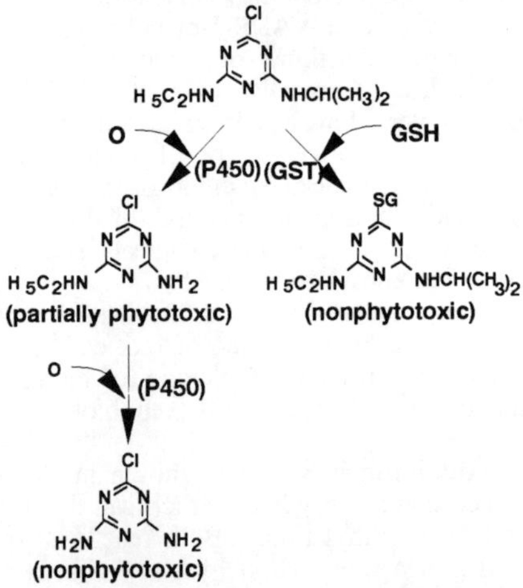

Figure 2 *Metabolism of atrazine by maize*

3 Plant P450 Species Metabolizing Herbicides

P450 enzymes are involved in the metabolism of herbicides in plants. For example, in the metabolism of chlortoluron in wheat and barley, both ring-methyl hydroxylation and N-demethylation were suggested to be catalyzed by P450 enzymes. The resistance of both crops to the herbicide has been attributed largely to the high capacity of these species to perform ring-methyl hydroxylation. On the other hand, N-demethylation may afford relatively low tolerance to the herbicide as demonstrated in cotton.[4]

When tobacco cultured S401 cells were treated with 2,4-D, the cells metabolized chlortoluron to give ring-methyl hydroxylated and N-demethylated metabolites, whereas the cells hardly produced these metabolites without 2,4-D treatment. Based on these results, we examined to clone P450 cDNAs involved in the metabolism of chlortoluron in tobacco cultured S401 cells treated with 2,4-D. As a result, four novel P450 cDNA clones were isolated. Based on their sequences, these were named as CYP71A11, CYP81B2, CYP81C1 and CYP81C2. Figure 3 shows a phylogenic tree of four P450 species together with plant P450 species reported. These P450 species cloned were found to be in type A, which were mostly related to the biosynthesis of phenylpropanoids and other secondary metabolites.

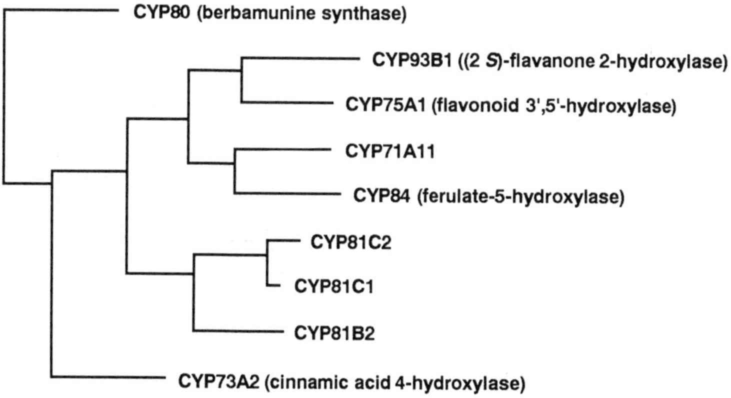

Figure 3 *A dendrogram showing the relatedness of novel tobacco P450 species with those of the plant P450 species reported*

Northern blot analysis with the cloned cDNAs revealed that both CYP71A11 and CYP81B2 were inducible in the S401 cells by 2,4-D treatment. Therefore, we examined to express each of both cDNA clones in the yeast *Saccharomyces cerevisiae*. When both CYP71A11 and CYP81B2 were expressed in the yeast together with yeast P450 reductase or tobacco P450 reductase, the yeast microsomal fractions showed high 7-ethoxycoumarin O-deethylase activity. The yeast cells expressing both CYP71A11 and yeast P450 reductase exhibited enhanced ring-methyl hydroxylation and N-demethylation, whereas the yeast cells expressing both CYP81B2 and tobacco P450 reductase showed a slightly enhanced ring-methyl hydroxylation toward chlortoluron. Therefore, both CYP71A11 and CYP81B2 were found to be involved in the metabolism of chlortoluron in the S401 cells treated with 2,4-D, as shown in Figure 4.[5] Recently, it was reported that CYP73A1, CYP76B1, CYP81B1 from plants also metabolized chlortoluron, although fatty acids were endogenous substrates for CYP81B1.[6]

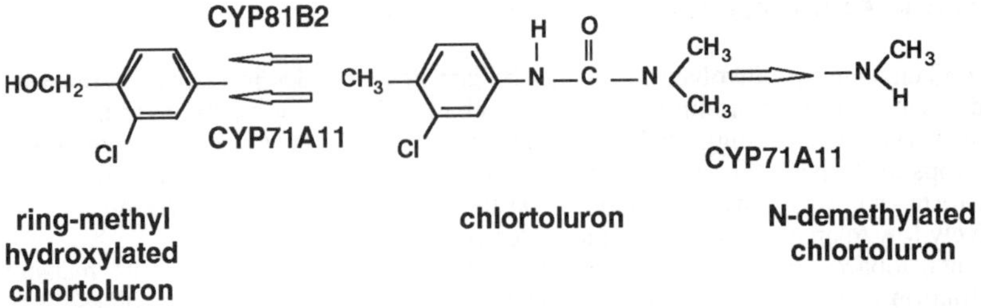

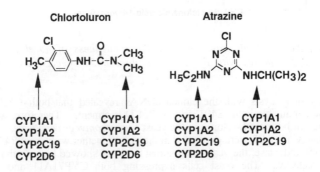

Figure 4 *Metabolism of the herbicide chlortoluron in tobacco cultured S401 cells treated with 2,4-D*

4 Human P450 Species Metabolizing Herbicides

In mammals, a number of microsomal P450 species are involved in the metabolism of xenobiotics. Many of them were characterized by molecular cloning and heterologous expression of the corresponding cDNA clones. Based on the results of these studies, it was suggested that individual P450 species showed overlapping broad substrate specificity and confer the ability to metabolize a number of unknown lipophilic compounds.

We attempted to express rat CYP1A1 and its fused enzyme with yeast P450 reductase in the yeast. The recombinant yeast showed 7-ethoxycoumarin O-deethylase as well as chlortoluron metabolizing activities, although the fused enzyme exhibited higher specific activities than the expression of rat CYP1A1 alone did.[7] Then, we examined the metabolism of herbicide chemicals in the yeast cells or the microsomes expressing human P450 species. It was found that human P450 CYP1A1 catalyzed both ring-methyl hydroxylation and N-demethylation toward chlortoluron as well as removal of both N-alkylgroups of atrazine. These metabolic reactions towards both herbicides were also found to be catalyzed by CYP1A2, CYP2C19 and CYP2D6, as shown in Figure 5.[8]

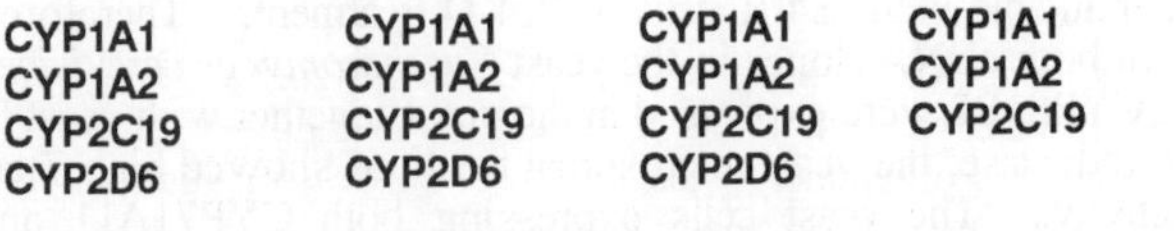

Figure 5 *Metabolism of herbicides in human P450 species*

5 Expression of Human P450 Species in Plants

It was reported that the transgenic potato plants[9] expressing rat CYP1A1, and tobacco plants[10] expressing the fused enzyme between rat CYP1A1 and yeast P450 reductase metabolized chlortoluron through ring-methyl hydroxylation and N-demethylation, exhibiting resistance to the herbicide.[11,12] Therefore, we attempted to express human CYP1A1 in potato plants for examination of cross-resistance to herbicides, since human CYP1A1 showed multiple metabolic reactions towards several herbicide chemicals with different structures. The potato plants transformed with the expression plasmid shown in Figure 6 by Agrobacterium method expressed human CYP1A1 in the microsomes. These transgenic plants metabolized chlortoluron through ring-methyl hydroxylation and N-demethylation, and atrazine through both N-dealkylations, resulting in resistance to the herbicides.[13]

Thus, it was found that the expression of single human CYP1A1 in potato plants conferred cross-resistance to the herbicide chemicals with different structures and modes of herbicide action.

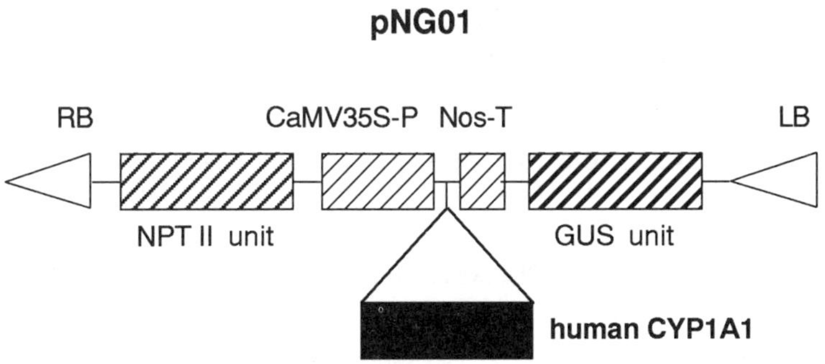

Figure 6 *An expression plasmid for human CYP1A1*

6 Concluding Remarks

Plant P450 species involved in the metabolism of the herbicide chlortoluron in tobacco cultured S401 cells treated with 2,4-D were identified as CYP71A11 and CYP81B2. On the other hand, human CYP1A1, CYP1A2, CYP2C19 and CYP2D6 were found to metabolize chlortoluron as well as atrazine. So, it was suggested that in plants and humans, different species of P450 enzymes catalyzed the metabolic reactions of the herbicide chemicals.

The yeast and plant expression systems for P450 species metabolizing xenobiotics from plants and human were useful for comparative metabolism of the chemicals between plants and human. Particularly, the systems expressing human P450 species are important models for human metabolism of pesticides.

The expression of human CYP1A1 in potato plants conferred cross-resistance to the herbicides chlortoluron and atrazine. The transgenic plants expressing P450 species metabolizing xenobiotics appears to be useful for crop breeding with herbicide resistance and low pesticide residues. These plants are also important for phytoremediation.

References

1. K. Kreuz, R. Tommasini and E. Martinoia, *Plant Physiol.*, 1996, **111**, 349.

2. K. A. Feldmann, R. Feyereisen, D. W. Galbraith and R. Winkler, "BRAIN SEMINAR: Cytochrome P450 and Plant Genetic Engineering" Ed. by H. Ohkawa and Y. Ohkawa, BRAIN, Tokyo, 1997, p. 13.

3. K. K. Hatzios, "Regulation of Enzymatic Systems Detoxifying Xenobiotics in Plants" Ed. by K. K. Hatzios, Kluwer Academic Publishers, Dordrecht, 1997, Chapter 1, p. 1.

4. J. W. Gronwald, "Herbicide Resistance in Plants", Ed. by S.B.Powles and J. A. M. Holtum, Lewis Publishers, Boca Raton, 1994, chapter 2, p. 27.

5. T. Yamada, PhD Thesis, Kobe University, 1998.

6. F. Durst, I. Benveniste, A. Lesot, F. Pinot, J. P. Salaün and D. Werch-Reichhart, "BRAIN SEMINAR: Cytochrome P450 and Plant Genetic Engineering" Ed. by H.Ohkawa and Y. Ohkawa, BRAIN, Tokyo, 1997, p. 1.

7. N. Shiota, H. Inui and H. Ohkawa, *Pestic. Biochem. Physiol.*, 1996, **54**, 190.

8. N. Shiota and H. Ohkawa, "BRAIN SEMINAR: Cytochrome P450 and Plant Genetic Engineering" Ed. by H. Ohkawa and Y. Ohkawa, BRAIN, Tokyo, 1997, p. 31.

9. H. Inui, N. Shiota, T. Ishige, Y. Ohkawa and H. Ohkawa, *Breeding Sci.*, 1998, **48**, 135.

10. N. Shiota, A. Nagasawa, T. Sakaki, Y. Yabusaki and H. Ohkawa, *Plant Physiol.*, 1994, **106**, 17.

11. H. Ohkawa, N. Shiota, H. Inui, M. Sugiura, Y. Yabusaki, Y. Ohkawa, and T. Ishige, "Regulation of Enzymatic Systems Detoxifying Xenobiotics in Plants" Ed. by K. K. Hatzios, Kluwer Academic Publishers, Dordrecht, 1997, Part 5, p.307.

12. H. Ohkawa, H. Imaishi, N. Shiota, H. Inui and Y. Ohkawa, *Rev. Toxicol.*, 1997, **1**, 1.

13. H. Inui, Y. Ueyama, N. Shiota, Y. Ohkawa and H. Ohkawa, *Pestic. Biochem. Physiol.*, 1998, Submitted.

Metabolism of Azoxystrobin in Plants and Animals

Robert S.I. Joseph

ZENECA AGROCHEMICALS, JEALOTT'S HILL RESEARCH STATION, BRACKNELL, BERKSHIRE RG42 6ET, UK

1 INTRODUCTION

Azoxystrobin is a new fungicide with a novel biochemical mode of action. Its synthesis was inspired by a group of natural products, the strobilurins, which are produced by several species of Basidiomycete fungi, which grow on decaying wood. The structures of azoxystrobin, Strobilurin A and the closely related Oudemansin A, are shown in Figure 1. The natural products and azoxystrobin share a common structural feature, namely the methyl ester of β-methoxyacrylic acid.

Synthesis chemists at Zeneca Agrochemicals first became interested in the strobilurins in the early 1980s. An extensive programme of research was initiated, with the aims of preparing analogues of the natural products with high levels of activity, and with suitable physical properties for an agricultural fungicide[1]. This effort culminated in the discovery of azoxystrobin in 1988, after the synthesis of some 1400 analogues. Following extensive biological characterisation, azoxystrobin was selected for development as a new fungicide.

Azoxystrobin shares its biochemical mode of action with the natural strobilurins. It demonstrates a very broad spectrum of activity and is active against fungal pathogens from all four taxonomic groups, the Oomycetes, Ascomycetes, Deuteromycetes and Basidiomycetes.[2,3]

The site of action of azoxystrobin is the fungal mitochondrion, where it binds to the cytochrome bc_1 complex, thereby preventing electron transport and energy production via oxidative phosphorylation. Azoxystrobin is not cross resistant to other site specific fungicides, such as the benzimidazoles, DMIs, or phenylamides. Azoxystrobin therefore represents an effective new tool for the management of resistance, which is frequently a significant factor in the choice of fungicide products.

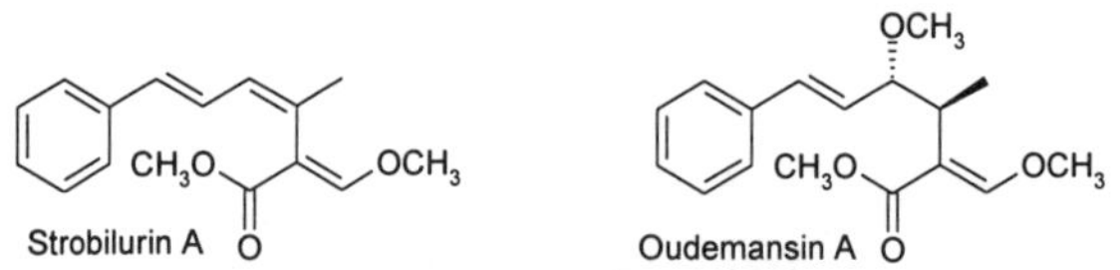

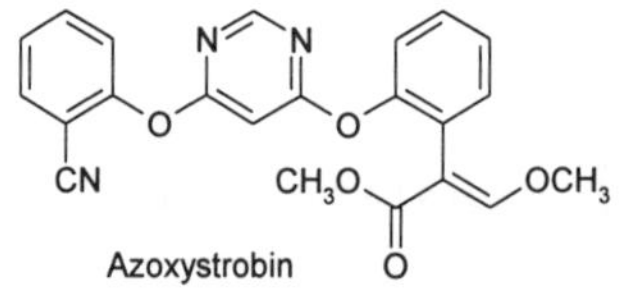

Figure 1 *Structures of Natural and Synthetic β-Methoxyacrylates*

This paper describes the results of studies into the metabolism of azoxystrobin in plants and animals conducted at Zeneca Agrochemicals during the period 1991 to 1995. The fate in soil is also considered, due to its importance in determining the nature of the terminal residue in plants.

2 METABOLISM IN ANIMALS

2.1 Excretion Studies

The metabolism and excretion of azoxystrobin in rats has been studied following oral administration at both low and high doses (1 and 100 mg kg^{-1}) and also following pre-dosing for 14 days with unlabelled azoxystrobin.

Initial whole body autoradiography (WBA) studies were conducted in glass metabolism cages. Azoxystrobin, radiolabelled individually in each of the three rings (Figure 2), was dosed at 1 mg kg^{-1} body weight and urine, faeces and expired air collected for up to 48 hours after dosing. Up to 75% of the radioactivity was excreted in the faeces and up to 21% in the urine. There was no significant radioactivity in the expired air from any of the three radiolabels. Analysis of tissues slices indicated a marked reduction of radioactivity in tissues between 24 and 48 hours. There were no significant differences between the rates and routes of excretion or tissue distribution between the three labelled forms of azoxystrobin, therefore subsequent WBA studies were conducted with the pyrimidinyl label only.

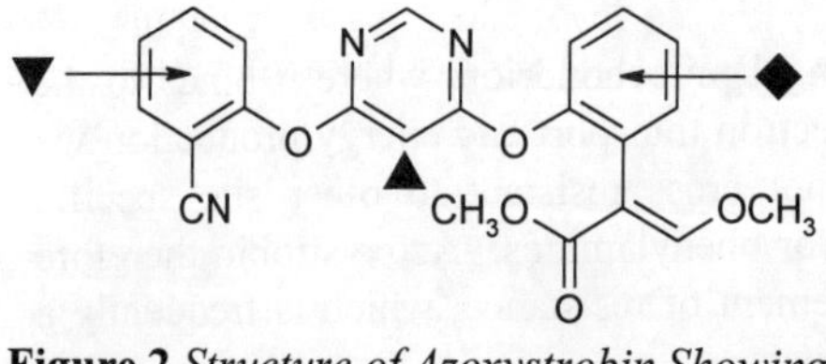

Figure 2 *Structure of Azoxystrobin Showing Position of Radiolabelling*

Further single and repeat dose excretion and tissue retention studies showed a similar pattern, with the majority of the dose (up to 89%) found in the faeces and up to 17% in urine. There was little radioactivity remaining in any of the tissues or the carcass 7 days after dosing. The highest tissue concentrations were found in the liver and kidney: the organs of metabolism and excretion. There were no significant differences between the rates or routes of excretion or tissue retention of radioactivity between the repeat dose study and either of the single dose studies or between males and females. Based on the above studies it can be concluded that azoxystrobin does not bioaccumulate in animal tissues.

2.2 Metabolism Studies

The principal metabolic pathways of azoxystrobin in the rat have been defined and are shown in Figure 3. Each of the three radiolabelled forms of azoxystrobin were dosed to bile duct cannulated rats at 100 mg kg^{-1}. The principal route of elimination was in the bile (between 56 and 74% of dose) with approximately 5% in urine and 20% in faeces. Again there were no marked differences between the rates and routes of excretion between the three radiolabelled forms. Azoxystrobin was extensively metabolised with at least 18 metabolites formed. All the metabolites, with the exception of Compound 54 (Compound 2 glucuronide), were present at levels of less than 10% of the administered dose.

There were five principal metabolic pathways:

- Hydrolysis of the methyl ester to form the β-methoxyacrylic acid (Compound 2) which then underwent conjugation to form the glucuronide (Compound 54). This metabolite was the most abundant, accounting for approximately 30% of dose.
- Cleavage of the acrylate double bond to form Compound 21. The proposed mechanism for this transformation is outlined in Figure 4. Compound 21 is then further metabolised by ring hydroxylation to give Compound 55 and also by ester hydrolysis to give the phenyl acetic acid, Compound 20. Compound 20 is further hydroxylated on the α-carbon to give Compound 35.
- Hydrolytic ether cleavage between the aromatic rings to give Compounds 28 and 18 and also Compounds 13 and 3. Compound 13 was identified principally as the glucuronide conjugate, Compound 53.
- Ring hydroxylation on the cyanophenyl ring to give the isomers Compounds 23 and 50. Both compounds were subsequently conjugated to form the glucuronides, Compounds 52 and 51.
- Glutathione conjugation on the cyanophenyl ring. The glutathione conjugate, Compound 46 was then catabolised through the corresponding glycinecysteine and cysteine conjugates (Compounds 47 and 48 respectively) to give the mercapturic acid conjugate, Compound 49.

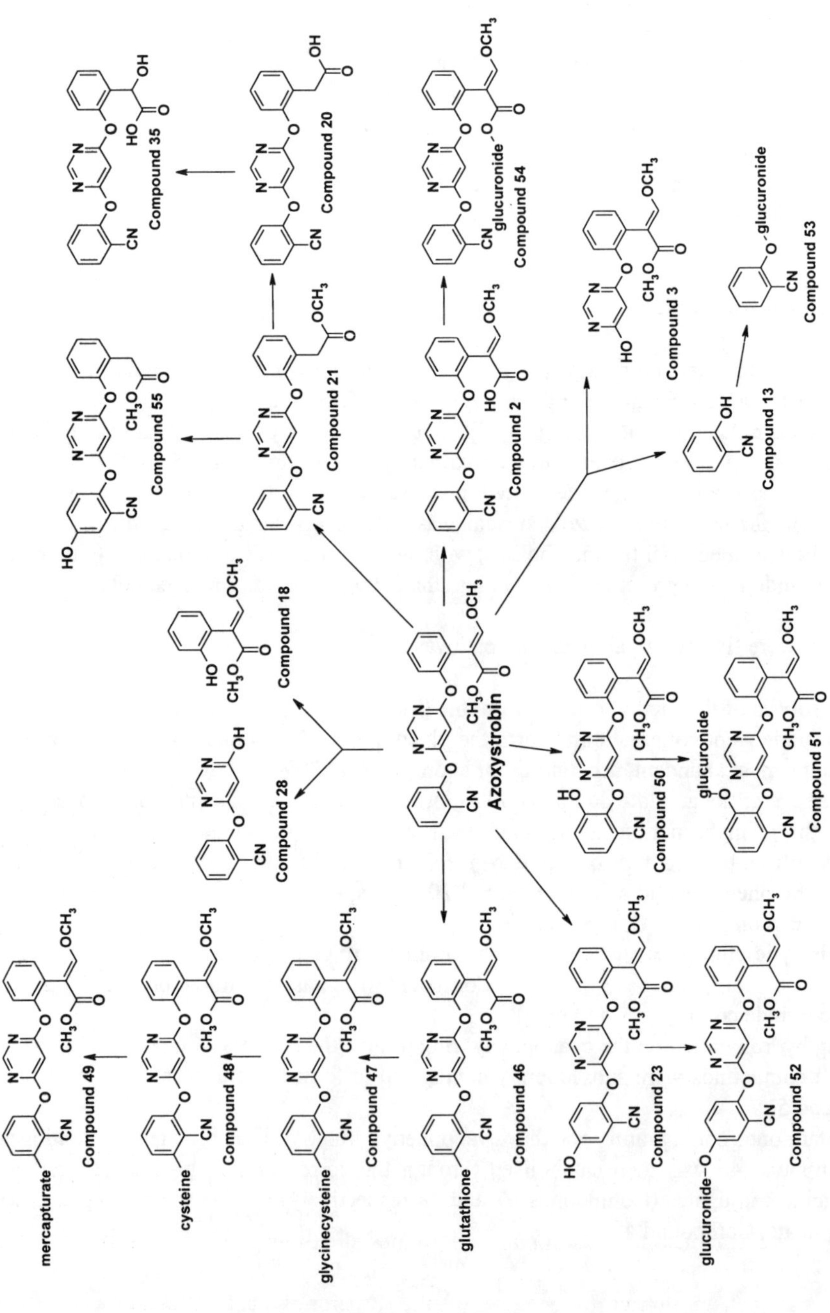

Figure 3 *Proposed Biotransformation Pathway in the Rat*

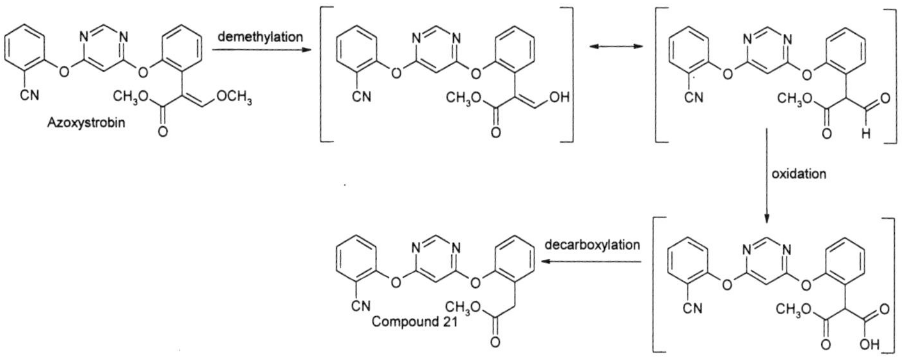

Figure 4 *Proposed Pathway to Compound 21*

Azoxystrobin has been established to be of low oral toxicity based on an extensive range of studies, and it is therefore considered that the metabolites present in the rat are not of toxicological concern.

The metabolism of azoxystrobin has been studied in farm animals, namely the goat and the hen. In both studies each of the three radiolabels of azoxystrobin was dosed orally at exaggerated dose levels for up to 10 days. Excretion was rapid and complete and there were no or very low residues present in meat, milk and eggs. The metabolic pathways were similar to that observed in the rat.

3 FATE IN SOIL

In order to understand the nature of the residue in plants and following crops, it is necessary to fully understand the fate of the pesticide in soil. Even for foliar applied compounds, significant quantities of active ingredient may still reach the soil either directly during application or subsequently by removal from the foliage, for example by wash off following rainfall. This residue in soil can contribute to the terminal residue in plants, either by direct root uptake of the parent compound and/or its soil degradates, or as the result of incorporation of radiolabelled carbon dioxide, which is produced from mineralisation of the pesticide in soil.

3.1 Laboratory Soil Photolysis

The soil surface photolysis of azoxystrobin was studied on a sandy loam soil for a period of 30 days. Azoxystrobin underwent rapid degradation with a DT_{50}, measured as a mean of the three radiolabels, of 11 days of natural summer sunlight. The degradation pathway was complex, producing many degradates all at low levels (<10% of the applied dose). The photodegradates themselves were all readily photolysed resulting in none of them building up during the course of the experiment and ultimately leading to mineralisation of the entire molecule to carbon dioxide. Thus the major product of degradation, for all three radiolabels, was carbon dioxide, reaching levels of up to 29% of applied by day 30. The evolution of carbon dioxide from soil is shown in Figure 5.

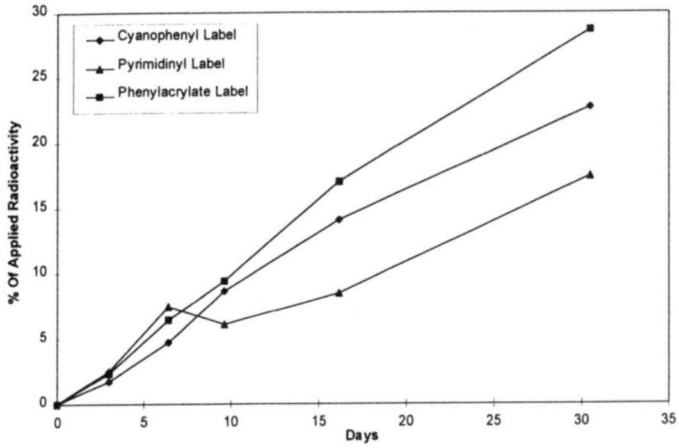

Figure 5 *Evolution of Carbon Dioxide from Soil Surface Photolysis of Azoxystrobin*

The results of this study indicate that soil surface photodegradation could be an important dissipation mechanism for azoxystrobin under field conditions and that if this process were to occur in the field, then carbon dioxide would be the major terminal degradation product.

3.2 Metabolism in Soil

The metabolism of azoxystrobin, separately labelled in each of the three rings, was studied under laboratory conditions in the dark in three soils incubated under aerobic conditions. In addition the degradation rate was studied in three soils following application of unlabelled compound. The DT_{50} of azoxystrobin was typically in the range of 8-12 weeks. There was no significant degradation under sterile conditions demonstrating that the degradation observed in the non-sterile soils was due to microbial processes only. The major metabolite in soil was Compound 2 reaching maximum levels of 20% of applied. There were no other metabolites in soil >10% of applied. Compound 2 itself was readily degraded through a series of transient intermediate metabolites resulting in mineralisation of the entire molecule. Carbon dioxide was the major product of degradation from all three radiolabels reaching levels of up to 50% of applied by one year after application. The evolution of carbon dioxide from soil is shown in Figure 6.

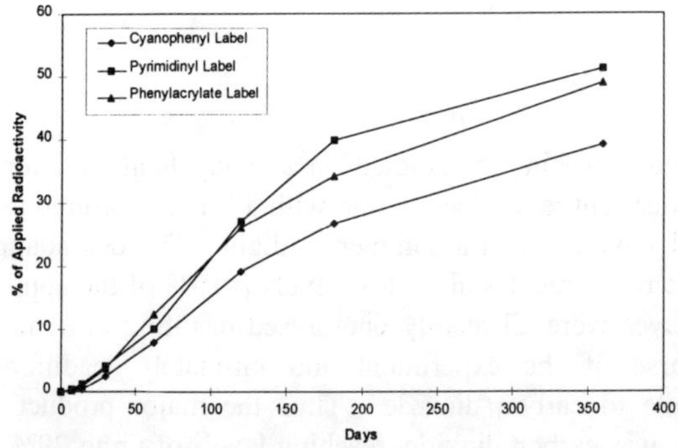

Figure 6 *Evolution of Carbon Dioxide from Aerobic Soil Metabolism of Azoxystrobin*

3.3 Degradation Pathway under Field Conditions

Based on the results of the laboratory soil photolysis and metabolism studies described above, it was necessary to determine what the significant degradates were under field conditions and also to establish the relative importance of photolytic and microbial degradation to the overall dissipation of azoxystrobin. To this end a radiolabelled field soil dissipation study was conducted at Visalia, California, USA. Azoxystrobin, radiolabelled in each of the three rings, was applied to soil, at 500 g ai ha^{-1} and samples taken at intervals up to one year after application.

Azoxystrobin dissipated rapidly, with a DT_{50} of 14 days. The only two significant degradates were Compounds 28 and 30, which reached maximum levels of 6 and 8% of applied respectively. Compound 2, the major microbial degradate, was detected at <1% of applied. Azoxystrobin remained as the major component in all samples. The overall degradation pathway for azoxystrobin in soil under field conditions is shown in Figure 7. There was no significant downward movement of radioactivity, despite heavy irrigation, indicating that azoxystrobin and its degradates are unlikely to be mobile in soil. There was a steady loss of radioactivity throughout the study, consistent with mineralisation of azoxystrobin to carbon dioxide, as seen in the laboratory soil studies.

These results indicate that under field conditions appreciable quantities of the applied substance which reach the soil will be mineralised to carbon dioxide. Previous experience with other pesticides in our laboratory suggests that, even under field conditions with significant dilution and dispersion in air, this radiolabelled carbon dioxide can be incorporated into natural products in the plant at significant levels. In addition these results indicate that the major residue in soil available for uptake into following crops will be parent azoxystrobin with lower levels of the two photodegradates Compounds 28 and 30 also present. Compound 2 the microbial degradation product is unlikely to be present at significant levels. These results were supported by an extensive programme of "cold" field dissipation studies conducted over several years in Europe and the USA, all of which showed a similar pattern of degradation.

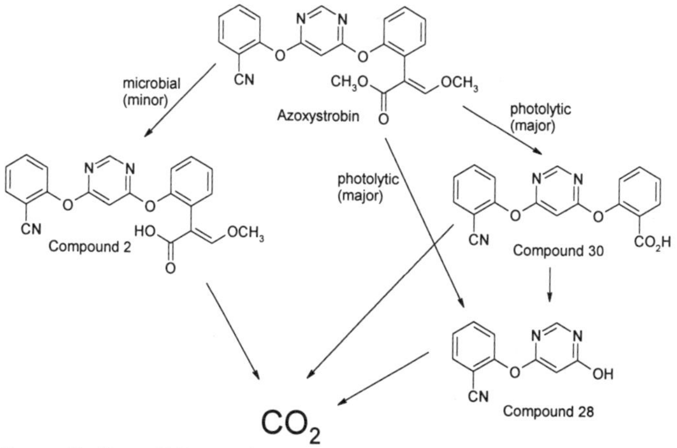

Figure 7 *Overall Degradation Pathway for Azoxystrobin in Soil*

4 METABOLISM IN PLANTS

The metabolism of azoxystrobin has been studied in wheat, grapes and peanuts. In all three studies, the three radiolabelled forms of azoxystrobin, formulated as suspension concentrates, were applied as multiple foliar sprays to separate plots. All three studies were conducted under field conditions in order to take account of all the biokinetic factors which can influence the nature of the terminal residue on the crop e.g. uptake, translocation, metabolism, foliar photodegradation, foliar wash off and soil uptake. In each of the studies, azoxystrobin was applied at rates equal to or greater than the labelled use rates as follows:

- Wheat - UK, 2 x 500 g ai ha^{-1}, PHI 61 days
- Grapes - France, 250, 1000, 1000 and 250 g ai ha^{-1}, PHI 21 days
- Peanuts - USA, 800, 900 and 300 g ai ha^{-1}, PHI 14 days

For each crop appropriate agricultural commodities were sampled and the total radioactive residues (TRR) determined (Table 1). Residues in human food items (grapes, peanut kernels and grain) were low. Residues on the foliage portions (potential animal feed items) of the crops were higher. These data are as expected given that azoxystrobin is not phloem mobile and the majority of the spray will be intercepted by the foliage. These higher residues in the foliage are of minimal concern from a dietary exposure perspective, given the low transfer of residues observed in the animal metabolism studies.

4.1 Plant Metabolites

The metabolic pathway in plants is shown in Figure 8. The metabolism of azoxystrobin was complex with a total of 17 metabolites identified, mostly at very low levels. All metabolites were present at less than 10% TRR.

Table 1 : *Total Radioactive Residues in Crops (mg kg^{-1})*

	Radiolabel Position		
Commodity	Cyanophenyl	Pyrimidinyl	Phenylacrylate
Grain	0.08	0.08	0.08
Grape	0.38	1.43	0.95
Nut (Kernel)	0.24	0.65	0.49
Wheat Forage	2.79	1.02	2.14
Wheat Straw	9.41	3.06	7.22
Peanut Vine	17.4	16.4	20.8

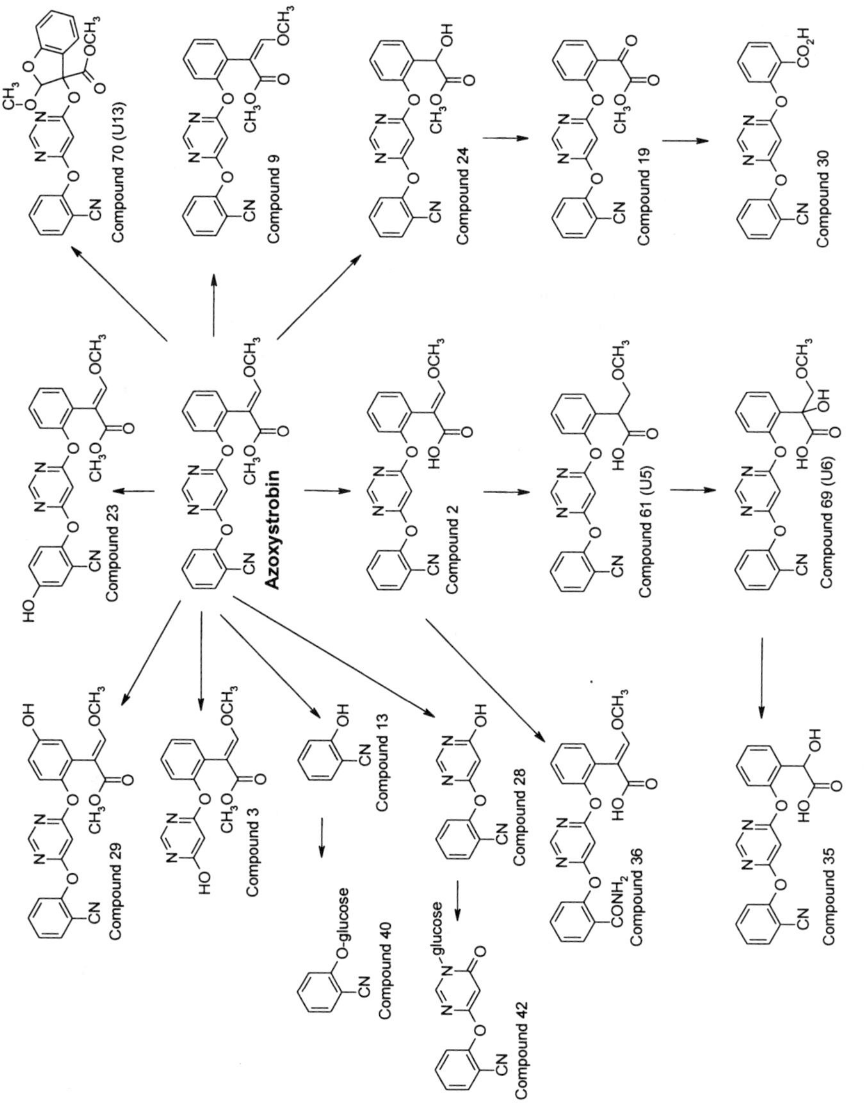

Figure 8 *Proposed Biotransformation Pathway for Azoxystrobin in Plants*

There were five principal metabolic pathways:

- Hydrolysis of the methyl ester to form the β-methoxyacrylic acid (Compound 2). The acrylate double bond of Compound 2 was then reduced to give Compound 61, which was than hydroxylated to give compound 69. Finally the methoxymethylene residue from the acrylate was cleaved to give Compound 35. Compound 2 also underwent nitrile hydrolysis to give the amide Compound 36 as a very minor metabolite.

- Oxidative cleavage of the acrylate double bond to form Compound 24 which was then further oxidised to give Compound 19 and ultimately Compound 30. It is also likely that compound 35 was formed from hydrolysis of the methyl ester of Compound 24.

- Hydrolytic ether cleavage between the cyanophenyl and pyrimidinyl rings to give Compounds 13 and 3. Compound 13 was also identified as its glucose conjugate, Compound 40. Similarly cleavage occurred between the other two rings to give Compounds 28 and its N-glucoside Compound 42. Compound 18,the other product that would be expected, was not detected. It is interesting to note that in nearly all metabolism studies conducted on azoxystrobin, no metabolites were detected that contained only the phenylacrylate ring. This is presumed to be due to the rapidity by which this fragment is further metabolised.

- Ring hydroxylation on the cyanophenyl and phenylacrylate rings to give Compounds 23 and 29.

- Photo-rearrangements on the leaf surface to give the Z isomer Compound 9 and Compound 70.

Many of the plant metabolites were also animal metabolites. Those that were not were generally present at extremely low levels and in the foliage only. The biotransformation pathways in the three crops were very similar, indicating that the metabolism of azoxystrobin is independent of crop type and is relatively insensitive to environmental factors.

In all samples, with the exception of the peanut kernel where there was no significant residue of either parent or metabolites (all <0.002 mg kg^{-1}), parent azoxystrobin was the major component of the residue representing between 17 and 65% of the total radioactive residue.

The two most abundant metabolites in plants were Compounds 13 and 28. Levels were below 0.01 mg kg^{-1} in wheat grain and peanut kernel, and around 0.01 to 0.02 mg kg^{-1} in grapes (3- 6% TRR). These metabolites were more significant in the foliage but still typically represented less than 10% TRR. Compound 9, the Z isomer of azoxystrobin, was present in most samples at 2-4% TRR. Compound 9 was included in the residue analytical method for crops due to its structural similarity to the parent. Compounds 13 and 28 were not included in the residue method for the following reasons:

- both fungicidally inactive
- both animal metabolites
- present at low levels
- structurally dissimilar to the parent compound

⇒ metabolites are not of toxicological concern

The majority of the other metabolites of azoxystrobin were screened *in vitro* for activity at the target site. Most showed no measurable activity, with only very weak activity associated with some of the metabolites which still retained the active moiety - the methyl ester of β-methoxyacrylic acid. A selected number of the more significant plant and soil metabolites were also evaluated for fungicidal activity in vivo and as expected showed no activity except for the Z isomer of azoxystrobin, Compound 9, which was weakly active.

4.2 Incorporation of Radiolabel into Natural products

In all three plant metabolism studies a significant portion of the total radioactive residue from each of the three radiolabels was found in natural products. This is consistent with the finding that azoxystrobin is extensively mineralised in soil, by both photolytic and microbial degradation, with the resulting radiolabelled carbon dioxide being incorporated into natural products via photosynthesis.

In grapes, radiolabel was found in the fruit sugars, glucose and fructose, which were separated and individually quantified by ligand exchange HPLC. In the wheat grain the radioactivity was found in starch which was hydrolysed by enzymes and the resulting glucose analysed by HPLC. In the peanut kernels significant radioactivity was found not only in polar fractions, representing incorporation of radiolabel into sugars and amino acids, but also in the non-polar hexane extracts which contained tri-glycerides and other fats. The tri-glycerides were base hydrolysed to fatty acids which were then derivatised with phenacyl bromide. The derivatised fatty acids were then identified by TLC co-chromatography against authentic standards of oleic and linoleic acids. A summary of the characterisation data for the peanut kernel is given in Table 2.

There were no significant residues of either parent or metabolites in the peanut kernels (all <0.002 mg kg^{-1}). The remainder of the residue consisted of unextracted residues, unanalysed fractions and minor unidentified residues (<0.01 mg kg^{-1}).

Table 2 *Characterisation of Natural Incorporation in Peanut Kernels*

Natural Product	Position of Radiolabel (%TRR)		
	Cyanophenyl	Pyrimidinyl	Phenylacrylate
Fatty Acids	42	49	44
Sugars/Amino Acids	21	20	21

5 UPTAKE AND METABOLISM IN FOLLOWING CROPS

The uptake and metabolism of azoxystrobin was studied in rotational crops following application of the compound to soil and ageing for different time periods. The three radiolabelled forms of azoxystrobin were separately applied to a sandy loam soil, in large pots, in a single application of 2 kg ai ha^{-1}. The soils were aged for intervals of 30, 200 and 365 days and following crops of lettuce, radish and spring wheat sown into the aged soils. Soil ageing and growing of the crops were conducted in a glasshouse. Soil cores were taken at each planting and appropriate agricultural commodities sampled for each crop.

This study design represents a worst case for the following reasons:

- exaggerated application rate compared to annual maximum use rates
- chemical applied as single dose of 2 kg ai ha^{-1} - maximum single application rate for crops is 0.25 kg ai ha^{-1}
- no crop interception
- dissipation reduced
- roots confined to the treated area

The above factors mean that the crops in this study were exposed to concentrations of azoxystrobin and its degradates in soil far in excess of what would be experienced under a realistic field situation. Thus in practice residues in following crops would be expected to be much lower.

The principal residues in soil were parent azoxystrobin, Compound 28 and Compound 30, consistent with results from the field soil studies. Residues in food items, grain, lettuce and radish root, were low. The residues in crops declined with each planting interval. The metabolism was complex (Figure 9). The metabolites were mostly conjugated forms of metabolites found in the target crops.

Azoxystrobin accounted for a significant percentage of the residue and was therefore selected as the analyte for limited field trials. As expected, based on the exaggerated nature of the confined study, residues in the field studies were mostly non detectable, even at very short intervals, with only very low residues ($\leq$0.04 mg kg^{-1}) in some forage, straw and hay samples. No residues, LOQ 0.01 mg kg^{-1}, were detected in any of the other commodities; wheat and millet grain, mustard greens or radish and turnip roots and tops.

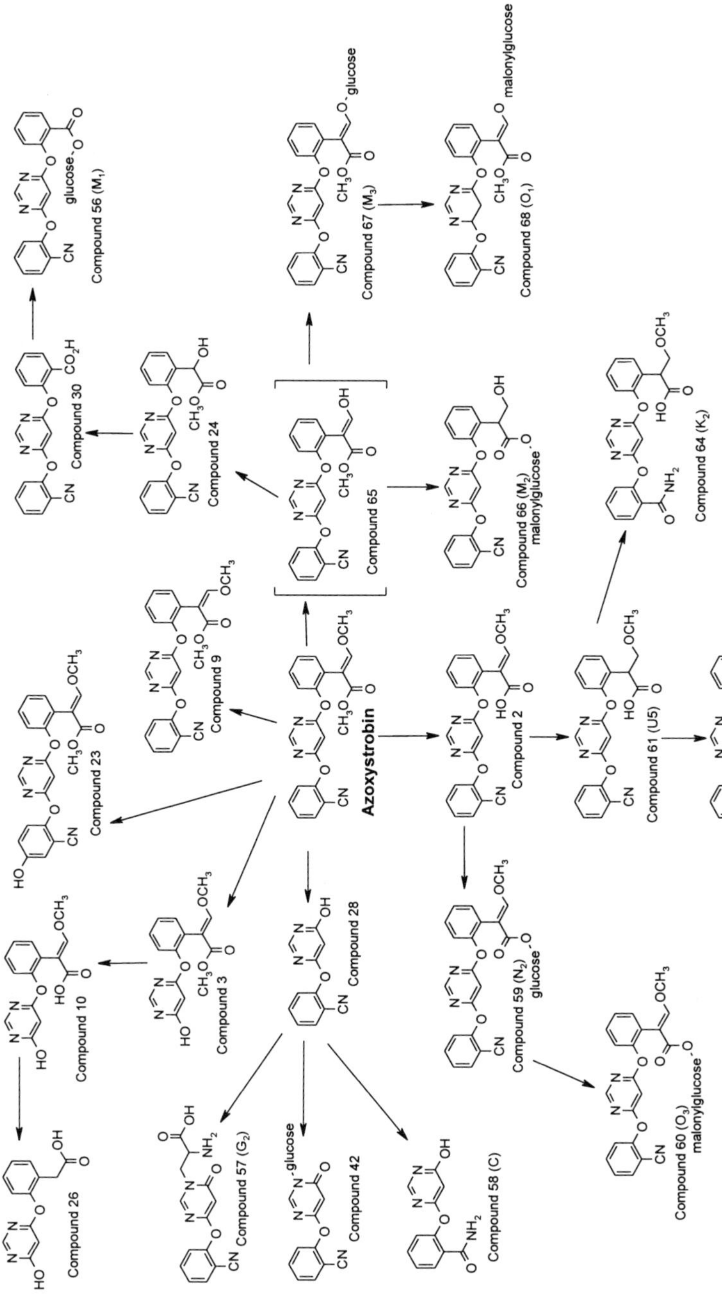

Figure 9 *Proposed Biotransformation Pathway for Azoxystrobin in Following Crops*

6 ACKNOWLEDGEMENTS

The work described in this paper involved many scientists within Environmental Sciences, Zeneca Agrochemicals. I would like to acknowledge the key contributors in each of the areas as follows:

Plant: Michaela Wilkinson, Derek Hepburn, Steve Hadfield, Gini Earl, Jackie Webb, Simon Emburey
Soil: Colleen Spurgeon, Louise Winter, Brian Harvey, Jacqui Warinton, Rob Mason, Martin Weissler, Mike Earl
Animal: Alex Gledhill, Graham Lappin - Zeneca, CTL; Simon Emburey, Derek Hepburn, Sue Mayes, Tim Steel, Jackie Webb, Yvonne Bramley and Julia Turner
Rotational Crop: Wendy Goldsby, Jane Barnes, Colleen Spurgeon, Mary Miller, Doug Tambling, Bruce Onisko - Zeneca, WRC, USA

References

1. K. Beautement, J. M. Clough, P. J. de Fraine and C. R. A. Godfrey, *Pest Sci,* 1991, **31**, 499.
2. J. R. Godwin, V. M. Anthony, J. M. Clough and C. R. A. Godfrey, Proceedings Brighton Crop Protection Conference: Pest and Diseases, 1992, pp. 435-442
3. S. P. Heaney and S. C. Knight, Proceedings Brighton Crop Protection Conference: Pest and Diseases, 1994, pp. 509-516

Herbicide Metabolism in Plants: Integrated Pathways of Detoxification

Klaus Kreuz and Enrico Martinoia

NOVARTIS CROP PROTECTION AG, CH-4002 BASEL, SWITZERLAND; AND INSTITUT DE BOTANIQUE, UNIVERSITE DE NEUCHATEL, CH-2007 NEUCHATEL, SWITZERLAND

1 SUMMARY

Herbicide metabolism and detoxification in plants is generally a multiphase process involving functionalisation (most frequently by oxidation or hydrolysis), conjugation (to glutathione, glucose, or an organic acid), processing of the conjugate (by degradation, secondary conjugation) and compartmentation (vacuolar or extracellular deposition). Many enzymes involved in the metabolism of herbicides in plants such as cytochrome P450-dependent monooxygenases, glutathione *S*-transferases, glucosyl and malonyl transferases have been investigated in the past. Transfer into crops of genes for herbicide-metabolising enzymes from bacteria or mammals has been successfully employed to increase herbicide tolerance. Recent research has revealed that both glutathione and glucose conjugates of herbicides are exported from the cytosol to the central vacuole of plant cells. Accumulation of glutathione conjugates in the vacuole is mediated by transporters in the vacuolar membrane that are directly energised by ATP. Evidence exists that glucose conjugates of herbicides are transported into the vacuole by ATP-stimulated transporters that are distinct from those involved in glutathione conjugate transport. While herbicide glutathionation and glucosylation takes place in the cytosol, subsequent compartmentation of these conjugates and possible further metabolism in the vacuole may be crucial for sustained detoxification. Herbicide safeners enhance herbicide oxidation and conjugation, as well as the activity of vacuolar transport ATPases involved in herbicide conjugate export. Thus, the various processes involved in cellular detoxification of herbicides in plants constitute integrated pathways that are apparently subject to coordinated regulation.

2 INTRODUCTION - SIGNIFICANCE OF PLANT HERBICIDE METABOLISM

Foreign compounds such as crop protection chemicals generally undergo extensive biotransformations in higher plants. Pesticide metabolism in plants has a crucial impact on the nature and amount of residues present in the harvested crop and on the environmental fate of pesticides. Moreover, in the context of herbicides, metabolism in plants has important implications for the herbicidal activity and crop tolerance of these agents. The majority of herbicides are transformed by plants leading to herbicide inactivation or "detoxification". Plants are able to metabolise herbicides by a variety of enzymatic reactions and with extraordinary species diversity.[1,2] In fact, the differential ability of plant species to detoxify a herbicide is, in most cases, the physiological basis for crop selectivity of herbicides. Compounds known as 'herbicide safeners', which are used to improve crop tolerance to particular herbicides, act by enhancing herbicide metabolism in the crop.[3] Inhibition of herbicide metabolism, on the other hand, is often the basis for synergistic interactions between herbicides and other agrochemicals.[4] Furthermore, enhanced capability of herbicide detoxification is one of the mechanisms of weed resistance to herbicides.[5] Finally, genetic engineering for enhanced herbicide metabolism is one strategy for the development of herbicide-tolerant crops.[6]

3 HERBICIDE METABOLISM PATHWAYS IN PLANTS

Herbicide metabolism in plants is generally described as a multiphase process. It often includes a primary step, most commonly an oxidation or hydrolysis, that introduces or reveals a functional group which is suitable for subsequent conjugation to an endogenous moiety such as glutathione (γ-glutamyl-cysteinyl-glycine; GSH), glucose or an organic acid.[1,2] Some herbicides contain suitable functional groups that are *per se* amenable to conjugation to those moieties. The resulting conjugates are usually subject to further processing which may include secondary conjugation, degradation, deposition in the vacuole or extracellular space, and formation of 'bound residues'. These biotransformations and many of the enzymes involved have been widely studied in the past. While the conjugations have been known to take place in the cytosol, only recent research has revealed that at least some of the subsequent processing reactions proceed in a different compartment of the cell, the large central vacuole. Thus, specific transport of conjugates across cellular membranes appears to be a pivotal step in pesticide metabolism pathways in plants. One aim of this paper is to discuss these recent findings which suggest that the various processes involved in cellular detoxification of herbicides in plants constitute integrated pathways. Furthermore, attempts to confer herbicide tolerance to crops by gene transfer of detoxifying enzymes will be discussed in light of these recent findings.

3.1 The Glutathione-dependent Pathway

Conjugation of herbicides to GSH or structurally related endogenous thiols is one of the major detoxification and selectivity factors in plants.[1,2] These reactions occur by nucleophilic attack of the thiolate anion of GSH to an electrophilic centre of the substrate and are catalysed by glutathione *S*-transferases (GSTs). The occurrence, function and regulation of plant GSTs has recently been reviewed.[7] In all plant species investigated so far, GSTs exist as multiple isoenzymes and are predominantly soluble proteins located in the cytosol of the cell. The cytosolic GSTs are either homo- or heterodimeric proteins with subunit molecular masses of 24 to 29 kDa. The role of microsomal and extracellular GST isoforms in xenobiotic metabolism is currently unknown. The broad substrate specificity of GSTs in a given plant species can be attributed in part to the presence of different isoenzymes. Moreover, individual GST isoenzymes may accept a fairly wide and sometimes overlapping range of mostly hydrophobic electrophiles as substrates. In contrast, the specificity toward the thiol donor is high with only GSH and very similar tripeptides, *e.g.* homoglutathione (γ-glutamyl-cysteinyl-β-alanine) found in leguminous plants, being accepted. These specificities of substrate and GSH binding sites can be explained on the basis of X-ray crystallographic studies.[8]

GSH conjugates of herbicides are more water-soluble than the parent herbicide and do not readily penetrate through biological membranes. They usually undergo rapid degradation by peptidases to yield the corresponding cysteine conjugates. The cysteine conjugates are further processed to *e.g.* *N*-malonylcysteine or thiolactic acid derivatives which are major herbicide residues in many plant species.[1]

Metolachlor (1, Figure 1) is a chloroacetanilide herbicide that is initially metabolised by conjugation of the chloroacetyl side chain to GSH.[9] Cytosolic GSTs from several plant species have been shown to catalyse this reaction.[7] In maize and other crops, thiolactic acid conjugates and their sulfoxides represent terminal products of the GSH-dependent metabolism of metolachlor.[9] While the principal metabolic reactions of the GSH-dependent pathway have long been known, the involvement of a specific transport process has only recently been discovered. Experiments with isolated plant vacuoles have revealed for the first time the existence of a transporter activity that mediates the uptake of the GSH conjugate of metolachlor (metolachlor-GS, 2) into the vacuole.[10] This transport across the vacuolar membrane is energy-dependent and results in the vacuolar accumulation of metolachlor-GS against the concentration gradient. Metolachlor-GS was shown not to be chemically modified during or shortly after vacuolar uptake. Uptake of metolachlor-GS and of glutathionated *N*-ethylmaleimide, used as a model probe for transport, is directly energised by hydrolysis of nucleoside triphosphates, preferentially the Mg^{2+} salt of ATP (MgATP). Uptake is insensitive to agents that dissipate transmembrane H^+-gradients, and is strongly inhibited by the phosphoryl transition state analogue vanadate. The transport system is not activated by the nonhydrolysable ATP analogue adenosine-5'-(β,γ-imino)-triphosphate.

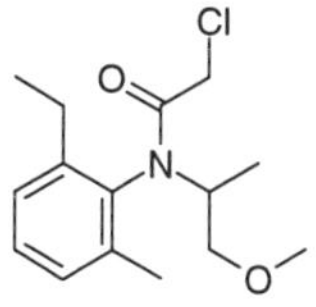

(1)

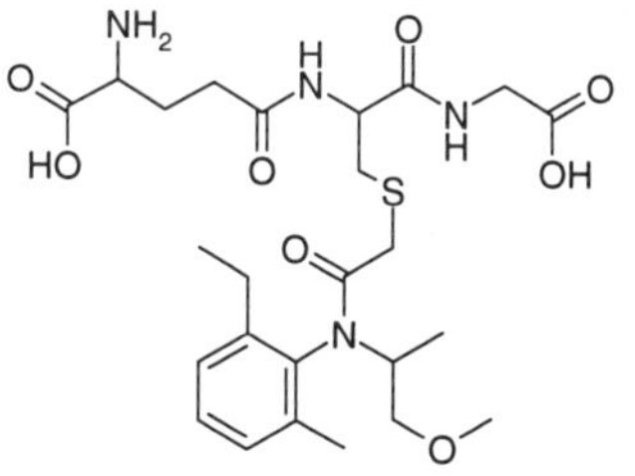

(2)

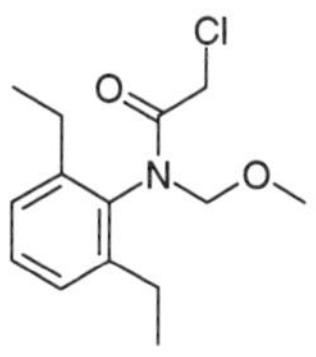

(3)

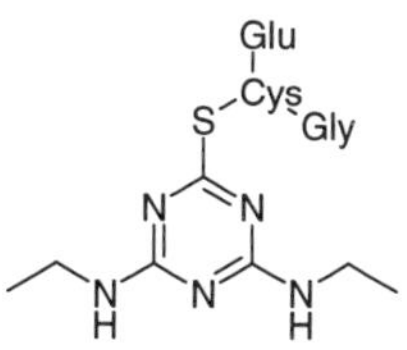

(4)

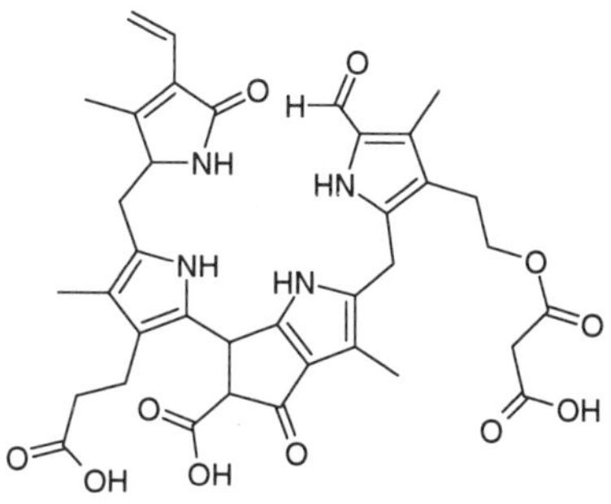

(5)

Figure 1 *Chemical structures of herbicides and substrates for vacuolar transporters*

Subsequent studies have demonstrated the uptake into isolated plant vacuoles and vacuolar membrane vesicles of *S*-(2,4-dinitrophenyl)-glutathione (DNP-GS), glutathionated monochlorobimane, and oxidised GSH (GSSG).[11] Evidence exists that the GSH conjugates of two other herbicides, the chloroacetanilide alachlor (3) and the *s*-triazine simetryn (simetryn-GS, 4), are also transported by the vacuolar ATPase (Figure 1).[10,11] Taken together, these data suggested the presence in plants of directly-energised transporters that functionally resemble the well-known GSH conjugate transporting ATPases of mammalian cells referred to as 'GS-*X* pumps'.[10,11] In both cases, the GSH conjugates of structurally unrelated xenobiotics as well as GSSG, but not GSH, are recognised, and transport is directly energised by MgATP.

A significant step toward the molecular investigation of these transport functions in plants was the recent isolation of several genes from *Arabidopsis thaliana* (thale cress) that encode functional GSH conjugate pumps.[11-13] These genes and their products are structural and functional homologues of the human multidrug resistance-associated protein, MRP1, and of the yeast (*Saccharomyces cerevisiae*) cadmium factor, YCF1.[11-13] Both MRP1 and YCF1 are directly MgATP-energised transporters for GSH conjugates of organic compounds. In addition, YCF1 was shown to mediate the vacuolar accumulation of Cd^{2+} after complexation with GSH to form bis-(glutathionato)-cadmium. To date, three genes for *A. thaliana* MRPs, designated *AtMRP1* to *AtMRP3*, have been characterised. They encode intrinsic membrane proteins of 170 to 180 kDa molecular mass with a domain organisation equivalent to those of MRP1 and YCF1. AtMRPs as well as MRP1 and YCF1 can be grouped into a subclass of the ATP-binding cassette transporter superfamily that has been extensively studied in animals and microbes. The *A. thaliana AtMRP* genes have been shown by expression in yeast to indeed encode functional GSH conjugate transporters. The detailed kinetic analysis *in vitro* of heterologously expressed AtMRP1, AtMRP2 and AtMRP3 revealed that they catalyse the MgATP-dependent transport of metolachlor-GS and DNP-GS as well as GSSG.[11-13] Interestingly, AtMRP2[12] and AtMRP3[13] are additionally capable for high-affinity transport of an endogenous nonglutathionated compound, *i.e.* the malonyl ester of a linear tetrapyrrole (5, Figure 1) derived from chlorophyll catabolism. This is in line with previous findings on mammalian MRPs that do not exclusively transport GSH conjugates but also other large amphipathic anions. These characteristics have been explained by the modular protein architecture of MRPs that encompasses two separate domains for substrate recognition and two ATP-binding folds.[11]

Alachlor (3) has recently been shown *in vivo* to accumulate rapidly as GSH conjugate in the vacuole of barley leaves.[14] Furthermore, the GSH conjugate of alachlor is degraded to the γ-glutamyl-cysteine conjugate by a carboxypeptidase which is exclusively localised in the vacuole. There is evidence that cleavage of the γ-glutamyl residue to yield the cysteine conjugate of alachlor also takes place in the vacuole. Using the fluorescent model substrate monochlorobimane which is amenable to both spontaneous and GST-catalysed GSH conjugation, Coleman *et al.* have subsequently demonstrated the cellular pathway *in vivo* starting with conjugation to GSH in the cytosol followed by transport of the conjugate into the vacuole.[15]

In summary, it can be concluded that xenobiotic GSH conjugates formed in the cytosol of the plant cell are exported by transport ATPases to the large central vacuole, where cleavage by peptidases to the γ-glutamyl-cysteine and possibly to the cysteine conjugates takes place (Figure 2). It is presently not known in which compartment(s) of the plant cell the further transformations of cysteine conjugates (rearrangement, oxida-

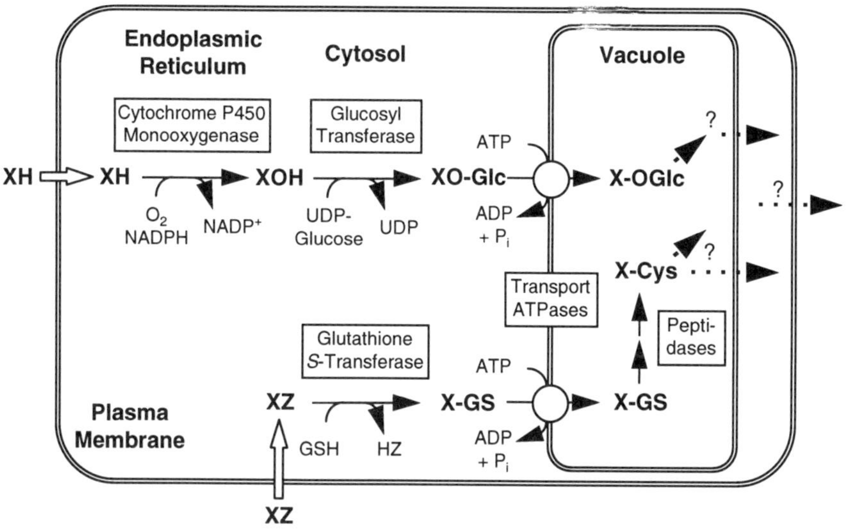

Figure 2 *Schematic representation of herbicide detoxification in a plant cell. XH and XZ are herbicides entering the oxidation/glucose conjugation pathway and the glutathione-dependent pathway, respectively. XOH is the hydroxylated herbicide; XO-Glc, X-GS and X-Cys are the glucose, GSH and cysteine conjugates, respectively*

tive deamination, malonylation *etc.*)[1] are accomplished.

The existence of vacuolar transporters recognising GSH conjugates of structurally unrelated compounds suggests that the GSH moiety serves as a specific tag that mediates the transfer of these conjugates into the vacuole. The biological rationale for energy-consuming transport could be twofold: (*i*) some GSH conjugates are potent inhibitors of GSTs and their efficient elimination from the cytosol could be crucial for sustained detoxification;[1,2] (*ii*) the plant vacuole is known to contain many hydrolytic enzymes and, thus, vacuolar transport could be a prerequisite for degradation of diverse GSH conjugates and for the recycling of amino acids.

It has been suggested that the terminal metabolites of GSH-dependent herbicide metabolism may be stored in the vacuole; on the other hand, metabolism through GSH-conjugation sometimes results in high levels of bound residues.[1] In the latter case, excretion of solutes across the plasma membrane into the extracellular matrix is required which involves yet-unknown mechanisms (Figure 2). It has been proposed earlier that the commonly observed addition of a malonyl residue to glucose conjugates (see below) or to cysteine or thiolactic acid conjugates facilitates their transport into the vacuole or extracellular space. As mentioned above, the plant vacuolar transporters AtMRP2 and AtMRP3 do not only transport GSH conjugates but also the malonyl ester of a tetrapyrrole chlorophyll catabolite (5, Figure 1). It might be interesting to investigate whether these or other GSH conjugate transporters also recognise other amphiphilic anions such as the commonly found malonyl, malonylglucose or malonylcysteine conjugates of herbicides.

3.2 The Oxidation/Glucose Conjugation Pathway

The first metabolic reaction of a herbicide in the plant is quite often an oxidation or hydrolysis. In recent years it has been firmly established that cytochrome P450 (Cyt P450)-dependent monooxygenases play a pivotal role in the oxidation of many herbicides in plants. The Cyt P450 monooxygenases found in plants are, like the well-characterised ones in the endoplasmic reticulum of mammalian liver cells, membrane-bound heme-thiolate proteins (molecular mass ~55 kDa) which require molecular oxygen for catalytic activity, NADPH and NADPH-cytochrome P450 reductase. The most common Cyt P450-mediated reactions with herbicides are hydroxylations of aromatic rings or alkyl groups, and heteroatom release (*O*-, *N*-dealkylation). Multiple Cyt P450 isoforms exist in plants, and the number of new isoforms being cloned and characterised is rapidly increasing. As is the case with GSTs, differences in Cyt P450 isoforms account for the conspicuous species-specific differences in the metabolism of herbicides and other xenobiotics. The occurrence, function and regulation of plant Cyt P450 monooxygenases has recently been reviewed.[16]

The primary metabolic attack through oxidation or hydrolysis, which usually serves to introduce a hydroxy, amino or carboxylic acid function, is often followed by rapid and extensive glucosylation to yield *O*-glucosides, *N*-glucosides, or glucose esters. These reactions are catalysed by the respective glucosyl transferases which utilise uridine-5'-diphosphoglucose (UDP-glucose) as sugar donor. Simple glucose conjugates are frequently subject to secondary conjugations to malonyl or carbohydrate (glucosyl, xylosyl, or arabinosyl) residues.[1] So far, the vacuolar deposition *in vivo* of a herbicide glucosyl conjugate has only been shown for the glucoside or 6-*O*-(malonyl)-glucoside of (2,5-dichloro-4-hydroxy-phenoxy)-acetic acid, a metabolite derived from 2,4-D.[17] Only recently, direct evidence was obtained that herbicide glucosides may be actively transported into the plant vacuole. The sulfonylurea herbicide primisulfuron (6, Figure 3) is metabolised in maize mainly by hydroxylation at the phenyl or pyrimidine ring followed by glucosylation. Both hydroxylations have been shown to be mediated by microsomal Cyt P450-dependent monooxygenases.[18] Glucosylation of (hydroxyphenyl)-primisulfuron (7) also proceeds *in vitro* with a partially purified glucosyl transferase and UDP-glucose, and yields the corresponding hydroxyprimisulfuron-glucoside (8) (Eberhard, Logel and Kreuz; unpublished results). Hydroxyprimisulfuron-glucoside is taken up by isolated plant vacuoles through a MgATP-stimulated, vanadate-inhibitable transport activity.[19,20] This transporter has not yet been characterised molecularly, but it appears to be distinct from the vacuolar ATPases mainly involved in GSH conjugate transport (Figure 2).

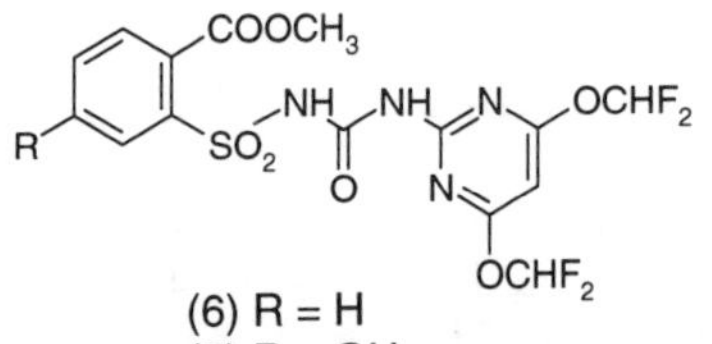

(6) R = H
(7) R = OH
(8) R = (Glucosyl)-O

Figure 3 *Chemical structures of the herbicide primisulfuron and some metabolites formed in the plant*

3.3 Chemical and Genetic Regulation of Plant Herbicide Metabolism

Even though herbicide metabolism in plants is accomplished largely by constitutive enzymes, a wide range of biotic and abiotic factors is known to influence herbicide metabolism. Herbicide safeners are a group of chemically diverse compounds with the ability to enhance crop tolerance to certain herbicides. Safeners act by accelerating the metabolism of herbicides through increased activities of Cyt P450 monooxygenases, GSTs and glucosyl transferases, and through raised GSH levels.[3] It is well documented that the GST and Cyt P450 families contain constitutive as well as safener-inducible isoenzymes.[3,7,16] Moreover, recent studies have shown that the activities of vacuolar ATPases for the transport of both GSH and glucose conjugates of herbicides are enhanced by the safener cloquintocet-mexyl (9, Figure 4) in barley.[19] Stimulation of conjugate uptake was the result of an increased uptake velocity whereas the K_m remained unaltered, suggesting that the higher activity was due to a higher expression of the transporter. Cloquintocet-mexyl has previously been shown to enhance Cyt P450-mediated herbicide oxidation and glucose conjugation, GST activity, and GSH levels in cereals.[3,21] Thus, a safener may induce several enzymes of cellular detoxification as well as vacuolar transporters involved in xenobiotic conjugate transport. It remains to be elucidated how this apparently coordinated induction is accomplished.

Herbicides are also likely to enhance the activity of vacuolar transporters. It has been shown that the *A. thaliana* MRP gene *AtMRP3* is strongly induced by treatment of plants with primisulfuron (6) or with the structurally unrelated experimental herbicide IRL 1803 (10, Figure 4).[22] In contrast, metolachlor (1) only scarcely induced this gene in *A. thaliana*.

Genetic engineering of plants for enhanced metabolic detoxification of herbicides has been employed as an approach to generate herbicide-tolerant crops.[6] Several classes of enzymes including hydrolases, monooxygenases, GSTs, and acetyl and glucosyl transferases have been used for this purpose. Up to now, the enzymes and their corresponding genes that have been reported to successfully confer resistance to target crops originated from either microbial or mammalian organisms. These enzymes include, *e.g.*, a bacterial acetyl transferase for phosphinothricin, bacterial nitrilase for bromoxynil, and bacterial carbamate hydrolase for phenmedipham;[6,23] a rabbit liver esterase for thiazopyr;[24] and a fused enzyme of rat cytochrome P4501A1 and yeast NADPH-cytochrome P450 oxidoreductase for chlortoluron detoxification.[25] Genetic engineering toward enhanced expression of herbicide-metabolising enzymes from plant origin has not yet been reported to yield acceptable levels of tolerance.[6] One conceivable reason for this could be that "detoxification" enzymes in plants may fulfil hitherto unknown endogenous functions resulting in the perturbation of physiological pathways upon

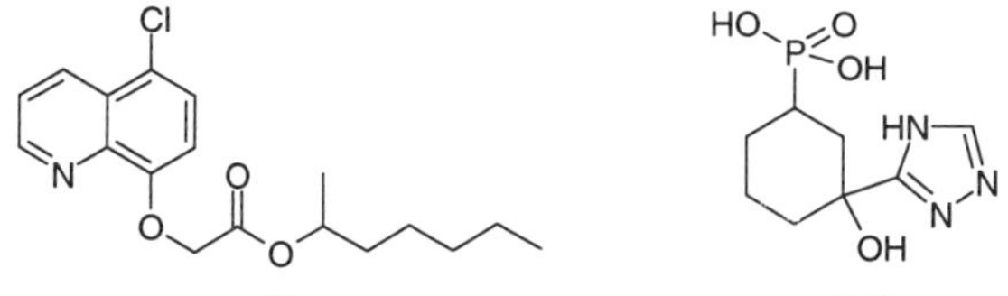

(9) (10)

Figure 4 *Chemical structures of the safener cloquintocet-mexyl and the experimental herbicide IRL 1803* increased expression of such enzymes in transgenic plants.[2,7,16] Another constraint for this latter approach could be that, in certain cases, increased levels of herbicide conjugates formed in the transgenic plant could exert deleterious effects due to a limited capacity of cells to transport conjugates out of the cytosol, as briefly discussed in Section 3.1.[11] Vacuolar transport ATPases have thus been proposed as targets for genetic manipulation of plant responses to herbicides and other xenobiotics.[11,26]

4 CONCLUDING REMARKS

Detoxification of herbicides in plants depends on the specialised functions of various subcellular compartments, *i.e.* the cytosol including membrane-bound enzymes of the endoplasmic reticulum, and the large central vacuole. The vacuolar membrane contains xenobiotic conjugate transporters that have been investigated only during the past five years. These transporters act as high-affinity "pumps" which efficiently eliminate the conjugates from the cytosol and lead them to further processing and/or final storage in the vacuole. Thus, the well-established scheme for plant xenobiotic metabolism probably needs to be amended in so far as transmembrane transport processes form a tightly integrated part of herbicide detoxification. The enzymology, regulation (*e.g.* by herbicide safeners) and genetic manipulation of plant herbicide metabolism have been, and will continue to be, attractive areas of both basic and applied research.

References

1. L. Lamoureux, R. H. Shimabukuro and D. S. Frear, in 'Herbicide Resistance in Weeds and Crops', J. C. Caseley, G. W. Cussans and R. K. Atkin (eds.), Butterworth-Heinemann, Oxford, UK, 1991, p. 227.
2. K. Kreuz, R. Tommasini and E. Martinoia, *Plant Physiol.*, 1996, **111**, 349.
3. K. Kreuz, in 'Proceedings of the Brighton Crop Protection Conference - Weeds', British Crop Protection Council, Brighton, UK, 1993, Vol. 3, p. 1249.
4. K. Kreuz and R. Fonné-Pfister, *Pestic. Biochem. Physiol.*, 1992, **43**, 232.
5. S. B. Powles and J. A. M. Holtum (eds.), 'Herbicide Resistance in Plants. Biology and Biochemistry', Lewis Publishers, Boca Raton, Florida, 1994.
6. R. Hain and P. H. Schreier, in 'Pflanzenschutz-Nachrichten Bayer, Special Issue', M. Esters (ed.), Bayer AG, Crop Protection Business Group, Leverkusen, 1996, Vol. 49, p. 25.
7. A. Marrs, *Annu. Rev. Plant Physiol. Plant Mol. Biol.*, 1996, **47**, 127.
8. T. Neuefeind, R. Huber, H. Dasenbrock, L. Prade and B. Bieseler, *J. Mol. Biol.*, 1997, **274**, 446.
9. H. M. LeBaron, J. E. McFarland, B. J. Simoneaux and E. Ebert, in 'Herbicides. Chemistry, Degradation, and Mode of Action', P. C. Kearney and D. D. Kaufman (eds.), Marcel Dekker, New York, 1988, Vol. 3, Chapter 7, p. 335.
10. E. Martinoia, E. Grill, R. Tommasini, K. Kreuz and N. Amrhein, *Nature*, 1993, **364**, 247.
11. P. A. Rea, Z.-S. Li, Y.-P. Lu, Y. M. Drozdowicz and E. Martinoia, *Annu. Rev. Plant Physiol. Plant Mol. Biol.*, 1998, **49**, 727.

12. Y.-P. Lu, Z.-S. Li, Y. M. Drozdowicz, S. Hörtensteiner, E. Martinoia and P. A. Rea, *Plant Cell*, 1998, **10**, 267.
13. R. Tommasini, E. Vogt, M. Fromenteau, S. Hörtensteiner, P. Matile, N. Amrhein and E. Martinoia, *Plant J.*, 1998, **13**, 773.
14. A. E. Wolf, K.-J. Dietz and P. Schröder, *FEBS Lett.*, 1996, **384**, 31.
15. J. O. D. Coleman, R. Randall and M. M. A. Blake-Kalff, *Plant Cell Environ.*, 1997, **20**, 449.
16. M. A. Schuler, *Crit. Rev. Plant Sci.*, 1996, **15**, 235.
17. R. Schmitt and H. Sandermann, Jr., *Z. Naturforsch.*, 1982, **37c**, 772.
18. R. Fonné-Pfister, J. Gaudin, K. Kreuz, K. Ramsteiner and E. Ebert, *Pestic. Biochem. Physiol.*, 1990, **37**, 165.
19. C. Gaillard, A. Dufaud, R. Tommasini, K. Kreuz, N. Amrhein and E. Martinoia, *FEBS Lett.*, 1994, **352**, 219.
20. M. Klein, G. Weissenböck, A. Dufaud, C. Gaillard, K. Kreuz and E. Martinoia, *J. Biol. Chem.*, 1996, **271**, 29666.
21. K. Kreuz, J. Gaudin, J. Stingelin and E. Ebert, *Z. Naturforsch.*, 1991, **46c**, 901.
22. R. Tommasini, E. Vogt, J. Schmid, M. Fromenteau, N Amrhein and E. Martinoia, *FEBS Lett.*, 1997, **411**, 206.
23. W. R. Streber, U. Kutschka, F. Thomas and H.-D. Pohlenz, *Plant Mol. Biol.*, 1994, **25**, 977.
24. P. C. C. Feng, T. G. Ruff, S. H. Rangwala and S. R. Rao, *Pestic. Biochem. Physiol.*, 1998, **59**, 89.
25. N. Shiota, H. Inui and H. Ohkawa, *Pestic. Biochem. Physiol.*, 1996, **54**, 190.
26. P. A. Rea, Y.-P. Lu and Z.-S. Li, 'Glutathione-S-conjugate transport in plants', Patent, University of Pennsylvania, PCT Int. Appl. WO 9821938, 1997.

Environmental Fate

Leaching Mechanisms

A.D. Carter

ADAS ROSEMAUND, PRESTON WYNNE, HEREFORD HR1 3PG, UK

Abstract

Losses of pesticides from soil by leaching may only amount for less than 1% of the application yet the impact of the process can cause considerable environmental concern. Increasing interest in groundwater protection and the need to protect non-target organisms has initiated more stringent data requirements from regulatory authorities. The magnitude and frequency of leaching events must be characterised in order that a full assessment of their potential impact can be made. Since leaching processes are dynamic in time and space it is necessary to understand the nature and properties of soils and their associated hydrological characteristics. Sorption and degradation kinetics determine the availability of pesticide for leaching but soil properties such as organic matter content, texture and structure determine the pathways that water, solutes and suspended particles follow. Cultivation practices such as tillage and drainage treatments which change the soil's intrinsic properties can also influence leaching mechanisms. Mathematical models are continuously being developed to account for complex leaching processes in order that pesticide fate can be predicted for all intended soil and agroclimatic scenarios.

1 INTRODUCTION

Leaching is a fundamental soil process whereby constituents (soluble elements or suspended particles) are lost from the soil profile by the action of percolating liquid water. It is a common and well documented process which is responsible for the transport of pesticides (and many other elements) from agricultural areas and can lead to the subsequent contamination of ground and surface water. Surveys in the USA and Europe have revealed a wide range of pesticides or their metabolites to be present in receiving waters.[1, 2, 3, 4] The amount of pesticide leached below the rooting zone depends on the physico-chemical properties of the pesticide, agroclimatic conditions and the time taken for sufficient rainfall to transport the pesticide. Under average conditions the amount is typically <1% to 1% of the applied mass but under certain circumstances can reach up to

5%.[5] Water pollution by pesticides can occur as a result of the approved use of a pesticide and the extent to which it does has initiated agroeconomic and environmental concerns for water quality and the sustainable use of pesticides. There are reports of contamination arising from point or semi-point sources, e.g. farmyard spillages and the use on non-agricultural surfaces [6, 7] where the soil layer is often by-passed and opportunities for degradation /dissipation are restricted causing a significant contamination incident. When pesticides are lost from agricultural fields losses tend to be diffuse but the potential for leaching losses varies at the local and national scale because of differences in soil properties, climate, cultivation and management systems.

The pesticide approvals process in Europe, the USA and within other regulatory systems requires that the leaching behaviour of an active substance and its metabolites are investigated prior to widespread use. There are a number of tiered data requirements for the investigation of the leaching process. Monitoring techniques including the use of laboratory leaching columns, thin layer chromotography, lysimeters and field investigations are all used to derive information concerning leaching. Predictive modelling has been developed to simulate the soil and pesticide processes for a number of potential usage scenarios. The accuracy and reliability of the models depends on their parameterisation, operation and their ability to describe the field leaching process. Appraisal and evaluation of predicted and measured data allows an assessment of leaching risk to be determined for registration.

2 SOIL HYDROLOGY AND LEACHING

Knowledge of the water content, energy status and fluxes within the soil is fundamental to the understanding of the leaching process. The mass or volume fraction of water in the soil is termed the soil water content and water is held within and around soil particles, aggregates, organic complexes or pore spaces by pressure forces or potentials. As the soil dries from saturation (0kPa) the pressure decreases (i.e. it becomes more negative) as large pores and voids empty of water. Progressive decreases in pressure continue to empty narrower pore spaces and the films of water held around soil particles are reduced in thickness. The decrease in pressure results in progressive decreases in soil water content. The relationship between water content and water pressure (the water retention characteristic) controls the amount of water retained and thus the volume and rate that water percolates through the soil. The water retention characteristic varies according to soil texture, structure, density and organic matter content.[8]

Water movement in soil occurs in response to two main forces which are applied by gravity or through differences in the soil pore water pressure. It is unusual for saturated conditions to persist in an agricultural soil, and water will normally drain from coarse pores (>60µm diameter) in response to gravity. Soil water will also flow from regions of high to low pressure i.e. from wetter to drier soil (e.g. 0kPa to -5kPa). The water retained after excess gravitational water has drained away is defined as that being held at field capacity (approximately -5 to -10kPa). When a soil is saturated water fluxes can be very rapid e.g. a saturated hydraulic conductivity of 0.40m day^{-1} can occur in a sandy subsoil. Once the coarse pores have drained, water movement continues as unsaturated flow through smaller soil pores at a much slower rate e.g. 0.01m day^{-1} in the same soil when at field capacity (-5kPa). In contrast, in a clayey soil matrix, flow is slower in the saturated state (e.g. 0.08m day^{-1}) since there are fewer coarse pores contributing to water movement than in a sand. Flow in the unsaturated state in a clayey soil can be greater than for a sand because there are a greater number of finer soil pores.[9] Water movement can also occur in response to evaporative processes and the flux will be upwards through the soil profile. Water movement can be lateral, sub-lateral or vertical through the soil profile.

3 SOIL STRUCTURE AND ITS EFFECT ON LEACHING

As the clay content of a soil increases there is a tendency for aggregation of soil particles into structural units call peds. Strong vertical structural development is characteristic of many clay soils particularly those with the potential to shrink and swell which also causes cracks to develop. Activity by soil fauna can also create coarse channels or biopores which carry flowing water. These structural and biological features, often referred to as macropores, can provide rapid by-pass flow routes for water movement. Consequently a large proportion of the soil matrix may not directly interact with flowing water and an equilibrium concentration of solute throughout the soil profile is not attained. Artificial drainage of slowly permeable soils aims to improve their drainage characteristics by providing channels for rapid water movement. Excess water is transported from the topsoil to systems within the subsoil or it can be removed directly from shallow water table conditions. Secondary treatments such as moling or subsoiling serve to increase the number of artificial channels which conduct water through the soil.

The movement of water between individual soil particles within the main soil matrix is the dominant flow process in sandy or light loamy soils and is know as chromatographic flow. There is however increasing evidence that preferential movement can occur in these soils particularly those which exhibit hydrophobic characteristics [10] or when hydrological discontinuities, such as a change in soil texture cause a change in flow pathways.

A comprehensive review of preferential flow mechanisms and the effect on solute transport is provided by Jarvis.[11] He states that 'preferential flow may significantly enhance the risk of leaching of surface-applied contaminants to surface and groundwater, since much of the buffering capacity of the biologically and chemically reactive topsoil may be by-passed. In this way, chemicals can quickly reach subsoil layers where degradation and sorption processes are generally less effective'. Where preferential channels connect directly with groundwater or with artificial drainage systems, the impact of transported solute on water quality can be significant.

The influence of soil structure on the hydrology of a heavy clay soil and associated patterns of solute movement was emphasised in a study by Brown *et al.* [12] Lysimeters were taken over mole drains from a heavy clay soil with either a standard agricultural tilth or a finer, deeper topsoil tilth. Leaching of the herbicide isoproturon was initiated by artificial irrigation either one day or 39 days after application. The finer tilth increased the water holding capacity of the topsoil and this resulted in slower wetting up of the subsoil, decreased flow volumes from the first events and delayed, by approximately four weeks, the breakthrough of a bromide tracer. There was a large effect of topsoil tilth on leaching of isoproturon independent of time to the first event after application. Maximum concentrations were reduced by 50-75% and total losses were approximately three times lower for the finer deeper tilth. The fine tilth decreased the incidences of bypass flow and provided a greater number of sites for sorption and degradation to occur. Subsequent experiments (Brown *pers. comm*) which aimed to reproduce the tilths using agricultural equipment have not shown the same differences between treatments. It was concluded that the soil moisture conditions at the time of cultivation precluded the generation of the finer, deeper tilth emphasising the difficulty of translating research into normal agricultural practice. Research in the USA has shown that pesticide leaching to groundwater from well established no-till soils (where soil structure was allowed to develop for several years) was greater than from under conventional tillage which destroyed natural structures. Increased leaching was thought to derive from preferential transport through macropores.[13]

Hardy *et al* [14] describe the importance of the sorbed phase in the leaching of pesticides. Large amounts of suspended sediment were observed in drainflow from a non-calcareous soil in contrast to the clear drainage from a calcareous soil. Topsoil, surface aggregates were found to be unstable during rainfall in the non-calcareous soil allowing fine soil particles and organic material to be transported through the soil profile to drains. The strongly sorbed herbicide diflufenican was transported by this process as was isoproturon. The latter was shown to desorb/diffuse into the aqueous phase of receiving waters. Concentrations of strongly sorbed pesticides in sediments of surface waters had previously thought to have derived from surface run-off or erosion but this study highlighted an alternative mechanism. It is thought that this process might be applicable to approximately 6% of soils in England and Wales.

4 SOIL/SOLUTE INTERACTION AND LEACHING

Leeds-Harrison [15] describes the principal mechanisms by which water and solute move through soil, emphasizing the link between water flow and solute transport. The physico-chemical interactions of the solute with soil particle surfaces will vary according to the nature of the solute and the surfaces themselves. Convective flow is the simplest form of solute movement as dissolved solute is transported by water when flow occurs. As water content changes within the soil profile concentration changes occur and the transport of solutes is subject to diffusion and dispersion because of the resulting concentration gradients in flowing water. Solute movement in sandy or light loamy soils where chromatographic type flow occurs is well characterised by simulation of these processes. Concentrations of a solute in a soil profile following leaching would exhibit a characteristic distribution whereby large pores caused leaching to deeper depths and the smallest pores caused retention at a shallower depth.[15] Classical breakthrough curves represent changing solute concentrations in leachate whereby the dominant processes controlling solute movement are convection and dispersion.

When preferential flow occurs in soils, solutes can be transported rapidly through the root zone and the convection dispersion processes may not dominate. Concentrations of solute may occur in soil aggregates which are not in equilibrium with the solute concentration in the macropores. An analysis of the hydrological characteristics of soils and their drainage response enabled an estimation of the relative importance of the different soil water flow mechanisms in UK soils.

Table 1 *The significance of soil water flow mechanisms in the UK*

% UK Soils	*Vertical single phase pore flow (chromato-graphic)*	*Vertical two-phase pore flow*	*By-pass flow (all via fissures/ biopores)*	*Overland flow (surface run-off)*	*Saturated lateral/sub-lateral flow (often to drains)*
13		***		*	
7	**	*		*	
1	**	*		*	**
4		**			**
4		**	?		**
1		***	?	*	
10		?		*	***
9		***	?	*	
9		**	*	*	**
6.5		*	***	*	**
17.5		?	**	*	***
3.5				*	***
9				?	***

after: J.M. Hollis and A.D. Carter, Soil Survey & Land Research Centre

***	Dominant flow mechanism
**	Major flow mechanism
*	Minor but significant flow mechanism
?	Importance not known

Table 1 shows that water flow in most soils is characterised by a range of mechanisms and that chromatographic flow is the dominant process in only 8% of UK soils.

Bettinson *et al* [16] reported leaching of isoproturon (IPU) from five soils with different structure, organic matter content and drainage characteristics (Table 2). Soils were contained in lysimeters, radioactive herbicide was applied and leachate collected.

Table 2 *Soil series and their description*

Soil series	*Description*
Fladbury	Slowly-permeable, strongly-structured, cracking clay soil in alluvium
Cuckney	Unstructured free draining sand
Sonning	Permeable, free-draining shallow loam over gravel
Ludford	Deep, weakly-structured loamy soil
Isleham	Shallow peat over permeable, free draining sand

No IPU was detected in the leachate from the Isleham series due to sorption by the organic matter in the topsoil. Moderate losses were seen from the Sonning soil with breakthrough observed after approximately 60 days, losses were attributed to its permeability and the low retention capacity of the subsoil. The Cuckney series showed a classical breakthrough pattern of loss which did not commence until approximately 75 days after the application of the herbicide. This pattern of loss was attributed to slow chromatographic flow in the soil profile which allowed maximum time for sorption and degradation processes to occur. The Fladbury series as expected, showed rapid breakthrough (approximately one month after application) with a greater amount of IPU lost overall and concentrations of up to seven times higher than from the Cuckney series. The weakly structured loam soil (Sonning series), suprisingly, also gave high losses with initial breakthrough concentrations greater than that of the Fladbury soil. On completion of the study the Sonning soil was excavated and a number of continuous worm channels were identified which accounted for the rapid bypass flow.

The implication of this study is that sandy soil (as often specified for regulatory studies) does not necessarily represent the worst case for leaching. However, the majority of clayey soils do not overlay aquifers and their drainage water usually contributes to surface water giving rise to alternative regulatory concerns. There are however examples of heavier textured soils which do overlay shallow groundwater which have the potential for preferential flow and where significant quantities of pesticide may be transported to underlying water.

The potential pathways which water and solute take within soil can be investigated using dye tracing. Flury *et al* [10] describe the use of the dye Brilliant Blue FCF in field studies whilst rhodamine-B has been applied in lysimeters studies.[17] If dye solutions are applied in a volume and intensity which is representative of normal practice, excavation will reveal the extent to which water and solute and thus a pesticide might interact with the soil profile.

Walker *et al.*[18] showed that adsorption of isoproturon to soil aggregates in a static laboratory system did not reach the equilibrium achieved with a conventional methodology of using sieved (<2mm) air dried soil even after 24 hours. The sorption rate was even slower with larger (9-12mm) soil aggregates. Adsorption isotherms measured in a static system with soil moisture content typical of winter field conditions also showed less adsorption than in a shaken laboratory system. The rate of decline in water extractable residues of the herbicides was more rapid than that of total extractable residues following application of isoproturon to a heavy clay soil in the field. The authors conclude that traditional laboratory derived sorption and degradation parameters which are used to predict pesticide leaching, particularly for highly structured soils, may give incorrect estimates.

5 CLIMATIC EFFECTS ON LEACHING

Water in excess of that required to maintain field capacity is known as excess winter rainfall and this water leaches through the soil profile and represents the recharge to water sources. The volume of excess winter rainfall varies each year according to factors such as precipitation amount, soil type, antecedent cropping, evaporation, temperature.

Table 3 *Rainfall in mm for selected durations and return periods for a location in the English Midlands*

	Return period (years)						
Duration	Twice a year	1	5	10	25	50	100
1 min	1.1	1.4	2.2	2.6	3.0	3.4	3.9
30 min	7.4	9.9	15.8	18.7	23.1	27.1	31.6
60 min	9.8	12.6	19.7	23.3	28.9	33.9	39.8
6 hrs	17.4	21.8	31.9	37.5	46.0	53.6	62.3
24 hrs	25.8	32.1	45.2	52.4	63.2	72.6	83.4
25 days	88.6	108.0	138.5	151.7	170.1	185.2	201.5

Data from the Meteorological Office, Bracknell, UK

In England and Wales the average annual excess winter rainfall varies from <100mm in the drier east to in excess of 500mm in the west of the two countries and in particular in upland areas. Leaching studies in lysimeter and field situations must therefore take into account the volume and timing of leachate loss and consider whether it is representative of the agroclimatic conditions where a pesticide is likely to be used.

The significance of rainfall during a field experimental study and of individual events can be assessed by comparison with long term statistical records. Table 3 shows the return periods for rainfall at specific intervals and intensities for a location in the English Midlands. For example, a rainfall event of 70mm during a 24 hour period is likely to occur approximately every 50 years. Should such an event occur during a leaching study, experimental results can be placed in context and in this case would be considered as an extreme worst case scenario. Events which occur every 10 years might be considered to be worst case whereas an event with up to a five year return period could be considered an average case.

Similarly statistical data may also be used to identify average/extreme times to rainfall occurring following the application of a pesticide on a given date. The uncertainty of rainfall occuring during the required phase of a field based or lysimeter leaching experiment may lead to the need for rainfall to be simulated. It will be important to ensure both the volume and intensity applied are realistic, uniform and are representative of the required climatic scenario

6 MONITORING TO INTERCEPT LEACHING

Results from laboratory leaching studies cannot be reliably used to predict the mobility of a pesticide in the field situation. [19] Sieved, air dry soil material is repacked into leaching colums and is not representative of true field conditions. Data do however provide indications of the relative mobilty of a pesticide. The analysis of sequential segments of soil cores sampled from the field situation to determine pesticide mobility may also provide misleading data since the component of interest is the soil water and not the solid phase. The trend towards deeper sampling for residue analysis by soil coring with the intention of determining pesticide leaching is inappropriate as it is time consuming, requires mechanised sampling equipment and does not provide an indication of the residue which is in the aqueous phase, i.e. the component which is leaching. Analytical determination of residues in soil water is usually less complex and limits of detection are often at least an order of magnitude lower than for soil. Direct analysis of soil water provides a cost effective option for which low levels of pesticide can be determined. Leachate samples may also be used in ecotoxicological testing.

The use of lysimeters overcomes these problems as leachate can be directly intercepted as it flows from the base of the lysimeter, whilst soil residues can be determined at the end of a study. It is however inappropriate to sample certain soils for use in lysimeters e.g. heavy clays or those with shallow bedrock unless special adaptations are made to the sampling and monitoring equipment but their use is generally confined to research projects. Field studies can provide a more realistic opportunity for determining leaching, and porous ceramic cup suction samplers are often used to collect soil water samples. Construction, installation and operation are described by Earl and Carter.[20] As suction samplers do not provide a measure of drainage this can be obtained by assessing the amount of excess winter rain (that rain falling when the soil is a field capacity) or by using a meteorological model to determine drainage losses.[21] The quantity of solute leaching can then be calculated from the concentration in the sample and the drainage volume. Concentrations from the lowest sampling depth should be applied to the volume of drainage water occurring between each sampling period. Cumulation of these data for the entire leaching period will allow the calculation of an average concentration in the recharge water and the mass of pesticide leaching. If comparison is made with the methodology used for interpretation of leachate concentrations from a lysimeter study then an average for a two year period could also be determined.

Samples should be taken after each 'significant' rainfall and site monitoring used to indicate the general direction and magnitude of the water flux. When matric pressures indicate a general downwards flux in the profile and are -10kPa or wetter, water is considered to be present in coarser pore spaces and leaching via drainage from the soil profile. Sampling when fluxes indicate upwards movement or pore water pressures are lower (drier) than -10kPa provides misleading data and can overestimate potential losses made, thus causing needless concern for regulators. Samplers are typically installed at approximately 100cm depth which is considered to be at or below the effective rooting zone for most crops. When a solute reaches this depth it is likely that it will continue to leach unless degradation or other processes cause it to disappear. Suction samplers can also be located at key depths in the soil profile, e.g. in different soil horizons or at the depth which drains occur.

7 MODELLING THE LEACHING PROCESS

A single model run with one combination of input parameters to determine leaching characteristics is of little use unless a benchmarking exercise is being carried out for comparison of the behaviour of one active substance versus that of another.

Table 4 *Flow processes simulated by a range of models used to predict pesticide leaching*

Model	Vertical single phase pore flow (chromato-graphic)	Vertical two-phase pore flow	By-pass flow (all via fissures/ biopores)	Overland flow (surface run-off)	Saturated lateral/sub -lateral flow (often to drains)
VARLEACH 2.0	*				
WAVE	*				
LEACHM 3.1	*				
PRZM 2.1/PELMO	*			*	
GLEAMS	*			*	
CRACK			*		*
PLM	*	*			
MACRO	*	*	*		*

after: J.M. Hollis and A.D. Carter, Soil Survey & Land Research Centre

Stochastic modelling enables a range of input parameters which might, for example, consider differences in soil type and texture, timing volume and intensity of rainfall and time to first rainfall event e.g. Wagenet *et al.*[22] Statistical evaluation of all input parameters will allow the production of probabalistic output which is used to determine the magnitude and return frequency of leaching events. However leaching models are still very restricted in their ability to simulate certain processes; for example, Table 4 compares the flow processes identified in Table 1 with the capabilities of a range of predictive models. Tables 1 and 4 show that most models can only simulate flow processes in 8% of UK soils. Whilst this fact might initiate concern, these models are specifically used in a regulatory context to provide supporting data, which is evaluated along side other data derived in the laboratory and from lysimeters or field studies. Models like MACRO, PLM and CRACK can simulate non-equilibrium processes but the validation status of the last two is poor. Whilst a number of standard scenarios and comprehensive data sets now exist there is still a requirement for more comprehensive field monitoring data to characterise input parameters and for validation purposes.[23] Variability in modelling expertise or interpretation can also influence the quality of model outputs with differences attributed to subjectivity when determining input parameter values.[24] Catchment scale models which integrate profile and field leaching processes are poorly developed and there is scope for their further development as demonstrated by a number of authors.[25, 26] Jarvis [11] states that 'one important task for the future is to encourage the transfer of these models to the realm of management applications. This would promote sound environmental decision-making and allow quantitative evaluation of alternative strategies to minimize environmental degradation'.

8 CONCLUSION

Leaching is the main mechanism responsible for the transport of pesticides in soils. Recognition of soil hydrology, structure, and agroclimatic conditions is critical to characterise the magnitude and frequency of leaching events whether there is concern for groundwater contamination or the potential impact on non-target aquatic organisms in surface waters. There are locations where water resources are known to be vulnerable to leaching and the use of pesticides is banned or restricted. In the future legislation may require that ground and surface water catchments are managed more sensitively with respect to their potential contamination. A wide range of management techniques are being investigated to minimise leaching losses but these are mainly at the research stage and their wider application in agriculture is not yet established.

Leaching mechanisms are largely known but are difficult to measure and quantify, particularly with reference to local and broad scale variability. A wider range of field monitoring data is urgently required for model parameterisation and validation particularly for those which simulate non-equilibrium flow processes. The use of the equilibrium type models will increase over the next few years in Europe in response to the registration directive 91/414/EEC [27] and their role in regulatory decision making procedures will become more important. It will be essential to consider the limitations of these models when interpreting their output.

References

1. A. D. Carter and A. I. J. Heather, Pesticides in Groundwater. In Pesticides - Developments, Impacts and Controls, Ed. G. A. Best and A. D. Ruthven, 112-123, Special publication No 174., Royal Society of Chemistry, Cambridge 1995.
2. K. R. Eke, *Pesticide Outlook,* **7,** 2, 15-20.
3. I. Heinz, A. Flessau, N. Zullei-Sebert, B. Kuhlmann, U. Schulte-Ebbert, M. Michels, J. Simbrey and G. Fleischer, Economic Efficiency Calculations in Conjunction with the Drinking Water Directive (Directive 80/778/EEC); Part III: The parameter for pesticides and related products. Research Report prepared for European Commission, DGXI Environment, University of Dortmund, Germany, 1996.
4. M. Isenbeck-Schröter, E. Bedbur, M. Kofod, B. König, T. Schramm and G.Attheβ, Occurrence of Pesticide residues in Water - Assessment of the Current Situation in Selected EU Countries. Berichte, Fachbereich Geowissenschaften, Universität Bremen, Nr 91 65 S, Bremen, Germany, ISSN 0931-0800, 1997.
5. M. Flury, *Journal of Environmental Quality*, 1996, **25,** 1, 25-45.
6. A. I. J. Heather and A. D. Carter, *Aspects of Applied Biology,* 1996, **44**, 157.
7. P. Fogg and A. D. Carter, Biobeds: the development and evaluation of a biological system for pesticide waste and washings, 49-58, In: Proceedings of the BCPC symposium proceeeding no. 70, Managing pesticide waste and packaging, Farnham, 1998.
8. D. G. M. Hall, M. J. Reeve, A. J.Thomasson, V. F. Wright, Water Retention, Porosity and Density of Field Soils, Technical monograph no. 9, Soil Survey of England and Wales, Harpenden, UK, 1977.
9. C. D. Brown, A. D. Carter and J. M. Hollis, Pesticide Mobility in Soils. 131-184, In: Environmental Chemistry of Pesticides, Eds: T. R. Roberts and P. Kearney, J. Wiley and Sons Ltd, Chichester, 1995.
10. M. Flury, J. Leuenberger, B. Studer, H. Fluhler, W. A. Jury, and K. Roth, Pesticide transport through unsaturated field soils: Preferential flow, ETH, Zurich, 1994.

11. N. J. Jarvis, Modelling the impact of preferential flow on solute transport in soils, In: Physical Non-equilibrium in Soils: Modelling and Applications, 195-222, Eds: H.M. Selim and L. Ma, Ann Arbor Press, 1998.
12. C. D. Brown, V.L. Marshall, A. D. Carter, A. Walker, D. J. Arnold and R. L. Jones, part I. Lysimeter experiment. *J. Soil Use and Management,* In press.
13. A. R. Isensee, A. M. Sadeghi, Soil Science, 1997, **161**, 6, 382-389.
14. I. A. J. Hardy, A. D. Carter, P. B. Leeds-Harrison, Abstract 6C-018, 9th International Congress of Pesticide Chemistry, The Food-Environment Challenge, Royal Society of Chemistry and the International Union For Pure and Applied Chemistry, London, 1998.
15. P. B. Leeds-Harrison, The movement of water and solutes to surface and ground waters. In: Pesticide Movement to Water, 3-12, Eds. A. Walker, R. Allen, S. W. Bailey, A. M. Blair, C. D. Brown, P. Günther, C. R. Leake and P. H. Nicholls. Monograph No 62 British Crop Protection Council, Farnham, 1995.
16. R. J. Bettinson, C. D. Brown and J. M. Hollis, Pesticides in Soil and the Environment Abstract for the COST 66 Workshop 1996, Stratford upon Avon , 169-170.
17. C. D. Brown, V. L. Marshall, A. Deas, A. D. Carter, A. Walker, D. J. Arnold and R. L. Jones, part II. Interpretation using a radiolabelled technique, dye tracing and modelling. *J. Soil Use and Management,* In Press.
18. A. Walker, I. J. Turner, J. E Cullington, S. J. Welch, Aspects of the adsorption and degradation of isoproturon in a heavy clay soil, *J. Soil Use and Management,* In Press.
19. J. M. Hollis and A. D. Carter, *Proceedings of the BCPC Conference*, 1990, **III**, 1005-1010.
20. R. Earl and A. D. Carter, *Agricultural Engineer*, 1991, Summer, 50-51.
21. E. I. Lord and M.A. Shepherd, *Journal of Soil Science* **44**, 435-451.
22. R. J. Wagenet and J L Hutson, Computer simulation models as an aid in estimating the probability of pesticide leaching, 786-794, In: Computers in Agriculture, Proceedings of the 5th International Conference, American Society of Agricultural Engineers, USA, 1994.
23. R. Calvet, European Journal of Agronomy, 1995, **4**, 4, 473-484.
24. C. D. Brown, U. Baer, P. Gunther, M. Trevisan and A. Walker, *Pesticide Science*, 1996, **47,** 3, 249-258.
25. D. J. Mulla, C. A. Perillo, C. G. Cogger, *J. Environmental Quality,* 1996, **25**, 3, 419-425.
26. A. Tiktak, A. M. A van der Linden, *Soil Science Society of America*, 1993, 259-281.
27. EC Directive 91/414/EEC Concerning the Placing of Plant Protection Products on the Market. O J L 265/87, 27. 9. 97.

Landscape–Scale Environmental Modeling

D.A. Laskowski

DOW AGROSCIENCES, 9330 ZIONSVILLE ROAD, INDIANAPOLIS, IN 46268, USA

1 ACTIVITY SINCE IUPAC 94

The 8th IUPAC International Congress of Pesticide Chemistry, held in 1994 in Washington DC, included a workshop/panel discussion on "Computer Models For Pesticide Risk Assessment: An International Progress Report". The primary topic of discussion was barriers to the use of models in a regulatory context, which at that time were identified as:

- unfamiliarity with modeling
- low level of confidence
- inadequate validation
- inadequate models
- poorly developed databases
- lack of tools that propagate variability

Since 1994 the pesticide modeling community has turned its attention to these barriers in order to improve modeling as a decision-making tool in the pesticide regulatory arena. This is driven by the expanding role of modeling in both the European and United States (U.S.) regulatory processes. In Europe modeling has become a part of the official regulatory process, and the U.S. uses modeling as an exposure assessment tool for ecological risk assessment, and now with passage of the Food Quality Protection Act (FQPA) for humans as well. FQPA dictates that pesticide concentrations in human drinking water must be assessed, and models formerly used for non-human aquatic exposure are now used to assess human drinking water exposure.

All of these things have caused a flurry of activity to improve models and their use, and the period since last IUPAC meeting can be characterized as a time of active development of modeling for purposes of better decision-making.

1.1 Activity in the EU

Activities within the European Union (EU) have proceeded largely through the "**FO**rum for the **C**oordination of pesticide fate models and their **US**e" (FOCUS). This body of modeling experts drawn from government, academia, and industry has turned its attention to the evaluation of models that deal with leaching and surface water. They also have worked to develop standard modeling scenarios for southern as well as northern Europe.

The FOCUS group dealing with leaching reported its findings in 1996.[1] This group dealt with model assessment, model limitations/deficiencies, European scenarios, and model validation. It also gave recommendations with regard to the correct use of models and their use in a regulatory context. The group concluded that current models are useful screening tools that can be used immediately in a tiered exposure assessment approach. No model was fully validated, but models could reliably predict the bulk of chemical mass movement. There was a need for EU modeling scenarios and better databases to support modeling, and a large source of error (uncertainty) arises from selection of input values by an individual modeler. Thus, there was a need for a Standard Operating Procedure (SOP) to provide guidance on the selection of model input. Many models exist and several were evaluated by the group, but the group concluded there must be selection of a "primary" model in order to provide more focus to development and improvement of modeling as a decision-making tool.

The FOCUS leaching work group currently is working on the development of standard modeling scenarios for the assessment of leaching potential in all of Europe.

The FOCUS work group for surface water modeling also published its report in 1996.[2] It concluded that current models can predict peak water concentration within a factor of 10. No model was fully validated, and modeling should be used in the context of a tiered approach to exposure assessment. Like the FOCUS leaching work group, the surface water group recognized the need to "harmonize" how models are used in a regulatory context, and recommended the development of a "community model" to focus limited resources more effectively into tool development, validation, and improvement. Current efforts are underway to develop standard scenarios for modeling the potential transport of plant protection products into EU surface waters.

For more details on the findings of the FOCUS work groups, the reader is directed to the work groups' reports.[1,2]

1.2 Activity in the U.S.

In the U.S., model development efforts have been centered within three activities. One is the Exposure Modeling Work Group (EMWG), comprised of government, academic, industry, and consulting experts working together to evaluate models, to improve them, and to develop new modeling tools. The EMWG evaluated several models and selected PRZM, GLEAMS, and EXAMS as primary models for modeling runoff, leaching, and surface water. It recognized that guidelines on modeling practices were needed and a guidance document outlining good modeling practices for regulatory purposes was produced by the group.[3] A series of standard modeling scenarios covering most areas within continental U.S. has been developed.

The EMWG began action on model validation by establishing the Federal Insecticide Fungicide Rodenticide Act (FIFRA) Environmental Model Validation Task Force (FEMVTF). This second task force is comprised of modeling experts from 15 agricultural chemical companies, with modeling experts from the U.S. Environmental Protection Agency (USEPA) serving as advisors. Its primary task is the conduct of the validation of primary models selected by the EMWG, its parent organization. Initially, the models under study were PRZM and GLEAMS, but due to lack of developer support for GLEAMS it was dropped from study and subsequently removed from the EMWG's list of primary models.

The FEMVTF's primary objective is to improve confidence in model predictions of pesticide behavior within the confines of FIFRA risk assessment. It also focuses on improvement of PRZM for runoff and leaching, and has created datasets and methodology for future model validation efforts. Like FOCUS, the FEMVTF found that a major source of uncertainty (error) in modeling comes from model input selection by a modeler, and an SOP for selection of model input was developed to standardize the modeling process. Sensitivity analysis is a part of the validation effort and to carry this out on PRZM, a Plackett/Burman sensitivity analysis tool was developed to assess relative importance of model inputs. The tool also can be used to assess inputs for GLEAMS.

The FEMVTF validation effort is a work in progress and details of the Task Force findings and conclusions are not yet available.

The third center of activity is an effort begun by USEPA to modernize its ecological risk assessment process in order to make it more probabilistic instead of continuing with older, more deterministic approaches used in the past. This effort is entitled the "Ecological Committee on FIFRA Risk Assessment Methods" and has become identified by the acronym ECOFRAM. Membership is comprised of USEPA, academic, consulting, and industry experts representing environmental chemistry, modeling, and ecotoxicology. Members are drawn from Europe and Canada, as well as the U.S. Two work groups have been formed, an aquatic group to deal with aquatic systems and a terrestrial group to address terrestrial systems. Primary focus of the working groups is to develop the capability to do probabilistic terrestrial and aquatic risk assessment.

The ECOFRAM work group for aquatic risk assessment has adopted a tiered approach to exposure assessment for water (surface water). Tier 1a is designed to be quite conservative and provides a worst case estimate of surface water concentrations immediately adjacent to the edge of a treated field. A simple screening-model approach (GENEEC) is recommended, which serves to identify need for further modeling. Tier 1b provides a more refined estimate of edge-of-field surface water concentrations by using PRZM linked to EXAMS. This adds additional chemistry to the simple screen in 1a by modeling actual high runoff/erosion scenarios. Tier 2 determines the complete distribution of concentrations in surface water immediately adjacent to treated fields by modeling a widespread range of scenarios encompassing an entire pattern of usage. In addition to identification of more sensitive areas, it looks at the impact of mitigation and management options on potential surface water concentrations. Tier 3 moves away from the field's edge and into a larger, watershed-scale assessment of surface water concentrations. It provides the most realistic concentration estimates, and it does this by modeling at the watershed scale and/or by carrying out watershed monitoring of surface waters.

Recommendations from the ECOFRAM aquatic work group emphasize clearly that the characteristics of environmental chemistry data must change from the single-point values derived from past and present-day studies to meet the future needs of probabilistic exposure assessment. Approaches to the evaluation of aquatic exposure will focus more clearly on assumptions, and they will propagate uncertainty throughout an assessment by means of probabilistic modeling techniques that use distributions of model input rather than single-point values. Modeling scenarios will be put in perspective through determination of their position on the scale of probability. The group also recommends the development of modeling tools that go beyond edge-of-field to watershed-scale, and this again translates directly into larger natural variation that must be dealt with. All of these things point to the need of an environmental chemistry database that has much greater range in data than are currently available.

The ECOFRAM terrestrial work group is also developing a probabilistic approach to exposure assessment. Figure 1 shows the conceptual, compartmentalized approach taken to address the heterogeneous nature of terrestrial systems. Terrestrial systems are different from aquatic ones because of the multitude of compartments and the several routes of exposure that must be considered in terrestrial exposure assessment. Aquatic environments have one primary route and that is through water; terrestrial systems must deal with vapor, vegetation, dust, direct deposition, and foods, as well as water. The

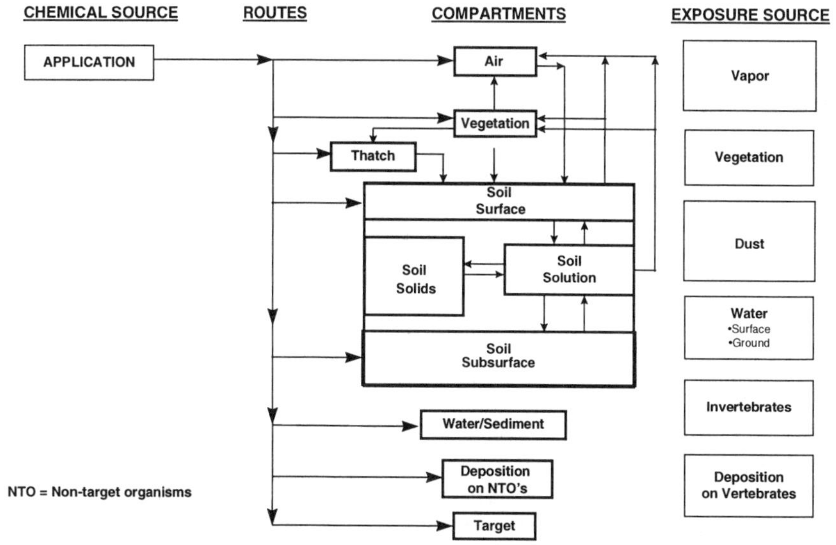

Figure 1 *Conceptual model of terrestrial environment (adapted from the work of the ECOFRAM Terrestrial Work Group)*

terrestrial work group is building models that consider these routes of exposure, and it is doing it in a probabilistic fashion that tries to capture the variation in concentration levels that exists naturally in such a heterogeneous system.

The terrestrial group's approach requires information regarding the fate of chemical for each of the relevant compartments in the figure; and if variability is carried throughout an exposure analysis, there also is need for information on the variation in concentrations of relevant compartments over time. These needs are different or at least shift the emphasis of environmental chemistry requirements that existed in the past.

2 A NEW ERA

Forces are operating to change the way environmental chemistry is carried out. There is increasing awareness of environmental complexity in terms of changing weather patterns, changing topography, and changing soil properties, and better perception of how these changes can be dealt with in terms of prediction of exposure of nontarget organisms to plant protection products. Probabilistic modeling tools are becoming available so that "all case" can be assessed instead of only "worst case" analysis of exposure. Computers have advanced in power so that they are no longer the limiting factor to probabilistic modeling as they once were. Other tools such as Geographical Information Systems (GIS) software and databases describing weather, landscape, and soil characteristics are becoming available to feed the probabilistic modeling approaches.

2.1 Probability Modeling – an Example

Figure 2 provides a simplified outline of probabilistic modeling in order to provide the reader with a clear understanding of this modeling process. In essence, ranges of model input that represent uncertainty are utilized to carry out the modeling of many potential real-world scenarios developed from different combinations of values from each input range. Output is then expressed in the form of frequency distributions that describe the likelihood of attaining various levels of output. The whole process serves to identify the effects that uncertainty in model input values have on the outcome of the model. Figure 2 depicts uncertainty in the two model input parameters (soil degradation rate and soil sorption constant, Koc) that environmental chemists supply.

To illustrate this process of uncertainty analysis further and to demonstrate the need for different data, it is posed hypothetically that an analysis of leaching potential for a plant protection product needs to be carried out. The leaching scenario chosen for this assessment is conservative, dealing with a vulnerable soil site (soil = loamy sand) receiving enough rain to cause movement of water through the soil's profile. Modeling with PRZM 3.12 indicates sufficient rain to give 66 cm of recharge below a 1-meter depth from the time of application of chemical in June to the end of the year. As typically the case, actual soil degradation rates and soil sorption partition coefficients have not been measured for the soil at this site, but there are data from studies with other soils that provide a number of measures of soil half-life and Koc (in this case, 24 soils providing the 24 values listed in Table 1). It is desired to assess how much chemical might leach out of this soil site's root zone, using the modeling tool PRZM 3.12 for assessment.

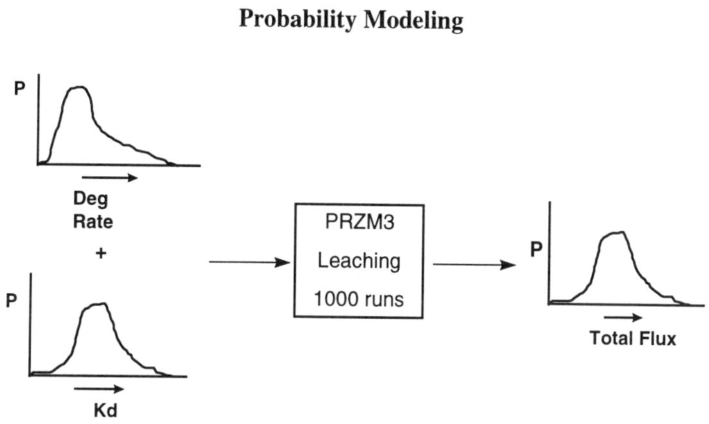

Figure 2 *Simplified conceptual view of probabilistic modeling*

The half-life and Koc numbers in Table 1 are real, taken from unpublished work carried out on an experimental compound by environmental chemists at Dow AgroSciences. Clearly no single "right" value exists for either of these important model input parameters because each number for Koc and half-life is a valid measure of the chemical's capacity to sorb and degrade in soil. This presents a dilemma because a common practice is to assign single values to half-life and Koc, which are then used to assess the leaching of chemical through the soil profile. But since all are valid measures of half-life or Koc, which value from Table 1 should be selected for this leaching assessment?

Some assessors err on the side of conservatism and select the maximum for half-life and minimum for Koc, intuitively knowing the chances of achieving greater leaching with this combination will likely be very small. Their attention focuses on "extreme worst case" — situations not likely to happen very often. Others do not wish to evaluate at the ultimate extremes; they select less extreme values and turn their attention to a more "typical worst case" scenario. Still others are interested in what most likely happens, and they use a measure of central tendency (e.g. mean values) to focus the assessment on "typical case", the case most likely to take place. But the latest trend in exposure analysis focuses on "all case" by carrying out the multi-run, probabilistic modeling illustrated in Figure 2. Here all available values for half-life and Koc are used to generate a multitude of potential modeling scenarios which are then used to describe all possible leaching results.

The following exercise demonstrates this by carrying out a probabilistic modeling assessment with the data in Table 1 and leaching scenario described earlier. It compares the total flux of chemical leaching beyond the root zone (depth of 67.5 cm) as predicted by PRZM 3.12 when (1) maximum/minimum, (2) mean + 90/10 percent confidence intervals, (3) mean values, or (4) samplings from cumulative distributions of the values from Table 1

Table 1 *Measured half-life and Koc values for 24 soils*

Koc						Half-life, days					
12	13.9	16	16	18	20	**13**	14	14	14	15	15.5
21	21.6	27	28	34	48.8	16	17	17	17	18	18
51	54	54.2	54.6	60	67	19	20	20.5	20.8	21	21
86.7	101	157	158	242	**262**	21	22	25.1	25.7	28	**29.2**

are assigned to half-life and Koc, respectively. For the sake of convenience, all other model input parameters including weather are held constant so that changes in leaching flux reflect only the impact of uncertainty/variability in half-life and Koc, even though it is recognized that other input parameters contribute their share of variation.

In order to generate a series of model input files reflecting the range in Koc and half-life in Table 1, the data were ranked and then cumulative percentile for each value within a rank was calculated with Excel's (v 7.0) rank and percentile data analysis tool. These transformed data were then used in a Monte Carlo add-in tool (@RISK for Windows, v 3.5e) for the Excel spreadsheet to generate a latin hypercube sampling of Koc and half-life values that reflect their distributions in Table 1.[4] One hundred combinations of Koc and half-life were generated and these were then used as input into PRZM 3.12 to generate 100 estimates of total flux of chemical out the bottom of the root zone.

Results from this analysis are summarized in Table 2 as percentages of applied material that leached beyond the root zone. Changes in Koc and half-life according to the distributions in Table 1 caused chemical flux to range from a low of 3×10^{-5} (essentially 0) to a high of 15.3 %. Initially this range appears very large, and one might conclude that uncertainties in Koc and half-life yield so much uncertainty in leaching that it is impossible

Table 2 *PRZM3 estimates of chemical leached beyond root zone*

Description	Flux, % of Applied
Range of 100 runs	3×10^{-5} – 15.3
Central Tendencies:	
Mean of 100 runs	4.0
Mean half life + mean Koc	1.6
50th percentile of 100 runs (median)	3.1
Extremes:	
Mean + 90 % CL of 100 runs	4.5
90 % half life + 10 % Koc[a]	3.1
Max half life + min Koc	18.1
90th percentile of 100 runs	9.6

[a]These are mean values of the data in Table 1 plus or minus the confidence interval at t = 0.1 for 1-tailed test assuming normal distribution of data.

to draw conclusions. But further analysis of the leaching data on the basis of statistical distribution and probability provides a great deal more information and allows the leaching to be put into perspective.

Table 2 presents data organized according to indicators of central tendency (typical case) and by indicators that focus on the outer tails of likelihood and represent the extremes of "typical worst case" and "extreme worst case". Central tendency is examined first, followed by a discussion on the measures of "extremes" in the table.

If it is assumed that flux values from the 100 simulations are distributed normally, a common assumption, their mean value is 4.0 %. If Koc and half-life are assumed to have a normal distribution and then averaged prior to estimation of flux (also a very common practice), PRZM3 estimates a flux of 1.6 %. The central tendency of the 100 simulations as measured by their median (assumes no statistical distribution) yields a value of 3.1 %. All measures of central tendency are similar to each other, ranging narrowly from 2 to 4 %.

A similar comparison at the extremes (typical worst case and extreme worst case) provides a different outcome. Here estimates of flux vary considerably according to the method of estimation. Assumption of normal distribution for the 100 values and then calculation of their mean plus the confidence interval to achieve a 0.1 single-tailed t-test value provides a flux estimate of 4.5 %. This changes to 3.1 % if the same approach is applied to Koc and half-life data, and these values are then used to model total flux. If the extreme worst-case is assumed by assigning maximum half-life and minimum Koc values from Table 1, the resultant PRZM3 estimate is 18.1 % flux. This is substantially different from all of the other measures of extreme. Finally, if the 100 results are ranked and transformed to percentiles, the 90th percentile value becomes 9.6%. Overall, extremes

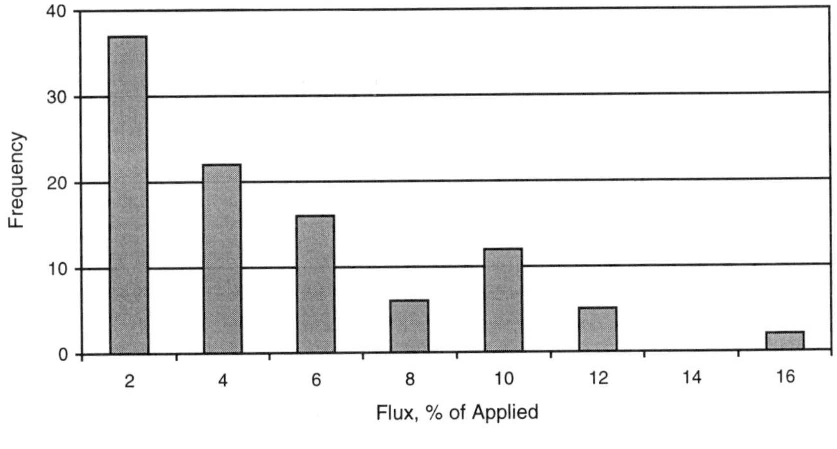

Figure 3 *Results from 100 leaching model simulations plotted as a frequency distribution*

analysis gives flux values that range from 3 % to 18 %, which is considerably more variation than the 2 to 4 % observed for central tendency analysis.

Examination of the 100 numbers as a frequency distribution plot, shown in Figure 3, suggests why there is not good agreement among the methods when extremes are evaluated. The numbers are skewed to the lower end of the distribution and are not distributed normally. Therefore, any estimators for extremes that are based on the assumption of normal distribution will not describe correctly the upper tail of the actual distribution of the data.

Data plotted as cumulative probability in Figure 4 show the entire spectrum of possibilities from uncertainty in the half-life and Koc, and help to place the uncertainty into perspective. It suggests a 50:50 chance of flux being 3.1 % or less, a 90 % chance of being less than 9.6 % and 10 % chance of being >9.6 %. Now it becomes more clear how the uncertainty in Koc and half-life impacted the assessment of leaching. This probabilistic approach that focuses on the distribution of data is more effective at recognizing uncertainty and puts its impact into better perspective than does a deterministic, single-point modeling analysis.

2.2 Observations

In this example it is seen that uncertainty in soil half-life and Koc can cause large changes in the forecast of chemical leaching through a soil profile. These changes were not described well by single modeling runs; in fact single modeling runs misled because they

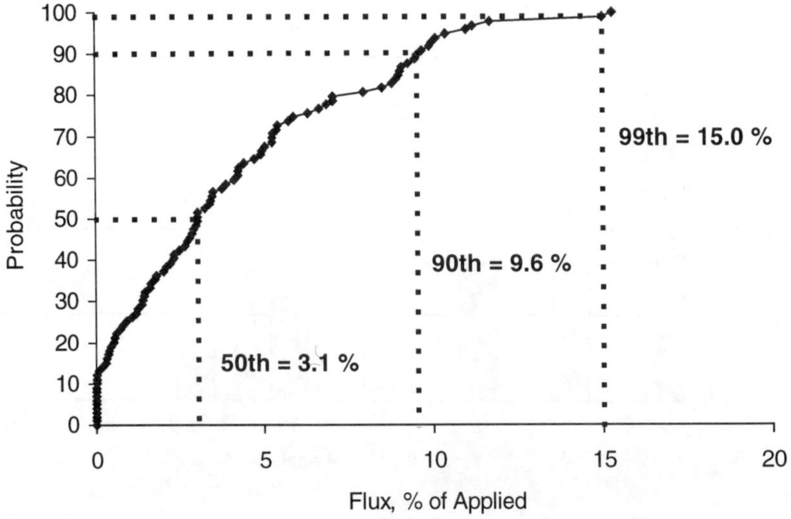

Figure 4 *Cumulative frequency plot of 100 estimations of flux beyond the root zone*

did not capture the pattern for all possible leaching results. Furthermore, the measures of leaching extremes that assume normal distribution did not describe the actual data properly because data were skewed to the lower end of the total range of possibilities. On the other hand, uncertainty in Koc and half-life could be characterized and put into perspective when a probabilistic modeling approach considering "all cases of outcome" was used to evaluate the uncertainty.

Uncertainty analysis is receiving much attention in the environmental arena and it is becoming a part of modern risk assessment methodologies. But the conduct of uncertainty analysis requires data, more data than the single-point kinds of information environmental chemists have supplied in the past. It also forces one to concentrate more heavily on the kinetic aspects of fate studies, and to evaluate how physical characteristics of the environment influence those kinetics. There must be experimentation that minimizes uncertainty through the development of correlation of such things as sorption or degradation rates with the properties of the surrounding medium so that variation as illustrated in Table 1 can be reduced through mathematical relationships. This perhaps will make the use of experimental design (such as factorial designs) a more common practice in environmental chemistry.

Uncertainty analysis requires that many modeling simulations be made (1000 or more) to delineate the real impact of model input uncertainty. Since it is necessary to do many runs, there must be tools developed to automate the modeling process for handling of input as well as output. The uncertainty exercise conducted in this paper used only 100 modeling runs, which perhaps is less than desired. But because tools for automation were not readily available for PRZM3, the process was too painful to carry out a greater number of runs. Tools for handling input and output of many runs are beginning to appear; and the two more popular ones are @RISK and Crystal Ball.[4,5] These are Monte Carlo tools that attach to Microsoft Excel or Lotus 123 spreadsheets and manage input and output in a Monte Carlo fashion for any model described in a spreadsheet setting.

Forces are at work to change the way environmental chemistry is conducted. They emanate from a shift in focus from field-scale, single-point exposure analysis to the larger more complex watershed-scale probabilistic approach that increases complexity and thus expands the range in uncertainty of exposure model input parameters. Computers have become powerful and affordable enough so that they no longer are limiting factors in the conduct of many modeling runs to construct probabilistic patterns of model output. And environmental databases such as weather and soil properties are going on-line to supply the model input needed by the probabilistic approach. Software tools such as GIS (Geographical Information Systems) to visualize this information in a comprehendible manner are also now available.

The next IUPAC Pesticide Chemistry Congress will be in the year 2002. It is of considerable interest to speculate and anticipate what will have transpired within the realm of environmental chemistry at that point in time.

References

1. 'Leaching models and EU registration. The final report of the work of the Regulatory Modeling Workgroup of FOCUS (FOrum for the Cor-ordination of pesticide fate models and their Use)', EU document 7617/VI/96.

2. 'Surface water models and EU registration of plant protection products. Final report of the work of the Regulatory Modeling Working Group on Surface Water Models of FOCUS (FOrum for the Co-ordination of pesticide fate models and their USe)', EU document 6476/VI/96.

3. T. Estes and P. Coody, 'Good modeling practices in chemical fate modeling', American Crop Protection Association WEB page, www.acpa.org/public/science_reg/misc/gmp-doc.html.

4. '@RISK. Risk analysis and simulation add-in for Microsoft Excel or Lotus 1-2-3', Windows version, 1997, Palisade Corporation. Newfield, NY, www.palisade.com.

5. 'Crystal Ball. Forecasting and risk analysis for spreadsheet users", Decisioneering Inc. Denver, CO, www.decisioneering.com.

Integrating Environmental Fate and Effects Information: The Keys to Ecotoxicological Risk Assessment for Pesticides

Keith R. Solomon

CENTRE FOR TOXICOLOGY AND DEPARTMENT OF ENVIRONMENTAL BIOLOGY, UNIVERSITY OF GUELPH, GUELPH, ON N1G 2W1, CANADA

The last 25 years have seen significant advances in analytical chemistry that have allowed more rapid, precise and accurate analysis of substances such as pesticides in environmental matrices. This increased capacity has generated fate and concentration data sets in the spatial and temporal dimensions. In this same time period, the science of environmental toxicology has grown prodigiously and our knowledge of the effects of many substances in a range of environmentally important species has multiplied. Traditional approaches to environmental risk assessment have not used these large data sets well and fail to recognize important differences between human health and environmental risk assessment. The recognition in several jurisdictions of the usefulness of distributional or probabilistic methods to characterize exposure and effects information has resulted in new approaches to ecological risk assessment that will increasingly be applied to pesticides in the environment. These approaches also provide useful planning and priority setting tools for analytical chemists seeking to maximize the utility of their data for risk assessment purposes and to focus their sampling on regions or use patterns where more data will better characterize high putative risks.

1 INTRODUCTION

Pests are those organisms perceived by humans to interfere with their own activities. This interference may have to do directly with our health, such as is the case with arthropod-borne diseases or infestation of foods with toxin-producing fungi; or with protection of our food from competition and consumption by pests during production, processing and storage; or with our own personal enjoyment and aesthetic pleasure. Pesticides are those substances, selected from nature or manufactured by humans, which are used to control and mitigate the effects of pests.

Pests have been known and described since the dawn of history and, along with pesticides, have played an important part in the development of modern society. The Greeks and Romans used pesticides to protect humans well before the birth of Christ and the Bible repeatedly refers to the ravages of pests (e.g., the plagues of Egypt [Exodus, 7] and the parable of the sower [Matthew, 13]).

The use of pesticides has brought many benefits to society. These include millions of lives saved from diseases such as malaria and yellow fever as well as an increased availability of inexpensive and high quality food to feed an ever-growing population. This increased efficiency of agricultural production has also indirectly freed up land for other uses, such as in the preservation of wildlife and habitats[1].

Against the benefits that pesticides have brought to humans, the use of these substances has also resulted in adverse effects in humans (accidental poisoning) and their environment (effects in non-target organisms). That pesticides produce both benefits and risks is well recognized and incorporated into many of the regulatory instruments, such as the US Federal Insecticide, Fungicide and Rodenticide Act, through which the sale and use of pesticides are controlled. The risks of pesticide use are borne by humans and the environment. Humans at greatest direct risk are those who handle the active ingredients during manufacture or those who apply or work with pesticides in the field. Because of smaller or negligible exposure, risks to the general public are considerably lower and, when weighed against the benefits of a high quality and varied diet, are essentially negligible[2]. This is not necessarily the case for organisms in the environment.

In crop production, the farmer's field is usually considered as a production unit that is maintained in as close to a monoculture state as possible. This is done through a variety of means, including cultural practices, selection of competitive varieties, and the use of pesticides. Because of habitat alteration, the environmental impact of agriculture, as a whole, is relatively great, however, the effects of pesticides are only a part of this total. The farmer's field is the target system for pesticides. Within the farmer's field, pesticides may have adverse effects on beneficial organisms, such as pollinators and natural enemies, that are useful or necessary parts of the agroecosystem. These within-field effects are considered in the choice of pesticides to use (or not) and are a component of many of integrated pest management (IPM) decision-making processes used today in conventional agriculture, but they are generally not regarded as effects on the environment.

Environmental concerns for pesticides are focused on their potential effects outside the agroecosystem. Because pesticides may move to non-target areas in the environment, or because organisms, such as birds and mammals, may enter the farmer's field, non-target species may be exposed to pesticides and suffer adverse effects as a result.

The presence of pesticides in non-target areas is well known and, in some cases, such as the volatile pesticides, can represent large quantities and considerable distances of transport [3,4]. Closer to the areas of use, pesticides may drift to non-target areas during application or move with water or sediment running off the field or leaching through the soil. At the same time, pesticides are subjected to a number of degradation processes (**Table 1**, from[5]) in soil, air, water, and biota that result in the production of usually less toxic or less environmentally relevant substances.

With the exception of those pesticides that bioconcentrate in organisms and biomagnify through the food chain and those that are transported to pristine environments,[3] these processes normally result in reductions in concentration. Thus, organisms exposed to pesticides outside the targeted farmer's field experience lower exposures which, depending on the inherent potency of the pesticide and on the degree to which the concentration has changed, may, or may not, cause adverse effects on non-target organisms.

This relationship between exposure and susceptibility is the key to conducting ecotoxicological risk assessments for pesticides. However, before dealing with the process of risk assessment, the framework within which ecotoxicological risk assessment takes place must be considered.

2 THE FRAMEWORK FOR ECOTOXICOLOGICAL RISK ASSESSMENT

Risk assessment provides the basis for risk-benefit and risk management decisions. As such, it is, ideally, a systematic means of quantifying, comparing and prioritizing risks that uses clear and consistent endpoints in a transparent manner that is understandable to all interested parties.

2.1 Differences Between Human and Ecotoxicological Risk Assessment

Risk assessment is not a new concept and has been routinely applied to assessing risks to humans from a number of sources. Risk assessment for human and environmental health is, in fact, mandated in a number of regulatory instruments in a number of jurisdictions. Human health risk assessment is older than ecotoxicological risk assessment and many of the procedures currently used in human health risk assessment are also used in ecotoxicological risk assessment.

Table 1 *Environmental processes, driving parameters, and matrices (from*[5]*)*

Process	*Driving parameter and primary matrix*
Transportation processes	
Volatilization	Henry's constant, surface texture (water and soil)
Movement	Wind velocity (atmosphere): Current velocity (water): Leaching (ground water): Particle and soluble transport (soil runoff)
Rain-out	Precipitation rate, sticking coefficient (atmosphere)
Sorption	Organic matter, lipid, clay content (soil, sediment)
Bioconcentration	Lipid content of organisms (biota)
Degradation processes	
Photolysis	Light intensity (water and atmosphere)
Oxidation	Oxidant concentration (water, air and soil)
Reduction	Reductant concentrations (water, air and soil)
Hydrolysis	Temperature, pH (water)
Biotransformation	Organism populations, nutrient concentrations, temperature, pH and other factors that affect the energetics of organisms (biota)

Despite the similarities between these sciences, it is important to recognize that there are significant differences between them. For many social and political reasons, human health risk assessment is based on the protection of individuals (humans) with a low degree of uncertainty. Human health risk assessment and its supporting science of toxicology involve the study of single species, whether these are the humans or the laboratory animals that act as surrogates for the humans. In contrast, ecotoxicology is the study of toxicity-mediated effects on assemblages of two or more species of organisms[6]. Unlike the situation for humans, individual organisms in the ecosystem are regarded as transitory and, because they are usually part of a food chain, are, in the ecotoxicological context, individually unimportant[5]. Thus, the objective of most ecotoxicological risk assessments is the protection of the population or the function of the population in the community. Only in the case of the protection of rare, endangered, or long-lived species are organisms in the environment afforded similar individual protection to that enjoyed by humans.

2.2 Objectives of Ecotoxicological Risks Assessment

The focus of ecotoxicological risk assessment on the functions of populations, communities and ecosystems acknowledges the fact that populations are less sensitive than their most sensitive member and, likewise, that functions of communities and ecosystems are less sensitive than their most sensitive components. This concept is well understood in ecology and has recently been the subject of experimental studies on the role of community diversity in the sustainability production of prairie ecosystems[7, 8]. Artificial manipulation of plant communities in North American Prairie ecosystems caused little change in function as the number of species was reduced from 24 to 10 and was only affected when the number of species was reduced to less than ten[7]. During this decrease in diversity, the function of the missing species was taken over by the remaining species. This phenomenon has also been observed in response to the effects of pesticides on algae and zooplankton in aquatic communities[9-12].

These observations support the concept that, in ecotoxicological risk assessment, some effects at the organism- and population-level can be tolerated, provided that these effects are restricted on the spatial and temporal scale and that keystone organisms are not adversely affected. This statement must be qualified in that some species may be affected to the point of a local extinction. However, this may not be considered ecologically relevant if the functions of these organisms (herbivory, predation or being prey) can be taken over by other organisms. Alternatively these organisms have the ability to repopulate the area from nearby unaffected populations or from special life stages (propagules) produced to allow the population to survive adverse environmental conditions, such as the drying up of an ephemeral pool. In this context, the functions of keystone organisms could not be replaced and the function of the community or ecosystem would be significantly changed in their absence. Keystone organisms should not be confused with organisms judged by society to be important. Organisms judged to be of value to society may not be essential for community or ecosystem function; however, for societal reasons they may be offered protection, sometimes, to the level normally only afforded to humans.

2.3 Direct and Indirect Effects and Keystone Organisms

The identification of keystone organisms requires a good knowledge of the ecology of the community, as their relationship to community structure may not be immediately obvious. Thus, removal of habitat (populations of other organisms) may be the root cause of the impact on the population designated for protection. Examples of this are seen in the case of the spotted owl in the Pacific NW[13] and the effects of pulp mill effluents on larval and juvenile fish in the Baltic. In the latter case, the presence of chlorate in the effluent had significant negative impacts on the populations of bladder wrack (*Fucus vesiculosus*), an aquatic macrophyte. The removal of this habitat and food supply had a significant secondary impact on crustaceans and, through the availability of food, on the fish[14].

2.4 The Frequency of Impacts

Another important issue in ecotoxicological risk assessment is the frequency at which the community is subjected to the stressor. Given that some organisms may have rapid generation times or produce large numbers of young (or both), the return frequency of an event (how often the event happens) is an important consideration in the risk assessment in relation to the ecological cost of recovery from the event[15]. In assessing risks, the return frequency protected against should be consistent with the resiliency of vulnerable populations. Low return frequencies, for example, one or fewer occurrences in thirty years may be necessary to protect some organisms but, for others, more frequent adverse events may be tolerated. For example, in phytoplankton, return frequencies of days to weeks will allow for recovery, even from high-impact events, especially if there is no persistent residue. Similarly, zooplankton and aquatic plants also would be protected by somewhat longer return frequencies, however, these would still be less than a season. In temperate regions many ecosystems undergo a period of dormancy and the system is, in a sense, reset seasonally by the winter and many organisms, possess mechanisms for propagation beyond the winter reset.

3 RISK ASSESSMENT, THE HAZARD QUOTIENT

The first real tier in the risk characterization process is the use of hazard quotients. These are simple ratios of exposure and effects and may be used to express hazard or relative safety. For example:

$$Hazard \approx \frac{Exposure\ concentration}{Effect\ concentration} \qquad (1)$$

or

$$Margin\ of\ Safety \approx \frac{Effect\ concentration}{Exposure\ concentration} \qquad (2)$$

The calculation of quotients traditionally has been conducted by utilizing the susceptibility of the most sensitive organism or group of organisms and comparing this to the greatest exposure concentration. This may be made more conservative by the use of an uncertainty (safety or application) factor[16] such as division of the effect concentration by a number >1. This is done to allow for unquantified uncertainty in the effect and exposure estimations or measurements; however, it does not factor in that worst-case exposure and effects measurements have likely been used to calculate the quotient.

Traditionally, risk assessors have used conservative assumptions rather than actually estimating or measure uncertainty. Worst case assumptions reduce the likelihood of a false negative (identifying a situation as having acceptable risk, when, in fact, it does not) and could be supported when the cost of error is great. However, as has been pointed out,[5] worst-case scenarios may not be multiplicative or additive. For example, high rates of rainfall (runoff) of a pesticide would seldom be accompanied by high rates of photoactivation. Worse case scenarios are inconsistent as it is always possible to conceive of a still worst case. Worst cases do not consider the probability of occurrence of a particular event; for instance, the worst case number in five measurements is very different from the worst case in a thousand measurements. Conservative assumptions are based on the axiom that there are no environmental costs resulting from the regulation of false positives. In fact, in many cases, risk decisions based on over-conservative assumptions result in the transfer of the risk to another medium or to another stressor. While this is certainly true for ecotoxicological risk assessment, the same situations apply in human health risk assessment[17].

The quotient approach is designed to be protective, not predictive, and is thus better used for the early tiers or preliminary steps of risk assessments where it can be employed to exclude substances that truly do not represent a risk for further consideration.

4 RISK ASSESSMENT, THE PROBABILISTIC APPROACH

By considering the probability of occurrence of an event (presence of a high exposure or a very sensitive species), risk assessment may be placed into a truly probabilistic framework (**Figure 1**).

The probability of occurrence of a particular event is, and has been, widely used in the characterization of risk from many physical events in human society. Common examples include those in the insurance industry and protection against failure in mechanical and civil engineering projects. Probabilistic approaches have also been suggested for establishing thresholds of concern for human health risk assessment[18]. This concept has been applied in ecotoxicological risk assessment for the characterization of both exposures and effects; however, this is being done with some qualifiers. These qualifiers relate to return frequencies of events and the ecological costs of replacement/recovery[15] in the same way as engineers constructing a high-cost bridge would opt for risk protection against rare flooding events such as the 100-year return flood level. Conservative approaches to ecotoxicological risk assessment may use low frequencies of return (for example, one-in-thirty years) to ensure protection of all organisms in situations where knowledge of mode of action or sensitivity of species is limited [19]. However, where more information is available, more realistic return frequencies may be used. The herbicide atrazine is a reversible inhibitor of photosynthesis in plants. Atrazine in surface waters primarily affects phytoplankton and aquatic plants, organisms with short life cycles and high rates of reproduction. Thus, return frequencies of the order of months will allow for recovery, even from high-exposure, high-impact events[11]. Similarly, zooplankton would also be protected by short return frequencies. Protection of longer lived species such as some fish may require consideration of return frequencies of several years but even these may be conservative because of repopulation from unexposed refugia. The example of the more rapid than expected recovery of the biota in the River Rhine from an endosulfan spill is a case in point[20].

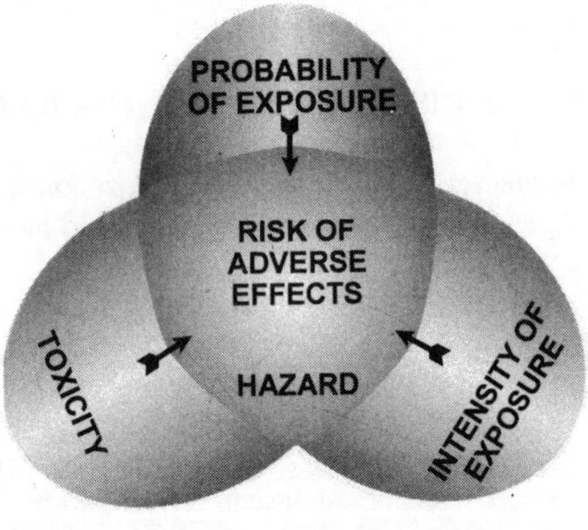

Figure 1 *Diagram illustrating the use of toxicity and intensity and probability of exposure into risk assessment*

4.1 Distribution Models

As discussed above, concentrations of substances in the environment can be affected by a large number of processes that relate to the amount released, the spatial and temporal distributions of the releases, and the results of the action of a large number of transportation and transformation processes (fate processes) on the substance. The likelihood and extent to which these myriad fate processes will affect a particular quantity of substance in the environment are essentially random and frequency distributions of exposure concentrations in the environment will therefore most likely be distributed according to a particular model[21,22] with the log-normal model being that most often observed [11, 19, 23, 24]. The same observations of log-normality apply to distributions of toxicological data. Many of the reactions through which fate and transformation processes, as well as toxicity mechanisms, are mediated are first-order or pseudo first-order. The logic for the log-normal model is supported by this knowledge. However, on occasions, data may be better described by other distributions or by distribution-free approaches.

The principle of the probabilistic approach (**Figure 2**) has been described before [11,23-26] and, when the distributions of exposure and toxicity are expressed as linearized cumulative distributions (**Figure 2**) exceedence profiles may be plotted as illustrated in **Figure 2**. The probabilistic approach can be applied in any situation where it is possible to express the exposure dose or the exposure concentration in a probabilistic manner. It has been most commonly applied to exposure concentrations in water but may also be applied to doses of pesticide consumed in food by birds and mammals.

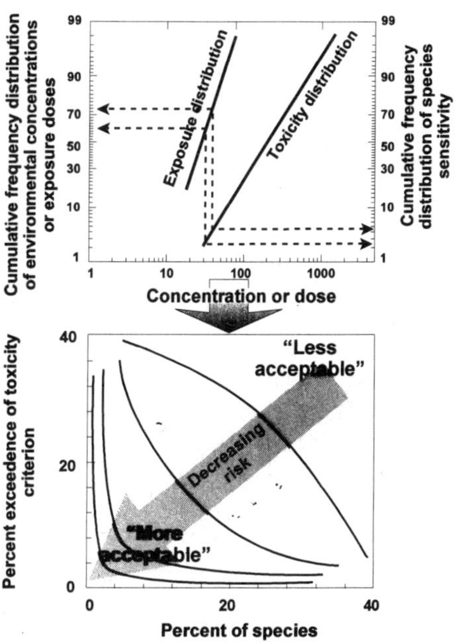

Figure 2 *Illustration of a linearize cumulative frequency plot of exposure and toxicity data (above) and the exceedence profile that can be derived from the relationship between the two distributions (below)*

4.2 Limitations to the Approach

There are some limitations to this approach [19, 27]. For example, the choice of protection level (e.g., 90% of species) may not be socially acceptable. However, the use of the exceedence profile (**Figure 2**) approach allows the significance of multiple protection criteria to be assessed at the same time as well as the rate at which the probabilities change with changes in exposure dose or concentration. It should be recognized that the probabilistic approach is a purely numerical method. It cannot identify keystone species or species that have social value. This remains the task of the ecologist and the ecotoxicologist and will probably require broader knowledge of ecology and ecotoxicological principles than is currently found amongst regulators[15].

Risks of persistent, bioaccumulative substances to species at the top of the food chain may not be sufficiently considered by this approach alone because toxicity tests used for developing the sensitivity distribution do not allow the maximal exposures that may occur through food chain exposure in the real environment. However, in most regulatory procedures, these substances would be identified through physical properties (K_{OW}) or specific tests for bioaccumulation[19].

The probabilistic approach to ecotoxicological risk assessment is data intensive and requires more data than the traditional approach[19]. For acute tests, the costs of generating the additional data are, relatively speaking, low, but not insignificant. However, these costs may be compensated for by reduced testing requirements in other areas, such as full life-cycle tests[19].

4.3 Analysis of Toxicity Data

In the probabilistic method, data are analysed as a distribution on the assumption that the data represent the species in the environment where the exposures occur. Obviously, it is neither practical nor possible to test all the species in an ecosystem and, for this reason, the distribution is used as an approximation or surrogate for all the species. In principle, this is not different from using a single species as a surrogate for all species (as in the quotient method) but it does at least consider the range of sensitivity that may exist across species for similar groups (insects and Crustacea) or those from very different groups (fish and plants).

In the case of pesticides, additional information may be available to allow more focused analysis of toxicity data. Many pesticides are designed to be selective in their toxicity and would thus be expected to be more toxic to species closely related to the target organisms than the non-target organism.

For example, herbicides may be selectively toxic to some groups of plants (weeds versus corn) as well as less toxic to animals and other organisms that do not possess the receptor system (say photosynthesis) for the pesticide. In the example of atrazine illustrated in **Figure 3**, fish are clearly less sensitive, as a group, than plants. When these organisms of inherently differing sensitivity were grouped together, the resulting distribution did not display a good fit to the log-normal model. However, when data were segregated into groups such as fish and plants, the data better fitted the log-normal model (**Figure 3**). Similarly, an insecticide that acts on the nervous system of insects is unlikely to be highly toxic to plants. However, organisms with well developed nervous systems (insects, other arthropods and vertebrates) are likely to be more sensitive. Specificity of action may not always be the case. For example some biocides, such as the chlorophenols, are similarly toxic to a wide range of organisms[28] — hence their use as biocides — and the grouping of all organisms together for distributional analysis may be appropriate. Thus, from a basic understanding of the mechanism of action of a pesticide, and from the toxicity data, it may be possible to identify and group sensitive organisms. Given a range of exposure concentrations in the environment, these organisms are the most likely to be adversely affected. This is helpful from the point of view of risk assessment as it allows the assessor to focus on the groups at greater direct risk and to devote less time and resources to groups that are exposed to very low or negligible direct risks. In addition, with a knowledge of the ecology of the potentially impacted system, it is possible to assess the likelihood that indirect effects will occur as a result of an effect on keystone groups of predator or prey/food organisms.

Using knowledge of food webs in risk assessment has been proposed as an adjunct approach to the assessment of ecotoxicological risks[29]. However, this approach requires detailed knowledge of the actual and the possible interactions in community, including redundancy of function, and the existence of good models.

While the mechanism of action of the pesticide is an important criterion for grouping of organisms, habitat may also be important. For example, there may also be good mechanistic reasons to separate data for freshwater (FW) and saltwater (SW) organisms where it is known that one group has an inherently different sensitivity because of interactions between salinity and the pesticide of concern[30].

It is also possible to group organisms together on the basis of their reproductive strategy and life cycle. Thus, organisms which are able to recover rapidly from an adverse effect at the population level (reduction in population caused by mortality) may be considered differently from another group of organisms that may require a longer period of recovery. This may also be important when deciding how the exposure data should be analyzed, for example whether the data are analyzed with return frequencies of weeks, months, or years.

4.4 Assessing Risks from Mixtures

The probabilistic approach can also be applied to mixtures of pesticides where it is known that the substances act in an additive manner, such as the organophosphorus insecticides. The likelihood of these interactions occurring can be assessed if the concentrations of the substances are normalized to the concentration of the most toxic substance present in the mixture and are expressed as toxicity equivalents or toxic units (TEs or TUs). TEs and TUs are usually calculated from the response of a single species to the range of substances being studied. In the case of the probabilistic approach to describing a toxicity profile, the point upon which the normalization could be based is the 10th or some other centile of the toxicity distribution for all organisms or, where sufficient data are available, for the groups of organisms relevant to the assessment. Then, for each sampling time, the concentrations of the pesticides present in the environment (determined from analyses) are converted to TEs and added together to give the total TE. These TEs can then be assessed probabilistically against the toxicity distribution of the substance to which the potencies of the components of the mixture were normalized. The advantage of this procedure is that it considers not only the probability that a particular concentration of a pesticide or substance will be found but that it will be present with other substances at certain concentrations.

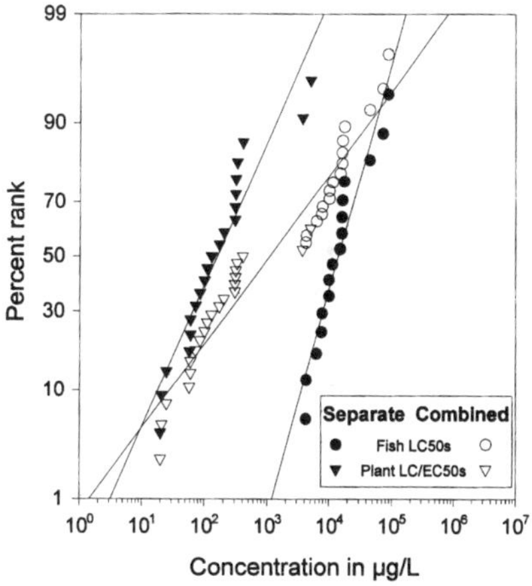

Figure 3 *Illustration of the distribution of toxicity values for atrazine in fish and plants analyzed separately and combined into a single data set (data from[11])*

An illustration of this is the example of atrazine and its metabolites. Hydroxylated and other atrazine metabolites were measured in Goodwater Creek, a small stream in the claypan soil regions of NE Missouri, USA[31]. Using toxicity data for some of the metabolites of atrazine,[32] TEs were calculated (**Table 2**). The concentrations of atrazine and metabolites on each day of measurement were multiplied by the TE and the TEs added together. These concentrations were plotted as a cumulative frequency distribution and compared to several probabilistic and deterministic measures of toxicity derived previously[11].

Table 2 *Toxicity of some atrazine metabolites measured in the green alga Chlorella pyrenoidosa*

Metabolite	*TEs*	*Source*
Atrazine	1.00	32
Hydroxy atrazine	1.00	Assumed
DEA	0.28	32
DIA	0.14	32
DEHA	0.28	Assumed
DIHA	0.14	Assumed

As might be expected, (**Figure 4**) the metabolites of atrazine contributed little additional toxicity to atrazine, except at low concentrations of atrazine. In neither case did the distribution exceed any of the measures of toxicity.

5 PROBABILISTIC RISK ASSESSMENT AND PESTICIDE RESIDUE CHEMISTRY

Increased application of probabilistic approaches to ecotoxicological risk assessment will, in all likelihood, require some changes in the way that pesticide residue chemistry is conducted.

5.1 Focusing Analytical Programs

Probabilistic risk assessment is more data intensive than other approaches and it is logical to suppose that this will increase the demand for analyses of pesticide residues in environmental samples. However, this need not be the case. Using existing distributions of exposure data collected from different sites and over different times and comparing these to toxicological responses allows identification of sites or times of the year/season when risks would be greater than others. These sites or use scenarios can then be ranked and the higher risk sites/scenarios identified by simple ranking of exceedence probabilities, as was done with atrazine[11]. These sites can then become the focus for more sampling and residue analysis to better characterize the risks. Alternatively the characteristics of these sites can be used to identify similar situations that may not have been studied. Sampling at sites with lower risks can be eliminated or reduced to better focus analytical resources on those sites with greater risks. The total number of analyses conducted would not necessarily increase, but the resources would be better focused.

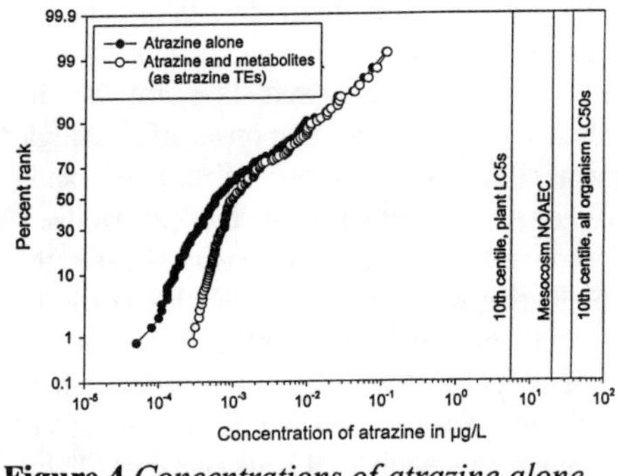

Figure 4 *Concentrations of atrazine alone and atrazine and its metabolites (expressed at atrazine TEs) in Goodwater creek*

Sampling protocols could be better designed to collect appropriate data for risk assessment. Traditionally, sampling of pesticide residues in the environment has been focused on misuse (spills, accidental contamination, etc.) or on testing hypotheses related to storm events, runoff and hydrology. While these data are useful for their designated purpose, they are less useful for probabilistic analysis. Truly random sampling of sites in the use environment may give a more realistic picture of the likelihood of any of these being contaminated. Time-varied sampling is inappropriate for use in probabilistic risk assessment as more closely spaced samples will contribute more weight to the distribution and less closely spaced samples. Although time-weighted-mean-concentrations can be calculated using an interval of time chosen from a knowledge of the exposures used in the toxicology testing, sampling at regular time intervals is more useful for proper description of the distribution. This is particularly relevant when the nature of variable flow rates and hydraulic events in rivers and streams results in exposure concentrations that vary over time.

Knowledge of the mechanism of action of the substance may also be useful in the selection of sampling procedures. For example, if it is known that the mechanism of action of the pesticide is fully reversible (competitive inhibition of an enzyme), then the average concentration that the organism is exposed to is a better descriptor and a time-integrating sample would be appropriate. For a substance that has little or no reversibility of action (covalent binding to an enzyme with a long reactivation/synthesis time), the peak exposure concentration may be more important and instantaneous samples would be the best approach.

Some substances have short degradation times or are rapidly sequestered into biologically unavailable forms. Knowledge of the length of exposures as they occur in the receiving environment may also be used to design appropriate toxicity testing procedures. This is particularly useful where environmental exposures shorter than those used in traditional toxicity testing (48 h, 96 h, 21 d, etc.) are found[19].

5.2 Modeling Exposures

Where well established pesticides have been used for a number of years, measurements of environmental concentrations may be relatively easy to come by. However, for substances not yet on in the market or in the process of registration, measurement of exposures may be impossible. In these cases, modeling is the only way to estimate these concentrations. This requires that suitable models for pesticide movement and dissipation be available and that these be run in a probabilistic manner. Monte Carlo simulations, using off-the-shelf software packages, can be employed for this purpose but they require that the range and distributions of the input values be well parameterized. While this has implications for the choice of appropriate weather and environmental conditions, it also requires that distributions of degradation rates in various matrices and other physicochemical properties of the substance be better characterized than at present. Better characterization will require more study and under more varied conditions. Studies to validate models also will be needed from the environmental chemist.

6 CONCLUSIONS

The greater use of probabilistic approaches in ecotoxicological risk assessment will require some changes in the way regulatory decisions are made and also in the way that environmental chemistry data are collected and used. Environmental chemists will have to cooperate more with ecotoxicologists both when measuring routine physicochemical properties of substances and when measuring environmental concentrations. The result of this cooperation will be a better understanding of the processes that determine the fate and the effects of pesticides in the ecosystem and better data to bring to bear on the questions posed by ecotoxicological risk assessment.

7 REFERENCES

1. R.D. Knutson, C.R. Taylor, J.B. Penson and E.G. Smith, Economic Impacts of Reduced Chemical Use, 1990. Knutson and Associates, College Station, TX. May 1990.

2. L. Ritter, for the *Ad Hoc* Committee on Pesticides and Cancer, 1997, *Cancer*, **80**, 2019.

3. T.E. Bidleman and D.C.G. Muir, *Chemosphere,* 1993, **27**, 1825.

4. L. Ritter, K.R. Solomon, J Forget M. Stemeroff and M. O'Leary, *Persistent organic pollutants: An assessment report on aldrin, chlordane, DDT, dieldrin, dioxins and furans, endrin, heptachlor, hexachlorobenzene, mirex, polychlorinated biphenyls, and toxaphene.* 1995 IPCS Interorganization Programme for the Sound Management of Chemicals. UNEP, ILO, FAO, UNIDO and OECD. WHO, Geneva November 1995, 48 pp.

5. G. Suter II, L.W. Barnthouse S.M. Bartell,T. Mill, D. Mackay and S. Patterson, *Ecological risk assessment*, 1993 Lewis Publishers, Boca Raton, 1993,538 pp.

6. J. Cairns Jr., *Environ. Toxicol. Chem.* 1989, **8**, 843.

7. D. Tillman, *Ecology*, 1996, **77**, 350.

8. D. Tillman, D. Wedlin and J. Knops, *Nature*, 1996, **379**, 718.

9. N.K. Kaushik, G.L. Stephenson, K.R. Solomon and K.E. Day, *Can. J. Fish. Aquat. Sci.*, 1985. **42**, 77.

10. G.L. Stephenson, N.K. Kaushik, K.R. Solomon, K.E. Day and P. Hamilton, *Environ. Toxicol. Chem.,* 1986, **5,** 587.

11. K.R. Solomon, D.B. Baker, P. Richards, K.R. Dixon, S.J. Klaine, T.W. La Point, R.J. Kendall J.M. Giddings J.P. Giesy L.W. Hall Jr., C.P. Weisskopf and M. Williams, *Environ. Toxicol. Chem.*, 1996, **15**, 31.

12. J.M. Giddings, R.C. Biever and K.D. Racke, *Environ. Toxicol. Chem.*, 1997, **16**, 2353.

13. F.C. James, Bull. Ecol. Soc. Amer.,1994, **June, 1994**, 69.

14. K-J. Lehtinen, B. Axelsson, K. Kringstad and L. Strömberg, *Nordic Pulp Paper Res. J.*, 1991. **2**, 81.

15. K.R. Solomon, *Risk Anal.*, 1996, **16**, 627.

16. CWQG, *Canadian Water Quality Guidelines and updates.* 1987 Task Force on Water Quality Guidelines of the Canadian Council of Resource and Environment Ministers, Ottawa, ON, 1987.

17. J.D. Graham and J.B. Wiener, (Eds), *Risk Versus Risk: Tradeoffs in Protecting Health and the Environment.*, 1997, Harvard University Press, Cambridge MA, 337 pp.

18. I.C. Munro R.A. Ford E. Kennepohl and J.G. Sprenger, *Food Chem. Toxicol.*, 1996, **34**, 829.

19. SETAC *Pesticide Risk and Mitigation. Final Report of the Aquatic Risk Assessment and Mitigation Dialog Group*, 1994, Society of Environmental Toxicology and Chemistry Foundation for Environmental Education, Pensacola, FL, 220 pp.

20. H.L. Friege, 1986, *Organic Micropollution in the Aquatic Environment* (CEC 5th European Sym, Rome, Oct 20-23) pp 132.

21. E.A. McBean and F.A. Rovers, *Ground Water Monitor. Rev.*, 1992, **12**, 115.

22. C.D. Carrington, *Human Ecol. Risk Assess.,* 1996. **2**, 62.

23. S.J. Klaine, G.P. Cobb, R.L. Dickerson, K.R. Dixon, R.J. Kendall E.E. Smith and K.R. Solomon, *Environ. Toxicol. Chem.*, 1996, **15**, 21.

24. K.R. Solomon and M.J. Chappel, *Triazine Herbicides: Ecological Risk Assessment in Surface Waters.* In: *Triazine Risk Assessment.* (Eds. L.G. Ballantine, J.E.McFarland and D.S. Hackett. ACS Symposium Series 683. 1998 American Chemical Society, Washington D.C. pp 357.

25. R.D. Cardwell, B.R Parkhurst, W. Warren-Hicks and J.S. Volosin, *Water Environ. Technol.* 1993, **5**, 47.

26. B.R. Parkhurst, W. Warren-Hicks, T.Etchison, J.B. Butcher R.D. Cardwell and J.S. Volosin, *Methodology for Aquatic Ecological Risk Assessment.* 1995. Final Report prepared for the Water Environment Research Foundation, Alexandria, VA. RP91-AER.

27. Health Council of the Netherlands. *Network,* 1993. **6(3)/7**(1), 8.

28. K. Liber, N.K. Kaushik, K.R. Solomon and J.H. Carey, *Env. Toxicol. Chem.*, 1992 **11**, 61.

29. Health Council of the Netherlands: *Standing Committee on Ecotoxicology. The food web approach in ecotoxicological risk assessment.* 1997. Rijswijk: Health Council of the Netherlands publication No. 1997/14E, 75 pp.

30. L.W. Hall and R.D. Anderson, *Crit. Rev. Toxicol.,* 1995, **25**, 281.

31. R.N. Lerch, W.W. Donald, L. Yong-Xi and E.E. Alberts, *Env. Sci. Technol.*, 1995, **29**, 259.

32. G.W. Stratton, *Arch. Environ. Contam. Toxicol.*, 1984, **13**, 35.

Environmental Fate, A Down Under Perspective

Jack Holland

MANAGER, RISK ASSESSMENT AND POLICY SECTION, ENVIRONMENT AUSTRALIA, BOX E305, KINGSTON, ACT 2604, AUSTRALIA

Abstract

The Risk Assessment and Policy Section of Environment Australia undertakes the environmental assessments of agricultural and veterinary chemicals on behalf of all the States and Territories of Australia. A broad perspective of the likely environmental impact of agvet chemical use in Australia has been obtained in the 12 years of performing this work, and the paper describes some of these experiences. The particular characteristics of the Australian environment that might lead to a chemical having a different fate on this continent are discussed, and the areas of highest intensity of pesticide use eg cotton, sugar cane, rice, horticulture and any particular sensitive areas adjacent to them identified.

Australia has both temperate and tropical parts, as well as very arid zones. What is known about differences in the fate of chemicals in tropical/temperate compared with colder regions is discussed and compared/contrasted briefly with what is known about the differences in ecotoxicity for these same regions. While Environment Australia has generally accepted Northern Hemisphere fate/toxicity data for our assessments, we have been more inclined to ask for locally generated fate rather than toxicity data, particularly if the pesticide use conditions are likely to be rather unique - eg in low organic matter or dry soils, or where differences in irrigation practices occur. This is illustrated by several examples of large scale environmental fate trials that have been/are being conducted in Australia, namely:

- The program examining the environmental fate of pesticides, particularly endosulfan, used in the cotton industry, which has generated some "real world" cotton field fate data; and

- The monitoring program set up to show that use of atrazine in forestry under reduced application rates does not lead to unacceptable contamination of the environment from this chemical.

In spite of this work and the 12 years experience in pre-registration assessments, there is still a major knowledge gap about the impact on the Australian environment resulting from the use of pesticides on the market, including how this may compare with predictions from the assessment. Environment Australia recently commissioned a consultant to investigate the extent of environmental contamination from pesticide use. The findings of the report, namely that there is little systematic investigation into the contamination of Australian soils, sediments, biota and waters (with some notable exceptions), and also little routine investigation of wildlife mortality incidents are discussed. One of the key recommendations is the need to have better information on pesticide use, ie what is used where and how much, to allow better policy making in this area. Some ongoing work in this area both in Australia and overseas is briefly mentioned.

1 INTRODUCTION

Along with those of many other developed nations, Australian registration processes for agricultural and veterinary chemicals have evolved significantly over the past two decades. The Department of Environment's formal involvement began in 1986 (under a different name and long before we were known as Environment Australia), when we were invited to join the voluntary system operating at the time between the Commonwealth (which did the assessments and provided "National Clearance") and the six States (which individually registered the chemicals), in order to fill a perceived gap ie the assessment of the potential impact of agvet chemicals on the non-human environment. Worksafe Australia was invited to join at the same time to cover occupational health aspects.

This voluntary system, which had operated for many years through an "informal agreement", was underpinned by legislation several years later. However, during the mid 1990s a more radical change occurred when after a couple of transitional years a truly national registration system (ie where the Commonwealth has responsibility for registering - two Territories had become involved in the interim) commenced on 15 March 1995.

This is the system that operates today on a 100% cost recovered basis from the agvet chemical industry. The National Registration Scheme is administered by the National Registration Authority for Agricultural and Veterinary Chemicals (NRA). Along with Worksafe Australia (mentioned above) and the Department of Health and Family Services, which assesses the public health aspects, Environment Australia evaluates the potential impact of these chemicals on the environment, and provides this advice to the NRA, with which we have a formal agreement to undertake this work on a fully cost recovered basis.

2 THE RISK ASSESSMENT AND POLICY SECTION'S EVOLVING ROLE

I have been part of the Section which has performed this function since the beginning, for the past 8 years as acting manager or manager, and in fact carried out the very first assessments. In this time these have become increasingly sophisticated as the art of environmental hazard/risk methodology grows (though we are by no means high tech) and the reports have become increasingly large, partly in response to the much greater environmental data packages we receive nowadays (this has increased from a maximum of several bound volumes to several boxes full for a large submission). The Section has also grown, though it is very small by comparison with most other developed countries. It is important to note that we conduct environmental impact assessment of both agricultural and veterinary chemicals, and as far as I know this is unique. Further the Section also conducts these assessments for industrial chemicals under Australia's regulatory system for these which commenced in 1990. In this way we have accumulated a very broad understanding of the likely impact of a wide range of chemicals on the Australian environment.

By necessity we had to adopt some priorities for agvet chemicals assessments when we began as a very small Section back in 1986, which were in the following order:

- all new active ingredients;
- cases where the proposed extension in use was likely to lead to a significant increase in the extent of environmental exposure; and
- where a new inactive constituent was intended to be used.

In the 12 years since we became involved we have assessed well over 100 new active constituents (as well as many further extensions of these new chemicals), but a much smaller number of the second kind (for example, extension of the registered use of abamectin from treating cattle internal parasites to application by air as a miticide on cotton) and very few new inactive ingredients (limited to a couple of herbicide "safeners" and some surfactants in glyphosate formulations when used to kill aquatic weeds). We have also used the second priority on occasions to undertake an assessment of chemicals where we have become aware of some environmental concerns either in the literature or after having been alerted by the public. However, it was not really until the NRA's Existing Chemicals Review Program (ECRP) began in the mid 1990s that we became part of a systematic review of the older chemicals which were put on the market well before we became involved in environmental hazard assessments back in 1986.

3 ASSESSMENT METHODOLOGY AND EXPERIENCE

We have adopted the same approach/methodology in conducting the assessments of these different types of chemicals, with a heavy emphasis on the extent of environmental exposure, and concentrating our assessment effort on those compartments where exposure is likely to be highest. We also use the traditional quotient approach to determine hazard, ie starting from the worst case and conducting a series of more refined iterations should the initial estimate indicate that an unacceptable hazard may occur.

For example for hazard to aquatic organisms we initially assume a direct overspray at the maximum rate to a shallow (15 cm deep) water body. If Q is < 0.1 based on the most sensitive fish, aquatic invertebrate, algal or aquatic macrophyte result, no further refinement is necessary and the assessment for this aspect stops. If Q is > 0.1, we move to the next level and assume 10% spray drift. If Q is still > 0.1 we assume a much more realistic figure for spray drift either from the literature or results obtained in Australia, such as from the program aimed at Minimising the Impact of Chemicals used in Cotton Production described below. This would also depend on the method of application (eg aerial, orchard or boom spraying). The potential for hazard from run off is often also considered at this time, followed by (if needed) other mitigating factors such as what is known about the crop situation (eg likely distance that water occurs from the crop boundary) and the rate of the chemical's degradation in water/dissipation to sediment. The latter is performed on a case by case basis depending on the information submitted on the properties and likely fate of the chemical. Where available (for example in our ECRP assessment of endosulfan - see below) environmental monitoring data are used. Industry sometimes includes environmental modelling details in their submissions, which we assess, though we do not have our own "in house" modelling capabilities.

While the main routes for environmental exposures may be different for these classes of chemicals, with veterinary drugs usually reaching the environment through excretion from the animal to which it is applied, and industrial chemicals either mainly reaching the environment through discharge in effluents to water, or by disposal to a landfill, this has generally worked well. It has meant that we have needed to be very flexible in our approach to data requirements as the extent required to be able to conduct an adequate assessment can vary widely, even between an agricultural chemical and a veterinary drug. An example is a peptide hormone which breaks down to its constituent amino acids when used to inject individual animals and where little, if any, environmental data would be required, compared with an insecticide proposed to be repeatedly sprayed from an aeroplane onto cotton, where a substantial package of data would be expected.

4 USE OF LOCAL VERSUS OVERSEAS DATA

In conducting these assessments, we have generally relied on the package of overseas data. Most people would agree that it could not be justified by reasons of duplication, let alone costs, that a new package of data should be generated for Australia as has been raised by some parties from time to time. Furthermore, in a number of States laws forbid or strongly discourage testing on local mammals and birds, and to a lesser extent, fish. However, ecotoxicological test methods using local species (generally aquatic invertebrates but also some fish species) have been, or are being, developed through various programs sponsored by government, industry and academia. Examples are the National Pulp Mills Research Program, effects of uranium and other toxicants on local aquatic fauna by the

Environmental Research Institute of the Supervising Scientist (ERISS), revision of the Australian and New Zealand Environment and Conservation Council's (ANZECC - which comprises the environment Ministers of the Commonwealth and each State and Territory) water quality guidelines.

Work in this area, initially funded by ANZECC (Johnston *et al.*, 1990) in the late 1980s and corroborated since that time, indicate that Australian species generally do not differ greatly in their sensitivity to a broad range of toxicants. Some unique situations might arise, however, due to differences in the physico-chemical characteristics of the water (eg the extremely soft waters of tropical northern Australia might exacerbate heavy metal toxicity when compared to the moderate to hard waters used in standard overseas test protocols). Higher water temperatures, which tend to increase toxicity, are also important.

Our experience with the assessment of chemicals indicates that generation of local toxicity data is generally not the issue. However, one area where we have requested additional or locally generated data on a number of occasions is in the exposure/hazard considerations. An example is a recent study we required as part of the registration of a new insecticide to be applied to pre-germinated rice, which is then aerially sown. We needed evidence to indicate that rice sown in this way was not consumed by birds either when it had landed on the banks or in the paddy itself. Luckily the insecticide had a rather favourable bird toxicity profile, being highly toxic only to Galliformes, and the registrant was able to demonstrate that very few birds in this family were likely to occur in the bare areas pertaining when rice is first sown.

Another difficulty we have encountered lies in drawing conclusions on fate in the Australian environment based on overseas (laboratory and field) data and we have requested the generation of a number of trials examining the persistence and movement of chemicals under local conditions.

5 THE AUSTRALIAN ENVIRONMENT AND AGRICULTURAL PRACTICES

Before we discuss a few examples in more detail it is worth examining why this may be the case. Australia is an island continent (see Figure 1) spanning from tropical areas close to the equator from about 11^{0}S to well below 44^{0}S in Tasmania, and stretching from aproximately 113^0 E to 153.5^{0}E. It contains both temperate and tropical parts, as well as very arid zones.

Australia is generally regarded as a relatively low user of chemicals compared with other OECD countries. What is clear from Figure 1 is that there are relatively few areas where intensive agvet chemical use is likely to occur, and that these are often close to the coast, or concentrated on our major river systems. For example, sugar cane is grown down the Queensland (QLD) and northern New South Wales (NSW) coasts, the upper half of which is in close proximity to the Great Barrier Reef, a well known World Heritage area. Cotton is mainly grown in north west NSW and stretching into QLD on the upper regions of the Murray Darling Basin, our major river system which drains about one seventh of the continent. On the other hand rice production is expanding rapidly in the irrigation areas of southern NSW, on the middle regions of the lower half of the Murray Darling Basin. Horticulture is scattered round the coast or again on the major river systems, and cereals are grown in a large crescent shape around the bottom half of the continent. What is also clear from Figure 1 is that there are very large areas where there are no farms, or sheep or beef cattle only are raised, and even lower levels of pesticide are used, generally restricted to those for animal health and locust control.

While the above considerations narrow down the range of possible locations that need to be considered, they still extend across the continent, albeit largely comprised of the coastal or near coastal region. One obvious fact is that much of Australia is hotter than Europe, or even the US, though as can be seen from Figure 1 areas of likely intensive pesticide use in tropical regions are relatively few, and largely confined to sugar growing on the east coast of QLD. However, other areas, in particular the cereal belt, are hot as well and can also be very dry as drought can occur anywhere.

Figure 1

Predominant Agricultural Landuse in Australia

Source: Aquatech (1997)

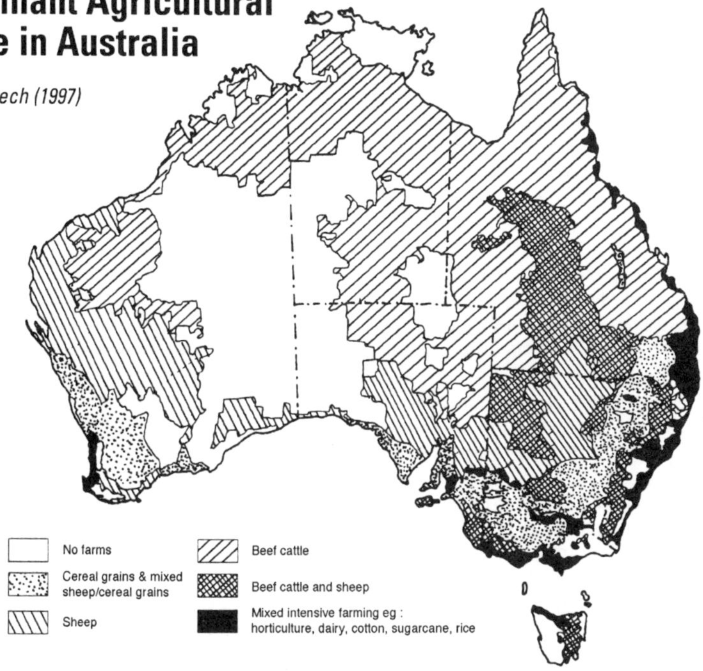

6 CONSIDERATION OF FACTORS LIKELY TO INFLUENCE PESTICIDE FATE IN THE AUSTRALIAN ENVIRONMENT

While there is literature support for significantly more rapid pesticide dissipation from tropical (and we assume by extrapolation sub-tropical soils) than under temperate conditions (Racke *et al.*, 1997), largely as a result of enhanced volatility and microbial degradation, soils are often dry which in our experience significantly delays, or even stops, degradation. Unfortunately there appear to be very few, if any, local studies that have specifically examined this aspect, and we must rely on the wider literature (eg Walker, 1991). However, we should note that the literature records that in some instances degradation proceeds more rapidly under dry soil conditions as with surface catalysed hydrolysis of organophosphate insecticides (Racke *et al.*, 1996).

One element that Australia has in abundance compared with Europe is sunlight, and industry has often tried to claim that aqueous photolysis is likely to be enhanced and to be a significant potential pathway for degradation of their chemical in Australia. Our fairly standard response has been to point out that compared with other countries Australian waters are often very turbid, which limits light penetration and is therefore likely to counteract this proposition. Unfortunately we have limited data on which to draw a definitive viewpoint, and are unaware of any work in Australia which may be investigating these aspects, as well as of any local work on the role of suspended solids/humic acid in acting as a photosensitiser.

One aspect that has come up in several recent pesticide assessments is the possibility of photolysis on soil being a major degradation pathway for pesticides. This has not accorded with our experience in assessing a large number of agvet chemicals as in the laboratory photolysis on soils seems to be significantly slower than in water in the vast majority of cases. Until recently our understanding has been that soil photolysis could only take place with that material that is present in the interstitial water (solution) phase.

However, our attention has recently been drawn to a recent IUPAC publication summarising the literature in this area (Racke *et al.*, 1997), which while noting that in the past photolytic degradation was not considered an important mechanism of pesticide loss from soil, gives some literature examples of where photo-induced transformations appears to have occurred. A number of different mechanisms have been advanced to explain this phenomenon, including one study raising the possibility of photo-induced oxidising species produced from the organic fractions of soil. On the other hand another study demonstrated that clay minerals seemed to be the dominant feature in catalysing oxidation reaction. Two studies are reported to indicate more rapid photodegradation of pesticides on moist versus dry soil surfaces. In contrast to this evidence other work has reported the apparent quenching effects of humic and mineral materials in soil.

It is clear that further investigation is required in this area, and we agree with Klöpffer (1993) who has highlighted the limits and gaps in knowledge of photochemical degradation testing: "Photochemistry in the adsorbed state (soils) clearly needs a better scientific foundation before a meaningful testing strategy can be developed." I hope that the Poster sessions in the Environmental Fate subtopic of this Congress will contain some work in this area.

Another feature of Australian soils is that many are very low in organic matter. These are often very sandy and either acidic (Western Australia) or basic (South Australia). This has implications for leaching into ground water, particularly with mobile herbicides and where the water table is very shallow. Sulfonylureas, which are used at low rates but are also very active are a good example. Ferris (1993) has reviewed the potential for sulfonylureas to persist in, and leach from, Australian soils and concluded that the alkaline sodic soils are the most vulnerable. These soils are characteristic of the cereal cropping areas in South Australia and therefore special attention needs to be paid to this concern in our assessments of this group of herbicides. In one case where this was a potential problem, the company was able to demonstrate through use of the PRZM model that this would not occur even after repeated applications at the maximum label rate over 10 years under worst case conditions, ie bare (no interception by plants), alkaline (pH 8) soils with low organic matter, and use of the longest known half-lives.

Another important difference in Australia compared with overseas seems to be in irrigation practices, with much more furrow irrigation appearing to be used here. A recent literature paper (Ebbert and Kim, 1998) has indicated that sprinkler irrigation leads to considerably much lower levels of suspended sediment concentrations in run-off compared with that from furrow irrigation. This has important implications particularly for cotton production in Australia, where the movement of chemicals adsorbed to eroded soil particles is an important transport mechanism for loss of activity from the cotton paddock. This is further discussed below.

It is appropriate at this point to discuss two large environmental fate studies which have been/are being undertaken in Australia, largely as the result of the registration process.

7 MINIMISING THE IMPACT OF CHEMICALS USED IN COTTON PRODUCTION

The first example relates to chemicals used in cotton production, which is one of the highest users of chemicals in Australia. In 1991 three organisations, the Land and Water Resources Research and Development Corporation (LWRRDC), the Cotton Research and Development Corporation (CRDC) and the Murray-Darling Basin Commission (MDBC) began to plan a co-operative study to minimise the impact of pesticides on the riverine environment using the cotton industry as a model (Schofield *et al.*, 1998). Work commenced in July 1993 for formal completion by December 1997. The goals of the program were to:

- assess the impact, if any, of the current cotton pesticide use on the riverine environment;

- develop practical and economic methods to reduce the transport of pesticides from application sites, and minimise the effects on the riverine environment; and
- provide a scientific basis for the development of management guidelines and regulatory codes.

In keeping with the above goals the program has consisted of three overlapping phases.

1. Determining and quantifying the major pathways of pesticide movement to rivers and their impact on aquatic biota.
2. Identifying and testing potential methods for ameliorating problems associated with pesticides.
3. Developing and implementing best management practices to minimise pesticide contamination.

In line with the theme of this Congress, we shall discuss only some of the results of the first phase. In practice this has focussed almost entirely on the insecticide endosulfan, due to its vital role in the protection of cotton from heliothis attack, particularly early in the season. These data proved to be central to our assessment of the environmental fate of endosulfan under the NRA's ECRP (National Registration Authority 1997a), examining various aspects of its transport and fate, namely spray drift, volatilisation, aerial transport on dust, movement in surface waters and suspended sediment, breakdown on plants and in soils and water. These were generated following application by standard methods, using typical application rates, aircraft and operating procedures. By sampling many elements at the same time, including the amount of drift, soils, irrigation water both in furrow and in tail waters, sediment etc, a complete budget of the endosulfan applied to the cotton field, including movement off site, has been able to be drawn up. Note that cotton is generally grown under furrow irrigation in areas with subtropical summer rainfall.

These studies have confirmed the importance of aerial transport, with around 70% of applied endosulfan lost through volatilisation and/or spray drift in the 7 days following aerial (ULV) application. The former seems most important though the latter may approach 10% of applied at a distance of 200-400 m downwind from target. Losses through volatilisation are particularly rapid over the first few days, especially if temperatures are high, and largely involve the more volatile α isomer. Off site contamination by dust movement seems relatively insignificant. For alkaline clay soils where cotton is typically grown, residues approaching 10% of applied may be expected in soil after a month, reducing to 1% after a year. Movement off site through furrow irrigation and into tailwaters are typically of the order of 1-2% and concentrations are in the order of 5-15 µg/L, declining to 2-3 µg/L late in the season. The balance (25-30%) is assumed to degrade.

The two isomers of endosulfan (α and β) both convert to the sulphate which typically persists at levels in the range of 100-200 µg/kg in the surface 2.5 cm. Residues can also be found at the tailwater outlet, at higher levels than in the soil from which they are transported. While residues in tailwater are mainly in the dissolved phase they appear to partition to sediment with a half-life in the order of about a week when ponded. Significant off-target movement of endosulfan sorbed to soil and suspended sediment occurs during major storms (levels of above 10% have been measured - these situations appear to have the strongest link with fish kills that occur from time to time).

As noted above these comprehensive studies seem to us to be unique, allowing a complete budget to be drawn up of what happens to a chemical once it is applied to the cotton field. They have set the benchmark for new insecticides for which registration has been sought in Australia over the past couple of years, and we have required the generation of similar data on several occasions, in particular when there have been concerns about persistence or potential off-field movement. It should be emphasised that these have shown different results from those described above, in particular in respect of being less volatile. The results from this very extensive series of studies have also been very useful to us in the assessment of new chemicals proposed for use on cotton.

The experimental work for this program has recently been concluded, and the results made public through a public conference entitled "Minimising the Impact of Pesticides on the Riverine Environment - Key Findings from Research with the Cotton Industry" held in Canberra from 23-24 July 1998. The proceedings of the conference will be published as a LWRRDC Occasional Paper before the end of 1998.

8 USE OF ATRAZINE IN FORESTRY

The second and quite different example involves the use of this rather controversial herbicide in forestry. In 1993 the major registrant in Australia approached the NRA with a proposal to limit the use of atrazine, as part of a decision to restrict registration world wide to the most essential uses on some major broadacre crops, due to concerns about the mobility of this herbicide.

This sparked a very significant response from both the private and public sector forestry agencies/interests in Australia, as they considered atrazine essential to the economic survival of plantation forestry (both pine and eucalypt) in this country, being the most cost effective way to suppress competing weeds in the first couple of years of growth. A series of negotiations led to a revised label with the conditions outlined below, and an agreement by forestry interests that a series of field experiments would be undertaken to show that atrazine could be safely used under these new restrictions.

- no mixing/loading or application within 20 m of any well, sink holes, intermittent or perennial stream;
- no application within 60 m of natural or impounded lakes or dams;
- no use in industrial and non-agricultural situations; and
- maximum annual use in plantation forestry to be 4.5 kg/ha except for 8 kg/ha in some special situations.

These restrictions are essentially identical to those for the remaining broad acre uses on the label, except that these are restricted to a maximum annual rate of 3 kg/ha.

A broadly based expert group named the Forest Herbicide Research Management Group (FHMRG), on which I represent the NRA, was set up, with half of its members independent of forestry interests. Its task was to design, commission and oversee the trials. The first named presented a real challenge as it soon became clear that practices in each State varied widely. Soil texture has a marked influence on practice, for example Tasmania generally applies atrazine by air, often at the highest allowable rate, whereas in contrast Western Australia applies it in a strip at the basic rate using a shrouded sprayer. In SA, NSW and Victoria it is mostly applied as core-coated granules.

Atrazine is not used in some States or Territories, or not on *Pinus* species or eucalypts. For example Queensland prefers to use simazine on pines, but does use atrazine on hoop pines, which belong to the genus *Araucaria.* Further, in some States such as WA and SA pines are grown on deep sandy soils, where leaching to ground water is the major concern, whereas in other areas such as parts of Victoria pines are grown on very steep slopes, where the potential to run off in surface water is the overwhelming consideration. As a result, a compromise protocol, which is scientifically robust but flexible enough, has had to be adopted, whereby as close as possible to the prevailing local conditions are used and trials are carried out on sites of "highest risk".

In this way a series of field experiments aimed at covering the "worst cases" of all the representative conditions (temperature, rainfall, slope, latitude and altitude) under which plantation forestry is practiced were undertaken. Sites were selected from those available for (re-)afforestation in 1996-7 and 1997-8. This has resulted in a total of five ground and 6 surface water field experiments, sometimes combined at the one site, depending on the nature of the catchment and which is of highest concern, with at least one trial in each State.

Only preliminary, and no published results are available as yet, but they do indicate that with close attention to application practices, atrazine appears to be able to be used safely as contamination appears to remain within guidelines for drinking water and those (yet to be finalised) for protection of aquatic species in the environment. One remarkable finding is that infiltration rates have been very high on all soil textures, even on steep slopes, and that even on sandy soils the movement of atrazine has been limited to the uppermost 60 cm. Atrazine metabolites have only been detected in the presence of atrazine itself.

Some of the analytical results do appear interesting in the context of this presentation, namely that the soil half-life in colder Tasmania is significantly longer than in warm temperate WA and SA, and longer still compared to semi-tropical Queensland, where three applications per year are needed over the first 2-3 years to adequately suppress the vigorous competing weed growth. It is also anticipated that through use of modelling, the threat to groundwater from the use of atrazine across Australia will be able to be clearly defined. This would also be desirable for surface water, but unfortunately the modelling capability for this aspect at the scale of forestry operations is not currently available in Australia.

9 POST REGISTRATION IMPACT OF CHEMICALS USE

In spite of the work described above,work and the 12 years experience in pre-registration assessments, participation in the NRA's ECRP has underlined that there is still a major knowledge gap about the impact on the Australian environment resulting from the use of pesticides once they are put on the market. It appears clear that apart from the work on endosulfan and atrazine mentioned above [Environment Australia was centrally involved in the review of atrazine under the NRA's ECRP which has now been published (National Registration Authority, 1997)], the only other chemical for which a substantial body of local research is being/has been undertaken is the insecticide fenitrothion by the Australian Plague Locust Commission, which uses this chemical as its major tool to control locusts in its area of responsibility. This chemical too is currently the subject of a review under the NRA's ECRP.

In order to begin to obtain a better understanding of the post-registration impact of the use of agvet chemicals, Environment Australia commissioned a consultant to investigate the level of investigation of environmental contamination of Australia. This was also prompted by a recommendation in a report of an inquiry into the use of these chemicals in Australia by a special Senate (the upper house of our Commonwealth Parliament) Committee. The findings of the report (Aquatech, 1997) make it clear that while work in this area is conducted in a number of States, this is generally carried out in an ad hoc manner and there is little systematic investigation into the contamination of Australian soils, sediments, biota and waters (with some notable exceptions being a couple of programs centred on the cotton areas of the Murray Darling Basin), and also little routine investigation of wildlife mortality incidents.

One of the key recommendations is the need to have better information on pesticides use, ie what is used where, how and how much, to allow better policy making. In Australia a joint agricultural/environmental working party has been formed to investigate this area. It is expected that the direction taken will rely heavily on initiatives in this area that have been started in the EU through EUROSTAT and are being promoted through the OECD Pesticides Forum.

Australia is also expected to consider over the next year or so the need for a more centralised environmental monitoring and incident reporting system, and what form this may take, taking into account a number of options put forward by the consultant. In this respect it should be noted that one of the recommendations of the atrazine ECRP review noted above was the need for an Australian monitoring program to confirm that the restrictions placed on atrazine, in particular the withdrawal of the very high rates in non-agricultural situations such as irrigation ditches, have been effective in increasing the margin of safety for use of atrazine in this country. This is being progressed through the NRA, and may prove to be a good pilot for a more extensive program.

10 CONCLUSION

The above is by necessity only a thumbnail sketch of the kind of environmental fate considerations Environment Australia needs to take into account when performing its assessments of the potential environmental impact of pesticides in Australia, and a brief outline of the kind of research work in this area that is being undertaken in this country. While the latter may be small by developed world standards, it is impressive and in some cases cutting edge, in particular the comprehensive program examining the fate of pesticides applied to cotton.

References

1. Aquatech, Monitoring of the Environmental Effects of Agricultural and Veterinary Chemicals in Australia - Preliminary Investigations, submitted to the Environment Protection Group of Environment Australia, Canberra, Australia, 1997.

2. J.C. Ebbert and M.H Kim, *Relation between Irrigation Method, Sediment Yields, and Losses of Pesticides and Nitrogen*, J. Environ. Qual., 1998, **27** (2), 372-380.

3. I.G. Ferris, *A risk assessment of sulfonylurea herbicides leaching to groundwater*, AGSO Journal of Australian Geology and Geophysics, 1993, **14** (2/3), 297-302.

4. N. Johnston, J. Skidmore and G. Thompson, *Applicability of OECD Test Data to Australian Aquatic Species: A report to the Advisory Committee on Chemicals in the Environment of the Australian and New Zealand Environment Council* ., July 1990.

5. W. Klöpffer, 'Photochemical Processes Contributing to Abiotic Degradation in the Environment. In "Chemical Exposure Predictions"', D Calamari (ed). Lewis Publishers, Boca Raton, 1993, (2), pp 13-26.

6. National Registration Authority for Agricultural and Veterinary Chemicals 1997. Technical Report on the NRA Review of ATRAZINE, Canberra Australia.

7. National Registration Authority for Agricultural and Veterinary Chemicals 1997a. Draft Technical Report for Endosulfan available at: http://www.dpie.gov.au/nra/prsendo.html (environmental chapter at: ftp://ftp.nra.gov.au/pub/prsendo7.pdf)

8. K. D. Racke, K. P. Steele, R. N. Yoder, W.A. Dick and E. Avidov, *Factors Affecting the Hydrolytic Degradation of Chlorpyrifos in Soil*, J. Agric. & Food Chem., 1996, **44** (6), 1582-1592.

9. K. D. Racke, M.W. Skidmore, D. J. Hamilton, J. B. Unsworth, J. Miyamoto and S. Z. Cohen, *Pesticide Fate in Tropical Soils* (Technical report). Pure & Appl. Chem., 1997, **69** (6) 1349-1371.

10. N. Schofield, V. Edge and R. Moran, *Minimising the Impact of Pesticides on the Riverine Environment using the Cotton Industry as a Model*, 1998, Water **25** (1), 37-40.

11. A. Walker, *Influence of Soil and Weather Factors on the Persistence of Soil-applied Herbicides*, Applied Plant Sci., 1991, **5**, 94-98.

Residues in Food and the Environment

Quality of Residue Data

Árpád Ambrus*

BUDAPEST PLANT HEALTH AND SOIL CONSERVATION STATION, BUDAPEST POB 340, HUNGARY

1 DEVELOPMENT OF QUALITY REQUIREMENTS OF CHEMICAL ANALYSIS

The results of measurements can be associated with very serious consequences and the responsible persons from a laboratory might have to defend their results in court. Therefore the results of measurements should be representative, and the laboratory should be able to prove the correctness of measurements with documented evidence. Analysts must be aware that their professional reputation and credibility could also be at stake. They are responsible for correct and timely analytical results, and are fully accountable for the quality of their work. Though the above argument seems self evident, considerable evidence exists in the literature that little attention is being paid by many professional analytical chemists to the question of the reliability of the analytical results they produce[1]. Expanding national and international trade, the responsibility of national registration authorities, who permit the use of various chemicals (e.g. medicines, pesticides) have long required reliable test methods, correct analytical data and complete reports reflecting all findings of complex studies.

Without attempting to give a complete historical background, the main activities performed to improve the quality of analytical data are summarised below.

1.1 Reliable analytical results

There was an early need to assess the available methods and test their performance under various laboratory and environmental conditions. It is worth mentioning that the Association of Official Analytical Chemists (AOAC) was already formed in 1884.

The primary objective of AOAC, as stated in its bylaws, is to "... obtain, improve, develop, test, and adopt uniform precise, and accurate methods for the analysis of foods, ...pesticides, ... and any other products, substances or phenomena affecting the public health and safety, the economic protection of the consumers, or the protection of the quality of the environment"[2].

The so-called Smalley program started in the U.S.A. in 1917 to promote better performance within and between laboratories testing cotton seed meal. The programme expanded by 1990 to include 32 fat/oil related categories with 450 certified participants world-wide[3].

1.1.1 Standardised Analytical Procedures. The first and probably the easiest way to improve the results of chemical analysis was to standardise the analytical methods. Many organisations, both national and international, had undertaken this task on a commodity by commodity basis for more than a century. It resulted in overlap and multiplication of effort and the difficulty in deciding which of two or more well authenticated methods, each properly established by collaborative study, should be selected for standardisation or referee or arbitration purposes. The analytical

*Current address: FAO/IAEA Training and Reference Centre for Food and Pesticide Control, FAO/IAEA Agriculture and Biotechnology Laboratory, Agrochemicals Unit, IAEA's Laboratories, Department of Research and Isotopes, A-2444 Seibersdorf, Austria

methods for various substrates were assessed and approved by several independent organisations competing for the dominance in providing referee methods.

The predecessor of FAO/WHO Codex Alimentarius Commission (CAC), the Joint FAO/WHO Committee of Government Experts on the Code of Principles Concerning Milk and Milk Products, established in 1958, was among the first who recognised the need for harmonised methods. Consequently, the Codex Secretariat formed a technical group consisting of experts from the International Dairy Federation (IDF), International Organisation for Standardisation and AOAC with the responsibility to supply jointly approved methods of analysis for supporting international CODEX standards. Later on, the Codex Committee on Methods of Analysis and Sampling (CCMAS) was established to assess and recommend methods for sampling and analysis of food products. The general criteria for the selection of methods of analysis for Codex purposes have been evolving during the years. They are published in the CAC Procedural Manual[4]. It should be pointed out that because of the specific requirements of trace analysis, the CCMAS does not deal with the methods for analysis of pesticide residues. This task was allocated to the Codex Committee on Pesticide Residues (CCPR) which publishes the lists of recommended methods for testing the compliance with the Codex Maximum Residue Limits.

1.1.2 Collaborative Studies. It was soon recognised that methods that were to be used by many laboratories had to be tested in several laboratories. Naturally, the isolated activities of various organisations led to independent guidelines for the design, conduct and interpretation of collaborative studies. There was an urgent need to harmonise the guidelines. This task was undertaken by the International Union of Pure and Applied Chemistry (IUPAC) inviting experts from the Analytical, Applied and Clinical Divisions and representatives of main international organisations (ISO, AOAC) sponsoring collaborative studies. Through a series of meetings a number of harmonised protocols have been developed[5,6,7].

The Codex Alimentarius Commission requires ideally methods that have progressed fully through a collaborative study in accordance with the internationally harmonised protocol[7]. This usually requires a study design involving minimum of 5 test materials, the participation of eight laboratories reporting valid data, blind replicates or split levels to also assess within laboratory repeatability parameters. Such studies are not practical for the validation of analytical methods for pesticide residues and veterinary drugs, because of the numerous possible combinations of residues and matrices. Following the recommendation of the Codex Committees concerned, a Joint FAO/IAEA Expert Consultation on Validation of Analytical Methods for Food Control was held in 1997[8]. The Consultation concluded that preferred methods for validation of analytical procedures are the collaborative studies or independent review of a multi laboratory study[9] (verification by a second laboratory). It further recommended that in those cases where collaborative studies or inter-laboratory studies are impractical or impossible to carry out, the validation of a procedure could be done in one laboratory operating under and internationally recognised quality system such as GLP, ISO/IEC (International Electrotechnical Commission) Guide 25, or Peer Review.

1.1.3 Proficiency Tests. In addition to the use of reliable methods the proficiency of the chemist and the testing laboratory are of vital importance in obtaining valid results.

The criteria for testing the proficiency of laboratories had been elaborated by several national and international organisations which were competing with each other. Again, the joint effort of ISO, AOAC and IUPAC resulted in harmonised protocols for testing the proficiency of laboratories[10,11,12,13].

1.1.4 Quality Guidelines for the Laboratories. Various guidelines and standards were introduced by different national and international organisations setting out the requirements for a laboratory to demonstrate that it operates according to them if it is to be recognised as competent to carry out the specified tests. The certification of the competence is performed through third party conformity assessment. A set of documents has been elaborated by ISO/IEC: ISO/IEC Guide 25[14],Guide 38,

Guide 43[a], Guide 45, Guide 49, Guide 58. Similar guidelines are being simultaneously produced by other organisations. For instance, the relevant Standards of the Joint European Standard Institution CEN/CENELEC are: EN 45001, EN 45002, EN 45003, EN 45011, EN 45012, EN 45013, EN 45014.

The basic principles of the ISO guides and EN standards are very similar and can be summarised as follows:

- for a laboratory to produce consistently reliable data it must implement an appropriate programme of quality assurance;
- analytical methods must be thoroughly validated before use, preferably with collaborative study which conforms to recognised protocol. These methods must be carefully and fully documented, staff adequately trained in their use and control charts should be established to ensure the procedures are under proper statistical control;
- where possible, all reported data should be traceable to reliable and well-documented standard materials, preferably certified reference materials. Note that the use of certified reference materials is very limited in the analysis of pesticide residues;
- accreditation of the laboratory by the appropriate national accreditation scheme, which itself should conform to accepted standards, indicates that the laboratory is applying sound quality assurance principles. It is anticipated that accreditation assessments will increasingly utilise the information produced by proficiency testing. Participating in proficiency testing schemes provides laboratories with a means of objectively assessing, and demonstrating the reliability of the data they are producing.

To assist laboratories in applying the general quality standards specific guidelines were prepared for validation of analytical procedures[9,15,16,17], and to apply good analytical practice in the analysis of pesticide residues[18,19,20,21]. A number of proficiency study programmes, such as CHECK[b], FAPAS[c], GTZ[d], regularly include samples for analysis of pesticide residues.

1.2 Reliable studies

The national registration authorities also found it necessary to introduce a system which ensures the reliability, traceability of the results obtained in long term and complex toxicological, metabolism and environmental fate studies involving a number of analysis of different substrates. Consequently the principles of Good Laboratory Practice (GLP) were elaborated first by the US FDA in 1976 followed by other national organisations. Since, medicines and pesticides are manufactured and distributed by multinational companies all over the word, and most of the countries' legislation requires the registration of such products, there was an immediate need for internationally acceptable harmonised guidelines. The Organisation of Economic Co-operation and Development (OECD) was found to be the best forum for the development of such guidelines. The first OECD GLP Guidelines were published in 1982. GLP principles were accepted rapidly and they were applied in many new fields such as short term studies, supervised field trials, and testing cosmetics. In order to meet the practical requirements, the OECD GLP Panel, consisting of the representatives of member countries, has elaborated and regularly updated the GLP Principles which are currently published in 11 monographs and guidance documents.

The GLP principles of OECD, EPA, FDA and Japanese Organisations are very similar; however, the minor differences in the requirements and the deviations in interpretation of certain terms by the national certification or accreditation bodies can cause many problems for the laboratories which should comply with all requirements. The need for harmonisation of the slightly different national requirements and the acceptance of each other's findings is well recognised, and actions

---

[a] Currently it is under revision

[b] Food Inspection Service, CHECK Secretariat, Eendrachtskade ZZ 2a 9726 CW Groningen NL

[c] Food Analysis Performance Assessment Scheme, CSL Food Science Laboratory, Colney, Norwich, NR4 7UQ UK

[d] Deutsche Gesellschaft fürTechnische Zusammenarbeit GmbH, Eschborn/Darmstadt, Germany

to achieve this goal are taken. Hopefully, the recently revised OECD GLP Principles[22] will be accepted by all countries.

1.3 Total quality management

Striving for gradual and continuous improvement of quality led to the development of Total Quality Management (TQM). It is a leadership style of management, which supports and cultivates the one team approach. It accepts risk taking, and recognises that mistakes and failures are inevitable but need to be learnt from. Emphasis is placed on continuous learning rather than rules. TQM in the laboratories means that all employees of the laboratory (from the top manager to each member) know their mission, their task and they work in full harmony with each other, with customers/clients to accomplish the task, the mission of the laboratory. TQM is particularly relevant to research and development and is one of the few approaches which addresses the creative part of the laboratory work. TQM places emphasis on good science, good management and empowerment of analyst. The process must ensure *that we do the right experiment as well as doing the experiment right*[23]. QA works when it is owned and controlled by the analyst and it facilitates quality rather than tries to impose simplistic rules.

Quality is a dynamic issue and the output of the process is continuous improvement. Systems alone cannot deliver quality. Staff must be trained, involved with the tasks in such a way that they can contribute their skills and ideas and must be provided with the necessary resources. We need to remember that a motivated person can move mountains[24].

Blumenthal adds that an analytical laboratory should define and communicate its mission as that of providing scientific evidence for important decisions and the resulting benefits. These mission viewpoints remove the analytical laboratory from its isolation as a generator of numbers, and place the products of the laboratory squarely in the centre of much of the corporate (enterprise's) "action".

The last aspect of TQM that I would like to draw your attention to is the relation of the laboratory and its staff to its customers. The customer relation has also been emphasised in the latest ISO 25[14] Guide and OECD GLP Principles[25]. The staff of the laboratory and the customers can only jointly achieve agreement as to what constitutes a quality service[26].

2 ROOM FOR IMPROVEMENT OF THE QUALITY OF RESIDUE DATA

The guidelines for quality control for residue analysis[19-20] and validation of analytical methods[21] cover most of the essential aspects of good analytical practice, which would result in high quality residue data if followed properly. In the following a few critical aspects are discussed which have not yet got the appropriate attention. The uncertainty of the analytical results (S_R) comprises the uncertainties of sampling (S_S), sample preparation (S_{Sp}) and analysis (S_A).

$$S_R = \sqrt{S_S^2 + S_{Sp}^2 + S_A^2} \tag{1}$$

Until recently method validation protocols placed the major emphasis on the determination of the performance of the analytical methods based on the analysis of fortified samples. In addition, information on the stability of residues during deep-frozen storage was a general requirement[27]. Very little attention has been paid to the effect of sampling and sample preparation.

It follows from Eqn. 1 that confidence intervals for the analytical data should be calculated from S_R taking into account the effect of sampling and sample preparation as well, and should not be estimated from the results of method validation performed with fortified samples which reflects only S_A.

2.1 Effect of sampling on the uncertainty of residue data

The residues are generally unevenly distributed within individual units and/or the bulk of any commodity. The ratio of the maximum and minimum residues found in treated lots is in the range of 100. Figures 1 and 2 show the relative frequency distributions of residues in individual apple fruit and composite random samples drawn with replacement[28] from the population of residues in individual apples.

Figure 1. Relative frequency polygon of chlorpyrifos-methyl residues in apple samples at day 0

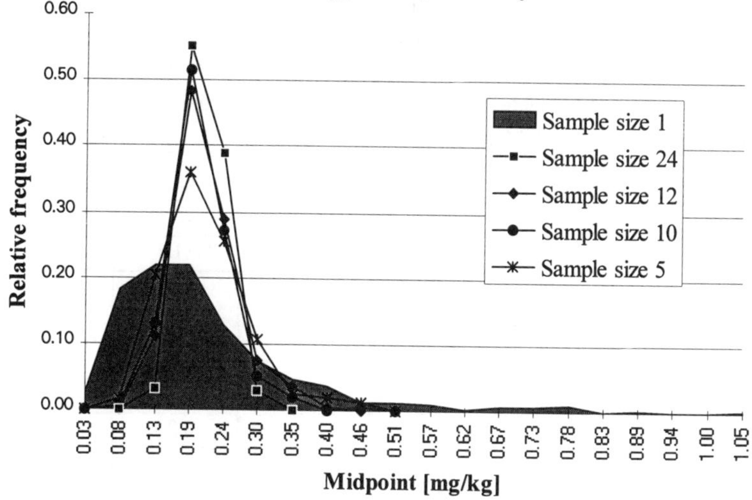

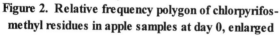

Figure 2. Relative frequency polygon of chlorpyrifos-methyl residues in apple samples at day 0, enlarged

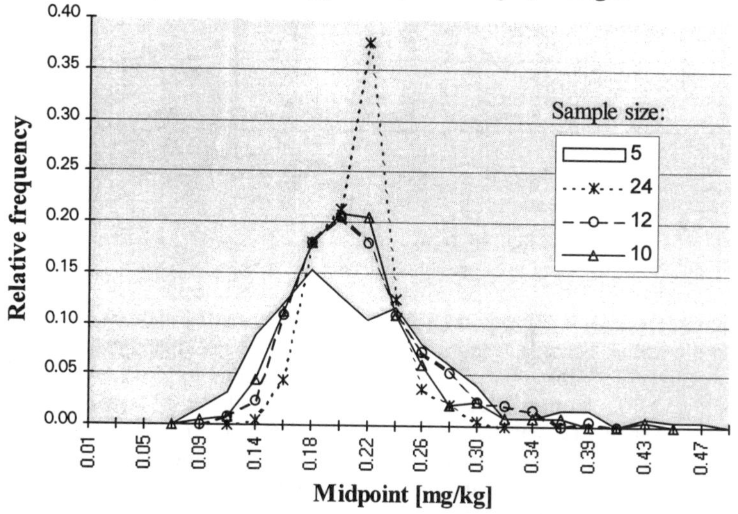

Table 1 summarises the results of simulated sampling experiments performed with residue data measured in unit crops or composite samples. These studies indicated that the coefficient of variation of residues being in composite samples of the same size (sample size = number of primary samples) is practically independent from the residue levels, provided that the natural units contained measurable residues. The Codex Sampling Method[29] requires a minimum of 5 and 10 primary samples for large and medium or small size crops, respectively. Provided that the whole laboratory sample is homogenised before the analytical portion is withdrawn, the uncertainty of sampling of such crops can be expected to be ≤ 40% and 30%, and respectively. The uncertainty of sampling of small crops is expected to be within the uncertainties of medium size ones. As there is no information on the residues in individual large crops, the estimated uncertainty is temporary. To be on the safe side, the recommended CV values for general use are slightly overestimated. These figures may be refined if more data will be available. Furthermore, the validation of sampling procedures applied following application of pesticides in row and furrow would be required to get information on the likely bias and uncertainty of sampling under those conditions.

Table 1. Variation of pesticide residues depending on the size of the samples

Sample size	1	5	10	12	24
Sample material	CV_S% of residues in samples				
Apple	72.2	33.6	22.9	21	13.7
	69.1	31.3	22.2	21.7	12.8
	63.2	32.8	19.9	17.8	11.9
Average apple	*68.2*	*32.6*	*21.7*	*20.2*	*12.8*
Average kiwi fruit[1]	*84.2*	*39*	*25.8*	*24.6*	*17.7*
Potato	0.89			0.22	
	0.82			0.24	
	0.82			0.23	
	0.94			0.26	
	0.32			0.1	
	1.12			0.31	
	1.53			0.4	
	0.41			0.11	
Average potato	*85.5*			*23.5*	
Soil	79.9	36.3	24.7	22.4	15.9
Average CV	*79*	*36*	*24*	*23*	*15*
Recommended CV for general use		**40**	**30**	**28**	**20**

Note: (1) average of 4 data sets obtained from residues of chlorpyrifos, diazinon, pirimiphos methyl and vinclozolin used for treatments during the growing season.

As it can be seen from Figure 3, if only a portion of the laboratory sample is processed (e.g. 1 or 2 apples) the spread of residues is much wider (both smaller and larger) than in composite samples of size 10-12. Consequently the results of such analysis cannot be used for checking compliance with Codex or national MRLs for plant products which are generally based on composite samples of size 10-12. This fact should be born in mind especially when quick field tests are used with bioassays and ELISA kits for the determination of pesticide residues from small portions of a single crop or soil increment. Furthermore, the results of preliminary screening for residues, based on the analysis of small portions of plant materials and soils, should be evaluated very carefully. It needs to be pointed out that, due to the wide variation of residues it could happen, when the test method is not sensitive enough, that the selected portion may not contain detectable residues while the residues in the tested commodity exceed MRLs.

Figure 3. Relative frequency polygon of phosphamidon residues in apple samples

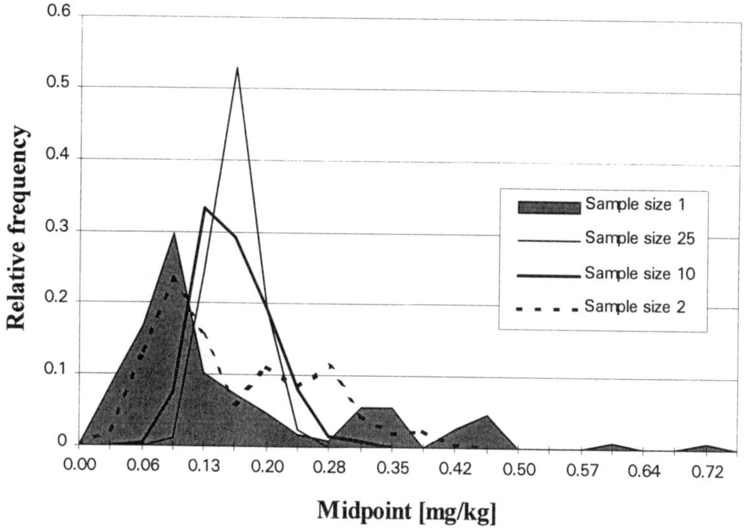

2.2 Effect of sample preparation

In order to obtain accurate and repeatable results, the laboratory sample should generally be cut into small pieces to obtain a statistically well mixed matrix for taking representative analytical portions for extraction. It is generally assumed that a sample comminuted in a chopper or blender is "homogeneous", and the uncertainty deriving from sample preparation is not taken into account. Experiments performed for determining the uncertainty of sample preparation do not support, however, the previous assumption.

The efficiency and the uncertainty of sample preparation can be tested utilising the "Sampling Constant", K_S,:

$$K_S = mCV^2 \qquad (2)$$

Where m is the mass of sample which should be withdrawn from a well mixed material to hold the relative sampling uncertainty to 1% with 68% confidence. Wallace and Kratochvil[30] applied the "Sampling Constant", defined by Ingamells and Switzer, to test whether a comminuted sample is statistically well mixed. If the sample is well mixed the sampling constant for small and large portions should be the same. Therefore:

$$m_L(CV_L)^2 = m_S(CV_S)^2 \qquad (3)$$

The authors recommended that ratio of $m_L/m_s \geq 10$ should be used in the tests to obtain reliable estimate. If Eqn. 3 holds, the *K* can be calculated. When the *K* is known, the uncertainty of sample preparation can be estimated for any analytical portion which is between m_S and m_L. It should be noted that the analysis of variance of a set of sub-samples of a given mass is suitable for the determination of the variation deriving from sample preparation, but it may not reveal any information on the fact whether the sample is well mixed or not.

Table 2. Sampling constants of "homogenised" plant materials

Sample material	Equipment	Homogenisation Aid	K_s [kg]
Apple, field treated[28]	UKM + B	none	11-21
Apple, surface treated[a]	S	none	15.5
	S	dry ice	2.8
	S + Wb	20% water for blending	0.2 -0.3
Cabbage, field treated[28]	UKM + B		1.6
Lettuce, surface treated	S		1.7
	S + Wb	20% water for blending	0.12

Notes: (a) about 20% of the surface of apples was treated with standard solution and dried before chopping; UKM: universal kitchen machine; S: Stephan UM-12; B: kitchen blender; Wb: Waring Blender

Table 3. Effect of the mass of the analytical test portion on the uncertainty of sample preparation

K_S [kg]	2.8[A]	15.5[B]
Test portion [g]	Uncertainty of sample preparation %	
5g	23.7	55.7
30g	9.7	22.7
50g	7.5	17.6
200g	3.7	8.8
5g	7.7[C], 8.6[D]	11.7[E]
25g	3.5[C], 5.1[D]	9.5[E]
50g	2.5[C], 4.5[D]	9.1[E]

Notes: A. Chopping with Stephan UM 8 universal machine in the presence of dry ice

B. Chopping with Stephan UM 8 universal machine without dry ice

C. Sampling constant after homogenisation of the pre-chopped sample in a blender with 20% water was 0.3 kg

D. and E.: Combined uncertainty of sample preparation following two steps homogenisation (further blending 200 g portion of pre-chopped sample) from A and B samples, respectively

Applying Wallace and Kratochvil's method, the efficiency of sample preparation was tested for apple and cabbage[31] and further studies were carried out in the Agrochemicals Unit of the FAO/IAEA Seibersdorf Laboratory which will be published elsewhere in detail. The findings of the experiments and the results, summarised in Tables 2 and 3, indicate that:

- the preparation of well mixed sample requires special equipment and attention;
- normal chopping procedure may not be sufficient for obtaining well mixed matrix, for instance, from lettuce and cabbage;
- adding dry ice to the sample improves the homogeneity of the matrix and simultaneously reduces the possibility of decomposition of pesticide residues during sample preparation;
- further blending in the presence of some water according to the method of Kadenczki and co-workers[32] improves the homogeneity of the sample and results in smaller K_S. The combined uncertainty of the two steps (chopping and consecutive blending) of sample preparation will have to be taken into account in such cases (Table 3);

- the smaller the analytical portion the larger is the uncertainty deriving from the sample preparation .

The latter fact should be taken into account when the size of analytical portion is reduced. Extraction of less then 30 g analytical portions may increase the uncertainty of the results substantially or the sample preparation procedure may not result in well mixed samples at all. Therefore, extraction of 1-5 g analytical portions by super critical fluid technique or according to miniaturised procedures such as described by Steinwandter[33] should only be applied if the appropriate efficiency of sample preparation had been previously confirmed.

2.3 Evaluation of calibration curves

Every guideline specifies the requirements for the performance of calibration. However none of them draws the attention to the fact that the calibration of chromatographic systems usually results in proportional residuals, and not constant residuals. Consequently for the correct evaluation of the calibration the weighed linear regression described in ISO/DIS 11095 Standard[34] should be applied. The difference is usually negligible, especially in the middle of the linear range, but may be substantial at the lower and higher ends of the calibration line. Figure 4 illustrates the calibration lines fitted with the two types of linear regression for the same data points.

Figure 4. Linear regression of calibration points

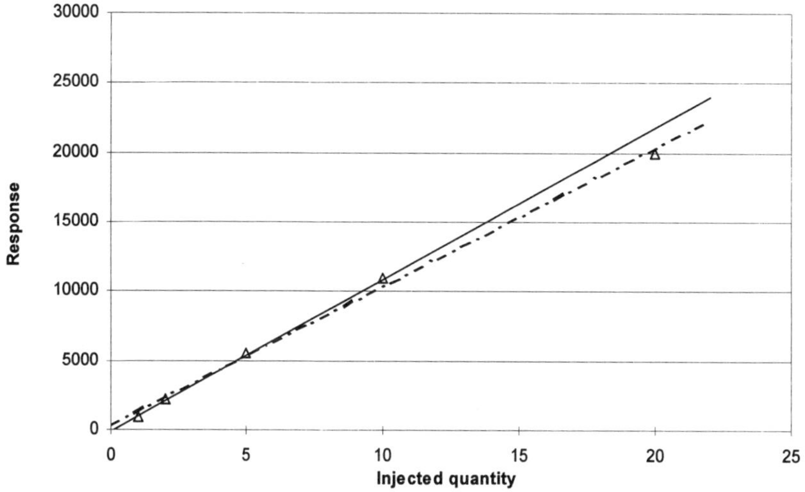

Notes: ——— Weighed linear regression; -·-·-·-·- Linear regression

2.4 Correction for the average recovery

The average recovery indicates a possible bias in the analytical results which should be considered in evaluation and use of the data. Currently there is no uniform internationally agreed approach for taking the results of recovery studies and blank values into account. Naturally, the recovery is also subject to random errors. Consequently, the correction for the recovery is justified only if the average recovery determined during method validation is significantly different from 100%, which should be tested with the Student Test. The performance of the method should be checked regularly with additional recovery tests. The method may be considered to be performed under statistical control if the other quality criteria are fulfilled and the recoveries fall within the acceptable range determined from the average recovery and its standard deviation obtained from the validation of the method.

2.5 Optimisation of residue analytical procedures

According to Eqn. 1 the overall uncertainty of the analytical results incorporates the uncertainties of all steps of the procedures from sampling to the final quantitative determination. Table 4 illustrates the effect of the three major steps of the procedure. Since the average residue can be assumed to be the same during an analytical procedure, the standard deviations can be replaced with the coefficient of variation, which enables general conclusions to be drawn independently from the actual residue levels.

As it was shown, in general, the uncertainty of sampling for Codex purposes cannot be expected to be lower than 30 – 40 % for small/medium and large size crops, respectively, there is little point to attempt to reduce the combined uncertainty of the sample preparation and analysis [$\sqrt{CV_{Sp}^2 + CV_A^2}$] below 12%, which allows 8%+9% or 5%+11% uncertainties for each step. Since the uncertainty of sample preparation can be more easily kept at lower level, a CV_{Sp} around 5% and a $CV_A \leq 11\%$ seem to be the optimum for small and medium size fruits and vegetables (sampling uncertainty is about 30%).

Table 4. Optimisation of analytical procedures

Sample size	CV_S %[a]	CV_{Sp} %	CV_A %	**CV_R %**	
5	40	15	15	45.28	**45**
	40	10	10	42.43	**42**
	40	5	12	42.06	**42**
	40	8	10	42.00	**42**
	40	1	10	41.24	**41**
	40	5	8	40.62	**41**
	40	1	5	40.32	**40**
10	30	15	15	36.74	**37**
	30	10	15	35.00	**35**
	30	2	12	32.17	**32**
	30	5	11	32.34	**32**
	30	8	9	32.33	**32**
	30	5	8	31.45	**31**
	30	5	5	30.82	**31**
	30	3	5	30.56	**31**

Note: Recommended CV for general use (see Table 1.)

3 SUMMARY OF CONCLUSIONS AND RECOMMENDATIONS

The ISO Guide 25 and OECD GLP Principles cover individual analytical tests and complex studies, respectively. The principles of their latest revisions are very similar and complementary to each other. They can be implemented simultaneously which would improve the quality of residue data as individual tests are regulated in more detail by ISO Guide 25.

Reports of studies should only contain the information specified in the GLP Principles. Indiscriminate inclusion of raw data into the report does not improve the quality of the study, but makes the evaluation more difficult.

A formal compliance with GLP or ISO 25 principles does not guarantee that a study is scientifically sound and that the results can be utilised. The responsibility of the Study Director and Scientific Management for the technical content and reliable implementation of the study is emphasised. Furthermore, laboratories performing regulatory control of pesticide residues should demonstrate their capability through regular participation in proficiency tests.

The sampling procedures should be selected according to the objectives of the study, and described in detail in the report. To provide data for the estimation of maximum residue limits and to control the compliance of an agricultural commodity with Codex MRLs the size of samples should be selected according to the Recommended Codex Sampling Procedure[29].

The sampling procedures used for supervised field trials should be validated, and the uncertainty associated with their application should be estimated. Since the applicability of a sampling procedure is independent from the pesticide applied, an international co-operation would substantially reduce the cost of their validation.

The efficiency of each sample preparation procedure should be tested by checking the homogeneity of the comminuted matrix. This procedure should be part of the method validation.

As the uncertainty of sample preparation depends only on the commodity and the equipment used, a well established and tested procedure can be applied for any pesticide without further testing. The stability of residues during a given sample preparation procedure should however be tested for each pesticide residue and typical commodity combination.

If the residuals of a calibration function are proportional to the injected amounts, the linear calibration function of chromatographic systems should be determined with weighed linear regression, and the confidence intervals for the calibration should be calculated accordingly.

The uncertainty of the residue analytical results should be estimated taking into account the uncertainty of sampling and sample preparation as well. The uncertainty of residue data cannot be lower than the combined uncertainty of all steps.

The estimated uncertainties of composite samples of size 5 and 10 are 40% and 30% respectively. The combined uncertainties of 14% and 12% for sample preparation and analysis would result in an total uncertainty of 42% and 32 %, respectively. Further improvement of the analytical procedures would decrease the uncertainty of the results marginally, which may not be worth in view of the substantially increased time and cost involved.

The reports on residue analysis should always contain the uncorrected residue values, blank values, fortification levels and corresponding average recovery values, their standard deviations and the number of recovery tests. The report should also clearly indicate if the results were corrected for recovery or not.

REFERENCES

1. W. Horwitz., `History of the IUPAC/ISO/AOAC Harmonization Programme`, 4th International Symposium on the Harmonisation of Quality Assurance Systems in Chemical Analysis, Geneva, 2-3 May 1991.
2. D.L. Brener, D. Friestone, `AOCS Certification Programs`, 4th International Symposium on the Harmonisation of Quality Assurance Systems in Chemical Analysis, Geneva, 2-3 May 1991.
3. E.J. Klesta and N. Palmer, `Proficiency Testing Needs as Seen by AOAC`, 4th International Symposium on the Harmonisation of Quality Assurance Systems in Chemical Analysis, Geneva, 2-3 May 1991.
4. Joint FAO/WHO Food Standards Programme, Codex Alimentarius Commission Procedural Manual 10th Ed., FAO, Rome, 1997, 57.
5. W. Horwitz, *Pure & Appl. Chem.*, **60**. 1988, 855.
6. W. Horwitz, *Pure & Appl. Chem.*, **66**, 1994, 1903.
7. W. Horwitz, *Pure & Appl. Chem.*, **67**,1995, 331.

8. FAO/IAEA Report of Expert Consultation on Validation of Analytical Methods for Food Control, FAO, 1998, http://www.fao.org/waicent/faoinfo/economic/esn/nutri.htm
9. AOAC International, AOAC Peer-Verified Methods Programme, 1997
10. W.D. Pocklington, *Pure & Appl. Chem.*, **62**, 1990,149.
11. M. Thompson, R. Wood, *Pure & Appl. Chem.*, **65**, 1993, 2123.
12. M. Thompson, R. Wood, *J. of AOAC International*, **76**, 1993, 926.
13. M. Thompson, R. Wood, *Pure & Appl. Chem.*, **67**, 1995, 649.
14. ISO/IEC Guide 25 General requirements for the competence of testing and calibration laboratories (4th edition) 1997.
15. Nordic Committee on Food Analysis, Validation of Chemical Analytical Methods, NMKL Procedure No.4, 1996.
16. D.G. Holcomb, Method Validation A laboratory Guide, Eurachem, EUUK Ex/96/05, 1996.
17. Bravenboer J, Putten A. v.d., Verwaal W., van Bavel-Salm M., Validation of methods, Inspectorate for Health Protection, Report No. 95-001, NL, 1995
18. Codex Alimentarius Commission, Pesticide Residues in Food Supplement to Vol. 2. 2nd Ed, FAO 1993. 151.
19. MAFF, Quality Control for Pesticide Residues Analysis, 2nd Ed., Ministry of Agriculture Fisheries and Food, UK, 1996.
20. A. Hill, Quality Control Procedures for Pesticide Residues Analysis, EU Document 7826/VI/97, 1997.
21. A. Hill, S. L. Reynolds, Guidelines for Validation of Analytical Methods for Monitoring Trace Organic Components in Food Stuffs and Similar Materials, Working Paper prepared for the Meeting of the Codex Committee on Pesticide Residues, 1998.
22. OECD, The OECD Principles of Good Laboratory Practice, ENV/MC/CHEM(98)17, Paris, 1998.
23. B. King, `Quality in the Analytical Laboratory`, 6th International Symposium on Quality Assurance and TQM for Analytical Laboratories, Melbourne, 4-11 December,1995. The Royal Society of Chemistry, 1995, 8.
24. T.K. Blumenthal, `Motivation and "Marketing" of the Analytical Laboratory for TQM: Pride of Place`, 6th International Symposium on Quality Assurance and TQM for Analytical Laboratories, Melbourne, 4-11 December, 1995, The Royal Society of Chemistry, 1995, 1.
25. OECD, The Role and Responsibilities of the Sponsor in the Application of Principles of GLP, ENV/MC/CHEM(98)16, Paris, 1998.
26. A. Nicholson, `Customer Satisfaction: A reality in Analytical Laboratories?`6th International Symposium on Quality Assurance and TQM for Analytical Laboratories, Melbourne, 4-11 December, 1995, The Royal Society of Chemistry, 1995, 232.
27. A. Ambrus (ed.) FAO Manual on the Submission and Evaluation of Pesticide Residues Data for the Estimation of Maximum Residue Levels in Food and Feed, Food and Agricultural Organization of the United Nations, Rome, 1997.
28. A. Ambrus, *J. Environ. Sci. and Health*, **B31**, 1996, 435.
29.Codex Committee on Pesticide residues, Revised Recommended Methods of Sampling for the Determination of Pesticide Residues for Compliance with MRLs, Alinorm 99/24, Appendix II.
30. D. Wallace, and B. Kratochvil, *Analytical Chemistry*, **59**, 1987, 226.
31. A. Ambrus, E.M. Solymosné, and I. Korsós, *J. Environ. Sci. and Health*, **B31**, 1996, 443.
32. L. Kadenczki, Z. Arpad, I. Gardi, A. Ambrus, and L. Gyõrfi, *J. Assoc. Off. Anal Chem.* **75**. 1992, 53.
33. H. Steinwandter, Development of Microextraction Methods in Residue Analysis, in T. Cairns and J. Sherma (eds.) "Emerging Strategies for Pesticide Analysis", CRC Press, 1992. 3.
34. ISO/DIS Linear Calibration Using Reference Materials, ISO/DIS 11095, 1996.

Advances in Pesticide Residue Methodology

Volker Bornemann

BASF AKTIENGESELLSCHAFT, PLANT METABOLISM AND RESIDUE STUDIES, AGRICULTURAL CENTER LIMBURGERHOF, PO BOX 120, D-67114 LIMBURGERHOF, GERMANY

1 INTRODUCTION

Residue studies are used to determine the magnitude of the residues in the harvested raw agricultural commodities (RACs) from supervised field trials, in the processed plant products from processing studies, in drinking water, and in the environment.

The proper design and conduct of the field portion of these supervised residue trials is a crucial prerequisite for obtaining accurate analytical data, since no analytical method, however well designed, will be able to rectify a distortion of the results caused by the improper application of the test substance, the harvesting, handling, shipping, storage and processing of the field samples as well as any inhomogeneities or contamination of the analysis sample. Field trials in the EU and in the US are now exclusively performed according to the principles of Good Laboratory Practice (GLP) which ensure that the formal aspects of these studies adhere to a uniform quality standard. However, the technical and scientific quality of the field trials still remain the exclusive responsibility of the Study Director (or Principal Investigator) of that study. Standardized protocol, study design and field data formats, potentially even in the form of electronic field notebooks, ensure not only the validity and a high quality standard but also a comparability of the residue data thus obtained for different trials and test substances.

The relevant pesticide residues are first determined in extensive Nature of the Residue Studies (metabolism studies). The results of these studies provide the foundation for the development of residue analytical methods. Registrants of pesticides develop these methods to determine the magnitude of the residues from supervised field trials in order to provide the basis for the setting of MRLs (tolerances). Residue analytical methods are also used by the various national agencies in residue monitoring programs, for the enforcement of the maximum residue levels (MRLs) as well as to seek out illegal pesticide residues in the ever faster moving food shipments or in the drinking water. In either event, the analytical data produced are used to protect the environment and the consumer from the potential risks of the exposure to pesticides in food of plant and animal origin and in the drinking water. Based on the questions that need to be answered by these various analyses, the residue analytical method must fulfill most or all of the following criteria:

1. specific analyte detection
2. unequivocal identification of one or multiple analytes
3. low limit of quantitation (LOQ)
4. low cost of analysis
5. short time of analysis
6. availability (short time interval from sampling to analytical results)
7. reliability of results

One of the main drivers of the evolution of methodology in pesticide residue analysis has been how to better meet these criteria. The classical process that describes an analytical method can be categorized into the following major sub-processes:

- extraction,
- clean-up,
- potentially derivatization,
- chromatographic separation, and finally
- detection of the analytes.

Needless to say, the potential for the introduction of errors or losses of analytes increases substantially with the complexity of the clean-up procedure and the matrix. The modern trend towards highly specific detection techniques, often combined with modern extraction and separation techniques, typically shortens the process to essentially the extraction followed directly by the detection (*shake and shoot*). So, losses due to clean-up steps and the introduction of potential errors are minimized. Thus, the results obtained are more accurate and reliable and provide a sounder base for the decisions derived from them.

Technical developments in the recent past have provided the analytical chemist with some very exciting, innovative, and powerful tools for this analysis. It is important to note that no single tool is capable of answering all the analytical questions. Instead, it is still the responsibility (and forte) of the analytical chemist to select from the array of the available techniques those which optimally suit the particular application.

2 NEW EXTRACTION TECHNIQUES

In most cases, the pesticide residues have to be extracted from the matrix before they can be further analyzed. The employed extraction technique should extract the desired analytes quantitatively and preferably selectively, i.e., only the analytes and as little as possible of the interfering matrix constituents that need to be removed by further clean-up procedures. In recent years, a number of extraction schemes utilizing Supercritical Fluid Extraction (SFE) have been reported in the literature. However, the routine use of this technology has been hampered by limitations (e.g., limited sample size and numbers, clogging of the apparatus by matrix constituents, polarity of the analyte) of the commercially available instrumentation. Improvements in the instrumentation have now overcome these limitations in many cases. The variable parameters of this extraction technique, like temperature, flow-rate, pressure, and most importantly the number, types, and amounts of modifiers which can adjust the efficiency and selectivity of the extraction, make SFE a very powerful tool. SFE has successfully been used with soil, plant matrices, and animal tissues. As an added advantage, the extraction of a sample series can be automated to a large extent.

Other approaches to automated extraction have been pursued by several instrument manufacturers and various designs of modified Soxhlet extractors, modified SFE, and Microwave Extraction equipment are now commercially available. The instruments use computer–controlled temperature programs to simultaneously or consecutively extract several samples in a semi-automated and controlled fashion. The currently available automated extraction instruments afford efficient extractions for a large spectrum of analytes, matrices, and sample sizes.

The automated extractors allow for a very standardized extraction protocol with high reproducibility. Compared to the conventional manual solvent extractions, these new extraction technologies not only save analyst time but have the following added advantages:

- efficiency,
- high reproducibility, and therefore
- accuracy of the extraction scheme.

3 CAPILLARY ELECTROPHORESIS (CE)

The development of new pesticides and the requirement to detect polar metabolites in residues have opened the spectrum towards more hydrophilic analytes. In many cases it has proven difficult to obtain chromatographic separation of these analytes with the high specificity required by environmental analysis. In recent years, capillary electrophoresis has increasingly been used to analyze polar pesticide residues in water, soil, and plant matrices[1]. Its separation principle is based on differences in electrophoretic mobility, which for the most part depends on the hydrophilicity (or the capability to ionize) of the analyte, the viscosity, the pH and ionic strength of the buffer solution, and the effective strength of the applied electric field. Therefore, the separation principle of CE is complementary to that of HPLC, which is generally based on hydrophobic interactions between the analyte and the separation system. CE has a very powerful selectivity for polar analytes[2], and in recent applications, the spectrum has even been extended to neutral analytes and non-aqueous, volatile buffer systems. The commercially available equipment is relatively easy to use. Compared to HPLC, few parameters have to be controlled. Thus, analyses with this methodology can be performed according to a detailed written protocol by analysts with only very little prior training in this technique. The sample sizes needed for CE analysis are in the nanoliter range and thus very small (however, current autosampler equipment still requires approximately 10 μL sample volumes to function properly). Therefore, CE can be a very powerful tool in conjunction with miniaturization of extraction and clean-up techniques as well as in those applications where sample sizes are limited.

The power of this highly selective separation technique is, of course, enhanced to a large extent when it is combined with the highly selective detection power of an electrospray ionization mass spectrometer (ESI-MS(/MS)), which not only allows the quantitative detection but also provides additional qualitative information about the analyte. Interfaces for the coupling of CE with ESI-MS have been commercially available now for several years. However, the technology is still in its infancy stage and currently not yet well enough developed to fulfill the ruggedness requirements of routine applications. In many cases, the quantitation values obtained by simple CE/UV will suffice.

The major current limitation of CE is the technique's relatively poor concentration sensitivity. For example, a (typical) 10 mm path length of an HPLC UV-detector's post-column flow cell compared to the 50 μm path length usually used in CE on-column UV detection gives the HPLC/UV detector a 200-fold sensitivity advantage over CE/UV. However, this problem can usually be overcome by appropriate extraction techniques as well as off-line or on-line pre-concentration techniques. Several publications on pesticide analysis with CE-based methods reported LOQ's at levels of 0.05 ppm. A.J. Krynitsky (US-EPA) and M.M. Safarpour (American Cyanamid Company, U.S.A.) reported good recoveries and adequate sensitivity at the 2.0 ppb level (LOQ) for several sulfonylurea herbicides in soil. They also concluded[3]: *CE/UV is a good complementary technique to existing HPLC/UV methods, but should not be interpreted as a replacement for HPLC.*

4 GAS CHROMATOGRAPHY (GC)

Gas Chromatography has been the method of choice for routine pesticide residue analysis and tolerance enforcement now for many years. More recently, the introduction of new sample introduction systems, e.g., on-column injection and loop-type and programmed temperature vaporizer (PTV) [4] injectors, and new detection systems, e.g., Atomic Emission Detectors (AED)[5,6,7] and various mass spectrometers, have dramatically widened the scope of this technology. Isotope dilution (ID), i.e., the addition of an internal standard which is labeled with a stable isotope (e.g., ^{13}C, ^{15}N, or ^{2}H), can significantly increase the sensitivity and accuracy of GC/MS(/MS) analysis. The isotope labeled analyte as an internal standard usually has identical chromatographic properties as the analyte and thus alleviates the decay of method accuracy and precision due to matrix interference.

Mass selective detectors are currently available in many laboratories and several multi-residue methods for tolerance enforcement now utilize this technology. The first generation of mass selective detectors were essentially limited to Electron Impact Ionization (EI) and could thus detect analytes only in the positive ion mode. The new generation of instruments now also allows for the Chemical Ionization (CI) and detection of the analytes in positive and negative ion mode. The instrumentation has also become smaller (bench top models) and less expensive. The sensitivity has increased by about one to two orders of magnitude compared to the first generation instrumentation. The equipment is now robust and relatively uncomplicated, easy to handle, and very well suited for the routine analysis of extensive sample numbers.

Parallel to the development of the mass selective detectors, the ion-trap technology has been increasingly improved. While the mass selective detectors have their maximum sensitivity in the *selected ion monitoring mode* (SIM), the ion-trap instruments have a comparable sensitivity also in the full scan mode (i.e. over the entire observed mass range). This results in an increased flexibility for the analyst, as the choices of masses for quantitation are wider and potential interferences by matrix constituents can be circumvented.

Today, the most important improvement for residue work is probably the selective reaction monitoring capability of the ion-trap instruments, This allows GC/MS/MS measurements at trace levels while combining the high resolution power of the GC with the orthogonal highly selective detection method of MS/MS at still significantly lower cost for the instrumentation in comparison[8] to LC/MS(/MS) methods. Some compound classes undergo ion/molecule-reactions in the ion-trap, which leads to poor reproducible results und thus renders the ion-trap mass spectrometer as an unsuitable detector for the quantitation of these particular analytes.

Due to the high resolution power of GC, the added specificity of the MS/MS detection is rarely needed in routine analysis. However, the GC based methodology often requires extensive clean-up procedures to eliminate interfering matrix constituents. Furthermore, it is still limited to the measurement of volatile polar and non-polar analytes, unless error prone and tedious chemical derivatization steps are included in the sample preparation procedure. Another drawback of the mass spectrometer detection system,as compared to conventional GC detectors, is the need for a high level of expertise and experience of well trained operators and analysts for these complex instruments in order to obtain reliable data.

5 LIQUID CHROMATOGRAPHY (LC)

The most significant advancements in LC instrumentation over the past four years have occurred, for the most part, in the area of LC/MS(/MS) detectors, in particular, the improvements in the Atmospheric Pressure Ionization (API) mass spectrometers.[9]

5.1 LC/MS

First, consider LC/MS. Several single quadrupole API instruments are now commercially available. These instruments lack the high selectivity of the ion-trap and triple quadrupole instruments described below. Their application in routine pesticide residue analysis is rather limited since they are very susceptible to matrix interferences from compounds with the same molecular mass as the pesticide analytes of interest. So, analytical methods utilizing this detection technique either require matrices with very few interfering compounds or an extensive clean-up procedure. Therefore, the application of single quadrupole mass spectrometers in pesticide residue analysis is rather limited to special cases. Typically then, this methodology is not very suitable for multi-residue analyses in tolerance enforcement.

5.2 Ion-trap LC/MS(MS)

The advancements of ion-trap MS-detectors for LC make them very powerful tools for pesticide residue analysis, because they combine the high sensitivity (> 100 × compared to single quad instruments in full scan mode) and selectivity (MS^n) of the ion-trap MS-detector with the separation power of liquid chromatography. Is this now to be the method of choice which promises to close the gap of polar, hydrophilic analytes, which cannot be analyzed by GC-based techniques without prior derivatization? When the ion-trap technology is compared to the recently advanced triple quadrupole–based LC/MS/MS detectors, the ion-trap has some significant differentiating features, which make it a very powerful analytical tool especially if it is regarded as complementary to the triple quadrupole instruments. The fragmentation in the ion-trap allows the sampling of MS^n, with n = up to 5 or more, thus providing extensive structural information which might allow the identification of unknown analytes at trace levels. However, the ion-trap is often less sensitive and has a smaller dynamic range than the triple quadrupole instruments. In comparison to the triple quadrupole instruments, the ion-trap mass spectrometers are less expensive. However, often a larger variation (c_V) of the quantitative results has been observed which makes the ion-trap less suitable for quantitative pesticide residue analysis, whenever a high degree of reliability of the residue results is demanded.

5.3 LC/MS/MS (Triple Quadrupole API-MS)

Recently, the main focus in the area of pesticide analysis by LC/MS/MS has been on the API triple quadrupole instrumentation. Applications of this technology have already been presented at preceding IUPAC meetings. Since then, the accessibility of LC/MS/MS has increased due to the dramatic decrease in the cost and size of the instrumentation over the past four years. In addition, the commercial instrumentation just introduced this year avows an increased sensitivity of at least a factor of 10.

Certain analytes are difficult to measure with the sensitivity required for pesticide residue analysis but the different modes of API-MS allow the analysis of a wide variety of analytes. API-MS with electrospray in the positive ion mode is favorably used for the detection of alkaline-type (proton-acceptor) analytes while negative ion spray is predominantly used for acidic-type (proton-donator). Those compounds, which are rather neutral and often non-polar and which show poor sensitivity in both positive and negative electrospray API-MS, can usually be successfully analyzed by *Atmospheric Pressure Chemical Ionization Mass Spectrometry* (APCI-MS). In this special case of API/MS, the quantifiable ions of the neutral analytes are formed by chemical ionization (ion/molecule reactions) in the gas phase. Even though several analytical runs might be necessary (APCI-MS requires a different source set-up than electrospray MS), the API-MS technology covers essentially the entire polarity range of analytes.

The main advantages of LC/MS/MS can be summarized as follows:

- suitable for polar and non-polar analytes
- separation power of HPLC, combined with
- high sensitivity and
- high specificity due to MS/MS detection
- individual multi-analyte detection

Both, ion-trap and quadrupole instruments are now available in a sufficient number of laboratories and a wide enough range of applications have been reported in that these technologies can be considered rugged enough for routine analysis. In summary, the features and advantages of modern LC/MS/MS often allow for a rapid, highly selective and reliable analysis, often alleviating most or all additional clean-up steps following the extraction.

However, despite the dropping costs for the instrumentation, the overall initial investment is still high in comparison to other analytical technologies (e.g. HPLC/UV, immunoassays) and the complexity of the instrumentation and the technique requires significant expertise and a very well–trained and experienced staff. Another drawback of the technique is the requirement of volatile buffer solutions (e.g. ammonium salts of formic acid or acetic acid). Interfering ions, like phosphates, halogenated organic acids, borates or most ion-pairing reagents have to be removed prior to analysis.

Despite the high selectivity of the MS/MS detection technique, interferences from matrix constituents are frequently observed and demand additional clean-up of the crude extracts, e.g., by solid phase extraction (SPE) techniques or on-line LC/LC-coupling.

Overall, the advances in LC/MS/MS technology have provided the analytical chemist with a very powerful tool to rapidly and accurately measure pesticide residues of a wide polarity range in essentially all matrices at very low quantitation levels (LOQ) without chemical derivatization. Several analytes can be detected in a single analysis. A complete chromatographic separation as with conventional chromatographic techniques (e.g., HPLC/UV) is not absolutely necessary, since overlapping analyte peaks can be potentially distinguished unequivocally by the mass selective detection in the MS/MS detector. These features make this methodology very attractive for the multi-residue testing used in tolerance enforcement.

6 IMMUNOCHEMICAL TECHNIQUES

Immunochemistry has provided the analytical chemist with a whole new battery of tools. It can be subdivided into two main areas: the immunoassays, which utilize antibodies as biochemical detectors, and secondly, the immunoaffinity techniques in which antibodies are immobilized on a carrier matrix allowing the selective separation of analytes, e.g., immunoaffinity chromatography. However, a detailed description of the principles and various facets of immunochemical methods is beyond the scope of this contribution. A recent review describing a broad variety of applications and contemporary aspects of the immunochemical technologies can be found in the *ACS Symposium Series* publication on Immunoanalysis of Agrochemicals.[10]

The main focus over the past 15 years has been on the ELISA (enzyme linked immunosorbent assay) format which has become well–established in the areas of medical diagnostics and medicinal chemistry as well as in the monitoring of pesticides in the environment. More recently, this technology has been increasingly utilized for the determination of pesticide residues in foodstuffs and environmental analyses for the registration of pesticides. This trend is reflected in the growing number of publications in the literature as well as the number of posters for this topic at this Congress.

A variety of different approaches can be used in the design of immunoassays, depending on the particular questions needed to be answered. The immunoassay utilizes the very selective and specific differentiation power of the mammalian immune system and is thus a truly orthogonal method to all conventional analytical detection techniques. The detection principle can be simply described as a lock and key principle, in which the specificity of the detection system is highly dependent on the design of the assay. Based on the objectives for the analytical method, an immunoassay can be designed to be either highly specific for a single analyte or it can equally detect or quantify a defined group of analytes with common structural features (e.g., a class of pesticides or a parent pesticide and its metabolites). However, since the specificity of an immunoassay cannot yet be unequivocally predicted, an immunoassay-based analytical method has to be tested and validated to fulfill a stringent set of quality criteria. Recently, Krotzky and Zeeh[11] proposed guidelines for the precision, standardization, and quality control of immunoassays which are used for the residue analysis of agrochemicals.

Immunoassays have been developed for quite a number of pesticides. They have been used as the detection system in residue analytical methods in a large variety of matrices. The high specificity and sensitivity of the detection system often allows the direct detection of the analytes in water samples or crude extracts without any or with little prior clean-up. Limits of quantitation of 0.01 mg/kg in plant and animal matrices and even lower LOQs in drinking water can routinely be achieved, usually without any additional and potentially error-prone preconcentration procedures. The advantages of immunoassay-based methods can be summarized as follows:

- high speed (little time from field sample to analytical result)
- high efficiency (number of samples compared to analysis time)
- low cost (analysis cost per sample)
- suitability for a wide polarity range of analytes
- high accuracy
- high sensitivity
- high selectivity (dependent on assay design)
- portability (analysis in the field is possible)

Thus, immunoassay-based methods are very powerful for selected applications in pesticide residue analysis.

Unfortunately, the initial euphoria in the pesticide immunoassay development produced a number of assays that did not live up to their promised expectations and hampered the further promotion of this technology. When validated appropriately, immunoassay-based methods are currently being accepted by an increasing number of regulatory agencies. This is due, in part, to the achievements of the dedicated work of a number of interest groups which have been promoting this technology over the past years. In the US, the Analytical Environmental Immunochemical Consortium (AEIC)[12], whose diverse membership includes research consultants, immunoassay equipment manufacturers, and pesticide manufacturers, has developed educational programs, furnished scientific expertise to regulatory agencies, and established performance standards for immunochemical methods. Similar groups are making comparable efforts in Europe.

Today, it can be assumed that this technology has matured from its infancy state to a powerful analytical technique which complements the array of conventional tools of the analytical chemist. Its use for tolerance enforcement can be envisioned to be very attractive due to its portability, speed, low cost, specificity, and sensitivity. However, in tolerance enforcement and residue monitoring programs methods which simultaneously can detect multiple analytes are typically preferred. Nonetheless, the emerging technology of miniaturized microspot multianalyte immunoassays[13] is set to become a powerful, fast alternative to conventional multi-residue methods.

7 CONCLUSIONS

Pesticide Residue Analysis of samples from supervised field trials determines the amount of residues likely to occur in foods. The results of these anlayses are an essential factor for establishing the Maximum Residue Levels (MRLs). Thus, Analytical Chemistry provides a foundation for decision making in the context of risk assessment during the registration process of pesticides, in tolerance enforcement, and in monitoring programs. The importance of these decisions demands a highly sensitive, specific, accurate, and reliable methodology.

In recent years, the main driver for the development in analytical instrumentation has been the demand for increased sensitivity, for very low Limits of Quantitation (LOQ). However, during the development of analytical methods with ever lower demanded LOQs, it has more and more frequently been found that the reliability of the results at or near the LOQ is concomitant with a decreasing level of confidence. Often the demand for very low LOQs could only be met by complex multi-step clean-up procedures and by additional derivatization steps to prepare the analytes for the highly sensitive detection techniques. Increases in the complexity of the clean-up procedure opens the door for the introduction of errors or losses of analytes. Other problems arise due to the variability of matrix constitutents between samples and the general variability of the detection system used. These problems are not always overcome by increased sensitivity of the detection method.

So even though the recent focus has been directed to greater sensitivity, the more significant aspects in Pesticide Residue Analysis are most probably the gain in accuracy, specificity, and reproducibility as well as the widening of the polarity range for the analytes. The trend toward hyphenated techniques in MS instrumentation (e.g., GC/MS/MS and LC/MS/MS) as well as the more frequent use of immunochemical techniques in the final quantitative step of analytical methods, lead to a higher degree of specificity. By increasing the specificity of the final detection technique, which is now often combined with a powerful separation system, less error-prone sample clean-ups are needed. State-of-the-art techniques now open the range toward more polar analytes, thereby alleviating the necessity for derivatization procedures. Still no single analytical tool can resolve all analytical problems. Instead, the multitude of available tools are complementary to each other and the most suitable technique has to be selected for each application. New emerging technologies in Mass Spectrometry and Capillary Electrophoresis (CE), as well as the maturing immunochemical techniques of Immunoassays and Immuno-Affinity Chromatography Separation provide the framework for obtaining more accurate, reliable, and judicially warranted analytical results. Thus, they provide a sounder foundation for Risk Assessment, the setting of MRLs, and Tolerance Enforcement.

8 REFERENCES

1. A. J. Kyrnitsky, D. M. Swineford, *J. AOAC Int.*, 1995, **78**, 1091.
2. K. Ohba, M. Minoura, M. M. Safarpour, G. L. Picard, H. Safarpour, *J. Pesticide Sci.*, 1997, **22**, 277.
3. A. J. Kyrnitsky, M. M. Safarpour, *111th AOAC International Annual Symposium*, 1997, Abstract **405**, 41.
4. H.-J. Stan, M. Linkerhäger, *J.Chromatogr. A*, 1996, **750**, 369.

5. K.-C. Ting, P. Kho, *J. Assoc. Off. Anal. Chem.*, 1991, **74**, 991.

6. N. Olsen, R. Carrel, R. Cummings, R. Rieck, S. Reimer, *J. AOAC Int.*, 1995, **78**, 1464.

7. J. L. Bernal, M. J. del Nozal, M. T. Martín, J. J. Jiménez, *J.Chromatogr. A*, 1996, **754**, 245.

8. I. Liska, J. Slobodník, *J.Chromatogr. A*, 1996, **733**, 235.

9. J. Slobodník, B. L. M. van Baar, U. A. T. Brinkman, *J.Chromatogr. A*, 1995, **703**, 81.

10. *Eds.*: J. O. Nelson, A. E. Karu, R. B. Wong, 'Immunoanalysis of Agrochemicals', ACS Symposium Series **586**, Washington, DC, 1995.

11. A. J. Krotzky, B. Zeeh, *Pure & Appl. Chem.*, 1995, **12**, 2065.

12. www.immunochem.org

13. R. P. Ekins, F. W. Chu, *Clin. Chem.*, 1991, **37**, 1955.

Pesticide Residues and the Consumer

Alastair Robertson

DIRECTOR OF TECHNICAL OPERATIONS, SAFEWAY STORES PLC, HAYES, MIDDLESEX UB3 4AY, UK

INTRODUCTION

Many may ask what food retailers know about Pesticide Chemistry. That is a fair question, and neither I nor any of my colleagues would pretend to be experts in pesticide discovery, synthesis or environmental profiling. However, our understanding of the pressures and forces at work in the industry today, and their effect in shaping policy and industry activities, is vital knowledge for all from pesticide chemists to producers supplying the major multiples such as Safeway.

Agriculture In Review

Paradoxically, in order to look forward, we need to look back and review where we've come from. Following the second world-war, the infrastructure of the UK was in dire need of a rebuild from manufacturing and service industries through to farming. The pressures of war had led to food shortages and rationing, and as well as rebuilding industry and the economy, there was a need to ensure the availability of good quality food to feed the nation. Agriculture rose to the challenge, and equipped with pesticide formulations that seem distinctly primitive when compared to today's compounds, farmers started to bridge the production gap. There was an optimism in post second word-war Britain that things could only get better, and that the science that had won the war would have spin-offs for the man in the street. Nuclear fission soon began to generate clean and efficient power from the UK's first commercial Magnox reactor which opened in Caldwell in 1956.

In agriculture, however, it was not long before questions started to be asked about the way we produce our food, and our over-reliance on pesticides. Could the availability of good, cheap food be too good to be true? Was this promise starting to ring hollow? It was not long before cracks started to show. In 1962 the landmark book 'Silent Spring' by Rachel Carson highlighted the adverse effects of pesticides on the environment. Maybe pesticides had a flip-side which needed to be considered? In the years which followed natural was good, and concern was voiced from some quarters that the potential risks of pesticides outweighed the benefits. This view has not diminished, but had been tempered in recent years with the emergence of other concerns allied to food production, and the realisation that food like life is not risk free.

Agriculture Today

Where does this leave us today? Recently, we've all been aware of the powerful forces at work that have impacted on our everyday lives and the livelihoods of those in UK agriculture. These forces have been instrumental in setting the scene for the legislative framework and consumer perception moving forward. Many of us and in particular our farmers have been on the BSE rollercoaster, experiencing constantly changing advice and confusion. The safety of genetically modified crops has been under the spotlight with debate from all quarters ranging from pressure groups, through manufacturers and retailers to regulators across Europe.

Moving specifically to pesticides, over the last two years we have seen the discovery of residues of organophosphate compounds in individual carrot roots in excess of the MRL. This was surprising as these compounds had been applied in accordance with label recommendations and Good Agricultural Practice. In the UK our Government has wasted no time in reacting to calls for accountability and transparency. Its promise of a body responsible for transparent regulation and with the role of consumers champion is emerging in the shape of the Food Standards Agency. Already we are seeing the framework of this body emerging, as brands associated with adverse results through surveillance are named and shamed.

Consumers And Food Risk

The important issue here is that we focus on the customer. As retailers, the customer is king. We exist to serve our customers, and in the evolving European and world agriculture, the realisation is finally dawning that we are all in the same extended food chain producing food for the final consumer who is ever more demanding.How do consumers in general deal with food risk? In a nutshell, food risks are not easily rationalised in the time our customers have to do their shopping, or without the background that comes with an in-depth knowledge of the subject in question, be it pesticide chemistry or microbiology. In our experience, food issues tend to break down into two main types. There are 'hard' issues such as *E Coli* 0157, BSE and pesticides, and 'soft' issues such as biodiversity, food choice and food miles. What we find is that when considering their personal health and welfare, the hard issues are consumers' number one priority. The softer issues such as animal welfare and the environment are also important, but are secondary, only being addressed once the consumers primary goals are satisfied. However, although there are clearly two tiers of consumer concerns here, both are vital in building and maintaining consumer confidence.

Being a successful retailer in todays' market is no longer a case of just selling food and making money. Indeed, being associated with the food chain in any sense carries certain responsibilities. This fact has not gone unnoticed by the major agrochemical companies some of which have established food chain teams to interface and build relationships with the prime movers in the food industry going forward.

Returning to the issue of food risk, trust is all important in building confidence and that warm feeling that comes with knowing all is well with food and the manner in which it has been produced. In recent times consumers feel that trust has been betrayed. It is the responsibility of the industry to work together to rebuild this trust, and through increasing transparency and accountability demonstrate we produce not only safe and wholesome food, but that we are working to address some of the softer secondary issues that are important to consumers in general.

The Role Of The Retailer

In setting the scene it is important to understand the retailers' role in the process. Retailers like Safeway are the primary interface between the industry and its customers. We receive over 5000 letters per month, and answer questions about all aspects of food production from the regulatory process for genetically modified foods, through food composition and intolerance, to the role of pesticides in agriculture and food production. We therefore have a good understanding of what our customers want and what is important to them. If we wish to continue to be their first choice in the competitive market moving forward, providing quality products at competitive prices is merely first base.

Our customers are increasingly more demanding and we are rising to this challenge, anticipating and developing initiatives which address and often exceed their expectations. This however is a considerable challenge. We trade with over 2000 Safeway Brand suppliers world-wide, and are investing significant effort in developing systems of control and best practice through a process of continuous improvement. We must exercise strong leadership, clear direction and encouragement for our suppliers, and ensure that initiatives we progress and support are co-ordinated across the industry for maximum benefit. Although challenging, we relish the opportunity of catalysing and driving forward positive change, and indeed our customers would expect nothing less.

Supplier Controls And Standards

So who supplies Safeway, how are their operating standards set and who monitors them? For our fresh produce business, we have over 900 supplying sites world-wide, with 300 in the UK, and 600 overseas. All suppliers hold detailed specifications for the products they supply under the Safeway brand. These cover not only key quality characteristics, but outline the production controls that must be in place to ensure products supplied are safe and wholesome.

In the case of crop protection practices, as well as complying with relevant legislation, suppliers must conform to standards as laid down in the UK Fresh Produce Consortium code of practice which covers procedure from the application of pesticides through to harvest procedure and record keeping. To ensure these standards are followed, trained and qualified Safeway Technologists make direct visits to suppliers both in the UK and overseas, examining production records, and working to ensure product packed for Safeway is of the highest quality. Our suppliers undertake their own pesticide residue testing, and this is supplemented by Safeway's own targeted pesticide residue testing programme which last year included a wide range of samples representing over 25000 pesticide/crop combinations. This is optimised on a year-on-year basis, and allows the identification of trends which can be addressed directly with our suppliers. Data generated from the Safeway programme also form part of the industry data submitted to the UK Pesticide Safety Directorate, and is published in the annual report of the Working Party on Pesticide residues.

Encouraging Best-Practice And Integrated Crop Management

For a number of years we've been looking for ways to encourage the adoption of Integrated Crop Management across our fresh produce supplier base. We've also recognised the potential vulnerability of UK horticulture to a crisis similar to that which has seen significant losses for beef farmers post-BSE. Recognising the value of investment in standards and systems, in 1992 we started work with the NFU and other retailers to develop crop-specific ICM protocols. These encourage farmers to reduce agrochemical inputs on fresh produce through alternatives which include the use of beneficial insects and crop rotations.

Some five years on, and recognising the importance of ensuring these standards are being followed and supported, the Assured Produce Scheme was launched at the UK Royal Show in July 1997. The Assured Produce Scheme is at the core of Safeways sourcing policy for UK-supplied fresh produce. Under the Assured Produce Scheme suppliers and their growers are visited not just by Safeway Technologists - who number 19 in our Produce Business Unit alone - but are also verified by independent auditors employed under the scheme itself. This means that over 6000 growers supplying both Safeway and the other UK retail multiples supporting Assured Produce will receive visits over the next 3 years, checking that standards as laid down under the scheme are being followed. This is a substantial task; however, it is clear is that standards are of little use unless they are checked, and that is exactly what this scheme sets out to do.

Key to the success of a scheme such as Assured Produce are clarity and focus, universal support, independent verification and continuous improvement. The universally supported crop protocols are clear in their requirement and provide a focus on the standards required. In having universal support from all the major UK multiple retailers, the scheme avoids the duplication that could have resulted from the myriad of different schemes that could have developed. The need for verification has been implemented with minimum cost, and the independent verification of standards at grower level has meant the scheme is open to challenge. A strong emphasis is placed on continuous improvement through yearly self-assessment audits conducted by growers, and this facilitates technology transfer to all growers and not just those with the resources to seek it out. Although a key component is a focus on reducing the use of pesticides, the scheme also provides a clear justification for the use of agrochemicals which it is clear are essential to sustainable crop production.

Progress Overseas

Having described our activities in the UK, what about the fresh produce that we source from overseas? Recognising that our customers expect the same high standards from all our fresh produce whether it be grown in the UK or overseas, Safeway looked for a way to establish an parallel initiative, recognising existing systems and encouraging the establishment of best-practice production methods for sources across Europe and beyond. What was clear from our experience in the UK was that co-operation and agreement on common standards was essential.

With this in mind and in conjunction with the EHI Institute in Köln, Germany, we established EUREP, the Euro-Retailer Produce working group. Since its inception and over the past 18 months, EUREP has grown to an association of over 25 European Retailers. Finding a way through the spectrum of retailer requirements has not been easy. In the UK our emphasis is on food safety, whereas the Dutch and Swedish have a strong focus on environmental protection. Some retailers have prescriptive lists of chemicals that must not be used on crops even though they may be approved, whilst others stipulate year-on-year reductions in pesticide applications. Standardisation of requirement and encouragement of ICM is the driving force behind EUREP, as well as a desire to support the common goal of establishing GAP and ICM for world-wide sources through co-operation. Safeway currently chairs this initiative, and in so doing has encouraged the development of a code of Good Agricultural Practice. This code will be shortly be issued to all suppliers, and will form a solid foundation for moving forward recognising the different approaches that currently exist across world-wide sources.

The Role Of Genetic Modification In Crop Production

In a future world agriculture, genetic modification offers a different approach to crop protection, but one that is surely here to stay. It does, however, raise certain issues across the chain from farmer through manufacturer and retailer to consumer. Over the past 10 years, we have seen an explosion in research on broad-acre crops such as rape, soybeans and maize, with a focus on agronomic qualities such as increased yield, improved weed management and pest and disease tolerance. There is no doubt that a genetic approach to crop protection is a valuable one, with the potential to reduce agrochemical inputs, increase yields and grow crops on marginal land previously unsuitable for crop production.

There are however some specific issues associated with the adoption of this genetic modification that must be considered. A challenge similar to that which we face with pesticides is the establishment of regulations for genetically modified crops across international boundaries. We now trade in a world market, and in this respect regulations must take a consistent approach across all sources with which we trade.

As each year goes by, the technical competency needed to farm to more stringent standards and to utilise new developments effectively and safely increases. This is also the case with genetically modified crops, where there is a very real need for increased management and knowledge at farm level of the benefits and the issues associated with genetically modified crops. Such crops are not currently being grown on a commercial scale in the UK, but as we have seen from experiences in other countries where GM crops are being grown, it is important that the farmer is sufficiently educated to recognise and deal with both technical and consumer issues as they arise.

Key crop protection agents such as Bt are vital not only to conventional but also organic agriculture, and farmers must ensure that appropriate resistance management strategies such as refugia are used to reduce the selective pressure on pests to develop resistance. This need for knowledge is being addressed by the Biotechnology companies themselves and organisations such as the British Agrochemicals Association in the UK, and it is vital that this knowledge is translated into practice at farm level.

Last but not least is the issue of segregation, traceability and consumer choice, and in this respect there is one point that should be carefully considered. Elsewhere in the food industry, and indeed in other industries, progress is normally associated with increased control over the operation in question. It seems strange, therefore, that in the case of genetically modified crops, their co-mingling at harvest and through transport with conventional counterparts means we are actually reducing our control and not enhancing it.

Agriculture And Sustainability

In the future, there is no doubt there will be an increasing pressure on the ability of world agriculture to feed a growing population. With the total global population projected to reach 9 billion by 2035, our capability to produce within ever more demanding environmental standards will be put to the test. Currently concerns centre around the sustainability of food production; however, the perspectives on this subject tend to differ from developed to developing countries. In the developed world we are concerned about our ability to continue to produce high quality food with minimal impact on the environment. However, developing countries are more likely to be concerned with continued food security, especially where the risk of crop failure is high. World-wide, increasing controls are being placed on agriculture from both legislators and industry. These are significant challenges, and challenges which we must face.

Organic production is not a serious proposition for the majority of the mouths that have to be fed; however, in the developed world we will surely see a continued increase in organic production, as growers become better structured, and are more able to respond to market requirements. In relation to conventional agriculture, there is no doubt that pesticides are here to stay. However, their use is likely to be on different terms and in a different framework to that which we have become accustomed. Those terms include justified and responsible use, and application a component of an integrated and sustainable approach to crop production. The use of pesticides will be increasingly controlled by pro-active industry-driven systems such as the Assured Produce Scheme in the UK. Legislation will set the starting point for agricultural practice, and will not in itself define best practice. What should be encouraged is the development of a 'toolkit' for agriculture. Different approaches may form part of this toolkit, and provided they are used in accordance with best practice, farm management should be the responsibility of the farmer who is responsible for his crop on a day-to-day basis.

Future Challenges

What are the challenges moving forward? There is no doubt that the market both for food and raw material commodities is now truly international. This brings benefits in terms of year round range and availability, and additional challenges as we seek to develop uniform international controls and standards. As these standards develop, there will be opportunities for pro-active producers who can meet those standards, and pressure placed on those where investment is required to improve. The issue of competitiveness is not restricted to growers, and all businesses which operate on the international stage must seek to invest wisely to ensure secure futures in their our own sectors, and opportunities in new and complementary markets.

In writing this paper, I sought to identify two simple messages that I considered important to the target audience. They are:

- Recognise that whatever your role you are part of the food chain. The food industry is larger and more complex than ever before, and increasingly interfaces with new organisations. Over the past 10 years, Safeway has moved from liaising with our suppliers, to increased dialogue with those spanning the food chain from Academics through Agrochemical and Biotechnology Companies to Legislators. We must all strive to improve our understanding of our role in this process, and the contribution we can make to the debate.

- Maintain your focus on the end consumer as your ultimate customer. This is critical, as without continued consumer acceptance and support we may win the battle but lose the war.

Evaluation of Pesticide Residues in Water

E. Jaskulké, L. Patty and A. Bruchet

SUEZ LYONNAISE DES EAUX-CIRSEE, 38, AV PRESIDENT WILSON, 78230 LE PECQ, FRANCE

Contamination of water resources by pesticide residues is one of the major problems to be addressed for the preservation and the sustainability of the environment. Since pesticide residues may sometimes occur in drinking water it is essential to set standards that will protect consumers' health. The first establishment of standards dealing with pesticides in waters has been carried out at the European level, with the implementation of the Drinking Water Directive (80/776/CEE). However, at that time, analytical tools did not allow the monitoring of more than one pesticide (atrazine) at the 0.1 microgram per litre level. Today, two new key factors have speeded up the R&D on the analysis of pesticide residues : the establishment of new directives, which may be strongly related to the WHO recommendations, related to individual toxicity of pesticides and the evolution of analytical tools, which could be useful for multiresidue analysis, with lower costs and higher sensitivity.

Because the validation of analytical methods for a large range of pesticides is complicated and very demanding in resources, seven major private and public European research centres[*] decided in 1996 to share their analytical capacities and expertise in order to develop common multiresidue methods. Supported by the Commission of the European Community (Standards Measurements and Testing programme) the STM4-CT96-2142 project aims at providing control laboratories with reliable and comparable analytical methods that allow the monitoring of priority pesticides in drinking and related waters. At present a common list of 38 priority pesticides has been established by consensus and two multiresidue methods based on Solid Phase Extraction in off-line with GC-MS and HPLC-DAD have been optimised. Both methods, under evaluation through an intra-laboratory exercise, will be evaluated as well through an inter-laboratory exercise, in 1999.

Many surveys dealing with the control and the occurrence of pesticides have been carried out in France. These show that a wide variety of pesticides can be detected in French natural waters in relation to the strong and diverse agricultural activities of this

[*]

CIRSEE-SUEZ LYONNAISE DES EAUX (FRANCE)	Dr A. BRUCHET
ANJOU-RECHERCHE (FRANCE)	Dr J.L. GUINAMANT
SVW (BELGIUM)	Dr F. VAN HOOF
DVGW-TZW (GERMANY)	Dr F. SACHER
KIWA (THE NETHERLANDS)	Dr I. BOBELDIJK
AGBAR (SPAIN)	Dr F. VENTURA
LNEC (PORTUGAL)	Dr M.H. MARECOS DO MONTE

country. If atrazine still remains the most frequent pesticide present in the rivers, new substances, such as atrazine metabolites (N-de-ethylatrazine) and substituted ureas are also frequently detected with a level higher than 0.1 µg/l.
Research carried out during the last decade has however allowed the water supply companies to implement appropriate treatment processes.

INTRODUCTION

In Europe, commercialisation and use of pesticides are authorised within a precise regulatory framework in order to limit human risks and environmental pollution. In addition, a regulation relating to water quality enforces the assessment of drinking water. This regulation, based on the European drinking water Directive (80/778/EEC), revised in October 1997, states that individual pesticide's concentration and the sum of all individual pesticide in drinking water should not exceed 0.1 and 0.5 µg/l, respectively. According to the latest version of this Directive, the analytical method used shall, as a minimum, be capable of measuring concentrations equal to the parametric value (100 ng/l) with a trueness, precision and limit of detection of 25 %. Consequently, it becomes necessary to provide control laboratories with analytical methods that allow the monitoring of pesticide residues at this trace level, with basic performance data in keeping with the revised drinking water Directive requirements.

In such a context, seven major private and public European Research Centres decided to share their analytical capacities and expertise to develop and validate a limited set of reliable and comparable multi-residue methods that allow the monitoring of priority pesticides in drinking and related waters.

This paper presents the state of work of this project and various results related to Lyonnaise des Eaux pesticides monitoring carried out in the last 15 years. Finally, a short presentation of the available treatment processes illustrates the different way pesticides can be removed from drinking water.

1 OPTIMISATION AND EVALUATION OF MULTIRESIDUE METHODS FOR PRIORITY PESTICIDES IN DRINKING AND RELATED WATERS

The European SMT4-CT96-2142 project, started in February 1997, aims at supporting the implementation of the drinking water Directive by providing control laboratories with a limited set of reliable and comparable validated analytical methods that allow the monitoring of priority pesticides in drinking and related waters. To bring this project to a successful conclusion, two main objectives have been defined : the establishment of a list of pesticides of common interest in European countries, and the development and the intra- and inter-laboratory evaluation of multi-residue methods based on Solid Phase Extraction in off-line with GC-MS, HPLC-DAD and HPLC-MS.

1.1 Materials and Methods

1.1.1 Establishment of the Common List of Priority Pesticides. The selection of priority pesticides of common interest in European countries has been carried out in various steps and the final list has been reached by consensus between partners. At first, the compilation of national lists coming from Germany, Belgium, France (ESU Ecotox List), Austria and U.K., gave a first global list of 148 substances. As partners agreed on the procedure used for the establishment of the French list, this one has been considered as a reference. So, after a first selection based on the presence of the 148 pesticides in at least two countries plus France, a second list of 68 pesticides has been obtained. At this stage, three new lists coming from Spain, The Netherlands and Germany (up-dated list) have been considered and a second selection has been done. This one was based on the occurrence of the 68 compounds in the different lists and on the exclusion of several substances. Pesticides like aminotriazole, for which extraction and analysis are too particular, chemicals for which a derivatisation is suitable before analysis and compounds for which an ISO standard is currently underway were excluded. After this second selection, a final list of 38 priority pesticides was obtained.

1.1.2 Chromatographic Analysis of Priority Pesticides. According to partners' experience in the analytical field and available information concerning the analysis of selected substances [1], screening exercises have been carried out in Gas Chromatography coupled with Mass Spectrometry Detection and in High Performance Liquid Chromatography coupled with UV-Diode Array Detection:

- GC-MS: standard solutions including the selected substances were analysed by GC-MS, using two injector types (Split-splitless or On-column) and two mass spectrometer types (Quadrupole or Ion trap) (Table 1)

- HPLC-DAD: 250 µl of a standard solution (at 250 µg/l) prepared in Methanol-water (10:90) and including pesticides that should be analysed by HPLC, were injected in a Hewlett-Packard 1090 Liquid Chromatograph equipped with a UV-Diode Array detector. Pesticides were separated on a Supelcosil ABZ + column (250 x 4.6 mm x 5 µm) at 40°C, using a water-acetonitrile elution gradient (Table 2).

Table 1. Operating conditions for GC-MS screening exercises

	CIRSEE	KIWA	SVW	DVGW-TZW
Chromatograph	Varian 3400	Carlo Erba HRGC	Varian 3400CX	HP 5890 Series II plus
Injector Temperature	On-column 50 °C to 260 °C at 180 °C/min 29 min at 260 °C	Split-splitless 270 °C	Split-splitless	Split-splitless 60 °C to 290 °C at 9 °C/min 5 min at 290 °C
Detector	Finnigan MAT Ion Trap ITS-40	Quadrupole QMD-1000	Varian Saturn 4D Ion Trap	Quadrupole MSD HP5972
Column (film thickness) Length x ID	PTE5 (0.25 µm) 30 m x 0.32 mm	BPX5 (0.25 µm) 50 m x 0.32 mm	DB5-MS (0.25 µm) 30 m x 0.25 mm	DB5 (0.33 µm) 12 m x 0.2 mm
Temperature programme	50 °C to 170 °C at 10 °C/min 170 °C to 180 °C at 1 °C/min 180 °C to 240 °C at 5 °C/min 240 °C to 260 °C at 15 °C/min 2 min at 260 °C	1 min at 40 °C 40 °C to 80 °C at 20 °/min 1 min at 80 °C 80 °C to 315 °C at 4 °C/min 10 min at 315 °C	50 °C to 150 °C at 10 °C/min 150 °C to 240 °C at 4 °C/min 240 °C to 270 °C at 15 °C/min 1 min at 270 °C	60 °C to 190 °C at 20 °C/min 5 min at 190 °C 190 °C to 210 °C at 1 °C/min 1 min at 210 °C
Injection volume	1 µl	2 µl	1 µl	2 µl

Table 2. Elution gradient for HPLC-DAD analysis (Anjou-Recherche)

Time (min)	Flow (ml/min)	Water (%)	Acetonitrile (%)
0	1	95	5
75	1	37	63
80	1	0	100
85	1	0	100
90	1	95	5

1.1.3 Solid Phase Extraction of GC-MS and HPLC-DAD Amenable Pesticides. Selection of the sorbent phases most suitable for extraction of GC-MS and HPLC-DAD amenable pesticides has been carried out in various steps. Preliminary tests have been performed on three sorbent types available in cartridges (RP-C18, styrene copolymer and graphitized carbon black) [2]. The 9 best promising phases identified were then tested under a common procedure. 500 ml of low DOC water samples (< 1mg/l) were spiked at two concentration levels (0.1 and 0.5 µg/l; 5 replicates by concentration level) with pesticides of the common list. Cartridge elution was performed with ethyl acetate and methanol and/or acetonitrile for GC-MS and HPLC-DAD amenable pesticides, respectively. Organic extracts were analysed by the corresponding technique. The 9 sorbent phases considered were: Isolute-C18 (IST), Envi 18 and LC 18 (Supelco), C18 Polar plus (J.T. Baker), LiChrolut EN (Merck), SDB-1 (J.T. Baker), Env + (IST), Oasis (Waters) and a home-made mix of RP-C18 and

LiChrolut EN phases. All data collected were processed [3] and the most suitable sorbent appeared to be the SDB-1 phase for Solid Phase Extraction of GC-MS and HPLC-DAD amenable pesticides. Both extraction procedures were optimised and the following protocols were adopted :

- for GC-MS amenable pesticides, the SDB-1 cartridge (6 ml - 200 mg) is conditioned successively with 3 ml of ethyl acetate, 3 ml of methanol and 6 ml of MilliQ water at 10 ml/min (without drying between steps). 500 ml of water sample are loaded onto the column at 5-10 ml/min and cartridge is rinsed with 6 ml of MilliQ water at 10 ml/min. After drying with gas (N2) for 15 min, elution is performed twice with 2.5 ml of ethyl acetate at 0.8 ml/min. The organic extract is concentrated to 500 µl under a gentle stream of nitrogen.

- for HPLC-DAD amenable pesticides, the SDB-1 cartridge (6 ml - 200 mg) is conditioned successively with 10 ml of methanol and twice with 10 ml of MilliQ water at 10 ml/min (without drying between steps). 500 ml of water sample are loaded onto column at 5-10 ml/min and cartridge is dried with gas (N2) for 5 min. Elution is performed in two steps, with 2 ml and 8 ml of a mixture of methanol-acetonitrile (60:40) at 3 ml/min. The organic extract is concentrated to 300 µl under a gentle stream of nitrogen and the volume adjusted to 1 ml with pure water.

1.2 Results

1.2.1 Common List of Priority Pesticides. The common list of 38 priority pesticides includes 27 herbicides, 3 metabolites of atrazine, 7 insecticides and one fungicide (Table 3). As stated before (Bruchet et al., 1997), 18 substances listed are cited in the French ESU Ecotox list considered as a reference for the establishment of this common list. 21 chemicals are herbicides cited in the priority list of Mediterranean countries (Barceló, 1993) and 7 pesticides correspond to priority substances included in the Annex I of the Directive 76/464/CEE and considered as priority pollutants of the European aquatic environment. Finally 10 compounds of the common list are mobile substances used in Europe at a level of more than 500 metric tons/year level.

The common list established is completed with a specific list for each partner country including at most 20 pesticides apt to pollute local water resources (Table 4).

1.2.2 Chromatographic Analysis. GC-MS screening exercises carried out proved that 24 pesticides of the common list will be analysed by this chromatographic technique (Table 5). In addition, five substituted ureas (chlortoluron, isoproturon, linuron, metabenthiazuron and metobromuron) could be detected by GC-MS with a chromatograph equipped with an on-column injector. But, as response and peak resolution are very low, those pesticides will be exclusively analysed by HPLC-DAD.

According to the HPLC-DAD screening exercise, 36 pesticides of the common list appear to be LC amenable (Table 5). However, the UV response of various substances represents less than 20 % of that for atrazine. Consequently, 25 pesticides of the common list will be analysed by HPLC-DAD because alachlor, chlorpyrifos-ethyl, dimethoate, ethofumesate, metazachlor, metolachlor and trifluralin would not be accurately quantified.

Finally, aldicarb, dicamba, fluroxypyr and pyridate not directly amenable to GC-MS and very badly detected in HPLC-DAD, will be analysed by HPLC-MS.

Table 3. Common list of priority pesticides

Priority pesticides		French Esu EcoTox list	Mediterranean priority list	List of 132 priority substances	Mobile pesticides used in Europe (metric ton/year)
Alachlor	Herbicide	X	X		> 500
Aldicarb	Insecticide	X			> 50
Atrazine	H	X	X	X	> 500
N-de-ethylatrazine	Metabolite				
N-de-isopropylatrazine	M				
Hydroxyatrazine	M				
Carbendazim	Fungicide	X			> 50
Chloridazon	H				> 500
Chlorpyrifos-ethyl	I	X			
Chlortoluron	H	X	X		> 500
Cyanazine	H	X			> 50
Dicamba	H				
Dichlobenil	H		X		> 50
Dimethoate	I			X	> 500
Diuron	H	X	X		> 500
Endosulfan	I	X		X	
Ethofumesate	H		X		> 50
Fluroxypyr	H	X			> 50
Isoproturon	H	X	X		> 500
Lindane	I	X		X	
Linuron	H	X	X	X	> 50
Metamitron	H		X		> 50
Metazachlor	H		X		> 500
Metabenzthiazuron	H		X		> 500
Methomyl	I	X			
Metobromuron	H		X		
Metolachlor	H		X		> 500
Metoxuron	H		X		
Metribuzin	H		X		
Pendimethalin	H	X	X		
Phenmedipham	H		X		> 50
Pirimicarb	I				
Propazine	H				
Pyridate	H				
Simazine	H	X	X	X	> 50
Terbuthylazine	H	X	X		> 50
Terbutryn	H		X		> 50
Trifluralin	H	X	X	X	

Table 4. Specific lists of pesticides for each partner country

Spanish specific list	
Bentazone	H
Diazinon	I
Fenitrothion	I
Methidathion	I
Methyl-Azinphos	I
Methyl-Parathion	I
Molinate	H
Propanil	H
Malathion	I
Thiobencarb	H
Triclorfon	I
Vinclozolin	F

Belgian specific list	
Bentazone	H
MCPA	H
Dichlorprop	H
2,4 D	H
Fenpropimorph	F
Aclonifen	H
Bromoxynil	H
Clopyralid	H
Chlorothalonil	F
Prosulfocarb	H
Lenacil	H

French specific list	
Oxydemeton-methyl	I
Triallate	H
Tridemorph	F
Captan	F
Cyproconazole	F
Dinoterb	H
Diquat	H
Fenpropimorph	F
Flusilazole	F
Folpel	F
Ioxynil	H

Portuguese specific list	
Molinate	H
Parathion	I

German specific list	
Ethyl-Azinphos	I
Bentazone	H
Bromacil	H
Carbofuran	I
Chlorfenvinphos	I
Clopyralid	H
2,4 D	H
Dichlorprop	H
Hexazinone	H
MCPA	H
Mecoprop	H
Monuron	H
Parathion	I
Disulfoton	I
Lenacil	H
Metalaxyl	F
Penconazole	F
Procymidon	F
Propetamphos	I
Vinclozolin	F

Dutch specific list	
Anilazine	F
Benazolin	H
Bromoxynil	H
Chlorothalonil	F
Diazinon	I
Dicloran	F
Furalaxyl	F
Lenacil	H
Oxadixyl	F
Propachlor	H
Triadimefon	F
Triadimenol	F
Triclopyr	H
Methyl-Tolclofos	F
Heptenofos	I

H: Herbicide
F: Fungicide
I: Insecticide

Table 5. Chromatographic analysis of priority pesticides of the common list

Priority pesticides	GC-MS	HPLC-DAD-UV	Response in % (/Atrazine)	λ max (nm)	HPLC-MS
Alachlor	X		13	222	
Aldicarb			5	246	X
Atrazine	X	X	100	222	
N-de-ethylatrazine	X	X	89	222	
N-de-isopropylatrazine	X	X	88	222	
Hydroxyatrazine		X	74	222	
Carbendazim		X	40	286	
Chloridazon	X	X	62	232	
Chlorpyrifos-ethyl	X		10	232	
Chlortoluron	*	X	53	246	
Cyanazine	X	X	87	222	
Dicamba	X(Me)		#	222	X
Dichlobenil	X	X	23	232	
Dimethoate	X		6	222	
Diuron		X	36	246	
Endosulfan alpha	X				
Endosulfan beta	X				
Ethofumesate	X		8	232	
Fluroxypyr	X(Me)		#	222	X
Isoproturon	*	X	46	246	
Lindane	X				
Linuron	*	X	46	246	
Metamitron	X	X	33	310	
Metazachlor	X		26**	222	
Metabenzthiazuron	*	X	43	272	
Methomyl		X	34	232	
Metobromuron	*	X	54	246	
Metolachlor	X		17	222	
Metoxuron		X	37	246	
Metribuzin	X	X	20	286	
Pendimethalin	X	X	28	246	
Phenmedipham		X	35	232	
Pirimicarb	X	X	49	246	
Propazine	X	X	100	222	
Pyridate			5	246	X
Simazine	X	X	100	222	
Terbuthylazine	X	X	100	222	
Terbutryn	X	X	90	222	
Trifluralin	X		10	272	

X: pesticide analysed by this chromatographic technique
*: pesticide detected with an on-column injector
X(Me): pesticide detected after methylation
**: this compound co-elutes with metobromuron and cannot be quantified in presence of metobromuron (but its presence allow the quantitation of metobromuron)
#: broad peak.

1.2.3 GC-MS and HPLC-DAD Multiresidue Methods. During optimisation of extraction procedures on SDB-1 phase, two multiresidue methods were defined. The SDB-1/GC-MS and SDB-1/HPLC-DAD methods will allow extraction and analysis of 24 and 22 priority pesticides, respectively. Mean recovery rates currently obtained

(Table 6) vary from 80 to 112 %, except for lindane and trifluralin. So, for the HPLC-DAD amenable pesticides and for the all majority of GC-MS amenable ones, recovery rates lie inside the acceptable range of 75 to 125 % defined in the revised Drinking Water Directive.

Table 6. Mean recovery rates obtained with SDB-1 phase during optimisation work

Analysis Sorbent phase Elution solvent	GC-MS SDB-1 Ethyl-acetate		HPLC-DAD-UV SDB-1 Methanol/Acetonitrile (60:40)
Conc° level (µg/l)	0.1	0.5	0.2
Alachlor	98 ± 3	103 ± 7	
Atrazine	111 ± 14	104 ± 7	95 ± 5
N-de-ethylatrazine	98 ± 5	100 ± 7	104 ± 6
N-de-isopropylatrazine	100 ± 10	102 ± 4	103 ± 11
Hydroxyatrazine			94 ± 13
Carbendazim			100 ± 5
Chloridazon	100 ± 6	103 ± 9	106 ± 12
Chlorpyrifos-ethyl	83 ± 5	85 ± 7	
Chlortoluron			104 ± 5
Cyanazine	98 ± 3	99 ± 5	112 ± 6
Dichlobenil	95 ± 5	98 ± 8	
Dimethoate	80 ± 11	84 ± 4	
Diuron			104 ± 4
Endosulfan alpha	91 ± 15	94 ± 8	
Endosulfan beta	94 ± 6	95 ± 5	
Ethofumesate	96 ± 3	104 ± 6	
Isoproturon			102 ± 5
Lindane	47 ± 18	61 ± 14	
Linuron			96 ± 5
Metamitron	91 ± 8	95 ± 7	102 ± 5
Metazachlor	98 ± 2	104 ± 6	
Metabenzthiazuron			96 ± 4
Methomyl			97 ± 12
Metobromuron			99 ± 7
Metolachlor	99 ± 3	104 ± 6	
Metoxuron			106 ± 4
Metribuzin	98 ± 4	98 ± 6	107 ± 4
Pendimethalin	85 ± 6	92 ± 6	
Pirimicarb	93 ± 3	98 ± 7	94 ± 3
Propazine	100 ± 2	102 ± 9	97 ± 6
Simazine	95 ± 6	90 ± 6	104 ± 5
Terbuthylazine	100 ± 3	103 ± 5	100 ± 7
Terbutryn	98 ± 4	102 ± 6	91 ± 6
Trifluralin	62 ± 6	67 ± 9	

In the latest version of this Directive, trueness, precision and limit of detection, related to the analytical method used, are precisely defined. Trueness is the systematic error and is the difference between the mean of the large number of repeated measurements and the true value, while precision is the random error and is usually expressed as standard deviation (within and between batch) of the spread of results about the mean. In addition, acceptable precision is two times the relative standard deviation. Finally, limit of detection is either, three times the relative within batch standard deviation of a natural sample containing a low concentration of the parameter, or, five times the relative within batch standard deviation of a blank sample. This Directive states also, that these performance characteristics should not exceed 25 % of the parametric value (100 ng/l).

Consequently, both multi-residue methods will be validated through an intra-laboratory test (between the seven partners) to determine, among others, these three performance data for all pesticides investigated and compare them to the criteria defined in the revised Drinking Water Directive. In spring 1999, validated multi-residue methods will be disseminated and then evaluated and compared through inter-laboratory tests organised between 23 laboratories, including the seven partners.

2 OCCURRENCE OF PESTICIDES IN PARIS AREA: RESULTS FROM VARIOUS STUDIES

2.1 Background

2.1.1 Results from the French Agrochemical Association Study. In 1987, the French Agrochemical Association (Union des Industries de la Protection des Plantes) funded a study which aimed at monitoring a wide range of active substances in French water resources. From June 1987 to June 1988, 38 substances have been detected on a seasonal basis in 13 sampling sites covering the main French agricultural areas. Pesticides monitored included 19 herbicides, 7 insecticides and 12 fungicides. Sampling sites were located on 11 ground waters with different depth and geological characteristics and on the Seine and the Marne rivers which account for 40 % of the water supply for the Paris area. Since then, four additional campaigns have been carried out every year in June.

Among the 38 pesticides monitored in the 128 water samples (including 16 surface water samples) analysed between June 1987 and June 1992, 13 substances presented concentration above the reportable level of 50 ng/l, with the following decreasing frequency (4): atrazine (33 positive results) > simazine (25 positive results) > aminotriazole (5 positive results) = bentazone (2 positive results for 50 samples examined) > mecoprop (3 positive results) > other compounds (1 positive result each).

It should be noticed that the limited number of surface water samples analysed (16 out of 128) accounts for 38 % of positive results. In particular, polar substances like aminotriazole, mecoprop, bentazone and 2,4-D were detected. Finally, the 100 ng/l level has been exceeded for various active substances in approximately 20 % of samples analysed.

2.1.2 Results from the CIRSEE Survey (1981-1991). In 1981, the Lyonnaise des Eaux company which supplies drinking water to 13 million people in France, set up the quality assessment of 2000 French drinking water resources, with respect to the EEC directive. As part of this quality control program, around 3000 water samples were analysed, between 1981 and 1991 for pesticide monitoring. In addition to raw surface and ground water samples, drinking water samples were investigated as well.

In 1989, the list of pesticides routinely monitored was upgraded from 20 to 32 compounds. This list covers a variety of active substances including triazines, organophosphorous insecticides, chloroacetanilide herbicides, anilines, triazole derivatives and miscellaneous herbicides. Pesticide residues were extracted from water samples using liquid-liquid extraction with methylene chloride and analysed by GC-MS on a quadrupole and an ion trap mass spectrometer.

During this period, 1183 different sampling sites scattered throughout France were investigated for pesticide monitoring. It appeared that the concentration of atrazine exceeded the 100 ng/l level at 360 sampling sites, while the concentration of simazine exceeded the same level at 140 sites. This Maximum Admissible Concentration of 100 ng/l has been exceeded as well for metolachlor (2 sites), lindane (4 sites), endosulfan (4 sites), alpha HCH (1 site), hexachlorobenzene (1 site) and terbuthylazine (2 sites). The highest individual concentration observed during this survey was close to 29 and 15 µg/l for endosulfan and simazine, respectively. The other following substances were also occasionally detected (<10 positive results each) above their detection limit but their concentration never exceeded the 100 ng/l level: ethofumesate, propazine, parathion, beta-HCH, 2,4 and 4,4 - DDT, dichlorvos, alachlor, vinclozolin and 4,4-DDE.

Because of the GC-MS analytical technique implemented, it should be kept in mind that more polar active substances were not taken into account during this survey.

2.1.3 Other Studies. A study has been carried out in the north of Paris (Oise), and between 1988 [3] and 1991 [4], 300 wells have been monitored for atrazine, simazine and lindane on a 600 000 hectare area. Pesticide residues were extracted from water samples using liquid-liquid extraction with methylene chloride and analysed by GC-MS.

It appeared that the concentrations of atrazine and simazine were above the detection limit (20 ng/l) for 50.7% and 31 % of wells, respectively. The same concentrations exceeded the 100 ng/l level for 18% and 4.3 % of wells, respectively. During this study, lindane has never been detected.

Finally, many of the wells monitored showed a significant decrease in atrazine and simazine concentrations between 1988 and 1991. This observation has been explained by a decrease of both corn farming area (from 38 000 to 26 700 hectares) and rainfall during the same period [5].

2.2 Current Situation on the Seine River (1990-1997)

In relation to the high concentration of some pesticides in water resources, as mentioned above, Lyonnaise des Eaux Company decided to intensify the monitoring of various compounds. Since 1990, 42 active substances have been analysed between 1 and 105 times, either at CIRSEE (using GC-MS) or at the Official Laboratory of Paris (using GC-MS as well). The sampling site is located at the Drinking Water Treatment Plant of Croissy - Le Pecq, approximately 15 km west and downstream of Paris.

It appears that 32 pesticides were never detected between 1990 and 1997 (Table 7).

Table 7. Pesticides analysed at CROISSY (1990-1997) with 100% of measurement under the detection limit

24 D Methyl Ester
Alachlor
Aldicarb
Aldrin
Alpha HCH
Arochlor 1254
Beta HCH
Carbofuran (HPLC/UV)
Cyanazine
DDE 2,4'
DDE 4,4'
DDT 2,4'
DDT 4,4'
Dichlorvos
Diclofop.Methyl
Dieldrin
Dimethoate
Endosulfan Alpha and Beta
Ethofumesate
HCB (Hexachlorobenzene)
Heptachlor
Heptachloreepoxyde
Malathion
Methyl Parathion
Metolachlor
Parathion
PCB (DP5)
Phosdrin
Promethrin
Triadimenol
Trifluralin
Vinclozolin

As shown in Table 8, among the 10 substances detected, 6 compounds were identified more than 20 times. Furthermore, the concentrations of atrazine, simazine and terbuthylazine exceeded the 100 ng/l level more than 20 times.

Table 8. Pesticide detections

	Number of determinations	Substances identified more than 20 times	Concentrations higher than the 100 ng/l level
atrazine	111	x	x
simazine	111	x	x
terbutylazine	84	x	x
propazine	81	x	
lindane	43	x	
alpha HCH	1	x	
diuron	1		
N-de-ethylatrazine	15		
isoproturon	15		
N-de-isopropylatrazine	1		

Case by Case Occurrence

- The concentration of atrazine exceeded the 0.1 μg/l level for approximately 50 % of measurements and remained above the 0.5 μg/l level for 10 % of them (Figure 1). Though the concentration of this compound usually increases in July after agricultural application, it appeared that this compound is detected also at a very high level out of agricultural periods of use [6]. Point source pollution could explain the peaks observed.
- Though the concentration of simazine was decreased in the Seine river since 1994, it exceeded the 0.1 μg/l level for 30% of measurements. This parameter is strongly dependent on season and weather : simazine concentration was always higher during dry years (1993-1996) than wet ones.
- The concentration of N-de-ethylatrazine was quite constant with values ranging from 0.08 to 0.1 μg/l (Figure 1), even if measurements were not as frequent as for atrazine. Indeed the presence of N-de-ethylatrazine was always related to the presence of atrazine. The presence of degradation products of atrazine in the environment will be one of the main topics to be discussed at the national level during the establishment of the priority list of « substances to be carefully monitored ». Measurements issued from alluvial groundwater in the Oise area, indicated that the concentration of N-de-ethylatrazine may represent 3 times the level of atrazine.
- Analyses carried out by GC-MS since July 1996 showed that the concentration of isoproturon exceeded the 100 ng/l level for 50 % of measurements (like atrazine) with regular peaks in November and December (> 0.25 μg/l) and other ones, out of agricultural periods of use, probably due to point source pollution.
- Since 1990, the concentration of terbuthylazine exceeded the 100 ng/l level for 9 measurements, and a slight increase has been observed since 1996. It is important to remember that in some cases, terbuthylazine is used in agriculture instead of atrazine.

The above data were statistically processed using the occurrence probability. Results obtained show that pesticides usually detected are firstly **atrazine**, secondly **N-de-ethylatrazine** and thirdly **simazine**. For the other substances, the probability of detecting them is lower than 15 %.

A great diversity of active substances has been already detected in French natural waters. Consequently, water treatment processes implemented to comply with the EEC directive should not allow the removal of one or two specific compounds only, but of a wide range of pesticides.

Figure 1 : Evolution of Atrazine and N-de-ethylatrazine during the period 1990-1997 at Croissy

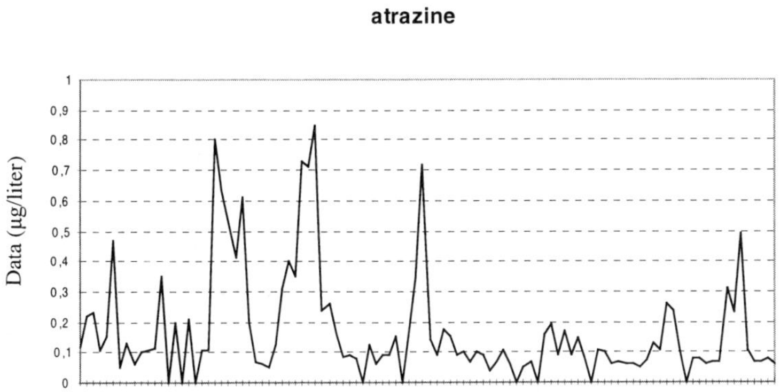

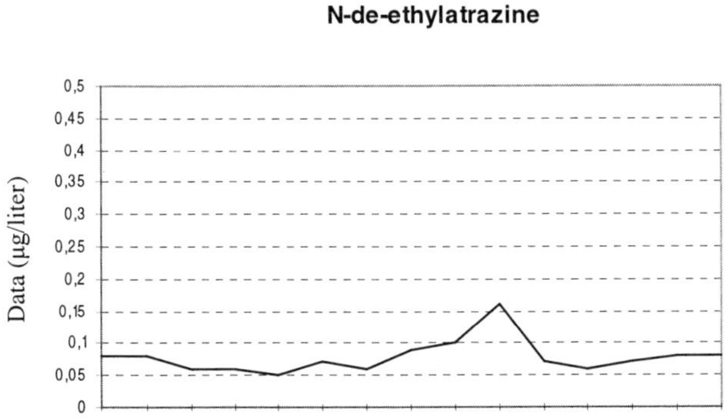

3 WATER TREATMENT PROCESSES FOR PESTICIDE REMOVAL

Most pesticides are poorly removed by coagulation-flocculation and sand filtration; as a consequence, compliance with the EEC standard often requires an upgrade of the existing treatment lines (addition of oxidation - adsorption stages for example) together with a change in their management Membrane technologies also represent a promising solution for pesticide or for micro-pollutant removal.

3.1 Adsorption

Adsorption onto granular or powdered activated carbon (GAC, PAC) has long been used for pesticide removal. The adsorption efficiency depends both on the structure of pesticides (the Freundlich coefficient for pesticides in pure solution is usually found in the 100-700 range) and on the GAC type. The capacity for atrazine in Seine river water has been found to vary from 0.04 to 0.22 mg/g as a function of the type of GAC used. The competition effect of the organic matrix present in natural waters drastically reduces the adsorption efficiency of pesticides. Isotherm experiments performed with Chemviron F400 GAC on different types of waters [7] indicate that the GAC capacity for atrazine which is equal to 20 mg/g in pure water is reduced to 0.6 - 0.15 mg/g when the organic matter load (DOC) in water increases.

In order to comply permanently with the EEC guideline, the GAC regeneration frequency will hence depend on the organic matter level and composition and pesticide load. It should also be kept in mind that a GAC filter acts as a chromatography column: depending on the state of GAC saturation and variations of pesticide concentration at the inlet, a peak of pesticide can be observed at the filter outlet a few days later.

A simulation performed using an adsorption model indicates that a GAC filter [8] is not an absolute filter and that it remains difficult to comply permanently with the low EEC pesticide guideline using a GAC filter alone. A drastic improvement in pesticide removal can be obtained by combining GAC adsorption with ozone or PEROZONE™ oxidation.

3.2 Oxidation

Ozone remains the most suitable oxidant for pesticide removal [9]. The efficiency of an ozonation process depends both on the chemical structure of the pesticides and the ozone species involved.

Ozone usually acts through two main species, molecular ozone and hydroxyl radicals. This last species is generated during ozone decomposition and hence the removal efficiency is controlled by pH, temperature, organic matter content and the concentration of hydroxyl radical scavengers such as bicarbonates. Specific ozonation conditions such as dissolved ozone concentration, contact time, applied ozone dose, concentration in the gas and reactor design also represent important parameters.

In many cases, molecular ozone has a limited effect but the efficiency can be improved by an increase in the hydroxyl radical production. Such an increase is usually obtained by the coupling of ozone with ultra-violet light or hydrogen peroxide. In contrast with molecular ozone which shows kinetic constants which can vary from less than 0.02 $M^{-1}.S^{-1}$ for endrin to $1.5x10^5$ $M^{-1}.S^{-1}$ for dinoseb, hydroxyl radicals react rapidly with pesticides which all show kinetic constants in the 10^8 - 10^{10} $M^{-1}.S^{-1}$ range. A process named PEROZONE™ has been developed and implemented since 1990 at more than 10 full-scale plants in our company.

Perozonation drastically increases the removal of atrazine, which results in GAC bed life improvement with respect to pesticide breakthrough [10]. Because of the non specific character of this process, similar results are expected with most pesticides, whatever their chemical structure.

3.3 Membrane Filtration

3.3.1 Ultrafiltration. Ultrafiltration efficiently removes turbidity and micro-organisms such as bacteria or viruses [11]. However the pore size (around 10 nm) does not allow the removal of micro-pollutants such as pesticides.

It has been shown that powdered activated carbon introduced in the recycling loop of the membrane process may efficiently remove pesticides. A 30 mg/l PAC dose allows the production of a permeate containing less than 20 nanograms per litter of atrazine.

The main advantage of such a process lies in the possibility of adjusting the PAC dosage as a function of the incoming pesticide concentration.

3.3.2 Nanofiltration. Initially, nanofiltration (N.F.) was based on modifications of reverse osmosis (R.O.) membranes in order to remove divalent ions (sulphates, calcium, magnesium,...) at low pressures (5-10 bar instead of 15-80 bar for R.O.) and with improved conversion yield. It was soon realised that due to the pore size (around 0.001 pm), organic matter and in particular trihalomethanes precursors could be extensively removed (up to 99 %).

Pilot studies have subsequently been performed in order to evaluate the efficiency of nanofiltration for pesticide removal. Tests indicate that the removal efficiency largely depends on the membrane and pesticide types. The removal efficiency usually varies between 45 and 99 % for a wide range of pesticides at an initial concentration of 3 to 5 µg/l. In the case of atrazine and simazine the removal was round to vary between 66 and 98 %. This shows that in order to comply with the EEC directive and depending on the type of membrane used, nanofiltration must be associated with another polishing treatment such as oxidation or adsorption, in addition to the treatment step used to reduce the Fouling Index. Nevertheless, nanofiltration is a promising technique for the simultaneous removal of hardness and pesticides.

CONCLUSION

Promising results issued from the European SMT Project indicate that it is possible at present to analyse more than 30 pesticides using a limited set of multi-residue methods. Additional work is necessary to validate this project and in particular collaboration with other laboratories (official control laboratories and pesticide manufacturer laboratories) will help the involved team to propose standards at the European level.

The various surveys summarised in this paper have shown that a wide variety of pesticides can be detected in French natural waters in relation to the strong and diverse agricultural activities of this country. Research carried out during the last decade has

however allowed the water supply companies to implement appropriate treatment processes.
Because of the very low EEC standard for pesticides and despite the increase in flexibility of the various processes available at the moment, additional research efforts are still necessary to further improve on-site measurements to control treatment processes in particular through the development of reliable continuous pesticides analysers.

ACKNOWLEDGEMENTS

Part of this work was supported by the Commission of the European Community - Standards, Measurements and Testing programme (project STM4-CT96-2142).

References

1. The Pesticide Manual (1994) - A world compendium : The Pesticide Manual (incorporating the Agrochemicals Handbook), Tenth edition. *Crop Protection Publications, Editor Clive Tomlin, 1341 p.*

2. Bruchet A., Patty L., Jaskulke E., Guynamant J.L., Acobas F., Van Hoof F., Van Wiele P.,Sacher F., Bobeldijk I., Ventura F. and Marecos do Monte M.H. (1997) - Optimization and Evaluation of Multiresidue Methods for Priority Pesticides in Drinking and Related Waters. *Workshop « Trends Analysis, Pre-normative Research in Environmental Measurements », November 05-07, JRC, Environment Institute - ISPRA,Italy.*

3. Patty, A. Bruchet and J.C. Alibar , J.L. Guynamant and F. Acobas, F. Van Hoof and P. Van Wiele, F. Sacher, I. Bobeldijk, F. Ventura and M.R. Boleda, M.H. Marecos do Monte - Optimization and Evaluation of Multiresidue Methods for Priority Pesticides in Drinking and Related Waters (European Project SMT4-CT96-2142)-, *8th Annual Meeting of SETAC-EUROPE, 14-18 April 1998, Bordeaux, France*

4. Duguet J.P. , Bernazeau F., Bruchet A.- Occurrence Of Pesticides In Natural Waters and removal during water treatment processes, *AIDE October 2/8 1993, Budapest.*

5. Gril J.-J., Mailloux-Jaskulké E., Fauchon N. - Ruissellement et transfert dans les bassins versants. *Colloque PHYT'EAU "Eau - Produits Phytosanitaires : Usages Agricoles et connexes" du 21-22 octobre 1992; Versailles; pp. 95-121.*

6. Narcy J.-B.. Le point sur les triazines. *Rapport Agence de l'Eau Seine-Normandie; Nanterre; 81 pages + annexes - 1996.*

7. Duguet J.-P., Anselme C., Wable O., Baudin I., Mallevialle J.- L'expérience industrielle de l'élimination des pesticides par de nouvelles techniques de traitement : les couplages ozone/peroxyde d'hydrogène et adsorption sur charbon actif en poudre/ultrafiltration. *Water Supply, Vol. 11; Amsterdam, pp. 141-148 (1993)*

8. Duguet J.-P., Feray C., Dumouttier N., Mallevialle J. - Efficacy of the combined use of ozone/UV or ozone/hydrogen peroxide combination for water disinfection. *Proceedings of the IOA Symposium on ozone + UV water treatment; Wasser, Berlin (1989).*
9. Reynolds G. and al.. Aqueous ozonation of pesticides. A review, *Ozone Science and Engineering, Vol. 11, N° 4; pp. 339-390 - 1989.*

10. Duguet J.-P., Wable O., Richard Y., Toffani G., Dalga N - Evaluation technico-économique de l'élimination de l'atrazine par le couplage ozone-peroxyde d'hydrogène/charbon actif en grains sur la station de traitement d'eau potable du mont-Valérien. *Conférence Européenne Spécialisée AIDE "L'atrazine et les autres pesticides"; Florence, 23 et 24 octobre 1991, pp. 105-110.*

11. Urbain V., Manem J. Les bioréacteurs à membranes pour l'élimination des pesticides et des nitrates. *Colloque "Contamination des eaux par les nitrates et les pesticides", 2 et 3 Juin 1994; à l'Ecole Nationale de la santé Publique, Rennes (France).*

Barceló D., - Official methods of analysis of priority pesticides in water using chromatographic techniques. *Environmental Analysis: Techniques, Applications and Quality Assurance. Chapter 5, 149-180.*

Pesticide Residues in Developing Countries – A Review of Residues Detected in Food Exports from the Developing World*

R.H. González

DEPARTMENT OF PLANT PROTECTION, FACULTY OF AGRICULTURAL SCIENCES & FORESTRY, UNIVERSITY OF CHILE, PO BOX 1004, SANTIAGO, CHILE

ABSTRACT

This paper describes the trend of international movement of pesticide residues in food products as reported by countries having a regular monitoring program. For the most part, detention of exports from developing countries refer to cases in which the illegal residues found in the produce pertain to either a non registered chemical or when the recipient country has not yet established a Maximum Residue Limit (MRL) for that particular crop/pesticide combination. Fewer violations are noted when residue levels exceed tolerance limits set by national or regional legislations. Information on violative cases has also permitted the study of high risk residue trends moving on the international front. In the past, persistent insecticides were responsible for most of the export rejections; at present, fungicides are becoming of greater concern due to the increasing development of disease resistance to fungicides, as well as the need to ensure better post-harvest protection practices. Risks are also foreseen on the use of a wide range of OP insecticides lacking registration and/or facing tolerance reductions in developed countries such as those being proposed by the recently implemented U.S. EPA 1996 Food Quality Protection Act. Actual residues detected in food exports from the various developing country areas are assessed with a view to highlight major actual or potential residues risks emerging from present world food trade practices.

INTRODUCTION

Developing countries, whilst still farming over 54 percent of the available land, consume scarcely one-fifth of the global amount spent world wide for pesticides (McDougall, 1994). In addition, pesticide consumption in the various countries varies considerably according to marketing conditions, cropping systems and plant health problems to be met. It is recognized, however, that international trade of agriculture produce is a major increasing factor of pesticide usage mainly due to phytosanitary and plant quarantine requirements. Approximate pesticide consumption rates by regions is estimated as follows (Waibel, 1994): East Asia, including China, 38 percent; Latin America, 30; Near East and North Africa, 15; South Asia 13 and Africa South of Sahara, 4 percent. It is worthwhile comparing these apparently high inputs with the

* See refs. 1 and 2

United States pesticide purchases in 1995 which accounted for 30% of the U.S. $ 37.7 billion world market for that particular year (Anonymous, 1997).

In terms of pesticide group usages, developing countries account for about 50 percent of world use of insecticides, 20 percent of fungicides and only 10 percent of herbicides (Klassen, 1995), a situation which reflects different plant protection priorities, subsistence cropping systems, hand labour availability and local economic constraints. Nevertheless, countries of the developing world which are increasingly acting as food producers are steadily increasing herbicide and fungicide uses whilst insecticide inputs are substantially decreasing as noted in the survey of pesticide residue trends moving in international food trade.

Notwithstanding current national and regional efforts in developed countries to update present registration schemes and food crop tolerances in the light of new toxicological and ecological assessments, it is noted that the use of undesirable pesticide groups has not yet declined in a very large part of the developing world in spite of public health and environmental concerns. In fact, and recognizing that economic problems, shortage of more adequate pesticide alternatives, and lack of efficient regulatory mechanisms may trigger the use of certain unwanted chemicals to protect domestic cropping systems, a more efficient protection technology to prevent residue risks in exported produce must be further sought and improved by the interested parties.

Monitoring programs established by developed countries to detect chemical residues moving in the international food commerce have permitted a map to be designed, and highlights the occurrence of unwanted residues which are still detected in a variety of imports such as legume crops, avian eggs, rice, biscuits and others (FDA, World Wide Import Detention Summary, years 1989-1993). The national and international implications of BHC/DDT residues in food after long term national usage in China is discussed by Li and Han (1995). On the other hand and because of lower costs, pesticides of WHO class 1a and 1b (extremely or very hazardous) are still frequently used by a large number of developing countries (Rola & Pingali, 1992).

Recent regulatory legislation which is expected to impact the world pesticide usage on food crops, is the "Food Quality Protection Act" (FQPA) of 1996 enacted by the Senate and House of Representatives of the United States of America to amend the former Federal Insecticide, Fungicide and Rodenticide Act to provide greater health and environmental protection, particularly for infants and children. It is scheduled to have all organophosphates and carbamate insecticides and carcinogens Class II reviewed by 1999, as well as all chemicals registered before 1984 and all Class I carcinogens to be evaluated by the year 2002. At the same time, a new safety factor will affect current residue tolerances since now there could be an additional 10x reduction in residue tolerances to protect infants and children, placing more stringent conditions on the consideration of benefits in setting pesticide residue limits. The final aim is a consolidated Registration Review Schedule to have all chemicals registered after 1984 and all biological pesticides assessed according to new parameters, at the same time ensuring that all pesticides are to be periodically re-evaluated for adherence to current safety standards supported by up-to-date scientific data (EPA, 1997).

DIVERSITY OF NATIONAL REGULATORY MECHANISMS AND RESIDUE PROBLEMS

The primary objective in regulating chemical pesticides is to provide man and the environment with maximum possible protection from potential adverse effects and facilitate international trade in food through the establishment of approved pesticide regulations, notwithstanding that these requirements may not always satisfy all importing countries thus creating potential barriers than facilitating the world commerce. However, considerations to set up regulatory mechanisms vary widely among developed and developing countries, a rather legal matter which has resulted in a primary reason for generating residue conflicts among producing and receiving countries. For example, the U.S. FDA monitors imported foods for pesticides for which there is no tolerance as well as those for which the U.S. EPA has set a Maximum Residue Limit (MRL) for a particular crop/pesticide combination. The majority of shipments classified by the FDA as violative results from those cases in which a tolerance has not been set rather than in detentions in which the residue tolerance has been exceeded (Wessel & Yess, 1991). On the other hand, exporting countries adhering to Codex Alimentarius standards may also be hampered by the diversity of MRLs established by the Codex guidelines and those set by national tolerance schemes.

Registration in the majority of developed countries enables authorities to exercise control over selection of chemicals, application and use levels, setting of tolerances and fixing pre-harvest intervals to meet internal market tolerances for that particular crop/pesticide combination. Nevertheless, and despite their expected efficiency on pesticide application technology and improved regulatory methods, detention reports show that a fairly large number of violative residues detentions are still produced by exports from developed countries (see Table 4). On the contrary, a large number of developing countries have enacted pesticide laws and registration procedures not always quite harmonized in their definitions and meanings, lacking analytical control mechanisms, to assist in the setting of appropriate tolerances and pre-harvest intervals. Furthermore, a number of materials are firstly registered in these countries since they serve as testing sites for the development of new pesticides which eventually could be later registered in other countries having less "soft" registration schemes.

As a consequence of the diversity of registration procedures and the slow pace of setting MRLs in regions such as the European Union group of countries and the increasing numbers of cancelled uses, violative residues of pesticides in food exports from both developed and developing countries may occur due to the following reasons:

a) the residue found pertains to a chemical which is not registered for use in the importing market regardless of its legal condition of use in the producing country;
b) the residue level exceeds the tolerance set for the registered chemical in the importing country for that particular pesticide/crop combination. Failures of this kind are expected to increase within the European Union and, in particular, in the United States of America, once the FQPA be fully implemented by EPA and the U.S. Food & Drug Administration in their search to diminish chemical risks in foods.

c) the residue level has been adopted from the Codex Alimentarius Limits (CXLs) which are normally higher than tolerances set by importing countries, thus leading to conflicts in international trade.

To overcome the above mentioned situations resulting from the lack of appropriate evaluation facilities and internal compliance mechanisms and, more importantly, on restraints relative to appropriate analytical facilities to monitor residue behaviour under own cropping systems, all efforts should be made by the exporting country to extrapolate data from related agroecosystems and yet obtain the industry support for assisting on analytical work where facilities are available.

Importing countries having regular residue monitoring programs should also make efforts to provide official information on findings to the exporting countries, with a view to correcting violative cases. In this respect it is worth noting that the U.S. FDA Center for Food Safety and Applied Nutrition contemplates the provision of the "FDA - CFSAN Regulatory Monitoring Data" listing all analyzed food samples under class groups (1 = in compliance; 2 = residues for which no action is taken; and 3 = violations). Class 3 applies when imported foods are found to be adulterated with illegal pesticide residues. In this respect the exporting countries should also be fully aware of the regulatory procedures which apply to rejected lots, i.e. destruction, re-exportation or recondition of the produce, and/or subject to follow up enforcement procedures such as "automatic detention" which is invoked by FDA when shipments of the same food commodity containing the same illegal residues are expected to continue; this procedure may be applied for a food commodity from a single shipper or grower, based on a single but reiterated violative findings.

National residue surveillance programs in importing countries vary to a very large extent in their decision actions and information procedures. For example, the Swedish compliance program determines that when a surveillance sample contains a residue above the national MRL or Guideline Level (GL), future imported lots from the same origin are detained pending analysis and clearance (Anderson & Bergh, 1991). Regular monitoring programs are likewise conducted in The Netherlands, Norway, United Kingdom, United States, Canada and Finland. From their annual reports as well as from other individual direct sources, the following information reflecting current trends on pesticide residues as well as highlights on major residue risks affecting the international market is discussed. It should however be noted that this is a problem not only affecting developing countries. Monitoring programs among European countries show a consistent fashion of violations in crops such as carrots, celery, endive, leek, lettuce, paprika, spinach, table grapes (van der Schee, 1997).

MAJOR PESTICIDE RESIDUES MOVING IN INTERNATIONAL TRADE

The kind and amount of pesticide residues circulating in food exports will be an indication of the appropriate pesticide technology utilized in plant protection and will also reflect the existence and properness of a national pesticide registration scheme. It should however be noted that occasional findings of a higher residue level in foods resulting from appropriate Codex Maximum Residue Levels (CXLs) approved by the exporting country, are not accepted by the importing country following their own regulations or adhering to a regional set of MRLs as those enforced by European Union Directives. With respect to European monitoring programs, over the past 7-8 years there has been an increase in the rate of exceadences (tolerances), due to the current use of EU-based MRLs which are lower than the Codex MRLs. The increase of violative cases is also due to an increase in the number of samples and the number of pesticides being monitored through multi-residue methods and the emphasis placed on certain target pesticides. Exporting countries must therefore comply with the residue level set at the point of destination of the food, a proposal which can not be easily met in practice when the food exports from a country are addressed to a diversified market in terms of non harmonized tolerances.

Tables 1 and 2 show the known frequency of dispersal of agricultural fungicides and insecticides compiled from information issued by the above mentioned importing countries. To maintain confidentiality on the individual origin of residues, only the relevant geographical area of the concerned country is presented. The detections listed do not necessarily imply violative cases.

With respect to higher risk pesticide residues leading to more conflicts in international trade, these are summarized in Table 4. Among determining factors for assessing risks, the following parameters have been considered; lack of overall registration and tolerances for a particular food/pesticide combination; unwanted old chemicals; residue persistence in particular food substrates; general world reduction of tolerances for a number of pesticides of toxicological and environmental concern and, increasing fungicide usage for post-harvest protection.

In Table 1, it is noted that the majority of findings have resulted from the application of post-harvest treatments, e.g. vinclozolin, thiabendazole, iprodione; or as a consequence of multiple preharvest applications of chlorothalonil, carbendazim (including benomyl), vinclozolin, iprodione, captan, and dithiocarbamates, the latter one of the most widely used chemicals. The dithiocarbamate metabolite ETU, refers to residues detected in processed foods (e.g. tomato paste, apple juice) resulting from repetitive pre-harvest treatments with reference to the calculation of total dithiocarbamate residues, these are determined as carbon disulfide (CS_2) and the parent EBDCs are all measured as CS_2 equivalents. Procymidone, commonly used in the European region has been much less used in the Latin American continent because of its lack of registration in the USA and Canada, hence data from this area are restricted.

It is predicted that due to the increased development of storage rot diseases and resistance to fungicides, the use of post-harvest applications as drenching, spraying or incorporated into cosmetic wax will substantially increase. There also exists the risk of expanding their usage to related crops in need of protection from storage diseases, for which there are no foreseeable tolerances. With respect to vinclozolin, a widely used fungicide for both pre- and post-harvest treatments, this chemical has recently been withdrawn by manufacturers from certain markets due to possible reproductive risks to applicators posed by dusting treatments.

Table 1. Major fungicide residues moving in international trade, listed by frequency of findings. Residue detections do not necessarily refer to violative cases. Regions refer to the general geographic location of the developing country of origin. (Compiled from various sources, period 1990-1996).

Fungicides	Major Food Groups	Main Regions of Origin
Chlorothalonil	Tomatillo, egg plant, bean cucumber, squash, tomato, pea, spinach, lettuce, hot pepper, strawberry, papaya, avocado, strawberries	North and Central America, Near East
Vinclozolin	Lettuce, parsley, tomato, carrot, kiwi Tomato Beans in pods, tomato, grape, kiwi Kiwi, grape	Southern Europe North East North America South America, Southern Europe
Iprodione	Peach, carrot, grape, kiwi, apple, papaya	Latin America, Southern Europe
Captan/Folpet	Pear, lemon, apple Tomato, lettuce Pepper, celery, peach, small fruits, grape, pear Melon, lemon, strawberry	North Africa Near East South America Near East
Thiabendazole	Citrus, apple, banana, potato, mango, papaya	South America, Near East Southern Europe
Dithiocarbamates (1)	Pear, lettuce, orange, cucurbits, tomato Papaya, tomato, potato, beans, pears, melons Processed stone fruits, tomato, grape, hops	South Africa Latin America Various
Carbendazim (2)	Pear, apple, papaya Pineapple Grape	South America, Southern Europe Caribbean, Latin America Near East
Imazalil	Banana, citrus fruits Melon	Central America, North & South Africa Near East
Procymidone	Green pea pods, bean in the pod, pepper, lettuce	Various

(1) Mancozeb, maneb and metiram are the most widely used in this group. Residues include ETU.
(2) Includes benomyl residues

Captan (and the sum of captan + folpet) is subjected to substantial reduction in tolerances and therefore it is recommended that its widespread use be cautiously watched. European countries are currently reassessing major captan-folpet tolerances and it is noted that major reductions are still under way on table grapes, strawberries and other horticultural produce. The same trend is noted for carbendazim and related benzimidazolic compounds.

Table 2. Major insecticide residues moving in international trade, listed by frequency of findings. Regions refer to the general geographic location of the developing country of origin. (Compiled from various sources, period 1990-1996).

Insecticide	Major Food Groups	Main Regions of Origin
Methamidophos	Pea pods, sweet pepper, frozen soybean, apple juice, bean, mustard, celery, pepper, passion fruit, lettuce, egg plant, apple, cantaloupe, cauliflower, bean, stone fruits, melon	Asia Near East, North Africa, Latin America
Acephate	Leaf vegetables, squash, parsley, mango, apple, melon, water melon Apples	Latin America Caribbean South Africa
Dimethoate	Pear, mushrooms, pear, cucumber, squash, snow peas, pineapple, citrus celery	Asia Latin America, Near East
Chlorpyrifos	Canned mushrooms clementine rice persimmon, citrus fruits, grape, lemon, string beans, cabbage, asparagus	Asia North Africa, & Middle East Far East Near East, Latin America
Profenofos	Sweet pepper tomato, squash, chayote pepper	Near East Latin America Caribbean
Endosulfan	Pepper parsley, tomato	Latin America Near & Far East
Omethoate	Pear, cucumber, squash, husk tomato	Latin America
Diazinon	Mushrooms, carrot, celery onion, fruit trees	Asia Latin America
Pirimiphis methyl	Dry banana, coconut, coconut products, pasta, noodles, sweet corn, rice, pears, beans, corn pepper	Asia Latin America Caribbean
BHC, Total	Egg plant, chicken & duck eggs, bean, peas, sesame, lentil,, rice spinach	Near East & Asia Central America
Parathion methyl	String beans, parsley, hot pepper grapefruit	Latin America Near East
Permethrin	Pepper, asparagus herbs	Latin America Asia
Monocrotophos	Okra snow peas, hot pepper cucumber	Asia Central America Near East
Azinphos-methyl	Celery, spinach, apple, peach, clementine	Latin America
Mevinphos	Lettuce, kale	Latin America
Phosmet	Apple, pear	Latin America
Esfenvalerate	Apple, pear	Latin America
Malathion	Rice	Far East
Methidation	Lemons	Southern Europe, North Africa

Most of the current insecticide residues moving in international trade pertain to the organophosphate group (OPs) which are widely used throughout Latin America, North and South Africa, the Near and the Far East. On the contrary, with the exception of endosulfan which is being used globally, the chlorinated insecticides such as BHC, lindane, and DDT reflect agricultural practices in Asia (China), ex-Soviet Union (CIS), and other eastern countries. The DDT usage in Russia has continued long after the official ban in 1970 and between 1970 and 1983, in Uzbekistan 70,000 metric tons of this insecticide were applied, and according to recent reports, poultry and egg samples, dry milk, soils and water reflect high residue levels above world standards (Anonymous, 1997, *op cit.* p. 7). Another chlorinated compound, BHC is still widely used in Asia as noted in the U.K. on Chinese imports (Anon., 1995).

Among OP s, residues most frequently moving in international trade include methamidophos (and acephate), profenofos, chlorpyrifos and parathion methyl, dimethoate, omethoate and assorted carbamates and few pyrethroids. Methamidophos, a systemic and contact insecticide is widely used in Latin America, because of efficacy, fairly long persistence and low application cost. EPA has announced the deletion of most food-use registrations. Profenofos, an alternative insecticide for vegetable crops is curtailed by its lack of registration in major importing countries; in addition, the California Department of Agriculture has added this material to its risks characterization priority list. Chlorpyrifos, is worldwide, one of the most used insecticides to control soil and aerial pest problems, and is also under careful global watch in view of its high toxicological profile. Still surprising is the continous use of monocrotophos in Central America and Asia, a chemical which has been voluntarily suspended elsewhere.

From Table 3, it can be concluded that several of the pesticides listed such as carbendazim, chlorfenvinphos, procymidone and tecnazene, which are fairly widely used within the Region concerned, could face problems in other areas due to the lack of tolerances and/or registrations, namely in North America. In terms of major residue findings, diazinon, chlorothalonil and methamidophos were detected in some 40 percent of analyzed assorted food samples, the highest findings being in carrots and strawberries (Norway), and peppers (Spain). The use of pre-or post-harvest fungicides such as iprodione, thiabendazole, vinclozolin, diclofluanid, imazalil and procymidone, have shown an outstanding increase in the past 5 years. While the levels found were of low profile from the consumers' viewpoint, the occurrence of recurrent residues of these chemicals is indicative of impending problems on vegetable crops produced in the southern part of Europe.

Table 4 shows the large number of fairly persistent organophosphates and a wide range of fungicides frequently detected in international trade. The recurrence of wide spectrum systemic materials such as methamidophos-acephate reflects their widespread use on leaf and fruiting vegetables, legumes, cabbage crops and fruit trees. Whilst this is perhaps the most common insecticide detected in the past 5 years, it must be recognized that few violative cases have ocurred. However, it is encouraged that methamidophos protection programs exclude its metabolite acephate (and vice-versa) to avoid conflicting residue problems. The occurrence of violative cases detected by FDA on apples imported from South America and South Africa (Roy *et al.*, 1997), due to either methamidophos and/or acephate treatments is a matter of concern.

Table 3. Major pesticide residue findings in produce from developing countries exported within the European Region. 1990-1996, listed in alphabetical order. From various sources, namely from Sweden, UK., The Netherlands and Norway.

Pesticide	Food Produce
Acephate	Grapes, apples
Azinphos-methyl	Apple, pear, citrus
Bupirimate	Berries, tomato
Captan/Folpet	Pear, cucumber, strawberry, apple, grape
Carbendazin	Apple, celery, pear, strawberry, tomato
Chlorfenvinphos	Root crops, carrot, lemon, celeriac, celery, clementine
Chlorothalonil	Celery, cucumber, parsley, lettuce, endive, tomato, strawberry, squash, onion, celery
Chlorpyrifos	Apple, olive, citrus, celery, sweet pepper, grape, persimmon, pear, kiwifruit
Chlorpyrifos-methyl	Pepper
Cypermethrin	Grape, lettuce
Diazinon	Carrot, parsley, celery, orange
Dichlofluanid (Tolyfluanid)	Current, strawberry, raspberry, other berries
Dithiocarbamates	Pear, leek
Folpet	Peppers
Imazalil	Melon, grapefruit, clementine, orange
Iprodione	Lettuce, kiwi, grape, endive, raspberry
Metalaxyl	Orange
Methamidophos	Cucumber, egg plant, tomato, celery, peapods, leek, pepper, melon
Methidathion	Orange, grapefruit, lemon
Omethoate	Apple
Parathion methyl	Grape, lemon, pear (Southern Europe)
Pirimicarb	Endive, lettuce, cole crops
Pirimiphos methyl	Gherkins, kiwi
Procymidone	Grape, wine, lettuce, strawberry, carrot, celery, paprika
Tecnazene	Potato
Thiabendazole	Apple, pear, potato, orange, clementine, grapefruit
Vinclozolin	Kiwi, strawberry, blackberry, lettuce, tomato, endive, cherries, carrot

Table 4. Major pesticide residue risks in international trade.

Pesticides	Reasons for Risk
Methamidophos	Wide use on horticultural crops; persistent residues; restricted tolerances in importing countries; protection programmes should not include joint uses with acephate to avoid excess of final combined residues.
Acephate	Wide use on fruit, vegetable and industrial crops; persistent residues; restricted tolerances in importing countries; limited number of Codex MRLs. Some uses have been cancelled voluntarily.
Omethoate	Persistent residues in fruits, seed crops, leafy vegetables, including minor crops treated with either this technical material or dimethoate; not registered in a major world market (e.g. USA). Codex recommending withdrawal of several MRLs.
Profenofos	A wide spectrum insecticide of increasing use in vegetable crops; restricted tolerances in several world markets; no Codex MRLs yet approved; subjected to targeted surveillance residue programmes; not registered in the USA.
Pirimiphos methyl	Increasing usage as aprotective material on grains and other stored products. Many importing countries lack tolerances for this insecticide on rice.
Dichlofluanid (Tolyfluanid)	Associated violations have lately been noted, particularly for lacking MRLs (e.g. grapes, strawberries)
Imazalil	A post-harvest protectant with rather high tolerances; however, application methods may lead to exceeding MRLs.
Endosulfan	Still widely used in several developing countries; easily detected in monitoring programmes due to its persistent residues, particularly on vegetables.
Monocrotophos	Insecticide of high residue risk in international food trade. Notwithstanding, its use is rapidly declining due to national cancellations, developing countries still make use of this product without due concern to acute toxicity.
Chlorpyrifos	Fairly persistent insecticide used as a soil and foliar material with international problems due to low MRLs, particularly on fruit trees.
BHC (HCH)	Although residues are slowly disappearing in the international trade, far eastern countries are still making a substantial use of this persistent chemical (e.g. meat products).
Captan/Folpet	Multiple pre-harvest treatments are yielding residues above currently approved MRLs. Either captan or folpet are being employed in a number of crops for which there are no Codex MRLs. The European Union has set new, lower MRLs, which are in conflict with Codex limits.
Aldicarb	A restricted chemical subject to targeted monitoring. Extremely persistent within the plant requiring long pre-harvest intervals for the existing low MRLs.
Carbendazim (Benomyl)	Most of the national and regional legislations have either cancelled most of its uses, or reduced post-harvest treatments. A number of alternatives are currently available for both pre- and post-harvest treatments.
Procymidone	Pre- and post-harvest treatment on lettuce, grapes, raisins, carrots, are generating persistent residues on a number of horticultural crops for which no tolerances are available.

Table 4 Major pesticide residue risks in international trade (continued)

Pesticides	Reasons for Risk
Triadimefon-Triadimenol, Penconazol, Tebuconazole	New generation fungicides for which very few tolerances are available on horticultural crops.
Heptachlor	Although of very limited usage, some countries are still making use of this chemical.
Thiabendazol	Notwithstanding its wide use as a post-harvest fungicide, violations due to this chemical are not of current concern; however, it is foreseen that uses on clementine. Grapefruit, apples and others are very much reaching MRLs.
Iprodione	Post-harvest treatments may exceed MRLs; to be watched when importing countries lack such tolerances (e.g. pome fruits, papayas).

Profenofos has rapidly emerged as a wide spectrum contact insecticide for the protection of vegetable crops, maize, potatoes and tomatoes. It is foreseen that due to the lack of Codex MRLs and the absence of tolerances in a large number of importing countries, violative cases will be expected. Countries must establish appropriate GAPs for this chemical and obtain from any available source of information data on dissipation curves in comparable crops to suit existing tolerances. An important volume of vegetable crops is detained in southern United States because of this chemical.

With respect to monocrotophos, the approach should be different since farmers have been familiarized with this material for the past 3 decades and existing Codex MRLs, although at very low levels, do not seem to prevent its use. Fortunately, a number of countries have already canceled all tolerances in view of its high acute toxicity and residue persistence.

Monitoring data from several countries are also indicating the heavy traffic of fungicide residues such as dichlofluanid, dithiocarbamates, iprodione, vinclozolin, thiabendazol, imazalil, bupirimate, procymidone, chlorothalonil, carbendazim and many others of the new generation chemical groups, which still lack tolerances such as chlozolinate, penconazol, pirimidyl and others. Captan/folpet tolerances have been drastically reduced in the EU and, therefore postharvest treatments, could represent a serious risks in that market area. On the other hand, several legislations (e.g. U.S. EPA) do not provide for tolerances for bupirimate, dichlofluanid, carbendazim, tebuconazole, procymidone and others of European or Japanese manufacture.

It is also worth noting that the quality of the substrate should also be taken into consideration in determining pesticide choices. Citrus fruits, in particular lemons and oranges, are apt to retain oil soluble insecticides such as parathion, chlorpyrifos, methidathion, clorfenvinphos, quinalfos and many other materials for which special GAPs must be designed to counteract the long residue persistence of these materials in the rind of sprayed fruits. Kiwifruit is another example of a highly retentive fruit due to its hairy surface which retains high deposit levels and fairly persistent residues of azinphos-methyl, chlorpyrifos, phosmet and carbaryl (González & Curkovic, 1994).

NEEDS FOR ADDITIONAL SETTING OF CODEX LIMITS FOR NEW PESTICIDE/ FOOD COMBINATIONS

The Codex Committee on Pesticide Residues (CCPR) at its 11th Session agreed to establish the *ad hoc* Working Group on Pesticide Residue Problems in Developing Countries to discuss main problems related to the export of commodities which have not yet been object of regulation by Codex. This is a most urgent matter since some Codex MRLs have been established without due consideration to GAPs utilized in developing countries, which are based on their own different cultural and environmental conditions. Such Codex limits, in some cases higher or lower than national tolerances of several importing countries, are creating barriers to international trade.

Legal MRLs are mostly based on supervised trials and toxicology data for estimation of these acceptable levels. On occasions, these data can be supplemented by selective surveys of crop/commodities where there is detailed information available on the use of that pesticide. A thorough study on the philosophy of setting MRLs, problems, procedures, toxicological implications and the Codex framework for the elaborations of world-wide Codex Maximum Limits was published by Bates in 1979, and the guidelines for submission and evaluation of residues data have recently been reviewed by FAO (1997).

The case of minor crops

Since the early 1970 s, when pesticide registration was made subject to more stringent environmental safety issues and new toxicological criteria, and consequently became more costly, the registration activities of the agrochemical industry were focused on the major commodity groups, namely cereals and maize, soybean, cotton and on selected crops of major trade value such as fruit trees and selected leafy and fruiting vegetables. Regretably, various other minor crops such as spices, pulses, herbs and small fruits, all of which fall in the category of "minor crops", were neglected for these purposes.
Minor crops of the herbs and spices group include export representative commodities such as basil (*Ocimum*), coriander (*Coriandeum sativum*), dill (*Anethum*), fennel (*Foeniculum*), marjoram (including oregano), parsley (*Petroselinum*) and perhaps tarragon (*Artemisia*). Among minor fruit crops of actual export interest, cherries, anonas or sugar apples, blackberries and other Rubus species, passion fruit (*Passiflora*), prickly pears (*Opuntia*) and tomatillo (*Physalis ixocarpa*), are examples of crops awaiting for Codex group residue limits. The prospective risks arise from their lack of tolerances which preclude their participation in the export bussiness.

From the economic standpoint the justification for establishing MRLs for minor crops is rarely profitable despite the fact that for a number of developing countries these are important export items regardless of their small acreages. However it is recommended that, since the setting of MRLs may not be practical for most of these crops, a pragmatic solution might be the extrapolation of tolerances set for similar major crops perhaps other berries, lettuce, celery or the like.

Candidate pesticides to suit general phytosanitary problems in the herb and spices groups include carbendazim, captan, chlorothalonil, dichlofluanid, dithiocarbamates and thiabendazol, just to suggest possible major fungicide materials. Insecticides of possible major use for these crops include wide spectrum insecticides having a diversity of Codex limits such as chlorpyrifos, cyhalothrin, diazinon, fenvalerate (and esfenvalerate), malathion, methomyl, permethrin and pirimiphos methyl.

A CASE EXAMPLE: THE CHILEAN APPROACH TO MINIMIZE RESIDUE RISKS IN EXPORTED COMMODITIES

Chile's great diversity of climates, a consequence of its long territory in the south western part of South America, has permitted the establishment of a wide range of temperate and subtropical fruit crops and assorted kinds of vegetables, table grapes, kiwifruit, pome and stone fruits, berries, lemons, asparagus, tomatoes, cole crops and cherries, are among the major horticultural exports. In the 1997-98 season over 150 million boxes are expected to be shipped to over 60 countries in all continents.

National MRLs follow Codex Limits which apply to domestic food crops. However, for export purposes, to meet the diversified MRL scheme adopted by importing countries, the task to comply with such wide array of individual MRLs is a challenging matter since most of the crops are not earmarked regarding the final destination of the produce. Therefore, several troublesome pesticides for Chilean's major markets, e.g. procymidone and dichlofluanid on table grapes with respect to the U.S. market, have been merely banned from the local table grape phytosanitary management. To achieve this and other limitations, the Chilean Exporters Association jointly with the University of Chile have developed a unique information system based on a supervised trial scheme to determine withholding periods (preharvest intervals, PHIs) to meet residue tolerances set by major export markets. These recommendations are published as deemed necessary in the series "PESTICIDE AGENDA: Registration, Tolerances and Preharvest Intervals for Fruits and Vegetable Exports" (González & López, 1996). The AGENDA identifies the pesticides permitted on foods to ensure that Chilean fruits and vegetables exported to all market areas comply with their tolerance and registration requirements.

At present, the AGENDA produces PHIs, and information on import tolerances for 30 horticultural export crops and some 60 chemical pesticides. These crop/pesticide/market combinations, yield over 300 recommendations, to apply approved pesticides on a limited or full scale range according to national good agricultural practices and relevant registration/tolerances in each major importing country. These recommendations are offered to growers through their respective exporting companies which are ultimately responsible for providing technical assistance and the provision of permitted pesticides to suit their needs.

For planning residue trials the FAO Guidelines on Residue Trials to obtain data for Pesticide Registration and Establishment of Maximum Residue Limits (FAO, 1987) is being followed. These guidelines refer to all necessary steps that should be carefully followed from the field installation of the plots, namely selection of crop sites, crop phenology, pesticide application systems, sampling and maintenance of samples, climatic data, analytical requirements and data assessment. Sampling dates are forecasted according to the frequency of applications, these being dictated by the local

Good Agricultural Practices. To extrapolate data from similar crop groups, care must be exercised not only with respect to botanical relationship of the crop group but also with the kind of pesticides used. Examples of crop comparability are presented by Hamilton (1994).

When the chemical is applied as a post-harvest treatment, the information to be obtained refers to the amount and nature of the residues during the normal course of storage and handling. In the planning of exports to the various geographical regions is of the utmost necessity to anticipate the fate of residues at arrival, according to temperature, humidity and other factors that may interfere during transit. Through the setting of adequate post-harvest supervised trials, these conditions can be adequately simulated for the various storage conditions and travelling periods.

The individual Chilean exporting companies do not routinely test fruit for residues before shipping the produce but solely rely on the recommendations issued by the Exporters' Association. These are closely controlled by the relevant Technical Department, particularly as to the timing and application systems for each chemical material, the latter being provided by the companies concerned. At the same time, and notwithstanding registration procedures which are officially enforced by the Ministry of Agriculture, some highly residue risk pesticides (e.g. methamidophos, omethoate, profenofos, procymidone) are not listed in the AGENDA and consequently they are not used by exporters. The outcome of this program has been very successful in terms of reducing residue violative cases. Information emerging from national monitoring programs is thus essential to verify the residue behaviour of exporting countries. Annual reports such as those issued by the Food and Drug Administration and USDA from the United States, the Swedish National Food Administration, the Dutch Inspectorate for Health Protection, the Italian Health Institute, the Working Party on Pesticide Residues, Ministry of Agriculture, Fisheries and Food of the U.K., the Finnish Customs Laboratory, and other agencies, are permitting a free and transparent access to of this valuable information to countries concerned.

CONCLUSIONS

Surveillance and compliance published monitoring programs implemented by selected food importing countries have become the only information source to assemble world data on the kind of pesticides and the extent of their residue ranges moving on the international front, albeit with a diversity of commodities sampled, multi-residue or specific analytical methods used, and targets in their enforcement actions. From this global database it can be concluded that, for the most part, pesticide residue detections are below established tolerance levels. In fact, overall reported levels were even below the 50 percent tolerances, a matter which is rarely known by the consumers' sector. Secondly, for certain commodities post harvest applications contribute significantly to the number of residues detected, namely thiabendazole, dicloran, imazalil and, by the same token, grain protectant insecticides.

An important component of presumptive or actual violations was found to occur for residues where no tolerances have been established, a matter which should easily be sorted out by the exporting country through warning notices to their producers and seek for better alternatives available.

REFERENCES

1 Andersson, A. 1996. Control of pesticide residues in fruits and vegetables on the Swedish Market. Paper presented at the 8th Annual California Pesticide Residue Workshop, San Francisco. 12 p.

2 Andersson, A., H. Palsheden, T. Bergh, & A. Jansson. 1995. Pesticide residues in food of plant origin, 1994. National Food Administration, Rapport 20/95, 44 p.

3 Andersson, A. & T. Bergh. 1991. Pesticide residues in fresh fruit and vegetables on the Swedish market, January 1985 - December 1989. Fressenius J. Anal. Chem. 339: 387-389.

4 Anonymous. 1997. Pesticide residues in food of plant origin 1996. The Swedish Nat'l. Food Admin. Livsmedelsv. rapport Nr. 25. 35 p.

5 Anonymous. 1997. Pesticide sales in the U.S. totales $ 11.3 billion in 1995. Pesticide & Toxic Chemical News 25(46): 18.

6 Bates, J.A.R. 1979. The Evaluation of Pesticide Residues in Food: Procedures and Problems in Setting Maximum Residues Limits. J. Sci. Food Agric. 30, 401-416.

7 Camoni, I., A. Di Muccio, M.S. Bellisai & P. Citti. 1991. Residui di pesticidi in campioni alimentari. Anno 1989. Prima parte. Istituto Superiore di Sanità Report, Roma, 44 p.

8 Environmental Protection Agency. 1997. EPA 1996 Food Quality Protection Act. Implementation Plan. US Env. Prot. Agency, 44 p. Wash. D.C.

9 FAO. 1981. Orientaciones para la experimentación de residuos de plaguicidas con vistas a obtener información para el registro de plaguicidas y el establecimiento de límites máximos de residuos. Bol. Fitosan. FAO, vol. 29, N° 1/2.

10 FAO. 1997. FAO Manual on the submission and evaluation of pesticide residues data for the estimation of maximum residue levels in food and feed. Rome, 158 p.

11 González, R.H. & T. Curkovic. 1994. Manejo de plagas y degradación de residuos de pesticidas en kiwi. Rev. Frutícola (Chile) 15(1): 5-10.

12 González, R.H. & J. López (eds.). 1996. AGENDA de Pesticidas: Registros, Tolerancias y Carencias en frutas y hortalizas de exportación. Asoc. Export. de Chile, Santiago. 130 p.

13 Klassen, W. 1993. Pest management and biologically based technologies: A look to the future. *In*: R.D. Lumsden & J.L. Vaughn (eds.) Pest Mngnt.: Biologically Based Technologies, pp. 410-422. Am. Chem. Soc., Wash. D.C., 435 p.

14 Klassen, W. 1995. World Food Security up to 2010 and the Global Pesticide Situation. 8th Int. Congress Pesticide Chemistry, Conference Proc. Series, Wash. D.C. pp.1-32.

15 Kok de, A. 1994. Multiresidue methods used in The Netherlands and Residue data obtained in 1993. *In*: Pesticide Residues '94 (Valverde & Fernández, eds.) pp. 87-113.

16 Hamilton, D. 1994. JMPR's Concept of Comparability IUPAC, 8th Internat. Congress of Pesticide Chemistry, Wash. D.C., pp. 7-12.

17 Li, W. & H. Han. 1995. A study of pesticides HCH and DDT residues in Chinese feed products (1987-1993 Monitoring Program). China Natl. Centre for Quality Supervision and Test of Feed. Presented at the 7th Annual California Pesticide Residue Workshop FDA/CDFA, 11 p.

18 Mc Dougall, J. 1994. Guarded optimism in agrochemicals. Chem. & Ind. 7, 264 p.
19 Nilsen, H.G.; A.L. Christiansen, G.T. Varran & B. Holen. 1997. Norwegian Monitoring of Pesticide residues in fruits and vegetables. Summary of Results 1994-1996. Norwegian Food Control Authoritary, Leaf., 8 p.
20 Rola, A. & P.L. Pingali. 1992. Pesticides and rice productivity: an economic assessment for the Philippines. International Rice Research Institute, Los Banos, Philippines.
21 Roy, R.R.; Wilson, F.; Laski, R., Roberts, J., Weishaar, J., Bong, R. and N. Yess. 1997. Monitoring of domestic and imported apples and rice by the U.S. Food and Drug Administration program. J.AOAC 80(4): 883-893.
22 van der Schee, H.A. 1997. The presence of pesticide residues in primary agricultural products of the Dutch market during the years 1994-1996. Inspectorate for Health Protection Rpt., Amst'dam, 16 p. + annexes.
23 United Kingdom. 1995. Annual Report of the Working Party on Pesticide Residues: 1994, Min. of Agric., Fish. & Food, London, 154 p.
24 Waibel, H. 1994. Global pesticide markets and future prospects for pesticides. Working Paper, 16th Session FAO/UNEP Panel of Experts on IPM, Rome, 25-29 April, 1994, 15 p.
25 Wessel, J.R. & N.J. Yess. 1991. Pesticide residues in foods imported into the United States. *In*: Reviews of Environmental Contamination & Toxicol. Vol. 120: 83-103.

e-mail: <rgonzale@abello.dic.uchile.cl>

Regulation and Risk Assessment

The Benefits of Pesticide Use

Sir Colin Spedding

COUNCIL OF SCIENCE AND TECHNOLOGY INSTITUTES, 20–22 QUEENSBERRY PLACE, LONDON SW7 2DZ, UK

The title of this paper does not call for a balanced assessment of benefits and costs or disbenefits, but some consideration of the latter cannot be avoided in an analysis of benefits. There are two reasons for this, one concerned with the relativity of benefits and the other with the identity of the beneficiary.

1 THE RELATIVITY OF BENEFITS

There is little point in listing benefits without indicating their relative importance and this varies with both the magnitude of the benefit (e.g. total elimination of a damaging pest) and the scale of it (*e.g.* only relevant to a minor crop grown only on a small area of one country).

But the *value* of a benefit (another way of assessing importance) is a *net* amount, representing the benefit minus the cost. This cost is not only that of achieving the benefit but also the costs and disbenefits consequent upon the use of a pesticide.

Obviously, if the cost of application exceeds the financial benefit, the process is unprofitable, will not occur and hardly qualifies as beneficial. But even when profitable, the application may have deleterious effects on the environment or other people: these effects may not always be known or quantified and may have no monetary significance but they have to be taken into account. Even so, it should be noted that attempts have been made[1] and the overall costs of pesticide pollution are discussed by Conway and Pretty.[2] Clearly, a benefit with few or no disbenefits is more worthwhile and more important than one with substantial consequential costs.

However, the recipient of the benefits may not be the bearer of the costs – which leads to the next question: "benefits to whom?"

2 IDENTITY OF THE BENEFICIARY

There is little point in listing even important benefits without specifying the beneficiary(ies), who may themselves be numerous or few, powerful or weak, commercially involved or not.

Benefits diminishing ↓

Disbenefits increasing ↓

		Numbers
A.	The Producer	Small or moderate
B.	The Processor	Small
C.	The Caterer	Small
D.	The Retailer	Small
E.	The Consumer	Large
F.	The Citizen	Vast

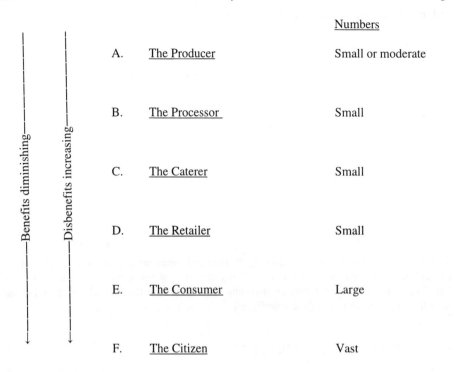

Figure 1 *The Beneficiaries*

In general, the likely – or possible – beneficiaries of pesticide use on a particular crop are the groups shown in Figure 1, listed as a sequence along the food chain (the picture is not significantly different for a non-food crop).

As indicated in the figure, however, the *numbers* involved will not be the same at each level and will vary with the crop grown and the commodity produced. But the likelihood is that the benefits will diminish, in both importance and scale, from A to F, and the disbenefits will probably increase over the same range.

It should be noted that we are *not* all consumers, even of major commodities, and, for some food products, the numbers of consumers may be quite low. We are all citizens, however, whether we consume a particular product or not, though not all citizens may be subject to every disbenefit.

Just by way of illustration, even a pesticide used on wheat, for example, does not concern the consumer of rice, *as a consumer*, but, if it had widespread environmental impact, could affect virtually all citizens of the world.

The identification of beneficiaries and those affected by disbenefits is even more complicated when a world view is taken.

However, using Figure 1 as a starting point, it is possible to attempt a list of the most probable benefits at each level. Since this paper is focussed on the benefits, the parallel task of listing the disbenefits in a similar fashion has not been undertaken. The list is illustrated in Table 1.

Table 1 *Benefits along the Food Chain*

Level	*Main benefits (will vary with the crop)*
Producer	Higher yields; Lower losses; Higher quality products; More uniform product; Higher profit.
Processor	Higher quality → less wastage; Lower costs; Uniform product → greater ease of handling.
Caterer	Lower risk of bio-contamination.
Retailer	More uniform product with fewer blemishes; Increased sales.
Consumer	Higher quality product; Possibly lower cost.
Citizen	Lower losses of the food product affected.

3 THE BENEFITS

As already indicated, it is easier to identify benefits at the level of production than for the citizen: indeed, for some products the citizen may not benefit at all and may nonetheless suffer from, or be aware of, the disbenefits.

It is hardly surprising that the benefits of pesticide use may not be so evident to the citizen or to the consumer, who may well suspect that the real benefits accrue, mainly in financial terms, to those involved in production, use and scale of the product.

Much depends here on whether the benefits are passed on to the consumer in terms of lower prices, or whether they are retained as profit at earlier stages.

Since the *perception* of the consumer is crucial, even if lower costs are passed on, what matters is whether this is believed to be the case. In this context, producers regularly damage their case by telling the consumer "of course, you can have what you want (food produced without pesticides, animal products from welfare-friendly systems, *etc.*) if you are prepared to pay x times as much for it."

This reinforces the quite false notion that the price paid is what the producer receives. But, of course, the cost of production is normally only a fraction of the retail price. In some cases, especially highly processed foods – such as potato crisps – the price per unit of potato may be fifty times the price paid to the producer. This means that if production does not involve a process or input that the consumer finds unacceptable, the extra cost of production may only have a marginal effect on the retail price.

The consumer is not usually told this by the producer, however. The benefit of lower cost to the consumer thus has to be analysed carefully and presented honestly. The case for better, or even tolerable, quality is more easily made. The argument that if you want a product free from blemishes/maggots/fungal rots, pesticides are necessary, is quite persuasive.

The benefits to the retailer are clear: higher quality, greater uniformity and lower cost. But increased sales depend either on lower prices or providing product qualities that the consumer wants. So it can be argued that the retailer would not benefit if the consumer did not also benefit.

The caterer is further removed from the consumer, in the sense that the product is mainly treated as a raw material and may be prepared and processed in ways that present the consumer with a much modified product. The benefits to the caterer are nonetheless similar to those of the retailer: lower cost, higher quality and greater uniformity (some of which may increase ease of preparation).

Much the same applies to the processor but for all three – retailer, caterer and processor – there may be qualities that affect storage and suitability for treatment and packaging that may have little significance for the consumer. Indeed, as an example, genetically-modified tomatoes that have a longer shelf life may even raise suspicions in the mind of the consumer that some deterioration is simply being hidden.

As stated earlier, however, the main and most obvious benefits are gained by the producer.

4 BENEFITS TO THE PRODUCER

4.1 Higher Yield

Worldwide, the losses from pests, during production and post-harvest, are often very substantial. They vary with the crop (see Table 2), the year, the location and many other factors. So it is not possible to give an accurate figure for the losses avoided by the use of pesticides or to generalise about them across the range of crops, years and locations.

Table 3 illustrates the losses from animal pests alone that have been quantified in the literature for rice production. The situation may be very different in developed and developing countries, especially in relation to storage and, to a lesser extent, transport.

Table 2 *Losses due to Pathogens, Animal Pests and Weeds on Different Crops, 1988–90* (after Oerke *et al.*, 1994)[3]

	% Loss			
Crop	*Pathogens*	*Animal Pests*	*Weeds*	*Total*
Rice	15.1	20.7	15.6	51.4
Wheat	12.4	9.3	12.3	34.0
Barley	10.1	8.8	10.6	29.4
Maize	10.8	14.5	13.1	38.3
Potatoes	16.4	16.1	8.9	41.4
Soybeans	9.0	10.4	13.0	32.4
Cotton	10.5	15.4	11.8	37.7
Coffee	14.9	14.9	10.3	40.0

Table 3 *Examples of Losses in Rice Production due to Animal Pests* (based on Oerke *et al.*,1994)[3]

Country	*% Loss*	*Period*
India	17.3	1985–90
Indonesia	12.6	1985–90
Africa (overall)	18.0	1988–90
America (overall)	14.0	1988–90
Asia (overall)	21.0	1988–90
Europe (overall)	13.0	1988–90
USSR (overall)	13.0	1988–90
Oceania (overall)	5.0	1988–90
World (overall)	21.0	1988–90

The highly sophisticated infrastructure of the developed countries may virtually eliminate storage and transport losses from pests (and from other causes). Some of the most vulnerable crops are moved very swiftly from field to a controlled-atmosphere store (including refrigeration) and very often the conditions during transport are not very different.

In developing countries, roads, vehicles and stores may be much less satisfactory and the time periods involved may be considerable.

It is not possible, as stated earlier, to generalise usefully about such losses. Suffice it to say that they can be very high and threaten both the farmers' livelihood and the local supply of food (and other commodities, *e.g.* cotton).

Table 4 gives examples of high estimated losses from specific crops and there are, of course, spectacular cases where production would probably not be feasible without the use of pesticides.

In the context of projected world population increases, it is argued that the sensible use of pesticides will be an essential tool in producing enough food.[4] However, food production does not of itself guarantee that hungry people (who are poor) will be fed.[5]

Table 4 *Examples of Estimated crop Losses due to Pests, Diseases and Weeds* (without pesticides) (after Oerke *et al.*, 1994)[3]

		% Loss		
Crop	*Region*	*Animal Pests*	*Pathogens*	*Weeds*
Potatoes	Central America	13.0	17.0	10.0
Barley	France	27.0	17.0	8.3
Wheat	USA	18.8	26.5	27.5

References

1. D. Pimental, D. Andow, R. Dyson-Hudson, D. Gallahan, S. Jacobson, M. Irish, S. Kroop, A. Moss, I. Schreiner, M. Shepard, T. Thompson and B. Vinzard, 'Environmental and social costs of pesticides: a preliminary assessment', OIKOS 34, 1980, pp. 126-40.
2. G.R. Conway and J.N. Pretty, 'Unwelcome Harvest', Earthscan Publications Ltd., London, 1991.
3. E-C. Oerke, H-W. Dehne, F. Schönbeck and A. Weber, 'Crop Production and Crop Protection', Elsevier Science B.V., Amsterdam, 1994.
4. D.T. Avery, 'Saving the Planet with Pesticides, Biotechnology and European Farm Reform', The Bawden Memorial Lecture, Brighton, 1997.
5. C. Spedding, 'Agriculture and the Citizen', Chapman and Hall, London, 1996.

Pesticide Risk Management and the United States Food Quality Protection Act of 1996

Stephen L. Johnson and Joseph E. Bailey

OFFICE OF PESTICIDE PROGRAMS (7501C), US ENVIRONMENTAL PROTECTION AGENCY, 401 MAIN STREET SOUTH WEST, WASHINGTON, DC 20460, USA

1 INTRODUCTION

The United States Environmental Protection Agency is charged with the responsibility of protecting human health and the environment from unreasonable adverse effects that may result from pesticide use. Agricultural production as we know it today depends on the use of pesticides to control a wide range of indigenous and exotic pests. The EPA must carefully review all pesticides, including those used in agriculture, for public health, and around the home, to ensure that their use does not cause unreasonable adverse effects. As a regulatory agency, the EPA is faced with challenges presented through the laws which provide the Agency with the authority to regulate pesticide use in the United States. Although many of these challenges arise from issues associated with the scientific aspects of pesticide toxicology and the rapidly advancing science in this area, many revolve around risk management issues and finding answers to questions and solutions to problems that arise about how to effectively mitigate risks based on the best science available. Most recently, changes to pesticide regulation were effected through the enactment of the Food Quality Protection Act of 1996[1]. Implementing this legislation has raised questions about how the Agency should evaluate potential pesticide risks, what information is necessary to accurately characterize risks, and how the Agency should manage risks when it is determined that unreasonable adverse effects may result from a pesticide's use. This paper provides an overview of the major provisions of the FQPA and the direction that the EPA is taking with regard to risk management. It also identifies some key issues facing pesticide risk managers in the United States.

2 MAJOR PROVISIONS OF THE FOOD QUALITY PROTECTION ACT

The FQPA was signed into law on August 3, 1996, fundamentally changing the way the EPA regulates pesticides. The FQPA greatly strengthens the regulations which protect the Nation's food supply from potentially unsafe pesticide residues and sets more strict standards that must be met in order to satisfy registration requirements for pesticides used on food commodities in the United States. Of particular interest is the emphasis that this legislation places upon the protection of infants and children from exposure to pesticides due to their increased sensitivity as compared to adults.

The EPA registers, and therefore regulates pesticides, primarily through statutory requirements mandated under the Federal Insecticide, Fungicide and Rodenticide Act (FIFRA). However, the changes that the FQPA has brought to pesticide regulation have been effected largely through the amendments that were made to the Federal Food, Drug and Cosmetic Act (FFDCA). The FFDCA gives the EPA the authority to establish tolerances for residues of pesticides in or on food commodities and the new FQPA safety standard has fundamentally changed the way the EPA sets tolerance levels. While it is the EPA's responsibility to determine what level of pesticide residue can be allowed to remain in or on foods from a health standpoint, it is the FDA's responsibility to monitor food items to ensure that the legally enforceable levels established by the EPA are not exceeded. The FQPA's new single health-based standard requires that the EPA can establish or maintain a pesticide tolerance if it is determined to be safe, and safe is defined in the law to mean that there is a reasonable certainty that no harm will result from exposure to the pesticide residue[2]. Other Federal Agencies, for example the U.S. Geological Survey's National Water Quality Assessment Program, monitor levels of pesticides in water resources in the United States.

The FQPA emphasizes the importance of considering extra sensitivities of infants and children to pesticide exposure. The 1993 National Academy of Sciences study, "Pesticides in the Diets of Infants and Children", reported the results of its research on what is known about the effects of pesticides in the diets of infants and children and evaluated current risk assessment methodologies and toxicological issues of concern[3]. In general, the report concluded that infants and children may react differently from adults when exposed to pesticides and that these differences should be considered when assessing potential risks from exposure. Further, the report recommended that, in instances where increased susceptibility to pesticides is believed to occur, an additional safety factor should be more routinely employed to adequately protect infants and children from potential risks. Legislators of the FQPA considered the recommendations of this report in developing pesticide regulatory policy reform, focusing particularly on the protection of infants and children.

Stemming from the recommendations of the National Academy of Sciences report, the FQPA requires that up to an extra 10-fold safety factor be utilized when it is determined that a pesticide may present risks to infants and children because of their increased sensitivity[4]. The EPA now places greater emphasis on its review of toxicological studies that provide insights to reproductive, developmental and neurological effects of pesticides that could indicate increased susceptibility to infants and children. In those cases where extra sensitivities are believed to be possible, after using a weight-of-evidence evaluation, the Agency will retain an additional 10-fold safety factor in order to adequately protect these more sensitive individuals unless there are reliable data that indicate some other safety factor, or no additional factor, is adequately protective. A standard 100-fold safety factor has routinely been used by the EPA in its risk assessments to account for intra- and inter-species variability and uncertainty.

In the past, the EPA has handled each chemical and each exposure scenario separately in its risk assessment determinations. Risks have been estimated for diet and residential uses, such as pest control in the home or on lawns and in gardens, but they were not added together to see what the combined risk estimate would be. Also prior to the FQPA, risk from drinking water was not quantitatively factored into the total dietary risk estimate. The FQPA has addressed the fact that people are not exposed to chemicals on an individual basis, but rather, may be subject to exposure to several different chemicals simultaneously and from a variety of sources. It is reasonable to approach risk assessment in such a way that reflects actual exposure in the real world, and the FQPA now requires that the EPA consider this issue through both aggregate exposure and through cumulative risk assessment[5]. The EPA is required to determine that a reasonable certainty of no harm will result when considering the additive effects of various exposure routes for a single pesticide; i.e., exposure from dietary sources (including food and drinking water) and all other non-occupational sources, largely residential exposure. Similarly, the EPA is required to consider the additive or cumulative effects of those pesticides which exhibit toxicological effects through similar mechanisms of action. This additive risk approach is not limited to the parent compound alone but includes degradates of toxicological concern as well.

At the time the FQPA was signed by the President, the EPA had about 9,700 tolerances on record. The FQPA requires that all existing tolerances be reviewed according to the new safety standard established by the legislation and an ambitious schedule for completion of the review of these tolerances is also established by the law. The FQPA requires that 33 percent of the tolerances on record be reviewed within 3 years, 66 percent within 6 years and 100 percent within 10 years[6]. In August 1997, the EPA published a notice in the *Federal Register* that outlined a schedule to meet this requirement[7]. Also, the law requires that the EPA gives priority to reassessing the tolerances for those chemicals which appear to pose the greatest risk. The EPA is advancing reviews of those pesticides that appear to pose the highest risks to the front of the queue and is developing an approach to review the organophosphate pesticides first, with review expected to be completed by August 1999.

Sharing a common toxicological endpoint, cholinesterase inhibition, it has been recommended that risk assessment for this class of pesticides be conducted cumulatively as the FQPA requires for chemicals that share a common mechanism of toxicity.

The FQPA has changed how pesticide benefits may be considered in determining the eligibility of a pesticide for registration. The new law allows tolerances to remain in effect for pesticides that might not otherwise meet the new safety standard based on benefits of the pesticide only under certain conditions. Pesticide residues would only be "eligible" for such tolerances if use of the pesticide prevents even greater health risks to consumers or the lack of the pesticide would result in "a significant disruption in domestic production of an adequate, wholesome and economical food supply"[8]. The new provision narrows the range of circumstances in which benefits consideration plays a significant role in determining whether or not a tolerance is appropriate for a particular pesticide use. Prior to passage of the FQPA, the EPA's risk management decisions considered both risks and benefits of a pesticide; in other words, benefits were considered in making decisions about whether or not to restrict a pesticide's use because of risk concerns. The FQPA removes the consideration of pesticide benefits from the risk management decision process for food uses by allowing tolerances to remain that might not otherwise meet the new FQPA safety standard if use of the pesticide in question will prevent a greater health risk to humans, for example in public health situations to control disease carrying pests, or if lack of the pesticide would result in disruption of the Nation's food supply. However, risk management decisions related to pesticide risks associated with ecological effects or worker effects do still involve a consideration of benefits.

A fundamental change to pesticide regulatory policy that the FQPA imposes is acknowledgment that science is not a static discipline and pesticide research, risk assessment and risk management methodology is a rapidly developing area. The EPA has undertaken reviews of registered pesticides previously under other programs. Most recently the 1988 amendments to FIFRA led to establishment of a program to review all pesticide active ingredients registered prior to November 1984. The intent of this reregistration program, which is currently underway, is to bring those active ingredients registered prior to November 1984 up to current day standards of testing as required by the EPA, but only on a one-time basis, with no further reviews required. Because science is continually evolving, the FQPA requires the EPA to periodically review all pesticide registrations with a review goal of every 15 years[9]. The EPA is required to establish, by regulation, a procedure for this periodic review. If the EPA determines that additional data are needed for any review, such data may be required under FIFRA section 3(c)(2)(B)[10]. This periodic revisiting of pesticide registrations ensures that a pesticide's physicochemical and toxicological characteristics are reevaluated according to current state-of-the-science and that risks can be managed as necessary based on the evolving science.

In response to an April 8, 1998 memorandum from United States Vice President Al Gore to the EPA and the USDA, in which the Vice President reaffirmed the

Administration's commitment to the FQPA and clarified how to fulfill the requirements of the law[11], an advisory group was created to ensure smooth implementation of the FQPA so that the important health aspects of the law are carried out and at the same time the Nation's important agricultural production is not impacted in a negative way. The advisory group, co-chaired by the EPA Deputy Administrator and the USDA Deputy Secretary, will serve to ensure that the implementation of regulatory processes flowing from the FQPA requirements are based on sound science, that all decisions are transparent, and that a reasonable transition period is provided for agriculture that reduces risk from pesticide use while not jeopardizing the level of agricultural food production. The advisory group has been requested to draft a framework for a process to review the first major class of chemicals the EPA is evaluating, the organophosphates. How best to implement risk management practices that may be indicated from risk assessment findings for these pesticides will be considered in light of providing a reasonable transition period for agriculture.

3 PESTICIDE RISK MANAGEMENT AND EPA'S DIRECTION

As mentioned earlier, the FIFRA provides the EPA with the authority to license, or register and re-register pesticides for use in the United States. The EPA grants this license to pesticide companies for new chemicals, or allows a license to continue subsequent to the review of an existing chemical, after it has reached a regulatory decision about the need for risk mitigation requirements based on science findings from risk assessments. The EPA's decision must ensure that unreasonable adverse effects will not be caused to human health or the environment by use of a pesticide. The EPA's Office of Pesticide Programs' risk assessors provide its risk managers with scientific analyses of the potential for pesticides to cause a risk of concern, based on the best available scientific information about a pesticide's toxicity and potential for exposure. The risk managers are provided with both qualitative and quantitative information that helps them understand as fully as possible, the magnitude, extent, and confidence of the potential hazards and exposures associated with a pesticide. The risk managers are faced with an extremely difficult task.... to understand the assessment and to weigh or balance it with socioeconomic benefits and political realities within the bounds of the framework of the FIFRA, the FQPA, the FFDCA, pesticide regulations, and established policies.

To clearly understand potential risk implications, risk managers must ask many questions before deciding whether or not risk mitigation measures are necessary and to what degree they may be necessary. The questions serve to fully inform the risk manager about the risks and to provide a forum for the risk manager and the risk assessor to exchange ideas and understand each other's position. For example, the risk manager must understand what the pesticide's effects of concern are and why they are of concern. Are the effects only seen in laboratory tests with a single species or do they occur in multiple species?

Have models been used to estimate exposures to ecosystems or to humans through drinking water from ground and surface water sources? If so, how reliable are these models and the data that have been utilized in them to predict the potential for pesticide exposure? Are actual monitoring data available to help predict more accurately the exposure in the real world? What extent of contamination with pesticide residues is considered acceptable? And once mitigation measures are put into place, how is their effectiveness evaluated? These are just a few of the examples of the types of inquiries necessary to develop effective risk management regimes. They all relate to one issue: what level of confidence does the risk manager have in the risk assessment's estimate of potential risk. Recently, the Agency has been using probabilistic assessments to provide risk managers with more refined information that will more accurately characterize the magnitude and probability of potential risks. The more refined characterization of the risk, the better position the risk manager will be in to make the most effective scientifically-based decision to protect human health and the environment.

Even for those toxicological endpoints of concern for which the EPA has established discrete levels or triggers above or below which risks are deemed to be unacceptable, it is not often a clear-cut, black and white decision about whether risk mitigation measures are needed. While this tends to be more of a statistical issue interpreting the significance of a risk calculation, it does impact the regulatory aspects. As technically precise as science can be, subjective interpretation of the science can lessen the objectivity of risk management decisions.

In the event that a pesticide's potential risks are determined to be unacceptable, a number of issues become important in the regulatory risk management decision process. For instance, if a pesticide should be restricted in some manner due to unacceptable risks, the risk manager must look at the impact of restricting that pesticide and decide if there are effective Integrated Pest Management principles that can come into play to help control the pest that is the target of the pesticide use. Similarly, the risk manager must consider what alternative pesticides are available and how do the potential risks posed by these alternatives compare with the risks posed by the pesticide of concern. In order to lower potential risks in some instances, difficult decisions may need to be made about which uses of the pesticide should be retained and which should not be allowed to continue. Such risk management decisions are often not without controversy.

The implications of the FQPA requirements on risk management are being realized by the EPA as decisions are made about what safety factor is appropriate to adequately protect public health, particularly infants and children. Further, aggregation of risks from multiple sources of exposure for a chemical is raising issues about how best to bring unacceptable aggregated risks into the acceptable range. Since the FQPA was passed, risk management decisions have been made that disallowed some pesticide uses because of the findings from risk assessments that included an additional safety factor or because of aggregate risk concerns from various exposure routes. In these instances, the EPA is carefully considering the data that are used in the risk assessment assumptions and has continued to refine such elements as drinking water contribution to the risk to ensure that the best possible data are used and that the resulting exposure estimates are realistic and as close to real world occurrences as possible. Refining the exposure and risk assessments will only become more crucial as cumulative risks are estimated for multiple pesticides with common mechanisms of toxicity. No doubt, some of these cumulative pesticide assessments will present risk management problems in the future; therefore, the need for realistic assumptions about potential exposure and risk is critical.

Considering risk management from a more general standpoint deserves a brief discussion of some of the activities within the EPA's Office of Pesticide Programs that exemplify the Agency's overall efforts toward pesticide risk management.

In cooperation with state agencies, the EPA has taken efforts to reduce risks from pesticide exposure to agricultural workers. In 1992, the Worker Protection Standard was issued to protect agricultural workers and pesticide handlers from poisonings and injuries that may result from unintentional pesticide exposure[12]. Although there are no specific provisions in FIFRA to protect this particular population, the legislative history of 1972 amendments to the law indicates an intent of Congress that farmers, farmworkers and others be afforded protection under the law. This set of regulations contains a number of occupational exposure reduction measures that serve to lower the risk of pesticide poisonings and injuries among agricultural workers and pesticide handlers on farms and nurseries and in forests and greenhouses. In general, the provisions of the Worker Protection Standard provide protection through regulations that prohibit applicators from applying a pesticide in such a way that will expose workers or other persons, require restricted-entry intervals on all agricultural product labels and prevent workers from entering a pesticide-treated area during the interval, with few exceptions. Personal protective equipment must be provided and maintained for handlers,and early-entry workers must be notified about treated areas to avoid inadvertent exposures. Adequate supplies of water, soap and towels must be maintained for handlers and workers for routine washing and emergency decontamination. Pesticide training for workers and handlers must be provided and safety posters must be displayed. Transportation must be provided in emergency cases to the nearest medical center and information must be provided about the pesticide to which the person may have been exposed.

In 1993, the Agency put its Reduced Risk Initiative into place that expedites the registration of conventional pesticides that the Agency believes pose less risk to human health and the environment than existing alternatives. This initiative provides an incentive for pesticide companies to develop reduced-risk pesticides, ensuring that they are available to the growers who need them as soon as possible. The EPA also expedites the review of biological pesticides in the same manner. The FQPA mandated that the Agency continue and enhance the Reduced Risk Initiative. Since the program was begun, the number of reduced-risk pesticides has steadily increased each year and has included fungicides, herbicides and insecticides for a variety of crop and non-crop uses. Conventional reduced-risk pesticides are advantageous from a risk management standpoint in that they may have low impact on human health, low toxicity to non-target organisms, low potential for water resource contamination, lower use rates, low pest resistance potential, and compatibility with Integrated Pest Management principles. Similarly, because biological pesticides are based on naturally occurring substances, they generally pose less risk to human health and the environment and the Agency has committed to fostering the registration of biological pesticides by forming a separate division devoted solely to such activities.

In 1994, the EPA established its Pesticide Environmental Stewardship Program which promotes environmental stewardship strategies and encourages the implementation of pest management practices designed to reduce pesticide use and potential risk. This program operates through a voluntary commitment by partners or supporters who agree to develop and implement strategies to reduce the risk of pesticides associated with specific use sites, crops, or regions of the country. Membership in the program is not limited to organizations that use or represent users of pesticides, but is open to organizations that are involved in pesticide issues or that may have influence over pest management practices of pesticide users such a food processors and public interest groups.

Currently ongoing, the FIFRA reregistration program actively pursues specific risk reduction measures for specific pesticides currently registered that are being re-evaluated. The FIFRA was amended in 1988, requiring the Agency to establish a pesticide reregistration program to review those chemicals registered prior to November 1984, ensuring that they meet the current requirements for registration. The FIFRA reregistration program ensures that pesticide active ingredients registered prior to 1984 reflect current advances in science, risk assessment and risk management practices. The Agency's intense review of these chemicals has frequently revealed the need for measures to reduce risk. Risk management decisions often require pesticide labeling changes that will reduce exposure of the pesticide to humans and the environment. If risks of concern are identified during the reregistration review, the Agency will have a dialogue with the pesticide company prior to reaching a final reregistration eligibility decision, to see what measures might be voluntarily adopted by the company to lower potential risks that may be associated with the use of the pesticide. If voluntary measures are not adopted, the Agency may impose specific risk mitigation requirements to ensure that the pesticide's use will not cause unreasonable adverse effects.

With regard to international activities and pesticide risk management, The EPA has taken a forward stance and become increasingly involved in the regulation of pesticides through cooperative, harmonization programs aimed at protecting the safety of the U.S. food supply, developing common international approaches to pesticide review, sharing work among governments to reduce the burden of regulatory review, supporting international activities to reduce pesticide risks and improve the safety of pesticide use, and to encourage open communications with the public on findings of pesticide evaluations. For example, through the Agency's involvement in the Organization for Economic Cooperation and Development's (OECD) Environmental Health and Safety Program and the North Atlantic Free Trade Agreement's (NAFTA) Technical Working Group on Pesticides, steps are taken to harmonize study guideline requirements for pesticide registration, and assistance is provided to other countries to manage the risks of pesticides as efficiently and effectively as possible.

4 CONCLUSION

Today, the use of pesticides plays a major role in maintaining food production necessary to sustain the world population. As a result, the EPA faces a rapidly changing, dynamic world as it carries out its mission to protect human health and the environment from potential unreasonable adverse effects that may be caused by pesticide exposure not only from a science standpoint but also from a risk management standpoint.

Along with ensuring a safer food supply for the public, the FQPA has presented challenges to everyone involved---growers, regulators, industry, environmental groups and government alike. Those aspects of the FQPA that require consideration of an additional safety factor, aggregate exposure, cumulative risk, and estimating risk from drinking water are complex science issues at the leading edge of evolving pesticide toxicology and risk assessment/risk management methodology. The EPA is carefully developing new exposure assessment tools and sound science policy to address these issues and is consulting with experts in these areas to allow the best scientific minds to contribute to the resolution of these very complex problems.

The evolution of pesticide risk management principles will continue on a parallel with the development and enhanced understanding of pesticide risk assessment methodology. A more robust understanding of the basis of our risk assessments and the toxicology of pesticides will only serve to further enhance realistic risk management policy and practices. The EPA firmly believes in the principles of pollution prevention and sustainable agriculture in order to preserve the standards of living to which we have become accustomed. Effective risk management will remain as a major tool that the EPA uses to protect human health and the environment from potential pesticide risks

Disclaimer: The views presented in this paper are those of the authors and not necessarily those of the U.S. Environmental Protection Agency (EPA or the Agency).

References

1 *Food Quality Protection Act of 1996*, 1996, Public Law 104-170.

2 *Federal Food Drug and Cosmetic Act*, 1996, § 408(b)(2)(A).

3 National Research Council. *Pesticides in the Diets of Infants and Children,* National Academy Press, Washington, DC, 1993, p. 2.

4 *Federal Food Drug and Cosmetic Act*, 1996, § 408(b)(2)(C).

5 *Federal Food Drug and Cosmetic Act*, 1996, § 408(b)(2)(D).

6 *Federal Food Drug and Cosmetic Act*, 1996, § 408(q).

7 Federal Register, August 4, 1997, Volume 62, Number 149, p. 42019.

8 *Federal Food Drug and Cosmetic Act*, 1996, § 408(b)(2)(B).

9 *Federal Insecticide, Fungicide and Rodenticide Act*, 1996, § 3(g)(1)(A).

10 *Federal Insecticide, Fungicide and Rodenticide Act*, 1996, § 3(c)(2)(b).

11 Memorandum from the United States Vice President to Secretary of Agriculture Daniel R. Glickman and EPA Administrator Carol M. Browner, April 8, 1998.

12 Federal Register, August 21, 1992, Volume 57, Number 163, p. 38102.

Pesticides in Food

Ian Shaw

CENTRE FOR TOXICOLOGY, UNIVERSITY OF CENTRAL LANCASHIRE, PRESTON PR1 2HE, UK

1 PUBLIC CONCERNS ABOUT PESTICIDE RESIDUES IN FOOD

There is very great public concern about food safety. Residues of pesticides in food are an important facet of this concern, even though the public's perception of the risks involved to their well-being are often grossly exaggerated. Perception of risk in this context is important, because consumers must be given the appropriate information to enable them to make their own risk assessment relating to the potentially harmful effects of pesticides in their food. It is very likely that if they were able to do this, that public concerns relating to pesticide residues in food would diminish considerably. On the other hand, if the effects of pesticide use upon the natural environment were added to this risk assessment equation, it is likely that the public would redirect their concerns towards the environmental effects of pesticides rather than the food-related effects.

2 PESTICIDE RESIDUES MONITORING

Most countries operate pesticide monitoring schemes, but the magnitude, reliability and scope of such schemes vary considerably from country to country. The UK has a Working Party on Pesticide Residues (WPPR) with a chairman who is independent of government. The WPPR carries out a surveillance programme which monitors the levels and frequency of residues of pesticides in food (both UK produced and imported) and reports its results to the Advisory Committee on Pesticides (ACP - the committee that advises Ministers on the effects of pesticides on people and the environment, and advises on the approval of pesticides for use in the UK) and to the Food Advisory Committee (FAC - the committee that advises Ministers on a broad spectrum of food safety and nutrition). In 1996, 4,000 samples from UK retail outlets were analysed for pesticide residues at a cost of approximately £2m. The sample number is small, in statistical terms, but it allows areas of concern to be identified with focused follow-up to assess the magnitude of the specific problem. This was illustrated well in 1995 when lindane (γ-hexachlorocyclohexane, γ-HCH) was found as a result of routine monitoring of the UK milk supply. Extensive follow-up illustrated the magnitude of the problem and, in turn, gave Ministers the information that they needed to decide whether milk was safe to drink or not (see later in this article).

Different countries have different sampling regimens and therefore collect widely varying numbers of samples (Table 1). Even when calculated on a *per capitum* basis the numbers between EU member States vary very considerably. The most intensive sampling is in Sweden and the least in the UK (Table 1). These sampling numbers, however, do not give any indication of the array of pesticides sought in each sample and therefore, perhaps, do not give a fair indication of the individual country's residues monitoring diligence.

Country	**Population ($x10^{-6}$)**	**N° samples analysed**	**N° samples analysed *per capitum* population ($x10^{6}$)**	**% with residues**	**% above MRL**
Belgium	10.2	932	91	52	1
Denmark	5.3	1273	240	23	1
Germany	83.5	4257	51	33	-
Greece	10.5	1132	108	18	1
Spain	39.2	3022	77	39	0.1
Ireland	3.6	505	140	43	3
Italy	57.5	7194	125	33	1
Luxemburg	0.42	212	506	30	1
Netherlands	15.6	11015	706	47	-
Portugal	9.9	600	61	-	1
Finland	5.1	3368	660	51	3
Sweden	8.9	8908	1001	39	2
Norway	4.4	2936	667	36	-
UK	58.5	878	15	34	<1

Table 1

Numbers of fruit and vegetable samples analysed for pesticides in Europe (EU,1996) showing the sample numbers analysed per capitum population.

2.1 Health risks of pesticide residues in food

In the UK the proportion of samples containing pesticide residues has been constant (at approximately 30%) for at least 3-years, similarly the number of samples exceeding the maximum residues level (MRL) has also remained constant (at <1%) (Table 2) (WPPR, 1994, 1995, 1996). Whether this level of contamination of food with pesticides presents a risk to health is an important question. Applying basic toxicological principles to the data suggests that most of the residues detected are extremely small and at least several orders of magnitude below the acceptable daily intake (ADI). Therefore, even if food containing such a residue was eaten every day of a consumer's life it would have no effect. This statement relates only to residues assessed on an individual basis; however, many foods contain multiple residues. It is possible that the effects of multiple residues might be at least additive; for example organophosphorus pesticides (OPs) which all have the same mechanism of action (inhibition of synaptosomal acetylcholinesterase) could have additive actions when the body is challenged with multiple residues. Very little work has been carried out on the potential toxicological effects of multiple pesticide residues; clearly research is necessary before we can decide upon the risk associated with consuming food containing several pesticides, albeit at very low concentrations.

Year	% with Residues	% Above MRL
1994	30	<1
1995	31	<1
1996	34	<1

Table 2

Frequency of residues of pesticides determined as part of the UK's monitoring programme (MAFF, 1994. 1995, 1996).

It is important to keep risks in perspective. There is no doubt that the health risks of pesticides are extremely low and that many of life's daily risks, which are accepted by most people, are significantly greater (Table 3). A good risk comparison is between pesticide residues and natural toxins that might be present in food. For example psoralens are found in parsnips, parsley and related vegetables. They are natural insecticides which are present as a natural defence against insect attack. Their concentrations in plants varies according to the situation. Stressing (eg by injuring) a parsnip will considerably increase the psoralen concentration in the parsnip's root. Most people would not worry about eating a parsnip even though it contained psoralens; however, if they knew that the parsnip contained $ng.g^{-1}$ concentrations of carbamate or OP insecticides used to control carrot root fly (which also affects parsnips) they might be a little more concerned. However their concern is ill-founded. Low levels of OPs taken in with food are metabolised and excreted quickly and have no measurable effect upon the body. On the other hand, the psoralens are photoactivated carcinogens. They are present in parsnips at about $40\mu g.g^{-1}$ and therefore might pose a greater long-term risk (albeit still extremely small) to health than the OPs.

RISKS	BENEFITS
Getting run over on the way to the shop	**Calories**
	Vitamins
Choking on a Brussels sprout	**Roughage**
	Anti-oxidants
Natural toxins	
Pesticide residues	

Table 3

A light-hearted look at the potential risks and benefits of eating fruit and vegetables. The adverse health effects of pesticides are very low indeed.

2.2 The frequency of pesticide residues in the EU and the USA

In the UK in 1996 (WPPR, 1996), 34% of food monitored contained measurable residues of pesticides, of these >1% exceeded MRLs. It is interesting to compare this figure with other countries (Tables 1 & 4). In the EU there are widely varying frequencies of pesticide residues from country to country. At the top is Belgium with 52% and at the bottom is Greece with 18%. It is, however, extremely misleading to compare these values directly, because they depend upon analytical limits of determination (LOD). If a country accepts high LODs they will find fewer pesticide residues than a country that insists on low LODs. On the other hand, MRLs are a good way of comparing different countries because MRL values are fixed. Within the EU MRL exceedences from state to state do not vary considerably. The greatest exceedance of 3% occurred in both Finland and Ireland; the lowest exceedance of 0.1% occurred in Spain. The other EU states have MRL exceedance frequencies of 1% or 2% and therefore show remarkable consistency.

It is interesting to compare the EU results with those from the USA (Table 3). MRL exceedences in the USA as a whole were 4.8% in 1996 (USDA, 1996). However MRL exceedances in California were significantly lower at 1.6% (CEPA, 1995). California is noted for its 'green' approach to life and it is possible that this philosophy has led to pesticides being used more carefully in this state when compared to other parts of the USA. Perhaps the most interesting comparison is between California and the EU whose MRL exceedences are comparable. If my conclusion about the careful use of pesticides is correct, this reflects very favourably on Europe.

	% with Residues	% >MRL
USA	72	4.8
California	35	1.6
EU mean	37	1.4

Table 4

A comparison of the frequency of pesticide residues in food from the USA as a whole (USDA, 1996), California (CEPA, 1995) and the EU (calculated from data in EU, 1996). The data show a striking similarity between the EU and California.

2.3 Examples of where the UK monitoring programme has shown residues problems

2.3.1 *Vinclozolin residues in winter lettuce*

Vinclozolin is a fungicide widely used throughout the world. It is not approved for use in the UK. The reason for vinclozolin not being approved for use in the UK is based on operator risks and not because of concerns about residues in food.

Vinclozolin has molecular structural analogy with the androgens. It appears to occupy the androgen receptor, blocking it and so inhibiting endogenous androgen activity. Relatively high doses of vinclozolin (ie far above residues in food levels) might inhibit the normal sexual development in males. The UK situation, where vinclozolin is likely to be used on winter lettuce grown in greenhouses, would lead to growers receiving greater vinclozolin doses when spraying within the close confines of a greenhouse than their counterparts in warmer countries where lettuces would be grown outside. It is largely for this reason that vinclozolin is not approved in the UK, but is approved in other EU member states.

The presence of residues of Vinclozolin (and chlorothalonil which is also not approved for use on lettuce in the UK) on UK winter lettuce has been a problem for several years (MAFF, 1996). Such residues indicate misuse of vinclozolin and chlorothalonil. The fact that vinclozolin is not approved in the UK, but is approved in other EU countries means that if residues (at whatever level) are found on lettuce of UK origin the grower had broken the law. Only residues above the MRL on lettuce from elsewhere would constitute a felony even if such lettuce were imported into the UK. Therefore by using legislation to protect the health of growers, a seemingly ridiculous residues anomaly has been introduced.

2.3.2 *Lindane in the UK milk supply*

Lindane is an organochlorine (OC) pesticide which has a shorter environmental $t_{1/2}$ than most other OCs. Its use is controversial in the UK because there have been widely publicised suggestions (which are scientifically unfounded) that it might be involved in the development of breast cancer. Its use in the UK is limited to a few very specific applications, for example use in sugar beet growing at the two cotyledon stage. Residues in food do not normally present a problem because of its use at a very early stage in the growing cycle.

In July 1995 the UK pesticide monitoring programme detected unexpectedly high levels of lindane in milk (MAFF, 1996). By September 1995, lindane concentrations in some milk samples had significantly exceeded the MRL, and there was considerable discussion at Ministerial level regarding the appropriate course of action. Since the ADI was never exceeded it was decided that the milk was safe to drink, but that a very close watching brief must be mounted to ensure that values did not continue to rise. Levels began to steadily fall until by December 1995 they were below the MRL. By January 1996 lindane levels in milk had returned to the normal very low levels that would be expected, based on many years of monitoring, and the knowledge that lindane is persistent in fat deposits (eg mammary fat).

A follow-up intensive sampling regimen was instituted in 1996. This showed that there was not a repeat of the 1995 situation, but interestingly that there was a marginal (and statistically insignificant) rise in lindane levels (all values were well below the MRL) which mirrored the 1995 results.

The lindane problem of 1995 therefore appears to have been a one-off event. The question is, what caused it? There are three possible explanations.

- The summer of 1995 was unusually hot and dry and therefore grass and cereal yields were low. A reduced food supply and intake might have led to cattle re-mobilising fat reserves. Such fat would contain lindane residues which when mobilised would have become available for excretion in milk.
- The poor grass and cereal yields might have resulted in the import of lindane-contaminated feed components.
- Cattle had chewed wood treated with lindane-containing wood preservatives.

It is impossible to determine retrospectively which, if any, of these possible explanations contributed to the problem. However cattle feed components were monitored for lindane residues in 1997 (MAFF, 1997). None were found, which shows that even if this contributed to the lindane in milk in 1995, it is no longer a problem.

2.4 High risk groups - Children

2.4.1 *Ethylenethiourea (ETU) residues in tomatoes*

ETU is a food processing breakdown product of dithiocarbamate fungicides (eg Mancozeb). It is possible that its residues might occur in tomato products, because dithocarbamates are used in the tomato production process. Tomato products (eg purees) are consumed in large quantities by children (eg in baked beans in tomato sauce). ETU has a low ADI and therefore if it were present in tomato puree would present a particularly high risk to children (who consume more food per kg than adults). No detectable residues (reporting limit 0.01 mg.kg^{-1}) were found in a total of 60 tomato product samples (both UK origin and imported)(MAFF, 1996). This is an excellent example of there being no risk to a particularly vulnerable group of consumers from food which might be predicted to be problematic.

2.5 Organochlirine (OC) pesticide residues in food from developing countries

2.5.1 p,p'-*DDE residues in rabbit meat from the Peoples Republic of China*

DDT is a cheap and effective pesticide which has a long environmental half-life, even in hot climates. For these reasons it is difficult to persuade poorer countries not to use it. Even if governments ban the use of OCs generally, and DDT in particular, it is almost impossible to prevent poor farmers taking advantage of clandestine sources of the pesticide. This is illustrated well by the UK's monitoring of rabbit meat imported from China in 1995 and 1996 (Table 5; MAFF, 1995 & 1996) when 3% and 7% of imports were contaminated with the DDT metabolite *p,p*'-DDE. The residues levels were all well below the MRL (0.1 mg.kg^{-1}), but suggest that DDT is still in use in China.

Year	N° of Samples	% Containing *p*,p'-DDE	DDE Level
1996	30	7	0.02, 0.009
1995	34	3	0.08

Table 5

Residues of p,p'-DDE in rabbit imported into the UK from China (MAFF, 1995 & 1996).

2.5.2 *Pesticides in fruit and vegetables in the Republic of Indonesia*

Indonesia has implemented an excellent pesticide approvals legislative process which is very similar to that operated in the EU and USA. To back up the legislation a monitoring programme has been introduced which, although a ridiculously small number of commodities are sampled, has demonstrated a remarkable improvement in pesticide residues over the past 6-years (Table 6). This improvement is particularly impressive for DDT which was not found in 1997 (Table 6). Even though these results are encouraging, knowledge of the sampling system is essential to fully assess the meaning of the results. It is important that Indonesia increases the sample number in future years. However, following the recent political unrest and currency collapse there are financial constraints upon this ideal.

Year	N° of Samples	% with Residues	% Above MRL	% with DDT
1997	40	42.5	2.5	0
1991	143	93	Not Known	20

Table 6

Residues of pesticides in food sampled as part of the Indonesian government's monitoring programme (Data from the Indonesian Pesticide Committee).

3 CONCLUDING REMARKS

This brief foray into the world of pesticide residues in food illustrates the importance of residues monitoring in both assessing the potential risks to consumers' health and as a means of monitoring the use of pesticides, particularly in developing countries.

The value of pesticide residues monitoring is great, even though the sample size *per capitum* is small. This value is both from a legislative point of view, but, perhaps more importantly, it gives the consumer confidence in the food that they eat.

References

EU,1996 Monitoring for Pesticide Residues in The European Union and Norway, European Commission Directorate General XXIV, Brussels

MAFF, 1994, 1995, 1996, 1997 Annual Reports of the Working Party on Pesticide Residues, MAFF Publications, London.

USDA, 1996 Pesticide Data Program Annual Summary, USDA, Washington

CDPR, 1995 Residues in Fresh Produce, Pesticide Enforcement Branch, Sacramento California

Professor Ian Shaw is chairman of the UK government's Working Party on Pesticide Residues. The views expressed in this article are his and not necessarily those of the WPPR.

E-mail: i.c.shaw@uclan.ac.uk

Subject Index

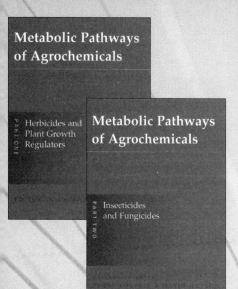